Biomedical Informatics

Andreas Holzinger

Biomedical Informatics

Discovering Knowledge in Big Data

Andreas Holzinger
Medical University Graz and
  Graz University of Technology
Graz, Austria

Additional material to this book can be downloaded from http://extra.springer.com.

ISBN 978-3-319-04527-6   ISBN 978-3-319-04528-3 (eBook)
DOI 10.1007/978-3-319-04528-3
Springer Cham Heidelberg New York Dordrecht London

Library of Congress Control Number: 2014932669

© Springer International Publishing Switzerland 2014
This work is subject to copyright. All rights are reserved by the Publisher, whether the whole or part of the material is concerned, specifically the rights of translation, reprinting, reuse of illustrations, recitation, broadcasting, reproduction on microfilms or in any other physical way, and transmission or information storage and retrieval, electronic adaptation, computer software, or by similar or dissimilar methodology now known or hereafter developed. Exempted from this legal reservation are brief excerpts in connection with reviews or scholarly analysis or material supplied specifically for the purpose of being entered and executed on a computer system, for exclusive use by the purchaser of the work. Duplication of this publication or parts thereof is permitted only under the provisions of the Copyright Law of the Publisher's location, in its current version, and permission for use must always be obtained from Springer. Permissions for use may be obtained through RightsLink at the Copyright Clearance Center. Violations are liable to prosecution under the respective Copyright Law.
The use of general descriptive names, registered names, trademarks, service marks, etc. in this publication does not imply, even in the absence of a specific statement, that such names are exempt from the relevant protective laws and regulations and therefore free for general use.
While the advice and information in this book are believed to be true and accurate at the date of publication, neither the authors nor the editors nor the publisher can accept any legal responsibility for any errors or omissions that may be made. The publisher makes no warranty, express or implied, with respect to the material contained herein.

Printed on acid-free paper

Springer is part of Springer Science+Business Media (www.springer.com)

To my family and friends

Preface

Amongst all the challenges which medicine and health care in all parts of our world are facing, including exploding costs, finite resources, aging population, etc., maybe one is *the* grand challenge: How to deal with the increasingly large, high-dimensional, weakly structured, and complex data sets and the growing masses of unstructured information produced by modern biomedical sciences. The future trend towards personalized medicine has resulted in an explosion in the amount of generated biomedical data sets from various sources, for example genomics, proteomics, metabolomics, lipidomics, transcriptomics, epigenetics, microbiomics, fluxomics, and phenomics.

This "Big Data" rapidly exceeds the capacity a single medical doctor may be able to handle manually. In the daily routine today, a medical doctor is no longer able to keep pace with these developments and to remember all the thousands of facts of the many diverse patients she or he has. Although human cognitive capacities are excellent in certain areas, in dealing with complex data sets the human learning is limited to a certain extent. This makes the application and further development of sophisticated machine learning approaches a pressing need.

For quite some time now, I have worked persistently on a synergistic combination of methodologies and approaches of two areas, which offer ideal conditions towards solving these challenges towards finding new, efficient, and user-centered algorithms and tools to deal with the "Big Data" challenge: Human–Computer Interaction (HCI) and Knowledge Discovery and Data Mining (KDD), with the goal of supporting human learning with machine learning to interactively discover new, previously unknown *insights into the data*.

Imagine a future where medical doctors would be able to ask questions (e.g., "What are the similarities/differences between patients with symptom X and patients with symptom Y") to their patient data *and get relevant answers*. The advantageous possibilities are endless if the doctors have an integrated overview on all *relevant* data at their clinical workplace and could find and diagnose diseases well in advance before they might become symptomatically apparent.

This book constitutes the lecture notes to my one-semester course where I try—and one can never be perfect—to provide undergraduate students of biomedical engineering, software engineering, and computer science a broad overview on "Biomedical Informatics" with emphasis on the aforementioned challenges, consequently focusing on a blend of aspects on data, information, and knowledge. Academic freedom lets teach in an egocentric way; consequently, I bring in what I find most interesting and which is amongst my research interests, following a research-based teaching style: I am passionate on extending advanced methods including time (e.g., information entropy) and space (e.g., computational topology), along with user-centered software engineering methods to create interactive software for mobile applications and content analytics techniques, and I follow three promising research streams: Graph-Based Data Mining, Entropy-Based Data Mining, and Topological Data Mining. Naturally, in the limited time of this course I can only scratch the surface and it is impossible to touch each and every interesting aspect—but as a first taster the goal is to open the eyes of the students for the current problems and to encourage them to study further.

Such a foreword is always an opportunity to thank people. This is not an easy task as many people contribute to the development of such a book, either directly or indirectly, and because there are so many, it is always the risk of forgetting somebody, so I will use the plural: I thank my Institutes both at Graz University of Technology and the Medical University of Graz for the academic freedom, the intellectual environment, and the opportunity to teach these fascinating topics to the students. I thank all my colleagues in academia, the clinical domain, and industry for their ongoing and constant fruitful support. I cordially thank all my team and the members from my research group and all my students for their valuable criticism and feedback, and my family and friends for their nurturing encouragement. Last but not least I thank the Springer management and production team for their great and smooth support!

Graz, Austria   Andreas Holzinger

Overview

The Journey Through This Course

The goal of this one-semester course is to provide the students with a broad overview on "Biomedical Informatics" with focus on data, information, and knowledge. We will follow the hypothesis that information can bridge the gap between scientific biomedical research and clinical health practice. The course consists of the following 12 lectures (see Slide 0-1):

Slide 0-1: Roadmap Through This Course

The course consists of the following 12 lectures:

1. **Introduction: Computer Science Meets Life Sciences**. We start with the basics of life sciences, including biochemical and genetic fundamentals, some cell-physiological basics, and a brief overview of the human body; we answer the question "what is biomedical informatics," and we conclude with an outlook into the future.
2. **Fundamentals of Data, Information, and Knowledge**. In the second lecture we start with a look on data sources, review some data structures, discuss standardization versus structurization, review the differences between data, information, and knowledge, and close with an overview of information entropy.
3. **Structured Data: Coding, Classification (ICD, SNOMED, MeSH, UMLS)**. In the third lecture we focus on standardization, ontologies, and classifications, in particular on the International Statistical Classification of Diseases, the Systematized Nomenclature of Medicine, Medical Subject Headings, and the Unified Medical Language.

(continued)

(continued)

4. **Biomedical Databases: Acquisition, Storage, Information Retrieval, and Use**. In the fourth lecture we get a first impression of a hospital information system, we discuss some basics of data warehouse systems and biomedical data banks, and we concentrate on information retrieval.
5. **Semi-structured, Weakly Structured, and Unstructured Data**. In the fifth lecture we review some basics of XML, before we concentrate on network theory and discuss transcriptional regulatory networks, protein–protein networks, and metabolic networks.
6. **Multimedia Data Mining and Knowledge Discovery**. In the sixth lecture we determine types of knowledge, focus on the basics of data mining, and close with text mining and semantic methods, such as Latent Semantic Analysis, Latent Dirichlet Allocation, and Principal Component Analysis.
7. **Knowledge and Decision: Cognitive Science and Human–Computer Interaction**. In the seventh lecture we review the fundamentals of perception, attention, and cognition; discuss the human decision-making process, reasoning, and problem solving; and learn some principles of differential diagnosis and a few basics on human error.
8. **Biomedical Decision Making: Reasoning and Decision Support**. In the eighth lecture we start with the question "Can computers help doctors to make better decisions?" and apply the basics from Lecture 7 to the principles of decision support systems and case-based reasoning systems.
9. **Interactive Information Visualization and Visual Analytics**. In the ninth lecture we start with the basics of visualization science; review some visualization methods, including Parallel Coordinates, Radial Coordinates, and Star Plots; and learn a few things about the design of interactive visualizations.
10. **Biomedical Information Systems and Medical Knowledge Management**. In the tenth lecture we discuss workflow modeling, some basics of business enterprise hospital information systems, Picture Archiving and Communication Systems, and some standards, including DICOM and HL-7.
11. **Biomedical Data: Privacy, Safety, and Security.** In the eleventh lecture we start with the famous IOM "Why do accidents happen?" report and its influence on safety engineering, and concentrate on aspects of data protection and privacy issues of medical data.
12. **Methodology for Information Systems: System Design, Usability, and Evaluation**. Finally in the twelfth lecture we slip into the developer perspective and have a look on design standards, usability engineering methods, and on how we evaluate such systems.

Contents

Lecture 1 Introduction: Computer Science Meets Life Science 1
1   Learning Goals 1
2   Advance Organizer 1
3   Acronyms 3
4   Key Problems 4
5   Our World in Data 4
6   What Is Life? 10
    6.1   The Building Blocks of Life 11
    6.2   Proteins 13
        6.2.1   From Amino Acids to Protein Structures 13
        6.2.2   Tertiary Structure of a Protein 14
        6.2.3   Protein Analytics 15
    6.3   DNA and RNA 18
    6.4   Cell Physiology 20
    6.5   Organ Systems 21
        6.5.1   Tissue 22
        6.5.2   Organs 22
        6.5.3   Cardiovascular System 23
        6.5.4   Anatomical Axes 23
7   What Is Biomedical Informatics? 23
    7.1   Medicine Versus Informatics 23
    7.2   Computer Machinery 24
    7.3   Medical Informatics 30
    7.4   Biomedical Informatics 31
8   Future Outlook and Research Avenues 35
    8.1   Grand Challenges 35
    8.2   Personalized Medicine 36
    8.3   Biomarker Discovery 39

9	Exam Questions		40
	9.1	Yes/No Decision Questions	40
	9.2	Multiple Choice Questions (MCQ)	41
	9.3	Free Recall Block	43
10	Answers		46
	10.1	Answers to the Yes/No Questions	46
	10.2	Answers to the Multiple Choice Questions (MCQ)	47
	10.3	Answers to the Free Recall Questions	49
References			51

Lecture 2 Fundamentals of Data, Information, and Knowledge ... 57

1	Learning Goals		57
2	Advance Organizer		57
3	Acronyms		59
4	Key Problems		60
5	Data in the Biomedical Domain		60
	5.1	Data Sources in Biomedical Informatics	60
	5.2	Levels of Data Structures	61
	5.3	Abstract Data Structures	63
	5.4	Big Data Pools in the Health Domain	65
	5.5	Standardization Versus Structurization	67
	5.6	Data Dimensionality	69
		5.6.1 Multivariate and Multidimensional	69
		5.6.2 Point Cloud Datasets	71
6	A Clinical View on Data, Information, Knowledge		75
7	A Closer Look on Information		78
	7.1	What Is Information?	78
	7.2	Information Entropy	81
8	Future Outlook		90
9	Exam Questions		94
	9.1	Yes/No Questions	94
	9.2	Multiple Choice Questions (MCQ)	95
	9.3	Free Recall Questions	97
10	Answers		100
	10.1	Answers to the Yes/No Questions	100
	10.2	Answers to the Multiple Choice Questions (MCQ)	101
	10.3	Answers to the Free Recall Questions	103
References			105

Lecture 3 Structured Data: Coding and Classification ... 109

1	Learning Goals	109
2	Advance Organizer	109
3	Acronyms	111
4	Key Problems	111

5	Standardization		112
	5.1	The Need for Standardization in Medicine	112
	5.2	Inaccuracy of Medical Data	113
	5.3	Data Standardization	115
6	Modeling Biomedical Knowledge		120
7	Ontologies		122
	7.1	Ontology Languages	127
	7.2	OWL	128
8	Medical Classifications		129
9	Future Outlook		137
10	Exam Questions		140
	10.1	Yes/No Questions	140
	10.2	Multiple Choice Questions (MCQ)	141
	10.3	Free Recall Questions	143
11	Answers		146
	11.1	Answers to the Yes/No Questions	146
	11.2	Answers to the Multiple Choice Questions (MCQ)	147
	11.3	Answers to the Free Recall Questions	149
References			151

Lecture 4 Biomedical Databases: Acquisition, Storage, Information Retrieval, and Use ... 153

1	Learning Goals		153
2	Advance Organizer		153
3	Acronyms		156
4	Key Problems		157
5	A First View on Hospital Information Systems		158
	5.1	Goals and Challenges of Hospital Information Systems	158
	5.2	Workflows	160
	5.3	Architecture of HIS	161
6	Databases		162
	6.1	Data Warehouse	164
	6.2	Data Marts	164
	6.3	Biomedical Databanks and Cloud Computing	165
	6.4	Cloud Computing	165
	6.5	Biomedical Databases	167
7	Information Retrieval		170
	7.1	Data Retrieval vs. Information Retrieval	171
	7.2	Text Retrieval	172
	7.3	IR Process	172
	7.4	Formal Notation	172
	7.5	Taxonomy of IR Models	175
		7.5.1 Set Theoretic Example: Boolean Model	177
		7.5.2 Example Algebraic Model: Vector Space Model	178
		7.5.3 Example: Probabilistic Model (Bayes' Rule)	182

8	Future Outlook		187
9	Exam Questions		188
	9.1	Yes/No Decision Questions	188
	9.2	Multiple Choice Questions (MCQ)	189
	9.3	Free Recall Block	191
10	Answers		194
	10.1	Answers to the Yes/No Questions	194
	10.2	Answers to the Multiple Choice Questions (MCQ)	195
	10.3	Answers to the Free Recall Questions	197
References			199

Lecture 5 Semi-structured, Weakly Structured, and Unstructured Data 203

1	Learning Goals		203
2	Advance Organizer		203
3	Acronyms		205
4	Key Problems		205
5	Review on Data		206
	5.1	Well-Structured Data	208
	5.2	Semi-structured Data	209
	5.3	Weakly Structured Data	211
	5.4	On the Topology of Data	212
6	Networks = Graphs + Data		213
	6.1	Networks in Biological Systems	213
	6.2	Network Theory	214
		6.2.1 Basic Concepts of Networks	214
		6.2.2 Computational Graph Representation	214
		6.2.3 Network Metrics	215
		6.2.4 Graphs from Point Cloud Datasets	219
	6.3	Network Examples	225
		6.3.1 The Human Brain as Network	225
		6.3.2 Systems Biology and Human Diseases	225
		6.3.3 Gene Networks	227
	6.4	The Essence: Three Types of Biomedical Networks	228
		6.4.1 Transcriptional Regulatory Networks	228
		6.4.2 Protein–Protein Networks	229
		6.4.3 Metabolic Networks	231
	6.5	Structural Homologies	233
7	Future Outlook		234
8	Exam Questions		235
	8.1	Yes/No Decision Questions	235
	8.2	Multiple Choice Questions (MCQ)	236
	8.3	Free Recall Block	238

9	Answers		241
	9.1	Answers to the Yes/No Questions	241
	9.2	Answers to the Multiple Choice Questions (MCQ)	242
	9.3	Answers to the Free Recall Questions	244
References			247

Lecture 6 Multimedia Data Mining and Knowledge Discovery 251

1	Learning Goals		251
2	Advance Organizer		251
3	Acronyms		253
4	Key Problems		253
5	Knowledge Discovery		256
	5.1	What Is Knowledge?	256
	5.2	Implicit vs. Explicit Knowledge	258
	5.3	Differences Between KDD and DM	258
6	Data Mining Methodologies		260
	6.1	Definitions	260
	6.2	Example Tasks	261
	6.3	Taxonomy of Methods	261
	6.4	Supervised Learning	263
		6.4.1 Artificial Neural Networks	264
		6.4.2 Clinical Example: Model for End-Stage Liver Disease	268
		6.4.3 Bayesian Network	271
		6.4.4 Alterative Approach: Support Vector Machines	275
	6.5	Text Mining and Semantic Methods	276
		6.5.1 Latent Semantic Analysis	277
		6.5.2 Latent Dirichlet Allocation	278
		6.5.3 Principal Components Analysis	279
7	Future Outlook		280
8	Exam Questions		285
	8.1	Yes/No Decision Questions	285
	8.2	Multiple Choice Questions (MCQ)	286
	8.3	Free Recall Block	288
9	Answers		291
	9.1	Answers to the Yes/No Questions	291
	9.2	Answers to the Multiple Choice Questions (MCQ)	292
	9.3	Answers to the Free Recall Questions	293
References			295

Lecture 7 Knowledge and Decision: Cognitive Science and Human–Computer Interaction 299

1	Learning Goals	299
2	Advance Organizer	299
3	Acronyms	301
4	Key Problems	302

5	Human Information Processing		302
	5.1	Decision Making and Reasoning	302
	5.2	The Three-Storage Memory Model	304
		5.2.1 Example: Visual and Audial Information Processing	306
		5.2.2 Central Executive System	308
		5.2.3 Selective Attention	309
	5.3	Clinical Decision-Making Process	310
	5.4	Reasoning and Problem-Solving Procedures	312
		5.4.1 Hypothetico-deductive Model (HDM) vs. PCDA Deming Wheel	312
		5.4.2 Signal Detection Theory	316
		5.4.3 Differential Diagnosis	319
		5.4.4 Rough Set Theory (RST)	320
		5.4.5 Heuristic Decision Making	326
6	Human Error		327
7	Future Outlook		328
8	Exam Questions		330
	8.1	Yes/No Decision Questions	330
	8.2	Multiple Choice Questions (MCQ)	331
	8.3	Free Recall Block	333
9	Answers		336
	9.1	Answers to the Yes/No Questions	336
	9.2	Answers to the Multiple Choice Questions (MCQ)	337
	9.3	Answers to the Free Recall Questions	339
References			341

Lecture 8 Biomedical Decision Making: Reasoning and Decision Support 345

1	Learning Goals		345
2	Advance Organizer		345
3	Acronyms		347
4	Key Problems		347
5	Decision Support Systems		348
	5.1	Decision Models	349
	5.2	Evolution of DSS	350
	5.3	Design Principles of DSS	353
	5.4	Clinical Guidelines	355
6	Case-Based Reasoning		360
7	Future Outlook		364
8	Exam Questions		365
	8.1	Yes/No Decision Questions	365
	8.2	Multiple Choice Questions (MCQ)	366
	8.3	Free Recall Block	368

| Contents | xvii |

9	Answers	371
	9.1 Answers to the Yes/No Questions	371
	9.2 Answers to the Multiple Choice Questions (MCQ)	372
	9.3 Answers to the Free Recall Questions	374
References		376

Lecture 9   Interactive Information Visualization and Visual Analytics 379
1   Learning Goals 379
2   Advance Organizer 379
3   Acronyms 381
4   Key Problems 381
5   Fundamentals of Visualization Science 382
   5.1   Verbal Information Versus Visual Information 382
   5.2   Is a Picture Really Worth a Thousand Words? 385
   5.3   Informatics as Semiotics Engineering 386
   5.4   Visualization Process 388
   5.5   The Case of John Snow 392
6   Visualization Methods 393
   6.1   Overview 394
   6.2   Parallel Coordinates 395
   6.3   Radial Coordinate Visualization 399
   6.4   Star Plots 401
7   Visual Analytics 403
8   Future Outlook 406
9   Exam Questions 407
   9.1   Yes/No Decision Questions 407
   9.2   Multiple Choice Questions (MCQ) 408
   9.3   Free Recall Block 410
10 Answers 413
   10.1 Answers to the Yes/No Questions 413
   10.2 Answers to the Multiple Choice Questions (MCQ) 414
   10.3 Answers to the Free Recall Questions 416
References 418

Lecture 10   Biomedical Information Systems and Medical Knowledge Management 421
1   Learning Goals 421
2   Advance Organizer 421
3   Acronyms 423
4   Key Problems 423
5   Workflow Modeling 424
   5.1   Workflow and Decision Support 425
   5.2   Formal Modeling 425
   5.3   Example Clinical Workflow 428
   5.4   Workflows in Bioinformatics 432

6	Hospital Information Systems		432
	6.1	Architectures	433
	6.2	Process-Oriented Information Systems	434
7	Multimedia in the Hospital		434
	7.1	PACS	435
	7.2	Data Standards (DICOM, HL7, LOINC)	436
8	Future Outlook		442
9	Exam Questions		445
	9.1	Yes/No Decision Questions	445
	9.2	Multiple Choice Questions (MCQ)	446
	9.3	Free Recall Block	448
10	Answers		451
	10.1	Answers to the Yes/No Questions	451
	10.2	Answers to the Multiple Choice Questions (MCQ)	452
	10.3	Answers to the Free Recall Questions	454
References			456

Lecture 11 Biomedical Data: Privacy, Safety, and Security 459

1	Learning Goals		459
2	Advance Organizer		459
3	Acronyms		461
4	Key Problems		462
5	Standardization and Health Care		463
	5.1	What Is Risk?	463
	5.2	The IOM Report	464
	5.3	Medical Error	466
		5.3.1 Eindhoven Classification Model	466
		5.3.2 Adverse Event Reporting	467
		5.3.3 Human Error	468
		5.3.4 Risk Management	468
		5.3.5 Ubiquitous Devices	470
		5.3.6 Context-Aware Patient Safety	471
	5.4	Safety, Security and Technical Dependability	473
6	Patient Data Privacy		475
7	Private Personal Health Record		480
8	Future Outlook		486
9	Exam Questions		487
	9.1	Yes/No Decision Questions	487
	9.2	Multiple Choice Questions (MCQ)	488
	9.3	Free Recall Block	490
10	Answers		493
	10.1	Answers to the Yes/No Questions	493
	10.2	Answers to the Multiple Choice Questions (MCQ)	494
	10.3	Answers to the Free Recall Questions	496
References			498

Lecture 12 Methodology for Information Systems: System Design, Usability, and Evaluation 501
1   Learning Goals 501
2   Advance Organizer 501
3   Acronyms 505
4   Key Problems 506
5   A Framework for Understanding Usability 507
6   Standards 509
    6.1   EU Directive: Medical Device Directive 510
    6.2   ISO Standards 511
    6.3   Quality Management Process Cycle 514
    6.4   Software Product Quality Model 515
7   Usability Engineering 516
    7.1   Usability Engineering Methods 516
    7.2   How to Measure Usability? 517
        7.2.1   The System Usability Scale (SUS) 517
        7.2.2   The Software Usability Measurement Inventory (SUMI) 518
        7.2.3   Usability Measurement Metrics 518
    7.3   User-Centered Design and Development 519
8   Evaluation 525
9   Technology Acceptance 526
10 Future Outlook 527
11 Exam Questions 530
    11.1 Yes/No Decision Questions 530
    11.2 Multiple Choice Questions (MCQ) 531
    11.3 Free Recall Block 533
12 Answers 536
    12.1 Answers to the Yes/No Questions 536
    12.2 Answers to the Multiple Choice Questions (MCQ) 537
    12.3 Answers to the Free Recall Questions 539
References 541

Index 547

Lecture 1

Introduction: Computer Science Meets Life Science

1 Learning Goals

At the end of this first lecture, you:

- would be fascinated to see our world in data.
- would have a basic understanding of the building blocks of life.
- would be familiar with some differences between Life Sciences and Computer Sciences.
- would be aware of some possibilities and some limits of Biomedical Informatics.
- would have some ideas of some future directions of Biomedical Informatics.

2 Advance Organizer

Bioinformatics	A discipline, as part of biomedical informatics, at the interface between *bio*logy and *infor*mation science and mathema*tics*; processing of biological data
Biomarker	A characteristic (e.g., body-temperature (fever) as a biomarker for an infection, or proteins measured in the urine) as an indicator for normal or pathogenic biological processes, or pharmacologic responses to a therapeutic intervention
Biomedical data	Compared with general data, it is characterized by large volumes, complex structures, high dimensionality, evolving biological concepts, and insufficient data modeling practices
Biomedical Informatics	2011-definition: similar to medical informatics but including the optimal use of biomedical data, e.g., from genomics, proteomics, metabolomics

Classical Medicine	is both the science and the art of healing and encompasses a variety of practices to maintain and restore health
Cognitive Performance	Human capabilities, e.g., time to perform a task, number of errors per task, attention
Genomics	Branch of molecular biology which is concerned with the structure, function, mapping, and evolution of genomes
Medical Informatics	1970 definition: "... scientific field that deals with the storage, retrieval, and optimal use of medical information, data, and knowledge for problem solving and decision making"
Metabolomics	Study of chemical processes involving metabolites (e.g., enzymes). A challenge is to integrate proteomic, transcriptomic, and metabolomic information to provide a more complete understanding of living organisms
Molecular Medicine	Emphasizes cellular and molecular phenomena and interventions rather than the previous conceptual and observational focus on patients and their organs
Omics data	Data from for example genomics, proteomics, metabolomics
Pervasive Computing	Similar to ubiquitous computing (Ubicomp), a post-desktop model of Human–Computer Interaction (HCI) in which information processing is integrated into everyday, miniaturized and embedded objects and activities; having some degree of "intelligence"
Pervasive Health	All unobtrusive, analytical, diagnostic, supportive, etc. information functions to improve health care, e.g., remote, automated patient monitoring, diagnosis, home care, self-care, independent living
Proteome	The entire complement of proteins that is expressed by a cell, tissue, or organism
Proteomics	Field of molecular biology concerned with determining the proteome
P-Health Model	Preventive, Participatory, Preemptive, Personalized, Predictive, Pervasive (=available to anybody, anytime, anywhere)
Space	A set with some added structure [Note: the colloquial space most familiar to us is called the Euclidean vector space $\mathbb{R}^n$ which is the space of all n-tuples of real numbers $(x1, x2, \ldots xn)$]. The $\mathbb{R}^2$ is therefore called the Euclidean plane. In Special Relativity Theory of Albert Einstein, this Euclidean three-dimensional space plus the time (often called fourth dimension) are unified into the so-called Minkowski space. For us in data mining one of the most important spaces is the topological space

Technological Performance	Machine "capabilities," e.g., short response time, high throughput, and high availability
Time	A dimension in which events can be ordered along a time line from the past through the present into the future
Translational Medicine	Based on interventional epidemiology; progress of Evidence-Based Medicine (EBM), integrates research from basic science for patient care and prevention
Von Neumann Computer	A 1945 architecture, which still is the predominant machine architecture of today (opp: Non-vons, incl analogue, optical, quantum computers, cell processors, DNA and neural nets (in silico))

3 Acronyms

AI	Artificial intelligence
AL	Artificial life
CPG	Clinical Practice Guideline
CPOE	Computerized physician order entry
CMV	Controlled Medical Vocabulary
DEC	Digital Equipment Corporation (1957–1998)
DNA	Deoxyribonucleic acid
EBM	Evidence-based medicine
ECG	Electrocardiogram
EEG	Electroencephalogram
EMG	Electromyogram
EPR	Electronic patient record
GBM	Genome-based medicine
GC	Gas chromatography
GPM	Genetic polymorphism
HCI	Human–computer interaction
LC	Liquid chromatography
LNCS	Lecture notes in computer science
MS	Mass spectrometry
mRNA	Messenger RNA
NGC	New general catalogue of Nebulae and star clusters in astronomy
NGS	Next-generation sequencing
NMR	Nuclear magnetic resonance
PDB	Protein database
PDP	Programmable data processor (mainframe)
PPI	Protein–protein interaction
RFID	Radiofrequency identification device
RNA	Ribonucleic acid

SNP Single nucleotide polymorphism
TNF Tumor necrosis factor
TQM Total quality management

4 Key Problems

At the intersection of computational and life sciences some key problems are as follows:

- A zillion of different biological species (humans, animals, bacteria, virus, plants, ...).
- The enormous complexity of medicine per se (Patel et al. 2011).
- Big datasets in the life sciences.
- Limited time, e.g., on average a medical doctor in a public hospital has only 5 min to make a decision (Gigerenzer 2008).
- Limited computational power in comparison of the complexity of life (and the natural limitations of the von Neumann architecture).

5 Our World in Data

Let's start with a look at some macroscopic data: In Slide 1-1 we see the globular star cluster NGC 5139 Omega Centauri, discovered by Edmund Halley in 1677, with a diameter of about 90 light years, including several millions of stars, and approx. 16,000 light-years away from earth; look at the structure—and consider the aspect of time: when our eyes recognize this structure—it might even no longer exist. Time and space are the most fascinating principles of our world (Hawking et al. 1996).

5 Our World in Data

Fig. 1 See Slide 1-1

Slide 1-1: Our World in Data (1/2): Macroscopic Structures

Image credit: ESO. Acknowledgement: A. Grado/INAF-Capodimonte Observatory, Online available via: http://www.eso.org/public/images/eso1119b

---

From these large macroscopic structures let us switch to tiny microscopic structures: Proteins. These are organic compounds having large molecules composed of long chains of amino acids. Proteins are essential for all living organisms and are used by cells for performing and controlling cellular processes, including degradation and biosynthesis of molecules, physiological signaling, energy storage and conversion, formation of cellular structures, etc. Protein structures are determined with crystallographic methods or by nuclear magnetic resonance spectroscopy. Once the atomic coordinates of the protein structure have been determined, a table of these coordinates is deposited into a protein database (PDB), an international repository for 3D structure files (see Lecture 4). In Slide 1-2 we see such a structure and the data, representing the mean positions of the entities within the substance, their chemical relationship, etc. (Wiltgen and Holzinger 2005).

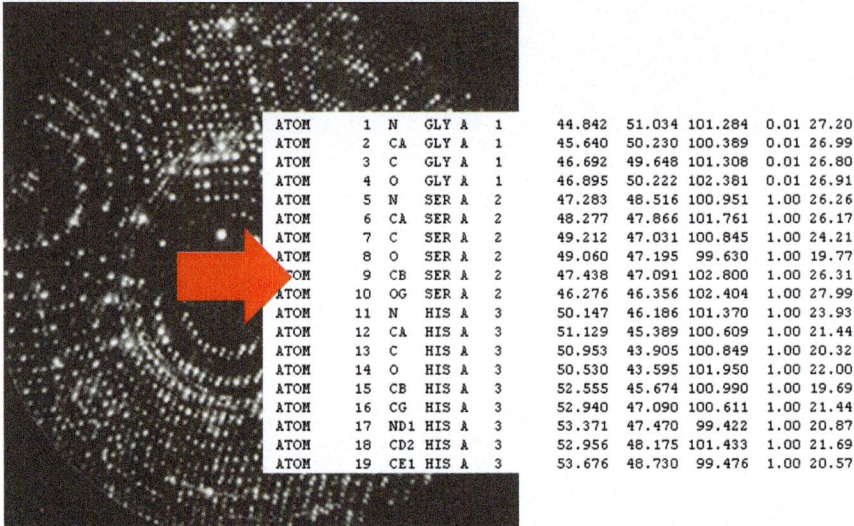

Fig. 2 See Slide 1-2

Slide 1-2: Our World in Data (2/2): Microscopic Structures

Structures of protein complexes, determined by X-ray crystallography, and the data stored in the PDB database (Wiltgen and Holzinger 2005). X-ray crystallography is a standard method to analyze the arrangement of objects (atoms, molecules) within a crystal structure. This data contains the mean positions of the entities within the substance, their chemical relationship, and various others. If a medical professional looks at the data, he or she sees only lengthy tables of numbers, the quest is now to get knowledge out of this data (see Slide 1-3).

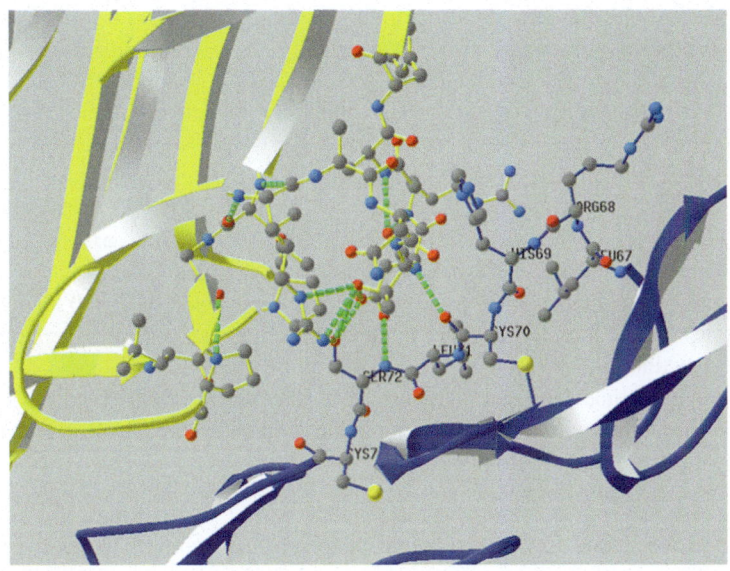

Fig. 3 See Slide 1-3

Slide 1-3: Knowledge Discovery from Data

It is essential to make such structures visible to the domain experts, so that they can understand and gain knowledge—for instance, it may lead to the discovery of new, unknown structures in order to modify drugs; the transformation of such information into knowledge is vital for prevention and treatment of diseases, consequently a contribution towards personalized medicine. This picture shows the 3D structure of the numbers seen in Slide 1-2. The tumor necrosis factor TNF (upper part—which causes the death of a cell) is "interacting" with the receptor (lower part). The residues at the macromolecular interface are visualized in a "ball-and-stick" representation. The covalent bonds are represented as sticks between atoms. The atoms are the balls. The rest of the two chains are represented as ribbons. Residue names and numbers of the TNF receptor are labeled. The hydrogen bonds are represented by these yellow dotted lines (Wiltgen et al. 2007).

Slide 1-4: First Yeast Protein–Protein Interaction Network

Good examples for data intensive, highly complex microscopic structures are yeast protein networks. Yeast is a eukaryotic microorganism (fungus), consisting of single oval cells, which asexually reproduces by budding, capable of converting sugar into alcohol and carbon dioxide. There are 1,500 known species currently, estimated to be only 1 % of all yeast species. Yeasts are unicellular, typically measuring 4 μm in diameter. In this slide we see the first protein–protein interaction (PPI) network (Jeong et al. 2001): *Saccharomyces cerevisiae*. It is perhaps the most useful yeast, used for brewing, winemaking, and baking since ancient times. This *S. cerevisiae* PPI network contains 1,870 proteins as nodes, connected by 2,240 identified direct physical interactions, and is derived from combined, nonoverlapping data, obtained mostly by systematic two-hybrid analyses. The nodes are the proteins; the links between them are the physical interactions (bindings), red nodes are lethal, green nodes are nonlethal, orange nodes are slow growing, and yellow are not yet known.

PPIs are essential for all biological processes, and compiling them provides insights into protein functions. Networks are relevant from a systems biology point of view, as they help to uncover the generic organization principles of functional cellular networks, when both spatial and temporal aspects of interactions are considered (Ge et al. 2003). In Slide 1-5 we see the first visualization of a human PPI (Stelzl et al. 2005). Most cellular processes rely on such networks (Barabási and Oltvai 2004), and a breakdown of such networks is responsible for most human diseases (Barabási et al. 2011).

Slide 1-5: First Human Protein–Protein Interaction Network

First visualization of a PPI structure of a human protein provided by researchers from the Charité in Berlin. Medical experts can gain knowledge out of this data, e.g., they can understand complex processes and understand illnesses (Stelzl et al. 2005); Light blue nodes are known proteins, orange nodes are disease proteins, and yellow nodes are not known yet. Contrast this image to Slide 1-1 and look at the similarities.

There is a very nice Video available, called Powers-of-Ten, which demonstrate the dimensions of both worlds very good; it is online available via: http://www.powersof10.com

A similar interactive visualization, focused on biological structures can be found here: http://learn.genetics.utah.edu/content/begin/cells/scale

Network biology aims to understand the behavior of cellular networks; a parallel field within medicine is called network medicine and aims to uncover the role of such networks in human disease. To demonstrate the similarity to nonnatural structures, we see in Slide 1-6 the visualization of the Blogosphere.

> **Slide 1-6: Nonnatural Network Example: Blogosphere**
>
> Hurst (2007) mapped this image as a result of 6 weeks of observation: densely populated areas represent the most active portions of the Blogosphere. By showing only the links in the graph, we can get a far better look at the structure than if we include all the nodes. In this image, we are looking at the core of the Blogosphere: The dark edges show the reciprocal links (where A has cited B and vice versa), the lighter edges indicate a-reciprocal links. The larger, denser area of the graph is that part of the Blogosphere generally characterized by sociopolitical discussion and the periphery contains some topical groupings.

A final example in Slide 1-7 shows the principle of viral marketing. The idea is to spread indirect messages which in return suggest spreading them farther. If you press the Like-button in Facebook, a similar process starts, similar to an epidemic in public health.

> **Slide 1-7: Social Behavior Contagion Network**
>
> Aral (2011) calls this *behavior contagion* and it is of much importance for research to know how human behavior can spread. We can mine masses of social network data in order to gain knowledge about the contagion of information, which is of interest for the health area, in particular for public health. A current trend of research is Opinion Mining, where you analyze such datasets (Petz et al. 2012, 2013).

A disease is rarely a consequence of an abnormality in a single gene, but reflects the perturbations of the complex intracellular network.

The emerging tools of network medicine offer a platform to explore systematically the molecular complexity of a particular disease. This can lead to the identification of disease pathways, but also to the molecular relationships between apparently distinct pathological phenotypes. Advances in this direction are essential to identify new disease-genes and to identify drug targets and biomarkers for complex diseases, see Slide 1-8 (Barabási et al. 2011).

> **Slide 1-8: Human Disease Network**
>
> In this slide we see a human disease network: the nodes are diseases; two diseases are linked if they share one or several disease-associated genes. Not shown are small clusters of isolated diseases. The node color reflects the disease class of the corresponding diseases to which they belong, cancers appearing as blue nodes and neurological diseases as red nodes. The node size correlates with the number of genes known to be associated with the corresponding disease (Barabási et al. 2011).

6 What Is Life?

The Austrian physicist Erwin Schrödinger provided in 1943 a series of lectures entitled "What Is Life? The Physical Aspect of the Living Cell and Mind" (Schrödinger 1944). He described some fundamental differences between animate and inanimate matter, and raised some hypotheses about the nature and molecular structure of genes—10 years before the discoveries of Crick and Watson (1953). The rules of life seemed to violate fundamental interactions between physical particles such as electrons and protons. It is as if the organic molecules in the cell have a kind of "knowledge" that they are living (Westra et al. 2007). It is both interesting and important to accept the fact that despite all external influences, this "machinery" is working now for more than 3.8 billion years (Schidlowski 1988; Mojzsis et al. 1996).

Fig. 4 See Slide 1-9

> **Slide 1-9: Living "Things" Are Able ...**
>
> Schrödinger's early ideas encouraged many scientists to investigate the molecular basis of life and he stated that information (negative entropy) is the abstract concept that quantifies the notion of this order amongst the building blocks of life.

Life is a fantastic interplay of matter, energy, and information, and essential functions of living beings correspond to the generation, consumption, processing, preservation, and duplication for information. Scientists in Artificial Intelligence (AI) and Artificial Life (AL) are interested in understanding the properties of living organisms to build artificial systems that exhibit these properties for useful purposes. AI researchers are interested mostly in perception, cognition, and generation of action, whereas AL focuses on evolution, reproduction, morphogenesis, and metabolism (Brooks 2001).

> **Slide 1-10: Life Is Complex Information**
>
> All complex life is composed of eukaryotic (nucleated) cells (Lane and Martin 2010). A good example of such a cell is the protist *Euglena gracilis* (in German "Augentierchen") with a length of approx. 30 μm.

It is also very interesting that in contrast to a few laws that govern the interactions between the few really elementary physical particles, there are at least tens of thousands of different genes and proteins, with millions of possible interactions, and each of these interactions obeys its own peculiarities. Consequently, in the life sciences are different processes involved including transcription, translation, and subsequent folding (Hunter 2009). Advances in bioinformatics generate masses of biological data, increasing the discrepancy between what is observed and what is actually known about life's organization at the molecular level. Knowledge Discovery plays an important role for the understanding, for getting insights and for sense making of the masses of observed data (Holzinger 2013).

6.1 The Building Blocks of Life

> **Slide 1-11: Building Blocks of Life: Overview**
>
> To get a "big picture" we have a first look on the human body: The average 70 kg adult contains approximately 3×10^{27} atoms and contains about

(continued)

(continued)

60 different chemical elements and is build up by approximately 10^{14} cells. The cell is the basic building block and consists of supramolecular complexes, chromosomes, plasma membranes, etc., consisting of macromolecules (DNA, proteins, and cellulose) and monomeric units, such as nucleotides, amino acids, etc. Very interesting is the large amount of human microbiome, which include microorganisms (bacteria, fungi, and archaea) that resides on the surface and in deep layers of skin, in the saliva and oral mucosa, in the conjunctiva, and in the gastrointestinal tracts. Studies of the human microbiome have revealed that healthy individuals differ remarkably. Much of this diversity remains unexplained, although diet, environment, host genetics, and early microbial exposure have all been implicated. Accordingly, to characterize the ecology of human-associated microbial communities, the Human Microbiome Project has analyzed the largest cohort and set of distinct, clinically relevant body habitats so far (Mitreva 2012). From the discovery of DNA to the sequencing of the human genome, the formation of biological molecules from gene to RNA and protein has been the central tenet of biology. Yet the origins of many diseases, including allergy, Alzheimer's disease, asthma, autism, diabetes, inflammatory bowel disease, multiple sclerosis, Parkinson's disease, and rheumatoid arthritis, continue to evade our understanding (Marth 2008).

Slide 1-12: The Dogma of Molecular Biology

To understand the big picture in Slide 1-11, we follow the central dogma of molecular biology, which states that DNA is transcribed into RNA and translated into protein (Crick 1970):

$$\text{DNA} \rightarrow \text{RNA} \rightarrow \text{Protein} \rightarrow \text{Cellular Phenotype} \qquad (1\text{-}1)$$

Similar to (1-1) there is a central dogma of genomics, which states (Pevsner 2009):

$$\text{Genome} \rightarrow \text{Transcriptome} \rightarrow \text{Proteome} \rightarrow \text{Cellular Phenotype} \qquad (1\text{-}2)$$

Three perspectives arise from those fundamentals:
(1) The cell, (2) the organism, and (3) the tree of life (evolution).

The cell is the basic building block of all organisms and forms organs and tissue. Before we look at this fantastic building block of life, let us first look at the fundamental building blocks of the cell.

6.2 Proteins

Proteins are large biological molecules consisting of one or more chains of amino acids and they vary from one to another mainly in their sequence of amino acids, which is dictated by the nucleotide sequence of their genes, and which usually results in folding of the protein into a specific three-dimensional structure that determines its activity. Proteins perform a variety of functions and they regulate cellular and physiological activities. The functional properties of proteins depend on their three-dimensional structures. The native structure of a protein can be experimentally determined using X-ray crystallography, nuclear magnetic resonance (NMR) spectroscopy, or electron microscopy. Over the past 40 years, the structures of 55,000+ proteins have been determined. On the other hand, the amino-acid sequences are determined for more than eight million proteins. The specific sequence of amino acids in a polypeptide chain folds to generate compact domains with a particular three-dimensional structure. The polypeptide chain itself contains all the necessary information to specify its three-dimensional structure. Deciphering the three-dimensional structure of a protein from its amino-acid sequence is a long-standing goal in molecular and computational biology (Gromiha 2010).

6.2.1 From Amino Acids to Protein Structures

> **Slide 1-13: Amino acid > Protein-Chain > Protein-Structure**
>
> 1. **Amino acid**: Protein sequences consist of 20 different amino acids serving as building blocks of proteins. Amino acids contain a central carbon atom, called Alpha Carbon (C_α) which is attached to a hydrogen atom (H), an amino group (NH_2), and a carboxyl group (COOH). The letter R indicates the presence of a side chain, which distinguishes each amino acid (Hunter 2009).
> 2. **Protein chain**: Several amino acids form a protein chain, in which the amino group of the first amino acid and the carboxyl group of the last amino acid remain intact, and the chain is said to extend from the amino (N) to the carboxyl (C) terminus. This chain of amino acids is called a polypeptide chain, main chain, or backbone. Amino acids in a polypeptide chain lack a hydrogen atom at the amino terminal and an OH group at the carboxyl terminal (except at the ends), and hence, amino acids are also called amino-acid residues (simply residues). Nature selects the combination of amino-acid residues to form polypeptide chains for their function, similar to the combination of letters of an alphabet to form meaningful words and sentences. These polypeptide chains that have specific functions are called proteins.

(continued)

(continued)

3. **Protein structure**: Depending on their complexity, protein molecules may be described by four levels of structure: primary, secondary, tertiary, and quaternary. Because of the advancements in the understanding of protein structures, two additional levels such as super-secondary and domain have been proposed between secondary and tertiary structures. A stable clustering of several elements of secondary structures is referred to as a super-secondary structure. A somewhat higher level of structure is the domain, which refers to a compact region and distinct structural unit within a large polypeptide chain.

3a. **Primary structure**: The linear sequence of amino-acid residues in a protein is described by the primary structure: It includes all the covalent bonds between amino acids. The relative spatial arrangement of the linked amino acids is unspecified.

3b. **Secondary structure**: Regular, recurring arrangements in space of adjacent amino-acid residues in a polypeptide chain are described by the secondary structure. It is maintained by hydrogen bonds (see the dotted lines in Slide 1-3) between amide hydrogens and carbonyl oxygens of the peptide backbone. The main secondary structures are α-helices and β-folding-structures (β-sheets). The polypeptide backbone is tightly coiled around the long axis of the molecule, and R groups of the amino-acid residues protrude outward from the helical backbone.

3c. **Tertiary structure**: It refers to the spatial relationship among all amino acids in a polypeptide; it is the complete three-dimensional structure of the polypeptide with atomic details. Tertiary structures are stabilized by interactions of side chains of non-neighboring amino-acid residues and primarily by non-covalent interactions.

3d. **Quaternary structure**: This refers to the spatial relationship of the polypeptides or subunits within a protein and is the association of two or more polypeptide chains into a multi-subunit or oligomeric protein. The polypeptide chains of an oligomeric protein may be identical or different. The quaternary structure also includes the cofactor and other metals, which form the catalytic unit and functional proteins (Gromiha 2010).

6.2.2 Tertiary Structure of a Protein

All biological mechanisms in the living cell involve protein molecules, consequently proteins are central components of cellular organization and function. The mechanistic view that the structure (the "shape" in the ribbon diagrams, see

Slide 1-14), manages the biological function in proteins has been confirmed in wet laboratory experiments (Anfinsen 1973). A unique set of atoms compose a protein molecule and determines to a great extent the spatial arrangement by these atoms for biological function. The state in which a protein carries its biological activity is called the protein native state. Microscopically, this macro-state is an ensemble of native conformations also referred to as the native state ensemble (Shehu and Kavraki 2012).

Proteins fold from a highly disordered state into a highly ordered one. The folding problem has been stated as predicting the tertiary structure from sequential information. The ensemble of unfolded forms may not be as disordered as believed, and the native form of many proteins may not be described by a single conformation, but rather an ensemble. For the quantification of the relative disorder in the folded and unfolded ensembles entropy (see Lecture 2) measures are suitable (Chirikjian 2011; Shehu and Kavraki 2012).

Slide 1-14: Tertiary Structure of a Protein

The tertiary structure of a protein is basically the overall, unique, three dimensional folding of a protein. In a protein folding diagram (ribbon diagram) we can recognize the beta pleated sheets (ribbons with arrows) and the alpha helical regions (barrel shaped structures). What makes proteins special is less a matter of their monomer types and more a matter of their specific sequences (Dill et al. 1995).

6.2.3 Protein Analytics

For Protein analytics a lot of different methods are known.

Slide 1-15: Protein Analytics

1. **X-ray crystallography** is the primary method and used for determining the atomic and molecular structures with the output of a three-dimensional picture of the density of electrons within a crystal structure. From this electron density, the mean positions of the atoms (see in Slide 1-2) in the crystal can be determined, as well as their chemical bonds, their disorder and various other information (Schotte et al. 2003).
2. **Gel-Electrophoresis** and in particular 2D-Gel-Electrophoresis had a major impact on the developments of proteomics (Southern 1975), although it is no longer the exclusive separation tool (See Mass-Spectrometry). Basically, electrophoresis separates molecules according

(continued)

(continued)

to their charge (mass ratio) and 2D-electrophoresis separates molecules according to both their charge and their mass.

The possibility of analyzing spots of interest coming from 2D gels was the real start of proteomics. At those times where no complete genome was published yet, such techniques provided enough information to look for homologs, or to devise oligonucleotides for screening DNA libraries (Rabilloud et al. 2010).

3. **Chromatography** is the collective term for the separation of mixtures, typical examples are Gas chromatography (GC) and Liquid chromatography (LC), the latter specifically a separation of proteins by size, i.e., gel filtration chromatography (Xiao and Oefner 2001).

    Gas chromatography is an excellent separation technique and detects the ions and generates mass spectrum for each analyte, and this structural information aids in its identification. The synergistic coupling of GC and MS (see below) renders the tandem technique a major analytical workhorse in metabolomics (Yip and Yong Chan 2013).

4. **Mass spectrometry (MS)**-based approaches involve either detection of intact proteins, referred to as top-down proteomics, or identification of protein cleavage products, referred to as bottom-up or shotgun proteomics. Mass spectrometry-based proteomics is an important tool for molecular, cellular biology and systems biology. These include the study of PPI via affinity-based isolations on a small and proteome-wide scale, the mapping of numerous organelles, and the generation of quantitative protein profiles from diverse species. The ability of mass spectrometry to identify and, increasingly, to precisely quantify thousands of proteins from complex samples can be expected to impact broadly on biology and medicine (Aebersold and Mann 2003).

    Top-down proteomic strategies retain a lot of information about protein sequence (protein isoforms). Recent advances in top-down proteomics allow for identification of hundreds of intact proteins in yeast and mammalian cells, however, clinical applications of top-down proteomics are still limited. **Bottom-up proteomic** approaches suffer from a loss of information about protein isoforms and posttranslational modification (PTM), especially for low-abundance proteins. Bottom-up proteomics greatly benefits from superior Liquid Chromatography (LC) separation of peptides prior to mass spectrometry, requires lower amounts of material, and provides better peptide fragmentation and higher sensitivity. Due to the very high number of routine protein identifications in biological samples, bottom-up proteomics remains the platform of choice for biomarker discovery. Process of protein identification by bottom-up

(continued)

(continued)

proteomic methods involves a set of consecutive steps, such as protein digestion, peptide separation by LC, peptide ionization, gas-phase peptide separation, peptide fragmentation, and detection of mass-to-charge ratios (m/z) and intensities of peptide ions and their tandem mass spectrometry (MS/MS) fragments (Drabovich et al. 2013).

5. **Nuclear magnetic resonance spectroscopy (NMR)** is a research technique that exploits the magnetic properties of certain atomic nuclei and determines the physical and chemical properties of atoms or the molecules in which they are contained. A goal is to obtain three-dimensional structures of the protein, similar to what can be achieved by X-ray crystallography. In contrast to X-ray crystallography, NMR spectroscopy is usually limited to proteins smaller than 35 kDa (Note: Dalton is the standard unit used for indicating the atomic mass: $1\ Da = 1\ g/mol = 1.6 \times 10^{-27} kg$) although larger structures have been solved. NMR spectroscopy is often the only way to obtain high resolution information on partially or wholly intrinsically unstructured proteins.

Slide 1-16: Comparison of Some Current Methods

This table shows some current Methods and some properties, e.g., whether they are minimally invasive, usable on live cells, in real time, etc. (Okumoto et al. 2012).

Abbreviations: Bq = Becquerel; MALDI = matrix-assisted laser desorption ionization; MRI = magnetic resonance imaging; NIMS = nanostructure initiator mass spectrometry; PET = positron emission tomography; SIMS = secondary ion mass spectrometry; TOF = time-of-flight mass spectrometry.

Slide 1-17: Enzymes

Enzymes are large biological molecules responsible for the thousands of chemical inter-conversions that sustain life. They are highly selective catalysts, greatly accelerating both the rate and specificity of metabolic reactions, from the digestion of food to the synthesis of DNA. Most enzymes are proteins, although some catalytic RNA molecules have been identified. Enzymes adopt a specific three-dimensional structure, and may employ organic (e.g., biotin) and inorganic (e.g., magnesium ion) cofactors to assist

(continued)

(continued)

catalysis. The tremendous potential of enzymes as practical catalysts is well recognized. The scope of industrial bioconversions, especially for the production of specialty chemicals and polymers, is necessarily limited by a variety of considerations. Most such compounds are insoluble in water, and water frequently gives rise to unwanted side reactions and degrades common organic reagents. The thermodynamic equilibrium of many processes are unfavorable in water, and product recovery is sometimes difficult from this medium (Klibanov 2001).

6.3 DNA and RNA

Slide 1-18: DNA–RNA Components: Overview

The Genome (Chromosomes) consists of Deoxyribonucleic acid (DNA), which is a molecule that encodes the genetic instructions used in the development and functioning of all known living organisms and many viruses. The DNA, the RNA, and the proteins are the three major macromolecules essential for all known forms of life.

The Ribonucleic acid (RNA) is a molecule that performs the coding, decoding, regulation, and expression of genes. The RNA is assembled as a chain of nucleotides, but is usually single-stranded. Cellular organisms use messenger RNA (mRNA) to convey genetic information (often notated using the letters G, A, U, and C for the nucleotides guanine, adenine, uracil, and cytosine) that directs synthesis of specific proteins, while many viruses encode their genetic information using an RNA genome.

Slide 1-19: DNA–RNA and the Five Principal Bases (A, G, T, C) and (A, G, U, C)

The five principal nucleobases are nitrogen-containing biological compounds cytosine (in DNA and RNA), guanine (in DNA and RNA), adenine (in DNA and RNA), thymine (only in DNA), and uracil (only in RNA), abbreviated as C, G, A, T, and U, respectively. They are usually simply called bases in genetics. Because A, G, C, and T appear in the DNA, these molecules are called DNA bases; A, G, C, and U are called RNA bases.

Slide 1-20: Nobel Prize 1957

Francis Crick, James D. Watson, and Maurice Wilkins were jointly awarded the 1962 Nobel Prize for Physiology or Medicine "for their discoveries concerning the molecular structure of nucleic acids and its significance for information transfer in living material"—they described the famous double helix structure of the DNA (Crick and Watson 1953).

Some landmarks in DNA (Trent 2012):

1953	Structure of DNA shown to be a double-stranded helix.
1972	Recombinant DNA technologies allow DNA to be cloned.
1977	Sequencing methods (Sanger and Gilbert awarded Nobel Prize).
1995	First complete bacterial sequence described for *H. influenzae*.
2007	Complete human diploid genome sequences publicly announced.
2009	First human genome sequence using single sequencing technique.
2010	Single molecule sequencing (third generation) has the potential to increase sequence data generated from Tb (terabyte) to Pb (petabyte).
2010	First publication of a human metagenome—an additional layer of complexity for bioinformatics.

Slide 1-21: Genetics (Genome, Chromosome)

Genetics concerns the process of trait inheritance from parents to progenies, including the molecular structure and function of genes, gene behavior in the context of a cell or organism, gene distribution, and variation and change in populations. Given that genes are universal to living organisms, genetics can be applied to the study of all living systems; including bacteria, plants, and animals. The observation that living things inherit traits from their parents has been used since prehistoric times to improve crop plants and animals through selective breeding. The modern science of genetics, seeking to understand this process, began with the work of Gregor Mendel (1822–1884). Genetics today is inexorably focused on DNA (Brown 2012).

Genes are molecular units of heredity of a living organism. It is widely accepted by the scientific community as a name given to some stretches of DNA and RNA that code for a polypeptide or for an RNA chain. Living beings depend on genes, as they specify all proteins and functional RNA chains. Genes hold the information to build and maintain an organism's cells and pass genetic traits to offspring. The Human Genome Project has revealed that there are about 20,000–25,000 haploid protein coding genes. The completed human sequence can now identify their locations. But only about 1.5 %

(continued)

(continued)

of the genome codes for proteins, while the rest consists of noncoding RNA genes, regulatory sequences, introns, and noncoding DNA (junk DNA). Surprisingly, the number of human genes seems to be less than a factor of two greater than that of many much simpler organism (earthworm, fruit fly). An example on how informatics can help in this field can be found in (Wassertheurer et al. 2003; Holzinger et al. 2008a).

Genomics is a discipline in genetics that applies recombinant DNA, DNA sequencing methods, and bioinformatics to sequence, assemble, and analyze the function and the structure of genomes (the complete set of DNA within a single cell). It includes topics such as heterosis, epistasis, pleiotropy, and other interactions between loci and alleles within the genome (Pevsner 2009; Trent 2012).

Epigenetics is the study of changes in gene expression or cellular phenotypes, caused by mechanisms other than changes in the underlying DNA sequence, and therefore, it is called epi- (Greek: επί- over, above, outer) genetics. Some of these changes have been shown to be heritable. Example: Looking beyond DNA-associated molecules, prions (infectious proteins) are clearly epigenetic, perpetuating themselves through altered folding states. These states can act as sensors of environmental stress and, through the phenotypic changes they promote, potentially drive evolution (Trygve 2011; Kiberstis 2012).

6.4 Cell Physiology

Slide 1-22: The Cell

Cells are the fundamental structural, functional and physiological units of organisms, and they are the smallest unit of life that is classified as a "living thing," and therefore are called the "building blocks of life."

There are two types of cells: eukaryotes, which contain a nucleus; and prokaryotes, without nucleus. Prokaryotic cells are usually single-celled organisms, while eukaryotic cells can be either single-celled or part of multicellular organisms.

Cells consist of protoplasm enclosed within a cell membrane, which contains biomolecules such as proteins and nucleic acids. Organisms can be classified as unicellular (consisting of only one single cell, e.g., bacteria) or multicellular (such as animals (=human) and plants). While the number of cells in plants and animals varies from species to species, humans contain about 100 trillion (10^{14}) cells. A simplified view on the cell lets us recognize

(continued)

> (continued)
>
> the basic physiological elements and we also see the similarity between animal cells and plant cells, for details please refer to one of the standard textbooks, e.g., Boal (2012). An example of how informatics may help in cell biology can be found in (Jeanquartier and Holzinger 2013).

> **Slide 1-23: The Cell Structure and Size**
>
> Just to get a feeling of the size of the cells, here a look at a microscopic image, the bar representing 25 µm and we are able to identify examples of the various cell types found in glomeruli, which is a part of the renal corpuscle in the kidney, part of the urinary (=renal) organ system (Sperelakis 2012).

The about 100 trillion cells make up the human body, which is the entire structure of a human organism. The anatomy describes the body plan and the physiology is the study of how the body works. So let us now move from the basic building blocks to the organ systems.

6.5 Organ Systems

> **Slide 1-24: Human Organ Systems**
>
> The major systems of the human body are as follows:
>   Cardiovascular system: the blood circulation with the heart, the arteries, and the veins;
>   Respiratory system: the lungs and the trachea;
>   Endocrine system: communication within the body using hormones;
>   Digestive system: esophagus, stomach, and intestines;
>   Urinary and excretory system: eliminating wastes from the body;
>   Immune and lymphatic system: defending against disease-causing agents;
>   Musculoskeletal system: stability and moving the body with muscles
>   Nervous system: collecting, transferring and processing information;
>   Reproductive system: the sex organs.
>   For a detailed view on these topics please refer to a classical text book, e.g., the Color Atlas and Textbook of Human Anatomy: three volumes, authored by Werner Kahle und Michael Frotscher (2010, 6th Edition,
>
> (continued)

(continued)

Stuttgart: Thieme) or "the Sobotta," the Atlas of Human Anatomy: Musculoskeletal system, internal organs, head, neck, neuroanatomy (with Access to www.e-sobotta.com), authored by Friedrich Paulsen and Jens Waschke (2011, 15th Edition, Amsterdam: Elsevier).

6.5.1 Tissue

Slide 1-25: Tissue

Tissue is a cellular organization between cells and a complete organism, it can be an ensemble of similar cells from the same origin, which in the collective carry out a specific function. Organs are formed by the functional grouping together of multiple tissues. The study of tissue is known as histology and in connection with disease, we speak about histopathology. As we can see, there is a range of different tissues, e.g., (a) the skin, (b) fibrous connective tissue, forming a tendon (or sinew, which connects muscles to bones, in German: "Sehne"), (c) adipose tissue (fat), (d) cartilage, at the end of a bone (in German: "Knorpel"), (e) bone, (f) blood (white cells, red cells, and plasma).

An area of research is tissue engineering, which involves the use of materials and cells with the goal of trying to understand tissue function and tissue or organ on the body to be made de novo (Rouwkema et al. 2008).

6.5.2 Organs

A collection of tissues joined in a structural unit to serve a specific function is called organ, we will just look at one single organ as an example:

Slide 1-26: Organ Example: The Human Heart

The heart is a hollow muscle pumping blood throughout the blood vessels in the circulatory cardiovascular system, by repeated, rhythmic contractions. The adjective cardiac (from Greek: καρδιά = heart) means "related to the heart." Cardiology is the medical subject that deals with cardiac diseases and abnormalities.

6.5.3 Cardiovascular System

> **Slide 1-27: Cardiovascular System**
>
> The essential components of the human cardiovascular system include the heart, the blood vessels (arteries and veins), and the blood itself. There are two circulatory systems: the pulmonary circulation, through the lungs where blood is oxygenated; and the systemic circulation, through the remaining body to provide oxygenated blood. An average adult contains 5 L of blood.

6.5.4 Anatomical Axes

> **Slide 1-28: Anatomical Axes**
>
> In Anatomy three reference planes are used: (1) a sagittal plane, which divides the body into sinister and dexter (left and right) portions; (2) a coronal or frontal plane divides the body into dorsal and ventral (back and front, or posterior and anterior); and (3) a transversal plane, also known as an axial plane or cross-section, divides the body into cranial and caudal (head and tail) portions.
>
> In standard anatomical position, the palms of the hands point anteriorly, so anterior can be used to describe the palm of the hand, and posterior can be used to describe the back of the hand and arm.

7 What Is Biomedical Informatics?

7.1 Medicine Versus Informatics

For a first understanding, let us compare the two fields: Medicine and Informatics.

Medicine is both the science and the art of healing and encompasses a variety of practices to maintain and restore health. Medicine is a very old discipline, having a more than 3,000 year long tradition (Imhotep, third millennium BC, Egypt) and is established since Hippocrates of Kos (460 BC–370 BC) the founder of western *clinical medicine* at the time of Pericles in Classical Athens and Asclepiades of Bithynia (124–40 BCE) the first physician, who established Greek medicine in Rome. Influenced by the Epicureans, he followed atomic theory, modification and evolution and can be regarded as the founder of *molecular medicine* (Yapijakis 2009).

Clinical medicine can roughly be separated into three huge areas:

1. Neoplasm (abnormal growth of cells).
2. Inflammation (=part of the biological response of vascular tissues to harmful stimuli, such as pathogens, damaged cells, or irritants, Note: this is not a synonym for infection).
3. Trauma (physiological injuries caused by an external source).

Note: You will often hear the term "Evidence-based medicine (EBM)." EBM has evolved from clinical epidemiology and includes scientific methods in clinical decision making (Sackett et al. 1996).

Informatics on the other hand is a very young discipline, existing for slightly more than 50 years. The word is stemming from the combination of the words *informat*ion + automa*tics* (Steinbuch 1957) and was a coined word for the science of automatic *information processing*. Automatic refers to the use of a machine which evolved slightly more than 10 years before that time and was called computer.

Informatics is the **science of information**, *not* necessarily of computers, but Informatics is researching in and with computational methods, therefore is using computers—so as Astronomers uses telescopes—but do not necessarily construct telescopes (Holzinger 2003a).

7.2 Computer Machinery

Computers are physical devices; and, actually what we understand as a computer is a general-purpose electronic digital programmable machine. This machine (hardware) responds to a specific set of instructions in a well-defined manner and executes pre-recorded lists of instructions (software or program). Up to date in nearly all our everyday computer systems, the so-called von Neumann architecture is used (Neumann 1945). This is a model which uses a central processing unit (CPU) and a single separate memory to hold instructions and data (see Slide 1-29).

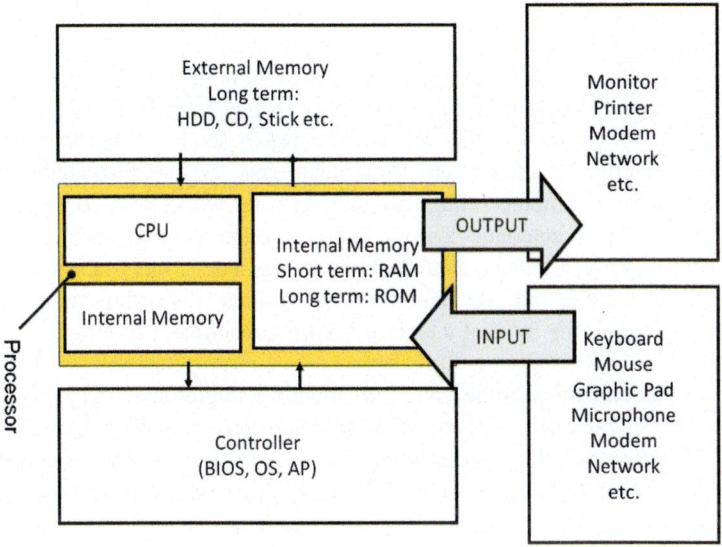

Fig. 5 See Slide 1-29

> **Slide 1-29: General-Purpose Digital Programmable Machine**
>
> In this image we can see the anatomy of a von Neumann machine: The processor executes the instructions with the central processing unit (CPU), the "heart" of the computer. The memory enables a computer to store, data and programs; the external memory are mass storage devices and allows a computer to permanently retain large amounts of data. The input device consists usually of a keyboard and mouse, but nowadays the finger replaces the mouse more and more (Holzinger 2003b).
>
> Finally the output device displays the results on a screen etc. Actually, the large mainframe computers until the 1970s had neither keyboard nor screens, they were programmed via punch-cards and the results were obtained via punch cards or tapes.
>
> The von Neumann Architecture is theoretically equivalent to a Universal Turing Machine (UTM). It was described by Alan Turing (1912–1954) in 1936, who called it an "A(utomatic)-machine," and was not intended as a practical computer—just as a Gedankenexperiment; a Turing machine can be used to simulate the logic of any computer algorithm, and is useful in explaining the functions of a CPU inside a computer (Holzinger 2002).

Turing's work is based on a more mathematical work by his doctoral advisor: Alonzo Church (1903–1995), whose work on lambda calculus intertwined with

Turing's in a formal theory of computation: The Church–Turing thesis, which states that Turing machines capture the informal notion of effective methods in logic and mathematics, and provide a precise definition of an algorithm (=mechanical procedure). All theory development reaches back to this format and is far reaching; here just one example: In 1980 Richard Feynman (Nobel Prize Winner 1965) proposed a quantum computer, actually an analogous version of a quantum system. Contrary to digital computers, an analogue computer works with continuous variables; however, such analogue computers have worked only for special problems. The digital computer, where everything is expressed in bits, has proven to be universally applicable and Feynman's innovative idea was without issue until the proposal was put in the form of a quantum Turing machine, i.e., a digital quantum computer.

The von Neumann machine (single instruction, single data stream) is still the basis for our digital world of today, although there are some other architectures of so-called "Non-vons": Single Instruction, Multiple Data streams (SIMD), Multiple Instruction, Single Data stream (MISD) and Multiple Instruction, Multiple Data streams (MIMD)—apart from different computational paradigms, e.g., evolutionary algorithms, swarm intelligence, or cellular automata (Burgin and Eberbach 2012; Cooper et al. 2008).

We may ask why digital technology generally is so successful. The answer is in the progress of technological performance, i.e., "Digital Power," which is mainly based on Moore's Law (Slide 1-30):

> **Slide 1-30: Moore's Law**
>
> In 1965, Intel co-founder Gordon E. Moore (1929–) noted that the number of components in integrated circuits had doubled every year from 1958 to 1965 and he predicted this trend would continue "for at least ten years" (Moore 1965). In 1975, Moore revised the Law to predict the doubling of computer processing power every 2 years—and this prediction has come true ever since, often doubling within 18 months. Argued several times, this law shows no sign of slowing down, and the prediction is beyond 2025 (Holzinger 2002).

Directly connected with the increasing performance of digital technology, another astonishing trend is visible, parallel to the raising processing speed and memory capacity: a vast reduction in cost and size (Slide 1-31).

> **Slide 1-31: Computer Cost/Size Versus Performance**
>
> In this slide we can observe a phenomenon which is directly connected with Moore's law: Whilst the digital power is increasing, both cost and size are

(continued)

> (continued)
>
> decreasing, making large computational power available and affordable today at pocket size (Smartphone).
>
> However, the limiting factor for continued miniaturization is that the structures will eventually reach the limits of miniaturization at atomic levels.

Consequently, there are international efforts to find alternatives for silicon-based semiconductors (see: http://www.itrs.net). The smallest size for both memory and logical devices depends on the mass of the information transporting particles, i.e., the smallest barrier width in Si devices is 5 nm (size of electrons in Si), and consequently there is a physical barrier that semiconductor designers will face. A big chance is biological computing: Due to the fact that the living cell is an information processor, which is extremely efficient in the execution of its functions, there are huge efforts towards creating of a Bio-cell, which outperforms the Si-cell in every aspect, see Slide 1-32.

> **Slide 1-32: Beyond Moore's Law → Biological Computing**
>
> Bio-Cell versus Si-Cell: Comparison between unicellular organism as information processing device with a modern Si-Cell; In the Bio-Cell (left) the components include L = Logic-Proteins; S = Sensor-Proteins; C = Signaling-Molecules, E = Glucose-Energy (Cavin et al. 2012).

Since the advent of computers, a dream of mankind was, to use such computers to augment human capabilities for structuring, retrieving, and managing information—however, this was and it is still not easy, although Human–Computer Interaction (HCI) has dramatically changed:

> **Slide 1-33: From Mainframe to Ubiquitous computing**
>
> The effects described in Slide 1-31 are directly connected with a change in HCI:
>
> 1. At the beginning (1960) there was one computer for many users, and the users were experts, who mostly worked in computer science.
> 2. The next idea was to provide every user a "personal" computer—the PC was born (2).
> 3. The reduction in size and the progress in wireless networking made computers mobile (3).
> 4. Finally, computers became pervasive and ubiquitous (4).

Ubiquitous computing (Ubicomp) was proposed by (Weiser 1991) and is a post-desktop model of HCI in which the information processing is integrated into everyday objects, see Slide 1-34:

Slide 1-34: Ubiquitous Computing: Smart Object

Ubicomp: Smart objects can exchange information, consequently can "talk" to each other (Holzinger et al. 2010b). Usually we speak from ubiquitous and pervasive technology: "ubiquitous" literally means omnipresent (in German "allgegenwärtig"); and the term pervasive implies that the technology is penetrating our daily life (in German "durchdringend"). Such devices are intended:

1. To support end users in their tasks without overwhelming them with complexity of networks, devices, software, databases (stay connected at every place in the world).
2. To ensure "context awareness" and knowledge about the environment in which the users focus on their tasks (have access to all your data from everywhere).
3. To provide natural interaction between user and technology, e.g., by using gesture, speech, and most intuitively by using multi-touch and advanced stylus technology (Holzinger et al. 2010c, 2011a, 2012a, b).

Ubicomp include Radio Frequency Identification (RFID), sensor networks, Augmented Reality, mobile, wearable and implanted devices. Since work in hospitals is formed by many cooperating clinicians having a high degree of mobility, parallel activities, and no fixed workplace, existing IT solutions often fail to consider these issues (Bardram 2004; Mitchell et al. 2000). Ubicomp in Health is promising (Varshney 2009) and may address such issues see Slide 1-35:

Slide 1-35: Pervasive Health Computing

An example for the application of Ubicomp in the area of Ambient Assisted Living: RFID technology for localizing elderly people, suffering from dementia (Holzinger et al. 2008b).

A further example for Ubicomp technologies can be seen in the future care lab of the RWTH Aachen (Slide 1-37): An integrated set of smart sensor technologies provides unobtrusive monitoring of a patient's vital health functions, such as:

(a) A smart floor, providing location tracking of multiple persons, fall detection, and weight measurement.
(b) An infrared camera, for noninvasive temperature measurement.
(c) Measurement equipment which is naturally integrated into the furniture, such as blood pressure or coagulation measurement devices.

Centerpiece of all HCI inside the future care lab is the 4.8 m × 2.4 m big multi-touch display wall.

Slide 1-36: Ambient Assisted Living: pHealth

Advances in Biomedical Informatics and Biomedical Engineering provide the foundations for modern patient-centered health-care solutions, health-care systems, technologies, and techniques. Ubicomp in the living lab, measures unobtrusively the vital parameters: blood pressure, pulse rate, body temperature, weight (Alagöz et al. 2010)—a further good example for "big data."

Slide 1-37: Example Pervasive Computing in the Hospital

Patients check in at the hospital—in addition to an ordinary wristband an RFID transponder is supplied. Patient data is entered via our application at the check-in point, any previous patient data can be retrieved from the HIS. From this information, uncritical but important data (such as name, blood type, allergies, vital medication, etc.) is transferred to the wristband's RFID transponder. The Electronic Patient Record (EPR) is created and stored at the central server. From this time on, the patient is easily and unmistakably identifiable. All information can be read from the wristband's transponder or can be easily retrieved from the EPR by identifying the patient with a reader. In contrast to manual identification, automatic processes are less error-prone. Unlike barcodes, RFID transponders can be read without line of sight, through the human body and most other materials. This enables physicians and nurses to retrieve, verify and modify information in the hospital accurately and instantly. In addition, this system provides patient identification and patient data—even when the network is crashed (Holzinger et al. 2005).

Slide 1-38: Smart Objects in the Pathology

A final example shall demonstrate how ubiquitous computer may help in the hospital. RFID devices are combined directly with the pathological specimen/example, so that they can be found again within the masses of specimens collected (Holzinger et al. 2005).

Slide 1-39: The Medical World Is Mobile

The usefulness of mobile computing applications in the area of medicine and health care is commonly accepted. Although many computer systems in the past have not paid off, retailers, service providers, and content developers are still very interested in mobile applications that are able to facilitate efficient and effective patient care information input and access, thereby improving the quality of services. A clinical example can be found in (Holzinger et al. 2011b).

7.3 Medical Informatics

The famous "Sputnik-Shock" in 1957 triggered an R & D avalanche, finally resulting in the 1969 Moonlanding, bringing computers as a "by-product," so at the same time the first computers were applied to the hospital. In Slide 1-40 we see the first computer in the University Hospital Graz in 1970 and this was the year when the term **Medical Informatics** was coined. However, the main use of a computer in the hospital at this time was accounting (Mahajan and Schoeman 1977), but other use cases were rapidly evolving, as the classical definition implies (Blum and Duncan 1990):

Slide 1-40: Turning Knowledge into Data

The PDP-10 was a mainframe computer manufactured by Digital Equipment Corporation (DEC) from the late 1960s on; the name stands for "Programmed Data Processor model 10." The first model was delivered in 1966, and made time-sharing common; the operating system was TENEX (Bobrow et al. 1972). It has been adopted to many university computing facilities and research labs (MIT, Stanford, Carnegie Mellon, etc., and Graz :-).

Slide 1-41: Four Decades from Medical to Biomedical Informatics

- 1970+ Begin of **Medical Informatics**
  - Focus on data acquisition, storage, accounting (typ. "EDV")
  - The term was first used in 1968 and the first course was set up 1978
- 1985+ Health Telematics
  - Health-care networks, Telemedicine, CPOE-Systems etc.

(continued)

(continued)

- 1995+ Web Era
  - Web-based applications, Services, EPR, etc.
- 2005+ Ambient Era
  - Pervasive and Ubiquitous Computing
- 2010+ Quality Era—**Biomedical Informatics**

Information Quality, Patient empowerment, individual molecular medicine, End-User Programmable Mashups (Auinger et al. 2009).

Slide 1-42: 2010: Turning Data into Knowledge

Today, we do not have the problem to have too little data—we have too much data and the main problem are the famous "5 minutes" to make a decision:

The problem now is that more and more such data is available, but the time available to make decisions is the same as before the advent of such technological advances. According to (Gigerenzer 2008) a typical medical doctor (in a public hospital) has approximately 5 min to make a decision. When everything turns out well no one complains; however, when something goes wrong, solicitors have nearly indefinite time to Figure out whether and why a wrong decision has been made. Whereas the technological performance is permanently increasing, the human performance is also increasing—but not at the same time rate: human cognition needed some millions of years for its development.

7.4 Biomedical Informatics

Slide 1-43: Definition of Biomedical Informatics

According to the American Association of Medical Informatics (AMIA) the term medical informatics has now been expanded to biomedical informatics and is defined as *the inter-disciplinary field that studies and pursues the effective use of biomedical data, information, and knowledge for scientific inquiry, problem solving, and decision making, motivated by efforts to improve human health* (Shortliffe 2011).

Note: Computers are just the vehicles to realize the central goals: To harness the power of the machines to support and to amplify human intelligence (Holzinger 2013).

Slide 1-44: Computational Sciences Meet Life Sciences

In this graphic we see that biomedical informatics is a perfect overlap of medicine, genomics, and informatics.

Example: Success in proteomics depends upon careful study design and high-quality biological samples. Advanced information technologies, and also an ability to use existing knowledge is crucial in making sense of the data (Boguski and McIntosh 2003).

Slide 1-45: In Medicine We Have Two Different Worlds ...

In the traditionally developed university system, there has been a cultural gap between the classical natural basic sciences (e.g., chemistry, biology, physics) and applied fields such as engineering or clinical medicine, the latter many believe to be more an art than a science (Kuhn et al. 2008). If we look at what both sides have in common, it is obvious that information and quality are in both areas considered as important. Consequently, modern information management can bridge the hiatus theoreticus, the gap between (scientific) knowledge and its application (Simonic and Holzinger 2010).

Slide 1-46: Information Quality as the Hiatus Theoreticus

Medical Information Systems of today are highly sophisticated; however, while we have seen that computer performance has increased exponentially, the human cognitive evolution cannot advance at the same speed. Consequently, the focus on interaction between human and computer is of increasing importance. The daily actions of medical professionals within the context of their clinical work must be the central concern of any innovation. Just surrounding and supporting them with new and emerging technologies is not sufficient if these increase rather than decrease the workload. Quality, actually, is a term which both medicine as well as informatics accept as an important issue (Holzinger and Simonic 2011), and must include the user-centered (human), the system-centered (computer), and process-centered (interaction) view.

Total Quality Management (TQM) provides a useful, simple yet important definition of quality: "consistently meeting customer's expectations" (Fisher et al. 2012). However, in medicine, this goal is not easy to accomplish, due to a number of problems, see Slide 1-47.

7 What Is Biomedical Informatics? 33

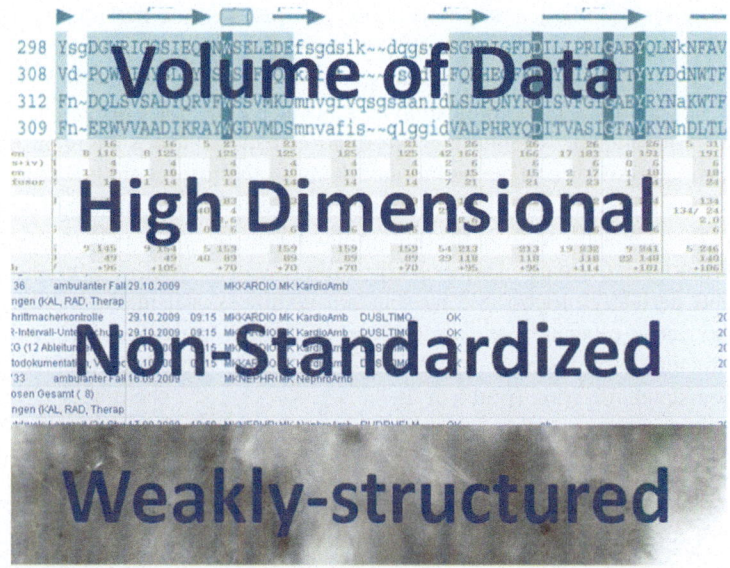

Fig. 6 See Slide 1-47

Slide 1-47: What Are the Problems?

1. **Volume of Data**: We are already aware of the fact that biomedical data covers various structural dimensions, ranging from microscopic structures (e.g., DNA) to whole human populations (disease spreading), producing "big data", so the first problem is in the volume of data.
2. **Complexity of the data**: Most of the data is multi-dimensional and multi-variate making it highly complex, so the second problem is in the curse of data dimensionality.
3. **Non-standardized Data**: Unfortunately, clinical-medical data is defined and collected with a remarkable degree of uncertainty, variability and inaccuracy. Komaroff (1979) stated that *medical data is disturbingly soft*. Three decades later, the data still falls far short of the exactness that engineers prefer. What did Komaroff mean with *soft*? The way patients "define" their sickness, questions and answers between clinicians and patients, physical examinations, diagnostic laboratory tests, etc., are far from the precision a computer scientist would expect. Even the definitions of the diseases themselves are often ambiguous; some diseases cannot be defined by any available objective standard; other diseases do have an

(continued)

(continued)

objective standard, but are variably interpreted. Another complication inherent in the data is that most medical information is incomplete, with wide variation in the degree and type of missing information. In both the development and the application of statistical techniques, analysis of data with incomplete or missing information can be much more difficult than analysis of corresponding data with all the information available—interestingly this was known before the term medical informatics was defined (Walsh 1960). So the third problem is in the non-standardized data.

4. **Weakly-structured Data**: Very little data is well-structured, and thus perfectly processable by standard computational methods. Most of the data—especially in clinical and molecular medicine—are weakly structured or even unstructured—combined with the problem of non-standardization (problem 3).

Slide 1-48: Big Data: We Need Machine Intelligence

Biomedical data is characterized not only by significant complexity but also by enormous size, making manual analysis by the end users often impossible. At the same time, medical experts are able to solve complicated problems almost intuitively, often enabling medical doctors to make diagnoses with high precision, without being able to describe the exact rules or processes used during their diagnosis, analysis and problem solving. Human thinking is basically a matter of the "plasticity" of the elements of the nervous system, whilst our digital computers, being von Neumann machines do not have such "plastic" elements (Holzinger 2013).

Slide 1-49: Big Data in Biomedical Informatics

The progress in medicine, covering all areas from drug development to personalized diagnostics and therapeutics, reflects the success of the most challenging disciplines involved: molecular biology and information science. Digital data is the key element for biomedical informatics. Particularly, the ongoing discoveries in molecular biology and their direct influence on the understanding of human diseases will have far-reaching consequences for the whole health-care system; hence influence prevention, diagnosis, and therapy. The progress in biomedical engineering will result in new diagnostic

(continued)

(continued)

and therapeutic options (e.g., image-guided minimally invasive surgery). HCI and Knowledge Discovery will play a central role in these advancements (Kuhn et al. 2008).

8 Future Outlook and Research Avenues

8.1 Grand Challenges

Slide 1-50: A List of Grand Challenges by Sittig (1994)

Sittig (1994) wrote an editorial for the Journal of American Medical Informatics Association (JAMIA) where he listed nine Grand Challenges in Medical Informatics:

1. A unified controlled medical vocabulary (CMV).
2. A complete computer-based patient record that could serve as a regional/national/multinational resource and a format to allow exchange of records between systems.
3. The automatic coding of free-text reports, patient histories, discharge abstracts, etc.
4. Automated analysis of medical records, yielding

    (a) the expected (most common) clinical presentation and course and the degree of clinical variability for patients with a given diagnosis;
    (b) the resources required in the care of patients compared by diagnosis, treatment protocol, clinical outcome, location, and physician;

5. A uniform, intuitive, anticipating user interface.
6. The human genome project (Brookhaven Protein Database, the Johns Hopkins Genome Database, and the National Center for Biomedical Information Genbank).
7. A complete three-dimensional, digital representation of the body, including the brain, with graphic access to anatomic sections, etc.
8. Techniques to ease the incorporation of new information management technologies into the infrastructure of organizations so that they can be used at the bedside or at the research bench.
9. A comprehensive, clinical decision support system.

Up to date much progress has been made, however, many goals have not been achieved.

> **Slide 1-51: An Update of the List: 20 Years Later**
>
> Grand new challenges from today's perspective are as follows:
>
> 10. Closing the gap between science and practice
> 11. Data fusion and data integration in the clinical workplace
> 12. To provide a trade-off between standardization and personalization
> 13. An intuitive, unified and universal, adaptive and adaptable user interface
> 14. Integrated interactive Knowledge Discovery Methods particularly for the masses of still "unstructured data"
> 15. Mobile solutions for the bedside and the clinical bench
>
> A consequence of 14 and 15 will be the vision of "Watson" on the smartphone. This goal was announced by IBM for the year 2020. The problem involved is the massive unstructured clinical datasets (Holzinger et al. 2013).

8.2 Personalized Medicine

Personalized medicine aims to optimize health and develop individualized therapy and personally designed drugs by combining a person's genetic and genomic data with information about their lifestyle and exposures. So far, the focus was on the prevention and treatment of conditions affecting adults, such as cancer and cardiovascular disease. Bianchi (2012) argues that a personalized medicine approach would have maximal benefit over the course of an entire individual's lifetime if it began in the womb or at least at birth. However, a major limitation of applying advanced genetic and genomic techniques in the prenatal setting is that genomic variation can be identified for which the clinical implications are not known yet. But as knowledge of the fetus and fetal development progresses by sequencing fetal DNA and RNA, new treatment opportunities will emerge.

Tanaka (2010) labels two future concepts in medicine and health care (Slide 1-52):

1. "Deeper" towards **individualized medicine**: Omics-based medicine and systems will realize a new personalized and predictive medicine.
2. "Broader" towards **ubiquitous health care**: preventive medicine, ambient assisted living, and ubiquitous home health care and wellness.

A new "translational informatics" plays an important role for deriving clinically meaningful information from the vast amount of omics data. To date, application of comprehensive molecular information to medicine has been referred to as "genomic medicine," which aims to realize "personalized medical care," based on the inborn (germ line) individual differences, or "polymorphisms," of the patient's genomic information (Tanaka 2010). A major challenge is in data fusion—to merge together the disparate, distributed datasets (confer with Slide 2-6 in Lecture 2).

8 Future Outlook and Research Avenues

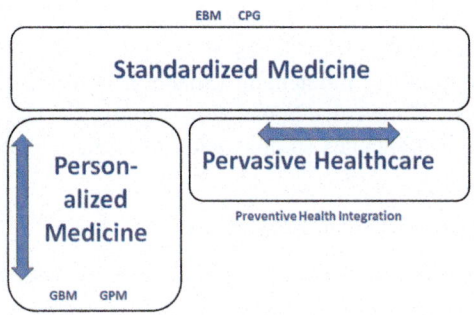

Fig. 7 See Slide 1-52

> **Slide 1-52: Between Standardization and Personalization**
>
> Future research must go towards an integration and trade-off of standardized medicine (EBM = Evidence-Based Medicine and CPG = Clinical Practice Guidelines) and Personalized Medicine (GBM = Genome-Based Medicine and GPM = Genetic Polymorphism) and Pervasive Health care, including Ambient Assisted Live and Wellness (Holzinger et al. 2010a). Big research issues include security, privacy, intimacy, and trust (Ziefle et al. 2011).

Personalized medicine is a typical "big data" innovation. The objective is to examine the relationships among genetic variation, predispositions for specific diseases, and specific drug responses and then to account for the genetic variability of individuals in the drug development process. Personalized medicine holds the promise of improving health care in three main ways: (1) offering early detection and diagnosis before a patient develops disease symptoms; (2) more effective therapies because patients with the same diagnosis can be segmented according to molecular signature matching (i.e., patients with the same disease often do not respond in the same way to the same therapy, partly because of genetic variation); (3) the adjustment of drug dosages according to a patient's molecular profile to minimize side effects and maximize response.

Public health is also a typical "big data" approach. For example, by using nationwide patient data and treatment database, public health officials can ensure the rapid, coordinated detection of infectious diseases, and a comprehensive outbreak surveillance and response through an Integrated Disease Surveillance and Response program.

Open Medical Data is a related idea which is not new. Many researchers in the past had followed the notion that Science is a public enterprise and that certain data should be openly available (Rowen et al. 2000) and it is recently also a big topic in the biomedical domain (Boulton et al. 2011; Hersey et al. 2012) and the British Medical Journey started a big open data campaign (Thompson and Heneghan

2012). The goal of the movement is similar to approaches of open source, open content, or open access. With the launch of open data government initiatives the open data movement gained momentum (Shadbolt et al. 2012) and some speak already about an Open Knowledge Foundation (Molloy 2011). Consequently, there are plenty of research challenges on this topic (Yildirim et al. 2013).

Slide 1-53: Towards Personalized Medicine

In the concept of (Tanaka 2010) we can see the pathway to an individualized and predictive personalized medicine: Omics data provide comprehensive molecular information of "diseased somatic cells." Hence, it varies during time, during the course of diseases and differs among sites of diseases, unlike the "germ line" genome sequences in tailored medicine which remain the same during the whole life. So, the Omics information lies at the intermediate level of the hierarchy so that the relation to the clinical phenotype is much closer to bring about the predictive medicine.

Note: Diseases are an integrated multi-hierarchical network system, comprising subcellular molecular network, cell–cell communication, tissue–organ linkage, and whole-body coordination. Diseases such as common complex diseases have a self-sustaining mechanism due to a bidirectional causative loop.

Direct-to-consumer (DTC) companies are already harnessing these discoveries by offering DNA tests that provide insights into personal genetic traits and disease risks. Genetic testing can improve lifestyle choices and increase preventive screening. However, understanding of the genetic contribution to human disease is far from complete. There is debate in the genetics community as to the usefulness of DTC testing (Ng et al. 2009).

Slide 1-54: Future p-Health Model: A 6 P's Paradigm

Pervasive care, in contrast to the current hospital-based medicine, aims to deliver health services beyond hospitals and into individual's daily lives.

More importantly, it supports individualization by providing means to acquire personal health information that are impossible to obtain inside hospitals.

Under the p-Health model, a collection of health information has to begin as early as possible, which can be starting at birth or even before birth. Types of information have to span multiple spatial scales of the human body, down from the genetic and molecular level and up to body system level. Information of different modality has to be captured by a variety of acquisition tools, e.g., sensing and imaging devices, and under different situations, e.g., during

(continued)

> (continued)
>
> day-to-day activities as well as irregular clinical visits. The set of information will eventually help to solve health issues arouse at different levels, from personal to global (Zhang and Poon 2010).
>
> In short the p-Health Model answers two questions: *What* decisions should be made and *how* the decisions should be made. The former relates to preventive, participatory, and preemptive measures, the latter to personalized, predictive, and pervasive actions.

8.3 Biomarker Discovery

Biomarkers are objectively measured molecule substances which indicate the presence of an abnormal condition within a patient. A biomarker can be a gene (e.g., SNP), protein (e.g., prostate-specific antigen), or metabolite (e.g., glucose, cholesterol) that has been shown to correlate with the characteristics of a specific disease. A biomarker in clinical and medical settings can be used for many purposes, including early disease detection, monitoring response to therapy, and predicting clinical outcome. Biomarkers can be categorized according to their clinical applications, e.g., in cancer, diagnostic markers are used to initially define the histopathological classification and stage of the disease, whilst prognostic markers can predict the development of diseases and the prospect of recovery. Based upon the individual cases, the predictive markers can be used for the selection of the correct therapeutic procedure. A potential biomarker should be confirmed that it is indeed specific to the disease state and is not a function of the variability within the biological sample of patients due to differences in diet, genetic background, lifestyle, age, sex, ethnicity, and so on (Issaq and Veenstra 2013).

> **Slide 1-55: Proteomic Samples for Biomarker Discovery**
>
> In this example we see proteomic samples used for biomarker discovery, along with their advantages and disadvantages/limitations. Tissue samples and proximal fluids are usually obtained through the highly invasive procedures such as surgery or biopsy, and require strict ethical approval by institutional review boards. A blood sample is minimally invasive and urine is a noninvasive collection and both are available in large quantities. Animal models and cell lines, on the contrary, are readily available already through commercial suppliers (Drabovich et al. 2013).

9 Exam Questions

9.1 Yes/No Decision Questions

Please check the following sentences and decide whether the sentence is true = YES; or false = NO; for each correct answer you will be awarded 2 credit points.

01	Regular, recurring arrangements in space of adjacent amino acid residues in a polypeptide chain are described by the secondary structure	☐ Yes ☐ No	2 total
02	Ubiquitous computing is a post-desktop model of Human–Computer Interaction in which the information processing is integrated into everyday objects.	☐ Yes ☐ No	2 total
03	The central dogma of molecular biology states that DNA is transcribed into RNA and translated into protein: DNA → RNA → Protein → Cellular Phenotype.	☐ Yes ☐ No	2 total
04	Proteins are molecules consisting of one or more chains of chromosomes and they vary from one to another mainly in their sequence of these chromosomes.	☐ Yes ☐ No	2 total
05	X-ray crystallography is a standard method to analyze the arrangement of objects (atoms, molecules) within a crystal structure.	☐ Yes ☐ No	2 total
06	The Church–Turing thesis states that Turing machines capture the informal notion of effective methods in logic and mathematics, and provide a precise definition of an algorithm	☐ Yes ☐ No	2 total
07	Within a Von Neumann Structure, the internal memory executes the instructions with help of the logic controller, the "heart" of the computer.	☐ Yes ☐ No	2 total
08	Biomarkers are measured molecules which indicate the presence of an abnormal condition within a patient, and can be a gene (e.g., SNP), protein (e.g., prostate-specific antigen), or metabolite.	☐ Yes ☐ No	2 total
09	Eukaryotic cells are usually single-celled organisms, while prokaryotic cells can be either single-celled or part of multicellular organisms..	☐ Yes ☐ No	2 total
10	Total Quality Management (TQM) provides a useful, simple yet important definition of quality: "consistently meeting customer's expectations.	☐ Yes ☐ No	2 total

Sum of Question Block A (max. 20 points)	

9.2 Multiple Choice Questions (MCQ)

The following questions are composed of two parts: the stem, which identifies the question or problem and a set of alternatives which can contain 0, 1, 2, 3, or 4 correct answers, along with a number of distractors that might be plausible—but are incorrect. Please **select the correct answers** by ticking ☒—and do not forget that it can be none. Each question will be awarded 4 points *only if everything is correct*.

01	Blood as biomarker has the advantage ... ☐ a) ... of having a wide range of protein concentrations. ☐ b) ... that it is available in large quantities. ☐ c) ... that it can be collected non-invasive. ☐ d) ... provides a good reflection of the physiologic state of the body.	4 total
02	Living "things" are able ... ☐ a) ... to reproduce. ☐ b) ... to grow. ☐ c) ... to process information. ☐ d) ... to self-replicate.	4 total
03	At the beginning of medical informatics in the 1970ies .... ☐ a) ... the focus was mostly on data acquisition, storage and accounting. ☐ b) ... there was an emerging trend to use Web based health applications. ☐ c) ... personalized medicine was already part of some health strategies. ☐ d) ... the first end-user programmable Mash-ups were introduced.	4 total
04	The "Quality Era" of biomedical informatics is mainly characterized by ☐ a) ... focus on data processing and storage. ☐ b) ... health care networks, telemedicine and CPOE-Systems. ☐ c) ... pervasive and ubiquitous computing technologies. ☐ d) ... patient empowerment and individual molecular medicine.	4 total
05	The problems in Biomedical Informatics are rooted mostly in the ... ☐ a) ... high dimensionality and complexity of the data. ☐ b) ... non-standardization of most of the data available. ☐ c) ... lacking memory capacity and computational power. ☐ d) ... heterogeneity and weak structurization of the available data.	4 total
06	Part of the definition of Biomedical Informatics is the ... ☐ ... effective use of biomedical data. ☐ ... motivation to improve computational capacities. ☐ ... effort to expand the technological capabilities. ☐ ... motivation to improve human health.	4 total
07	Tissue is defined as ... ☐ a) ... cellular organization between cells and a complete organism. ☐ b) ... an essential component of the human cardio-vascular system. ☐ c) ... part of the digestive system, e.g. esophagus, stomach and intestines. ☐ d) ... an ensemble of similar cells, carrying out a function as collective.	4 total
08	A Von Neumann Machine ... ☐ a) ... is theoretically equivalent to a Universal Turing Machine. ☐ b) ... is today present in nearly all of our "computers". ☐ c) ... uses a central processing unit and a single separate memory. ☐ d) ... can process continuous variables and multiple instructions.	4 total
09	Epigenetics ... ☐ a) ... is the study beyond protein folding, hence called epi-genetics. ☐ b) ... is the study of changes in gene expression or cellular phenotypes,. ☐ c) ... concerns mechanisms other than changes in the underlying DNA ☐ d) ... provides a genetical snapshot of the cell.	4 total

| 10 | Chromatography ...
☐ a) ... is the collective term for the separation of mixtures.
☐ b) ... is the primary method for determining the atomic structure.
☐ c) ... separates molecules according to their charge and their mass. | 4 total |

| **Sum of Question Block B (max. 40 points)** | |

9.3 Free Recall Block

Please follow the instructions below. At each question you will be assigned the credit points indicated if your option is correct (partial points may be given).

01	Please sketch the principle tertiary structure of a protein and indicate the α-helix and the β-sheet:	1-15 1 each 3 total
02	The Von-Neumann Architecture is the fundamental computer organization structure of nearly all of our todays computing systems (e.g. in your PC, smartphone, microwave oven, car, etc.), please roughly sketch the Von Neumann Architecture and indicate the main parts:	1-28 1 each 6 total

03	Please draw the overlapping circles of the three fields: Medicine – Genomics – Informatics, and indicate the areas of Bioinformatics – Medical Informatics – Biomedical Informatics and Clinical Genomics:	1/42 1 each 7 total
04	It is important to recognize the dimensions where the data comes from and the respective fields dealing with it. Please complete the following image: 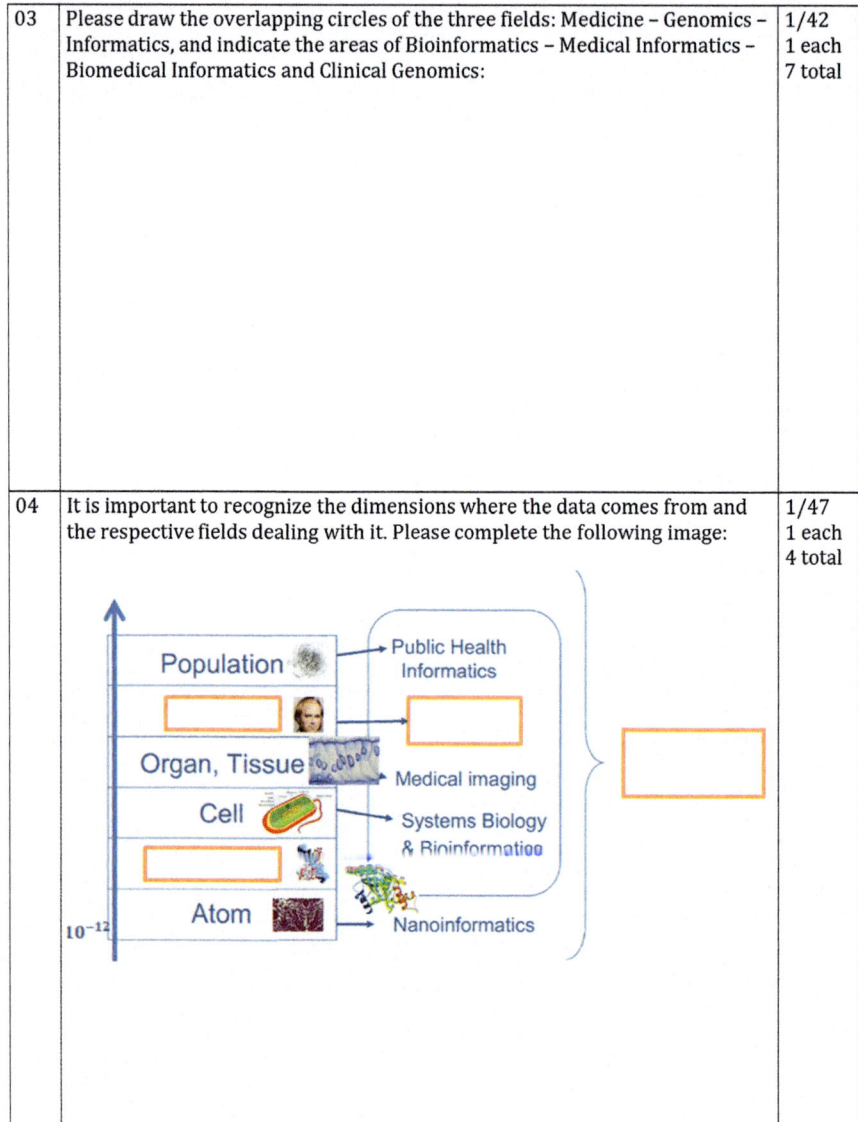	1/47 1 each 4 total

| 05 | Protein analytics is a primary source of data for research. Please assign the correct labels to the methods below (just by drawing a connecting line between the correct items): | 1/16 1 each 3 total |

	[gel image]	Liquid Chromatography
	[diffraction pattern]	Gel Electrophoresis
	[chromatogram peaks]	X-Ray Crystallography

Sum of Question Block C (max. 40 points)

10 Answers

10.1 Answers to the Yes/No Questions

Please check the following sentences and decide whether the sentence is true = YES; or false = NO; for each correct answer you will be awarded 2 credit points.

01	Regular, recurring arrangements in space of adjacent amino acid residues in a polypeptide chain are described by the secondary structure	☒ Yes ☐ No	2 total
02	Ubiquitous computing is a post-desktop model of Human–Computer Interaction in which the information processing is integrated into everyday objects.	☒ Yes ☐ No	2 total
03	The central dogma of molecular biology states that DNA is transcribed into RNA and translated into protein: DNA → RNA → Protein → Cellular Phenotype.	☒ Yes ☐ No	2 total
04	Proteins are molecules consisting of one or more chains of chromosomes and they vary from one to another mainly in their sequence of these chromosomes.	☐ Yes ☒ No	2 total
05	X-ray crystallography is a standard method to analyze the arrangement of objects (atoms, molecules) within a crystal structure.	☒ Yes ☐ No	2 total
06	The Church–Turing thesis states that Turing machines capture the informal notion of effective methods in logic and mathematics, and provide a precise definition of an algorithm	☒ Yes ☐ No	2 total
07	Within a Von Neumann Structure, the internal memory executes the instructions with help of the logic controller, the "heart" of the computer.	☐ Yes ☒ No	2 total
08	Biomarkers are measured molecules which indicate the presence of an abnormal condition within a patient, and can be a gene (e.g., SNP), protein (e.g., prostate-specific antigen), or metabolite.	☒ Yes ☐ No	2 total
09	Eukaryotic cells are usually single-celled organisms, while prokaryotic cells can be either single-celled or part of multicellular organisms..	☐ Yes ☒ No	2 total
10	Total Quality Management (TQM) provides a useful, simple yet important definition of quality: "consistently meeting customer's expectations.	☒ Yes ☐ No	2 total

Sum of Question Block A (max. 20 points)	

10 Answers

10.2 Answers to the Multiple Choice Questions (MCQ)

01	Blood as biomarker has the advantage ... ☐ a) ... of having a wide range of protein concentrations. ☒ b) ... that it is available in large quantities. ☐ c) ... that it can be collected non-invasive. ☒ d) ... provides a good reflection of the physiologic state of the body.	4 total
02	Living "things" are able ... ☒ a) ... to reproduce. ☒ b) ... to grow. ☒ c) ... to process information. ☒ d) ... to self-replicate.	4 total
03	At the beginning of medical informatics in the 1970ies .... ☒ a) ... the focus was mostly on data acquisition, storage and accounting. ☐ b) ... there was an emerging trend to use Web based health applications. ☐ c) ... personalized medicine was already part of some health strategies. ☐ d) ... the first end-user programmable Mash-ups were introduced.	4 total
04	The "Quality Era" of biomedical informatics is mainly characterized by ☐ a) ... focus on data processing and storage. ☐ b) ... health care networks, telemedicine and CPOE-Systems. ☐ c) ... pervasive and ubiquitous computing technologies. ☒ d) ... patient empowerment and individual molecular medicine.	4 total
05	The problems in Biomedical Informatics are rooted mostly in the ... ☒ a) ... high dimensionality and complexity of the data. ☒ b) ... non-standardization of most of the data available. ☐ c) ... lacking memory capacity and computational power. ☒ d) ... heterogeneity and weak structurization of the available data.	4 total
06	Part of the definition of Biomedical Informatics is the ... ☒ ... effective use of biomedical data. ☐ ... motivation to improve computational capacities. ☐ ... effort to expand the technological capabilities. ☒ ... motivation to improve human health.	4 total
07	Tissue is defined as ... ☒ a) ... cellular organization between cells and a complete organism. ☐ b) ... an essential component of the human cardio-vascular system. ☐ c) ... part of the digestive system, e.g. esophagus, stomach and intestines. ☒ d) ... an ensemble of similar cells, carrying out a function as collective.	4 total
08	A Von Neumann Machine ... ☒ a) ... is theoretically equivalent to a Universal Turing Machine. ☒ b) ... is today present in nearly all of our "computers". ☒ c) ... uses a central processing unit and a single separate memory. ☐ d) ... can process continuous variables and multiple instructions.	4 total
09	Epigenetics ... ☐ a) ... is the study beyond protein folding, hence called epi-genetics. ☒ b) ... is the study of changes in gene expression or cellular phenotypes,. ☒ c) ... concerns mechanisms other than changes in the underlying DNA ☐ d) ... provides a genetical snapshot of the cell.	4 total

| 10 | Chromatography ...
☒ a) ... is the collective term for the separation of mixtures.
☐ b) ... is the primary method for determining the atomic structure.
☐ c) ... separates molecules according to their charge and their mass.
☒ d) ... typically includes Gas-Chromatography and Liquid-Chromatography. | 4 total |

| **Sum of Question Block B (max. 40 points)** | |

10 Answers

10.3 Answers to the Free Recall Questions

01	Please sketch the principle tertiary structure of a protein and indicate the α-helix and the β-sheet:	1-15 1 each 3 total
	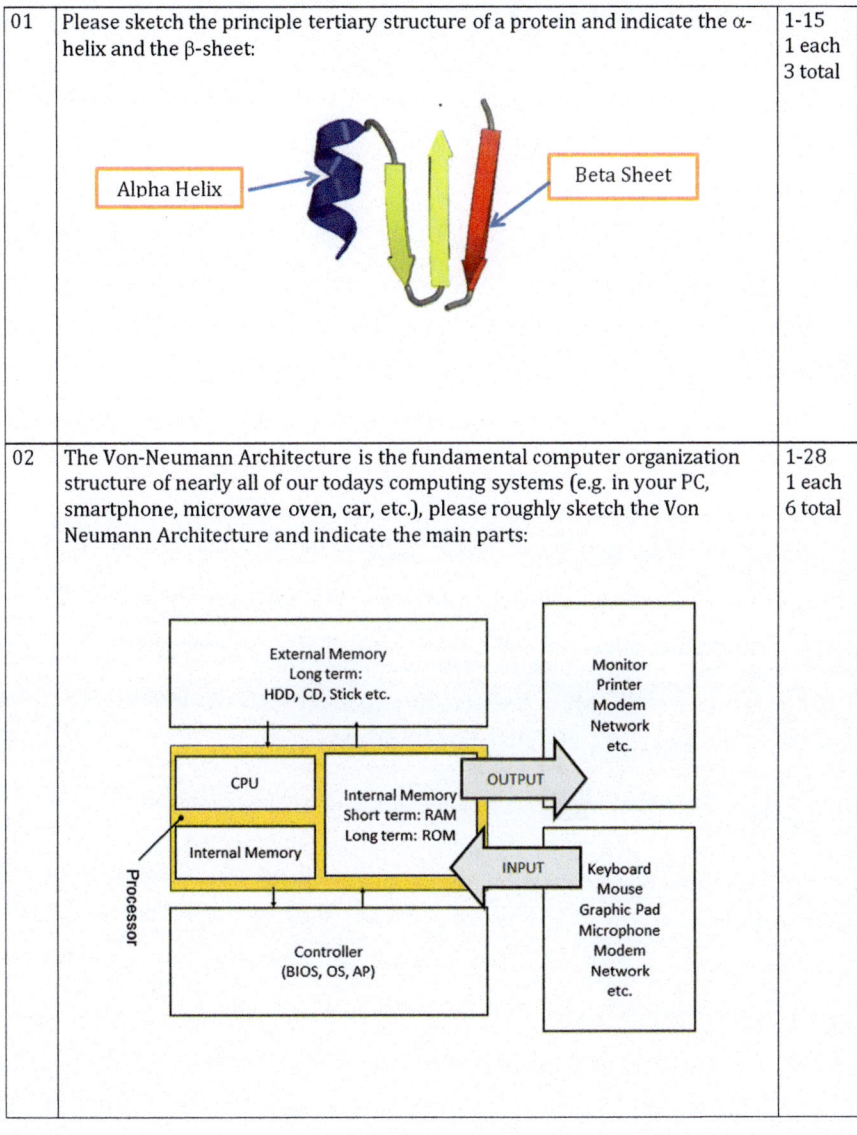	
02	The Von-Neumann Architecture is the fundamental computer organization structure of nearly all of our todays computing systems (e.g. in your PC, smartphone, microwave oven, car, etc.), please roughly sketch the Von Neumann Architecture and indicate the main parts:	1-28 1 each 6 total

03	Please draw the overlapping of the three fields: Medicine – Genomics – Informatics, and indicate the areas of Bioinformatics – Medical Informatics – Biomedical Informatics and Clinical Genomics:	1/42 1 each 7 total
04	It is important to recognize the dimensions where the data comes from and the respective fields dealing with it. Please complete the following image:	1/47 1 each 4 total

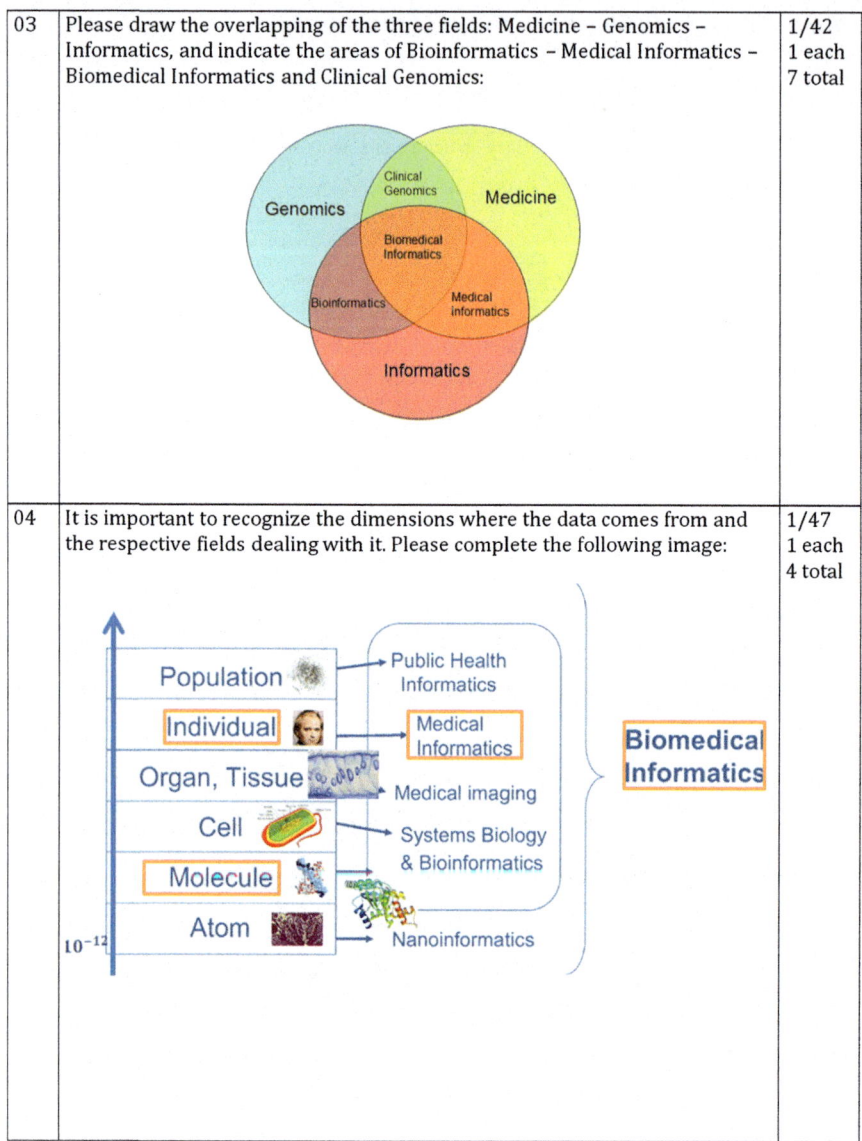

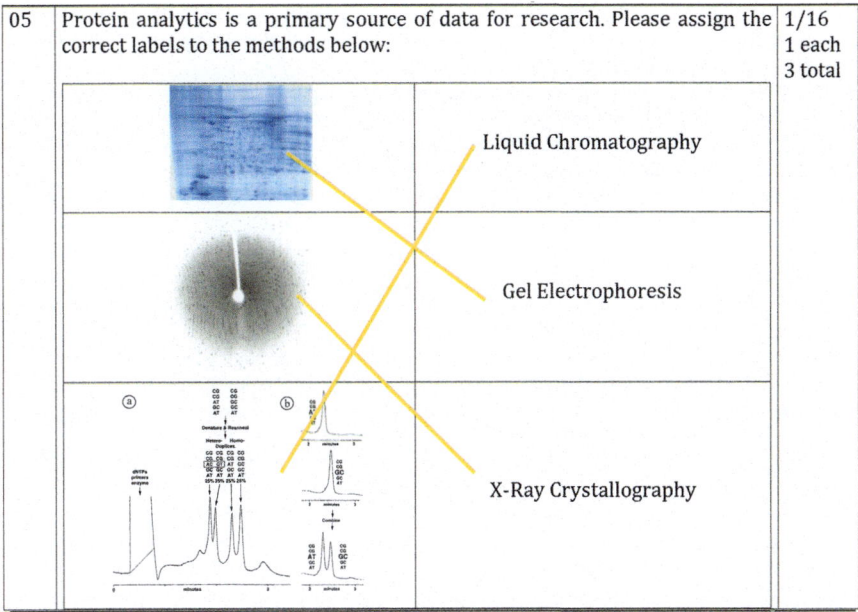

References

Aebersold R, Mann M (2003) Mass spectrometry-based proteomics. Nature 422(6928):198–207

Alagöz F, Calero Valdez A, Wilkowska W, Ziefle M, Dorner S, Holzinger A (2010) From cloud computing to mobile internet, from user focus to culture and hedonism—the crucible of mobile health care and wellness applications. ICPCA10 International Conference on Pervasive Computer Applications 2010, Maribor, Slovenia. IEEE, pp 1–9

Anfinsen CB (1973) Principles that govern the folding of protein chains. Science 181(4096):223–230

Aral S (2011) Identifying social influence: a comment on opinion leadership and social contagion in new product diffusion. Mark Sci 30(2):217–223

Auinger A, Ebner M, Nedbal D, Holzinger A (2009) Mixing content and endless collaboration—mashups: towards future personal learning environments. In: Stephanidis C (ed) Universal access in human-computer interaction HCI, part III: applications and services, HCI international 2009, vol 5616, Lecture notes in computer science (LNCS). Springer, Berlin, pp 14–23

Barabási AL, Gulbahce N, Loscalzo J (2011) Network medicine: a network-based approach to human disease. Nat Rev Genet 12(1):56–68

Barabási AL, Oltvai ZN (2004) Network biology: understanding the cell's functional organization. Nat Rev Genet 5(2):101–113

Bardram JE (2004) Activity-based support for mobility and collaboration in ubiquitous computing. In: Baresi L (ed) Second international conference on ubiquitous mobile information and collaboration systems. Springer, Riga, Latvia, pp 169–184

Bianchi DW (2012) From prenatal genomic diagnosis to fetal personalized medicine: progress and challenges. Nat Med 18(7):1041–1051

Blum BI, Duncan K (1990) A history of medical informatics. Addison Wesley, Reading, MA

Boal D (2012) Mechanics of the cell. Cambridge University Press, Cambridge

Bobrow DG, Burchfiel JD, Murphy DL, Tomlinson RS (1972) TENEX, a paged time sharing system for the PDP-10. Commun ACM 15(3):135–143

Boguski MS, Mcintosh MW (2003) Biomedical informatics for proteomics. Nature 422 (6928):233–237

Boulton G, Rawlins M, Vallance P, Walport M (2011) Science as a public enterprise: the case for open data. Lancet 377(9778):1633–1635

Brooks R (2001) The relationship between matter and life. Nature 409(6818):409–411

Brown T (2012) Introduction to genetics: a molecular approach. Garland, New York

Burgin M, Eberbach E (2012) Evolutionary computaion and the processes of life: opening statement. ACM Ubiquity 2012:1–13

Cavin R, Lugli P, Zhirnov V (2012) Science and engineering beyond Moore's Law. Proc IEEE 100 (13):1720–1749

Chirikjian GS (2011) Modeling loop entropy. Methods Enzymol 487:99

Cooper SB, Loewe B, Sorbi A (2008) New computational paradigms: changing conceptions of what is computable. Springer, New York

Crick F (1970) Central dogma of molecular biology. Nature 227(5258):561–563

Crick F, Watson J (1953) Molecular structure of nucleic acids. Nature 171(4356):737–738

Dill KA, Bromberg S, Yue K, Chan HS, Ftebig KM, Yee DP, Thomas PD (1995) Principles of protein folding—a perspective from simple exact models. Protein Sci 4(4):561–602

Drabovich AP, Pavlou MP, Batruch I, Diamandis EP (2013) Chapter 2—Proteomic and mass spectrometry technologies for biomarker discovery. In: Haleem JI, Timothy DV (eds) Proteomic and metabolomic approaches to biomarker discovery. Academic, Boston, pp 17–37

Fisher C, Lauría E, Chengalur-Smith S (2012) Introduction to information quality. AuthorHouse, Bloomington, IN

Ge H, Walhout AJM, Vidal M (2003) Integrating 'omic'information: a bridge between genomics and systems biology. Trends Genet 19(10):551–560

Gigerenzer G (2008) Gut feelings: short cuts to better decision making. Penguin, London

Gromiha MM (2010) Protein bioinformatics. Elsevier, Amsterdam

Hawking SW, Penrose R, Atiyah M (1996) The nature of space and time. Princeton University Press, Princeton

Hersey A, Senger S, Overington JP (2012) Open data for drug discovery: learning from the biological community. Future Med Chem 4(15):1865–1867

Holzinger A (2002) Basiswissen IT/Informatik Band 1: Informationstechnik. Wuerzburg, Vogel Buchverlag

Holzinger A (2003a) Basiswissen IT/Informatik. Band 2: Informatik. Vogel Buchverlag, Wuerzburg

Holzinger A (2003b) Finger instead of mouse: touch screens as a means of enhancing universal access. In: Carbonell N, Stephanidis C (eds) Universal access: theoretical perspectives, practice and experience, vol 2615, Lecture notes in computer science (LNCS). Springer, Berlin, pp 387–397

Holzinger A (2013) Human–computer interaction & knowledge discovery (HCI-KDD): what is the benefit of bringing those two fields to work together? In: Alfredo Cuzzocrea CK, Simos DE, Weippl E, Xu L (eds) Multidisciplinary research and practice for information systems, vol 8127, Lecture notes in computer science (LNCS). Springer, New York, pp 319–328

Holzinger A, Basic L, Peischl B, Debevc M (2011a) Handwriting recognition on mobile devices: state of the art technology, usability and business analysis. Proceedings of the 8th international conference on electronic business and telecommunications, INSTICC, Sevilla, pp 219–227

Holzinger A, Dorner S, Födinger M, Valdez A, Ziefle M (2010a) Chances of increasing youth health awareness through mobile wellness applications. In: Leitner G, Hitz M, Holzinger A (eds) HCI in work and learning, life and leisure, vol 6389, Lecture notes in computer science (LNCS). Springer, Berlin, pp 71–81

Holzinger A, Emberger W, Wassertheurer S, Neal L (2008a) Design, development and evaluation of online interactive simulation software for learning human genetics. Elektrotech Informationstechnik 125(5):190–196

Holzinger A, Kosec P, Schwantzer G, Debevc M, Hofmann-Wellenhof R, Frühauf J (2011b) Design and development of a mobile computer application to reengineer workflows in the hospital and the methodology to evaluate its effectiveness. J Biomed Inform 44(6):968–977

Holzinger A, Nischelwitzer A, Friedl S, Hu B (2010b) Towards life long learning: three models for ubiquitous applications. Wirel Commun Mob Comput 10(10):1350–1365

Holzinger A, Schaupp K, Eder-Halbedl W (2008b) An investigation on acceptance of ubiquitous devices for the elderly in an geriatric hospital environment: using the example of person tracking. In: Miesenberger K, Klaus J, Zagler W, Karshmer A (eds) Computers helping people with special needs, vol 5105, Lecture notes in computer science (LNCS). Springer, Berlin, pp 22–29

Holzinger A, Schlögl M, Peischl B, Debevc M (2010c) Preferences of handwriting recognition on mobile information systems in medicine: improving handwriting algorithm on the basis of real-life usability research (Best Paper Award). In: ICE-B 2010—ICETE the international joint conference on e-business and telecommunications, INSTICC, Athens, Greece, pp 120–123

Holzinger A, Schwaberger K, Weitlaner M (2005) Ubiquitous computing for hospital applications: RFID-applications to enable research in real-life environments. In: 29th Annual international conference on computer software & applications (IEEE COMPSAC), IEEE, Edinburgh, UK, pp 19–20

Holzinger A, Searle G, Peischl B, Debevc M (2012a) An answer to "Who needs a stylus?" on handwriting recognition on mobile devices. e-Business and telecommunications. Communications in computer and information science (CCIS), vol 314. Heidelberg, Springer, pp 156–167

Holzinger A, Simonic K-M (eds) (2011) Information quality in e-Health, vol 7058, Lecture notes in computer science (LNCS). Springer, Heidelberg

Holzinger A, Stocker C, Ofner B, Prohaska G, Brabenetz A, Hofmann-Wellenhof R (2013) Combining HCI, natural language processing, and knowledge discovery—potential of IBM content analytics as an assistive technology in the biomedical domain, vol 7947, Lecture notes in computer science (LNCS). Springer, Heidelberg, pp 13–24

Holzinger A, Stocker C, Peischl B, Simonic K-M (2012b) On using entropy for enhancing handwriting preprocessing. Entropy 14(11):2324–2350

Hunter L (2009) The processes of life: an introduction to molecular biology. MIT Press, Cambridge, MA

Hurst M (2007) Data mining: text mining, visualization and social media [Online]. http://datamining.typepad.com/data_mining/2007/01/the_blogosphere.html. Accessed May 10 2011

Issaq HJ, Veenstra TD (2013) Chapter 1—Biomarker discovery: study design and execution. In: Haleem JI, Timothy DV (eds) Proteomic and metabolomic approaches to biomarker discovery. Academic, Boston, pp 1–16

Jeanquartier F, Holzinger A (2013) On visual analytics and evaluation in cell physiology: a case study. In: Cuzzocrea A, Kittl C, Simos DE, Weippl E, Xu L (eds) Multidisciplinary research and practice for information systems, vol 8127, Lecture notes in computer science (LNCS). Springer, Heidelberg, pp 495–502

Jeong H, Mason SP, Barabasi AL, Oltvai ZN (2001) Lethality and centrality in protein networks. Nature 411(6833):41–42

Kiberstis PA (2012) All eyes on epigenetics. Science 335(6069):637
Klibanov AM (2001) Improving enzymes by using them in organic solvents. Nature 409 (6817):241–246
Komaroff AL (1979) The variability and inaccuracy of medical data. Proc IEEE 67(9):1196–1207
Kuhn K, Knoll A, Mewes H, Schwaiger M, Bode A, Broy M, Daniel H, Feussner H, Gradinger R, Hauner H (2008) From molecules to populations. Methods Inf Med 47(4):283–295
Lane N, Martin W (2010) The energetics of genome complexity. Nature 467(7318):929–934
Mahajan V, Schoeman MEF (1977) The use of computers in hospitals: an analysis of adopters and nonadopters. Interfaces 7:95–107
Marth JD (2008) A unified vision of the building blocks of life. Nat Cell Biol 10(9):1015–1016
Mitchell S, Spiteri MD, Bates J, Coulouris G (2000) Context-aware multimedia computing in the intelligent hospital. ACM Press, Kolding, Denmark, pp 13–18
Mitreva M (2012) Structure, function and diversity of the healthy human microbiome. Nature 486:207–214
Mojzsis SJ, Arrhenius G, Mckeegan KD, Harrison TM, Nutman AP, Friend CRL (1996) Evidence for life on Earth before 3,800 million years ago. Nature 384(6604):55–59
Molloy JC (2011) The Open Knowledge Foundation: open data means better science. PLoS Biol 9 (12):e1001195
Moore GE (1965) Cramming more components onto integrated circuits. Electronics 38 (8):114–117
Neumann JV (1945) First draft of a report on the EDVAC. University of Pennsylvania—techical report, 49 pp
Ng PC, Murray SS, Levy S, Venter JC (2009) An agenda for personalized medicine. Nature 461 (7265):724–726
Okumoto S, Jones A, Frommer WB (2012) Quantitative imaging with fluorescent biosensors. Annu Rev Plant Biol 63:663–706
Patel VL, Kahol K, Buchman T (2011) Biomedical complexity and error. J Biomed Inform 44 (3):387–389
Petz G, Karpowicz M, Fürschuß H, Auinger A, Stříteský V, Holzinger A (2013) Opinion mining on the Web 2.0—characteristics of user generated content and their impacts, vol 7947, Lecture notes in computer science (LNCS). Springer, Heidelberg, pp 35–46
Petz G, Karpowicz M, Fürschuß H, Auinger A, Winkler S, Schaller S, Holzinger A (2012) On text preprocessing for opinion mining outside of laboratory environments. In: Huang R, Ghorbani A, Pasi G, Yamaguchi T, Yen N, Jin B (eds) Active media technology, vol 7669, Lecture notes in computer science (LNCS). Springer, Berlin, pp 618–629
Pevsner J (2009) Bioinformatics and functional genomics. John Wiley & Sons, Hoboken, NJ
Rabilloud T, Chevallet M, Luche S, Lelong C (2010) Two-dimensional gel electrophoresis in proteomics: past, present and future. J Proteome 73(11):2064–2077
Rouwkema J, Rivron NC, Van Blitterswijk CA (2008) Vascularization in tissue engineering. Trends Biotechnol 26(8):434–441
Rowen L, Wong GKS, Lane RP, Hood L (2000) Intellectual property—publication rights in the era of open data release policies. Science 289(5486):1881
Sackett DL, Rosenberg WM, Gray J, Haynes RB, Richardson WS (1996) Evidence based medicine: what it is and what it isn't. Br Med J 312(7023):71
Schidlowski M (1988) A 3,800-million-year isotopic record of life from carbon in sedimentary rocks. Nature 333(6171):313–318
Schotte F, Lim MH, Jackson TA, Smirnov AV, Soman J, Olson JS, Phillips GN, Wulff M, Anfinrud PA (2003) Watching a protein as it functions with 150-ps time-resolved X-ray crystallography. Science 300(5627):1944–1947
Schrödinger E (1944) What is life? The physical aspect of the living cell. Dublin Institute for Advanced Studies at Trinity College, Dublin

Shadbolt N, O'hara K, Berners-Lee T, Gibbins N, Glaser H, Hall W, Schraefel M (2012) Open government data and the linked data web: Lessons from data. gov. uk. IEEE Intell Syst 27:16–24

Shehu A, Kavraki LE (2012) Modeling structures and motions of loops in protein molecules. Entropy 14(2):252–290

Shortliffe EH (2011) Biomedical informatics: defining the science and its role in health professional education. In: Holzinger A, Simonic K-M (eds) Information quality in e-Health, vol 7058, Lecture notes in computer science (LNCS). Springer, Heidelberg, pp 711–714

Simonic K-M, Holzinger A (2010) Zur Bedeutung von Information in der Medizin. OCG J 35(1):8

Sittig DF (1994) Grand challenges in medical informatics. J Am Med Inform Assoc 1(5):412–413

Southern EM (1975) Detection of specific sequences among dna fragments separated by gel-electrophoresis. J Mol Biol 98(3):503–517

Sperelakis N (2012) Cell physiology sourcebook: essentials of membrane biophysics, 4th edn. Elsevier, Amsterdam

Steinbuch K (1957) Informatik: Automatische Informationsverarbeitung. Standard Electric Lorenz (SEL) Nachrichten (4), p 171

Stelzl U, Worm U, Lalowski M, Haenig C, Brembeck FH, Goehler H, Stroedicke M, Zenkner M, Schoenherr A, Koeppen S, Timm J, Mintzlaff S, Abraham C, Bock N, Kietzmann S, Goedde A, Toksöz E, Droege A, Krobitsch S, Korn B, Birchmeier W, Lehrach H, Wanker EE (2005) A human protein-protein interaction network: a resource for annotating the proteome. Cell 122 (6):957–968

Tanaka H (2010) Omics-based medicine and systems pathology a new perspective for personalized and predictive medicine. Methods Inf Med 49(2):173–185

Thompson M, Heneghan C (2012) BMJ OPEN DATA CAMPAIGN. We need to move the debate on open clinical trial data forward. Br Med J 345:e8351

Trent RJ (2012) Molecular medicine: genomics to personalized healthcare, 4th edn. Elsevier, Amsterdam

Trygve T (2011) Handbook of epigenetics. Academic, San Diego

Varshney U (2009) Pervasive healthcare computing: EMR/EHR, wireless and health monitoring. Springer, New York

Walsh JE (1960) Analyzing medical data: some statistical considerations. IRE Trans Med Electron ME-7(4):362–366

Wassertheurer S, Holzinger A, Emberger W, Breitenecker F (2003) Design and development of interactive online-simulations for e-learning. In: Hohmann R (ed) Frontiers in simulation—17th symposium Simulationstechnik in Magdeburg (Germany). SCS European Publication, Delft, pp 97–105

Weiser M (1991) The computer for the twenty-first century. Sci Am 265(3):94–104

Westra R, Tuyls K, Saeys Y, Nowé A (2007) Knowledge discovery and emergent complexity in bioinformatics. In: Tuyls K, Westra R, Saeys Y, Nowé A (eds) Knowledge discovery and emergent complexity in bioinformatics. Springer, Berlin, pp 1–9

Wiltgen M, Holzinger A (2005) Visualization in bioinformatics: protein structures with physicochemical and biological annotations. In: Zara J, Sloup J (eds) Central European multimedia and virtual reality conference (available in EG Eurographics Library). Czech Technical University (CTU), Prague, pp 69–74

Wiltgen M, Holzinger A, Tilz GP (2007) Interactive analysis and visualization of macromolecular interfaces between proteins. In: Holzinger A (ed) HCI and usability for medicine and health care, vol 4799, Lecture notes in computer science (LNCS). Springer, Berlin, pp 199–212

Xiao WZ, Oefner PJ (2001) Denaturing high-performance liquid chromatography: a review. Hum Mutat 17(6):439–474

Yapijakis C (2009) Hippocrates of Kos, the father of clinical medicine, and Asclepiades of Bithynia, the father of molecular medicine. In Vivo 23(4):507–514

Yildirim P, Ekmekci I, Holzinger A (2013) On knowledge discovery in open medical data on the example of the FDA Drug Adverse Event Reporting System for alendronate (Fosamax). In:

Holzinger A, Pasi G (eds) Human-computer interaction and knowledge discovery in complex, unstructured, big data, vol 7947, Lecture notes in computer science (LNCS). Springer, Berlin, pp 195–206

Yip LY, Yong Chan EC (2013) Chapter 8—Gas chromatography/mass spectrometry-based metabonomics. In: Haleem JI, Timothy DV (eds) Proteomic and metabolomic approaches to biomarker discovery. Academic, Boston, pp 131–144

Zhang YT, Poon CCY (2010) Editorial note on bio, medical, and health informatics. IEEE Trans Inf Technol Biomed 14(3):543–545

Ziefle M, Rocker C, Holzinger A (2011) Medical technology in smart homes: exploring the user's perspective on privacy, intimacy and trust. 35th Annual IEEE computer software and applications conference workshops (COMPSAC) 2011, IEEE, Munich, pp 410–415

Lecture 2

Fundamentals of Data, Information, and Knowledge

1 Learning Goals

At the end of this second lecture you:

- would be aware of the types and categories of different datasets in biomedical informatics.
- would know some differences between data, information, knowledge, and wisdom.
- would be aware of standardized/non-standardized and well-structured/unstructured data.
- would have a basic overview of information theory and the concept of information entropy.

2 Advance Organizer

Abduction	A cyclical process of generating possible explanations (i.e., identification of a set of hypotheses that account for the clinical case on the basis of the available data) and testing those explanations (i.e., evaluation of each generated hypothesis on the basis of its expected consequences) for the abnormal state of the patient at hand
Abstraction	Data are filtered according to their relevance for the problem solution and chunked in schemas representing an abstract description of the problem (e.g., abstracting that an adult male with hemoglobin concentration less than 14 g/dL is an anemic patient)

Artifact/surrogate	Error or anomaly in the perception or representation of information through the involved method, equipment, or process
Data	Physical entities at the lowest abstraction level which are for example generated by a patient (patient data) or a (biological) process; data contains per definition no meaning
Data quality	Includes quality parameter such as: accuracy, completeness, update status, relevance, consistency, reliability, accessibility
Dataset	A collection of data, each single data object is called datum (singular of data)
Data structure	Way of storing and organizing data to use it efficiently
Dirty data	Data which is incorrect, erroneous, misleading, incorrect, incomplete, noisy, duplicate, etc.
Deduction	Deriving a particular valid conclusion from a set of <u>general premises</u>
DIK Model	Data–information–knowledge <u>three-level model</u>
DIKW Model	Data–information–knowledge–wisdom <u>four-level model</u>
Disparity	Containing different types of information in different dimensions
Entropy	Generally the degree of the disorder of a system; Information Entropy
Gedanken experiment	Thought experiment; "Gedanken" is the German word for thought; such an experiment enables us to prove a concept in the case that a physical experiment is not feasible
Heart rate variability (HRV)	Measured by the variation in the beat-to-beat interval
HRV artifact	Noise through errors in the location of the instantaneous heart beat, resulting in errors in the calculation of the HRV, and thus HRV is highly sensitive to artifact and errors in as low as even 2 % of the data will result in unwanted biases in HRV calculations
Induction	Deriving a likely general conclusion from a set of particular statements
Information	Derived from the data by interpretation (with feedback to the clinician)
Information overload	Difficulty of end users to make decisions in the presence of too much and too complex information
Information quality	Can be a means to bring technology and medicine closer together; information quality <u>combines</u> aspects understood by both fields (medicine and technology)
Information entropy	A measure for uncertainty: highly structured data contains low entropy, if everything is in order there is no uncertainty, no surprise, ideally $H = 0$

Knowledge	Obtained by inductive reasoning with previously interpreted data, collected from many similar patients or processes, which is added to the "body of knowledge" (explicit knowledge). This knowledge is used for the interpretation of other data and to gain implicit knowledge which guides the clinician in taking further action
Large data	Consist of at least hundreds of thousands of data points
Multidimensional	Containing more than three dimensions and data are multi-variate
Multi-modality	A combination of data from different sources
Multi-variate	Encompassing the simultaneous observation and analysis of more than one statistical variable (antonym: univariate = 1D)
Reasoning	Process by which clinicians reach a conclusion after thinking on all facts
Spatiality	Contains at least one (non-scalar) spatial component and non-spatial data
Structural complexity	Ranging from low-structured (simple data structure, but many instances, e.g., flow data, volume data) to high-structured data (complex data structure, but only a few instances, e.g., business data)
Time dependency	Data is given at several points in time (time series data)
Voxel	Volumetric pixel = volumetric picture element

3 Acronyms

ApEn	Approximate entropy
$\mathbb{C}_{data}$	Data in computational space
DIK	Data–information–knowledge three-level model
DIKW	Data–information–knowledge–wisdom four-level model
GraphEn	Graph entropy
H	Entropy (general)
HRV	Heart rate variability
MaxEn	Maximum entropy
MinEn	Minimum entropy
NE	Normalized ENTROPY (measures the relative informational content of both the signal and noise)
$\mathbb{P}_{data}$	Data in perceptual space
PDB	Protein data base
SampEn	Sample entropy

4 Key Problems

- Heterogeneous data sources (need for data fusion)
- Complexity of the data (high-dimensionality)
- Noisy, uncertain data (challenge of preprocessing)
- The discrepancy between data, information, and knowledge (various definitions)
- Big datasets (manual handling of the data is impossible)

5 Data in the Biomedical Domain

5.1 Data Sources in Biomedical Informatics

Our first question is: Where does the data come from? The second question: What kind of data is this? The third question: How big is this data? So let us look at some biomedical data sources (see Slide 2-1).

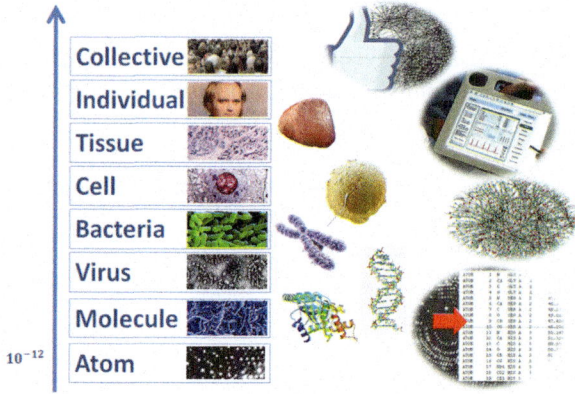

Fig. 1 See Slide 2-1

Slide 2-1: Data Sources in Biomedical Informatics

Due to the increasing trend towards personalized and molecular medicine, biomedical data results from various sources in different structural dimensions, ranging from the microscopic world (e.g., genomics, epigenomics, metagenomics, proteomics, metabolomics) to the macroscopic world (e.g., disease spreading data of populations in public health informatics). Just for orientation: the Glucose molecule has a size of 900 pm = 900×10^{-12} m

(continued)

(continued)

and the Carbon atom approx. 300 pm. A hepatitis virus is relatively large with 45 nm = 45×10^{-9} m and the X-Chromosome much bigger with 7 μm = 7×10^{-6} m.

Here a lot of "big data" is produced, e.g., genomics, metabolomics, and proteomics data. This is really "big data"—the datasets enormously large—whereas in each individual we estimate many Terabytes (1 TB = 1×10^{12} Byte = 1,000 GByte) of genomics data, we are confronted with Petabytes of proteomics data and the fusion of those for personalized medicine results in Exabytes of data (1 EB = 1×10^{18} Byte).

Of course these amounts are for each human individual; however, we have a current world population of 7 Billion (1 Billion in English language is 1 Milliard in European language) people (= 7×10^9 people). So you can see that this is really "big data." This "natural" data is then fused with "produced" data, e.g., the unstructured data (text) in the patient records, or data from physiological sensors—these data is also rapidly increasing in size and complexity. You can imagine that without computational intelligence we have no chance to survive in these complex big datasets.

5.2 Levels of Data Structures

Most of our computers are von Neumann machines (see Lecture 1), consequently at the lowest physical layer, data is represented as patterns of electrical on/off states (1/0, H/L, high/low); we speak of a **bit**, which is also known as Bit, the **B**asic **i**ndissoluble **i**nformation **u**nit (Shannon 1948). Do not confuse this Bit with the IEC 60027-2 symbol bit[1]—in small letters—which is used as an SI dimension prefix (e.g., 1 kbit = 1,024 bit, 1 Byte = 8 bit). Beginning with the physical level of data we can determine various levels of data structures (see Slide 2-2).

Slide 2-2: Levels of Data Structures: Towards a Taxonomy of Data

1. **Physical level**: In a von Neumann system: bit; in a Quantum system: qubit.
   Note: Regardless of its physical realization (voltage, or mechanical state, or black/white, etc.), a bit is always logically either 0 or 1 (analog to a light-switch). A qubit has similarities to a classical bit, but is overall very different: A classical bit is a scalar variable with the single value of either 0 or 1, so the value is unique, deterministic and unambiguous. A

(continued)

---

[1] Refer to: http://physics.nist.gov/cuu/Units/binary.html.

(continued)

qubit is more general in the sense that it represents a state defined by a pair of complex numbers (a, b), which express the probability that a reading of the value of the qubit will give a value of 0 or 1. Thus, a qubit can be in the state of 0, 1, or some mixture—referred to as a superposition—of the 0 and 1 states. The weights of 0 and 1 in this superposition are determined by (a, b) in the following way: qubit $\triangleq (a, b) \triangleq a \cdot 0_{bit} + b \cdot 1_{bit}$. Please be aware that this model of quantum computation is not the only one (Lanzagorta and Uhlmann 2008).

2. **Logical Level**:
    1. Primitive data types, including:
        (a) Boolean data type (true/false)
        (b) Numerical data type (integer ($\mathbb{Z}$), floating-point numbers ("reals"), etc.)
    2. Composite data types, including:
        (a) Array
        (b) Record
        (c) Union
        (d) Set (stores values without any particular order, and no repeated values)
        (e) Object (contains others)
    3. String and text types, including:
        (a) Alphanumeric characters
        (b) Alphanumeric strings (=sequence of characters to represent words and text)

3. **Abstract Level**: including abstract data structures, e.g., queue (FIFO), stack (LIFO), set (no order, no repeated values), lists, hash table, arrays, trees, graphs
4. **Technical Level**: Application data formats, e.g., text, vector graphics, pixel images, audio signals, video sequences, multimedia
5. **Hospital Level**: Narrative (textual, natural language) patient record data (structured/unstructured and standardized/non-standardized), Omics data (genomics, proteomics, metabolomics, microarray data, fluxomics, phenomics), numerical measurements (physiological data, time series, lab results, vital signs, blood pressure, CO_2 partial pressure, temperature, ...), recorded signals (ECG, EEG, ENG, EMG, EOG, EP ...), graphics (sketches, drawings, handwriting, ...); audio signals, images (cams, X-ray, MR, CT, PET, ...), etc.

5.3 Abstract Data Structures

In biomedical informatics we have a lot to do with abstract data types (ADT), and consequently we briefly review the most important ones here. For details please refer to a course on Algorithm and Data structures, or to a classic textbook such as (Aho et al. 1983; Cormen et al. 2009), or in German (Ottmann and Widmayer 2012; Holzinger 2003) and please take into consideration that data structures and algorithms go hand in hand (Cormen 2013).

Slide 2-3: Example Data Structures (1/3): List

List is a sequential collection of items $a_1, a_2, \ldots, a_n$ accessible one after another, beginning at the head and ending at the tail z. In a von Neumann machine it is a widely used data structure for applications which do not need random access. It differs from the stack (last-in-first-out, LIFO) and queue (first-in-first-out, FIFO) data structures insofar, that additions and removals can be made at any position in the list. In contrast to a simple set S the order is important. A typical example for the use of a list is a DNA sequence. The combination of GGGTTTAAA is such a list, the elements of the list are the nucleotide bases.

Nucleotides are the joined molecules which form the structural units of the RNA and the DNA and play the central role in metabolism.

Slide 2-4: Example Data Structure (2/3): Graph

Graph is a pair $G = (V, E)$, where $V(G)$ is a set of finite, non-empty vertices (nodes) and $E(G)$ is a set of edges (lines, arcs), which are 2-element subsets of V. If E is a set of ordered pairs of vertices (arcs, directed edges, arrows), then it is a **directed graph** (digraph). The distances between the edges can be represented within a distance-matrix (2D array).

The edges in a graph can be *multidimensional objects*, e.g., vectors containing the results of multiple Gen-expression measures. For this purpose the distance of two edges can be measured by various distance metrics.

Graphs are ideally suited for representing networks in medicine and biology, e.g., metabolism pathways.

In bioinformatics, distance matrices are used to represent protein structures in a coordinate-independent manner, as well as the pairwise distances between two sequences in sequence space. They are used in structural and sequential alignment, and for the determination of protein structures from NMR or X-ray crystallography. Evolutionary dynamics act on populations. Neither genes nor cells nor individuals evolve; only *populations evolve*.

(continued)

(continued)

This so-called **Moran process** describes the stochastic evolution of a finite population of constant size: In each time step, an individual is chosen for reproduction with a probability proportional to its fitness; a second individual is chosen for death. The offspring of the first individual replaces the second and individuals occupy the vertices of a graph. In each time step, an individual is selected with a probability proportional to its fitness; the weights of the outgoing edges determine the probabilities that the corresponding neighbor will be replaced by the offspring. The process is described by a stochastic matrix W, where w_{ij} denotes the probability that an offspring of individual i will replace individual j. At each time step, an edge ij is selected with a probability proportional to its weight and the fitness of the individual at its tail. The Moran process is a complete graph with identical weights (Lieberman et al. 2005).

Slide 2-5: Example Data Structures (3/3): Tree

Tree is a collection of elements called nodes, one of which is distinguished as a root, along with a relation ("parenthood") that places a hierarchical structure on the nodes. A node, like an element of a list, can be of whatever type we wish. We often depict a node as a letter, a string, or a number with a circle around it. Formally, a tree can be defined recursively in the following manner:

1. A single node by itself is a tree. This node is also the root of the tree.
2. Suppose n is a node and $T1, T2, \ldots, Tk$ are trees with roots $n1, n2, \ldots, nk$, respectively. We can construct a new tree by making n be the parent of nodes $n1, n2, \ldots, nk$. In this tree n is the root and $T1, T2, \ldots, Tk$ are the subtrees of the root. Nodes $n1, n2, \ldots, nk$ are called the children of node n.

Dendrogram (from Greek dendron "tree," -gramma "drawing") is a tree diagram frequently used to illustrate the arrangement of the clusters produced by hierarchical clustering. Dendrograms are often used in computational biology to illustrate the clustering of genes or samples. The origin of such dendrograms can be found in Darwin (1859).

The example by Hufford et al. (2012) shows a neighbor-joining tree and the changing morphology of domesticated maize and its wild relatives. Taxa in the neighbor-joining tree are represented by different colors: parviglumis (green), landraces (red), improved lines (blue), mexicana (yellow), and Tripsacum (brown). The morphological changes are shown for female inflorescences and plant architecture during domestication and improvement.

5.4 Big Data Pools in the Health Domain

Now that we have seen some examples of data from the biomedical domain, we can look at the "big picture". Manyika et al. (2011) localized four major data pools in the US health care and describe that the data are highly fragmented, with little overlap and low integration. Moreover, they report that approx. 30 % of clinical text/numerical data in the USA, including medical records, bills, laboratory and surgery reports, is still not generated electronically. Even when clinical data is in digital form, they are usually held by an individual provider and are rarely shared (see Slide 2-4).

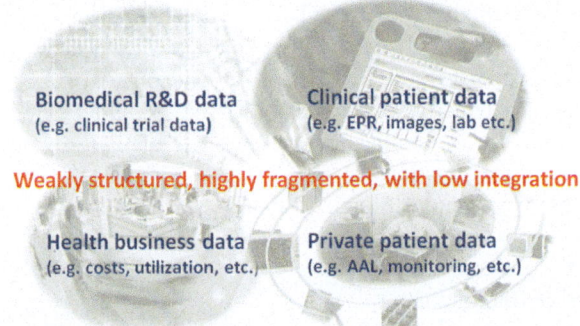

Fig. 2 See Slide 2-6

Slide 2-6: Big Data Pools in the Health Domain

Biomedical research data, e.g., clinical trials, predictive modeling, is produced by academia and pharmaceutical companies and stored in data bases and libraries. Clinical data is produced in the hospital and is stored in hospital information systems (HIS), picture archiving and communication systems (PACS) or in laboratory data bases, etc. Much data is health business data produced by payers, providers, insurances, etc. Finally, there is an increasing pool of patient behavior and sentiment data, produced by various customers and stakeholders, outside the typical clinical context, including the growing data from the wellness and ambient assisted living domain.

A major challenge in our networked world is the increasing amount of data—today called "big data." The trend towards personalized medicine has resulted in a sheer mass of generated (-omics) data, (see Slide 2-7). In the life sciences domain, most data models are characterized by complexity, which makes manual analysis very time-consuming and frequently practically impossible (Holzinger 2013).

Slide 2-7: Omics-Data Integration

More and more Omics-data are generated, including:

1. Genomics data (e.g., sequence annotation).
2. Transcriptomics data (e.g., microarray data); the **transcriptome** is the set of all RNA molecules, including mRNA, rRNA, tRNA, and noncoding RNA produced in the cells.
3. Proteomics data: Proteomic studies generate large volumes of raw experimental data and inferred biological results stored in data repositories, mostly openly available; an overview can be found here (Riffle and Eng 2009). The outcome of proteomics experiments is a list of proteins differentially modified or abundant in a certain phenotype. The large size of proteomics datasets requires specialized analytical tools, which deal with large lists of objects (Bessarabova et al. 2012).
4. Metabolomics (e.g., enzyme annotation), the **metabolome** represents the collection of all metabolites in a cell, tissue, organ, or organism.
5. Protein–DNA interactions.
6. Protein–protein interactions; PPI are at the core of the entire **interactomics** system of any living cell.
7. Fluxomics (isotopic tracing, metabolic pathways).
8. Phenomics (biomarkers).
9. Epigenetics, is the study of the changes in gene expression—other than the DNA sequence, and therefore the prefix "epi-."
10. Microbiomics.
11. Lipidomics.

Omics-data integration helps to address interesting biological questions on the biological systems level towards personalized medicine (Joyce and Palsson 2006).

A further challenge is to integrate the data and to make it accessible to the clinician. While there is much research on the integration of heterogeneous information systems, a shortcoming is in the integration of available data. **Data fusion** is the process of merging multiple records representing the same real-world object into a single, consistent, accurate, and useful representation (Bleiholder and Naumann 2008).

An example for the mix of different data for solving a medical problem can be seen in Slide 2-8.

Slide 2-8: Example of Typical Clinical Datasets

A good example for complex medical data is RCQM, which is an application that manages the flow of data and information in the rheumatology outpatient

(continued)

(continued)

clinic (50 patients per day, 5 days per week) of the Graz University Hospital, on the basis of a quality management process model. Each examination produces 100+ clinical and functional parameters per patient. This amassed data are morphed into better useable information by applying scoring algorithms (e.g., Disease Activity Score, DAS) and are convoluted over time. Together with previous findings, physiological laboratory data, patient record data, and Omics data from the Pathology department, these data constitute the information basis for analysis and evaluation of the disease activity. The challenge is in the increasing quantities of such highly complex, multidimensional, and time series data (Simonic et al. 2011).

5.5 Standardization Versus Structurization

Do not confuse structure with standardization (see Slide 2-9). Data can be standardized (e.g., numerical entries in laboratory reports) and non-standardized. A typical example is non-standardized text—imprecisely called "Free-Text" or "unstructured data" in an electronic patient record (Kreuzthaler et al. 2011).

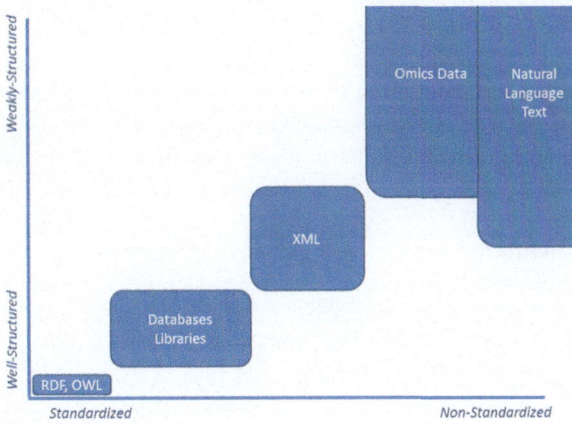

Fig. 3 See Slide 2-9

Slide 2-9: Standardization Versus Structurization

Standardized data is *the* basis for accurate communication. In the medical domain, many different people work at different times in various locations.

(continued)

(continued)

Data standards can ensure that information is interpreted by all users with the same understanding. Moreover, standardized data facilitates comparability of data and interoperability of systems. It supports the reusability of the data, improves the efficiency of health-care services, and avoids errors by reducing duplicated efforts in data entry.

Data standardization refers to

(a) The data content
(b) The terminologies that are used to represent the data
(c) How data is exchanged
(d) How knowledge, e.g., clinical guidelines, protocols, decision support rules, checklists, standard operating procedures, is represented in the health information system (refer to IOM[2])

Technical elements for data sharing require standardization of identification, record structure, terminology, messaging, privacy, etc. The most used standardized dataset to date is the International Classification of Diseases (ICD), which was first adopted in 1900 for collecting statistics (Ahmadian et al. 2011), which we will discuss in Lecture 3.

Non-standardized data is the majority of data and inhibit data quality, data exchange, and interoperability.

Well-structured data is the minority of data and an idealistic case when each data element has an associated defined structure, relational tables, or the resource description framework RDF, or the Web Ontology Language OWL (see Lecture 3).

Note: **Ill-structured** is a term often used for the opposite of well-structured, although this term originally was used in the context of problem solving (Simon 1973).

Semi-structured is a form of structured data that does not conform with the strict formal structure of tables and data models associated with relational databases but contains tags or markers to separate structure and content, i.e., they are schema-less or self-describing; a typical example is a markup language such as XML (see Lectures 3 and 4).

Weakly Structured data is the most of our data in the whole universe, whether it is in macroscopic (astronomy) or microscopic structures (biology)—see Lecture 5.

Non-structured data or *unstructured data* is an imprecise definition used for *information* expressed in natural language, when no specific structure has been defined. This is an issue for debate: Text has also some structure: words,

(continued)

---

[2] Institute of Medicine, http.//iom.edu.

(continued)

sentences, paragraphs. If we are very precise, unstructured data would mean that the data is completely randomized—which is usually called noise and is defined by Duda et al. (2000) as any property of data which is not due to the underlying model but instead to randomness (either in the real world, from the sensors or the measurement procedure).

Note: In this course we follow the definitions of Boisot and Canals (2004). They describe data as originating in discernible differences in physical states of the world. Significant regularities in this data constitute information. This implies that the information gained from data depends on the expectations called hypotheses. This set of hypotheses is called knowledge and is constantly modified by new information. Based on these definitions, the commonly used term "unstructured data" refers to complete randomness and unpredictability—noise. In informatics, particularly, it can be considered as unwanted nonrelevant data without meaning—or, even worse, with a not detected wrong meaning (artifact) (Holzinger 2012)—see also more details in Lecture 5.

5.6 Data Dimensionality

5.6.1 Multivariate and Multidimensional

"Multivariate" and "multidimensional" are modern words and consequently overused in literature. Each item of data is composed of **variables**, and if such a data item is defined by more than one variable it is called a **multivariable data item**.

Variables are frequently classified into two categories: dependent or independent.

Slide 2-10: Data Dimensionality

- 0D data = A data point existing isolated from other data, e.g., integers, letters, Booleans
- 1D data = Consist of a string of 0D data, e.g., Sequences representing nucleotide bases and amino acids, SMILES
- 2D data = Having spatial component, such as images, NMR-spectra, etc.
- 2.5D data = Can be stored as a 2D matrix, but can represent biological entities in three or more dimensions, e.g., PDB records
- 3D data = Having 3D spatial components, e.g., image voxels, e-density maps
- H-D Data = Data having arbitrarily high dimensions

In Physics, Engineering, and Statistics, a variable is a physical property of a subject, whose quantity can be measured, e.g., mass, length, time, temperature.

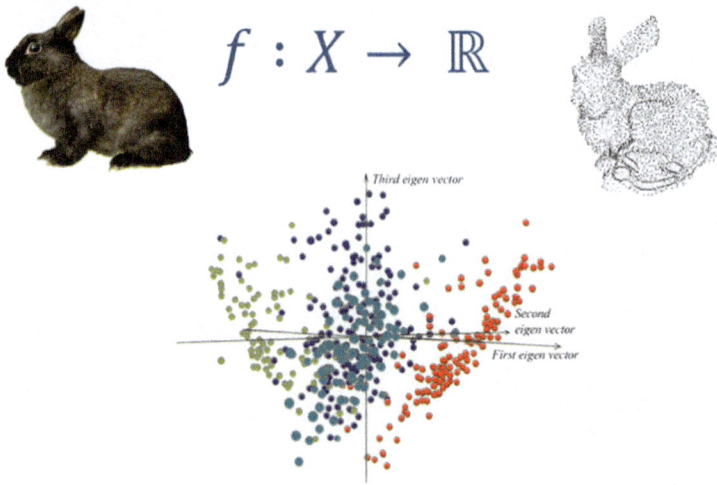

Fig. 4 See Slide 2-11

Slide 2-11: Space

In mathematics, and hence in informatics, however, a variable is associated with a *space*—often an *n*-dimensional Euclidean space $\mathbb{R}_n$—in which an entity (e.g., a function) or a phenomenon of continuous nature is defined. The data location within this space can be referenced by using a range of coordinate systems (Cartesian, Polar-coordinates, etc.): The dependent variables are those used to describe the entity (for example the function value) whilst the independent variables are those that represent the coordinate system used to describe the space in which the entity is defined. If a dataset is composed of variables whose interpretation fits this definition our goal is to understand how the "entity" is defined within the *n*-dimensional Euclidean space $\mathbb{R}_n$. Sometimes we may distinguish between variables meaning measurement of property, from variables meaning a coordinate system, by referring to the former as **variate**, and referring to the latter as **dimension** (Dos Santos and Brodlie 2002; dos Santos and Brodlie 2004).

A space is a set of points. A metric space has an associated metric, which enables us to measure distances between points in that space and, in turn, implicitly defines their neighborhoods. Consequently, a metric provides a space with a topology, and a metric space is a topological one. Topological spaces feel alien to us because we are accustomed to having a metric.

(continued)

5 Data in the Biomedical Domain

(continued)

Biomedical example: A protein is a single chain of amino acids, which folds into a globular structure. The Thermodynamics Hypothesis states that a protein always folds into a state of minimum energy. To predict protein structure, we would like to model the folding of a protein computationally. As such, the protein folding problem becomes an optimization problem: We are looking for a path to the global minimum in a very high-dimensional energy landscape (Zomorodian 2005).

5.6.2 Point Cloud Datasets

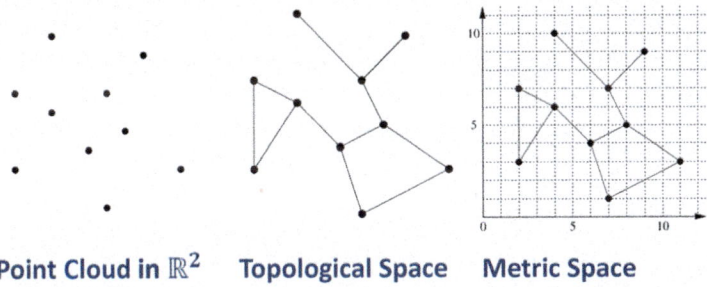

Point Cloud in $\mathbb{R}^2$ **Topological Space**   **Metric Space**

Fig. 5 See Slide 2-12

Slide 2-12: Point Cloud Datasets

Let us collect n-dimensional i observations in the Euclidean vector space $\mathbb{R}^n$ and we get:

$$x_i = [x_{i1}, \ldots, x_{in}] \qquad (2\text{-}1)$$

A cloud of points sampled from any source (medical data, sensor network data, a solid 3D object, surface, etc.). Those data points can be coordinated as an unordered sequence in an arbitrarily high dimensional Euclidean[3] space, where methods of algebraic topology can be applied. The main challenge is in mapping the data back into $\mathbb{R}^3$ or to be more precise into $\mathbb{R}^2$, because our retina inherently perceives data in $\mathbb{R}^2$. The cloud of such data points can be used as a computational representation of the respective data object.

(continued)

---

[3] In Einstein's theory of Special Relativity, Euclidean 3-space plus time (the "4th-dimension") are unified into the Minkowski space.

(continued)

A temporal version can be found in motion-capture data, where geometric points are recorded as time series. Now you will ask an obvious question: "How do we visualize a four-dimensional object?" The obvious answer is: "How do we visualize a three-dimensional object?" Humans do not see in three spatial dimensions directly, but via sequences of planar projections integrated in a manner that is sensed if not comprehended. Little children spend a significant time of their first year of life learning how to infer 3D spatial data from paired planar projections, and many years of practice have tuned a remarkable ability to extract global structure from representations in a strictly lower dimension (Ghrist 2008). Because we have the same problem here in this book, we must stay in $\mathbb{R}^2$ and therefore the example in Slide 2-12 (Zomorodian 2005).

The common type of space is the Euclidean space (left). The most general type is the topological space; A metric space (right) is a set S with a global distance function—the metric g (Zomorodian 2005).

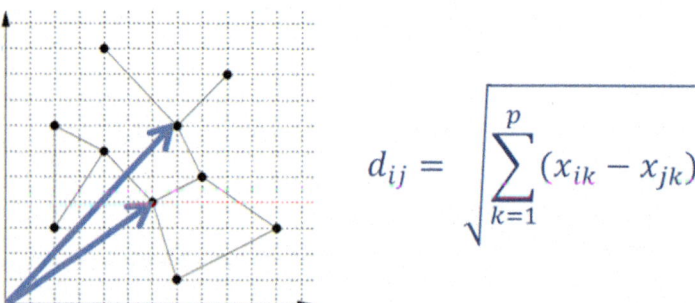

$$d_{ij} = \sqrt{\sum_{k=1}^{p}(x_{ik} - x_{jk})^2}$$

Fig. 6 See Slide 2-13

Slide 2-13 Example Metric Space

A metric space has an associated metric, which enables to measure the distances between points in that space and, implicitly define their neighborhoods. Consequently, a metric provides a space with a topology, hence a metric space is a topological space.

A set X with a metric function d is called a metric space. We give it the metric topology of d, where the set of open balls.

(continued)

(continued)

Most of our "natural" spaces are a particular type of metric spaces and the Euclidean spaces: The Cartesian product of n copies of $\mathbb{R}$, the set of real numbers, along with the Euclidean metric:

$$d(i,j) = \sqrt{\sum_{k=1}^{p}(x_{ik}-x_{jk})^2} \quad (2\text{-}2)$$

is the n-dimensional Euclidean space $\mathbb{R}_n$.

We may induce a topology on subsets of metric spaces as follows:

If $A \subseteq X$ with topology T, then we get the relative or induced topology T_A by defining $T_A = \{S \cap A | S \in T\}$.

For more information refer to Zomorodian (2005) or Edelsbrunner and Harer (2010).

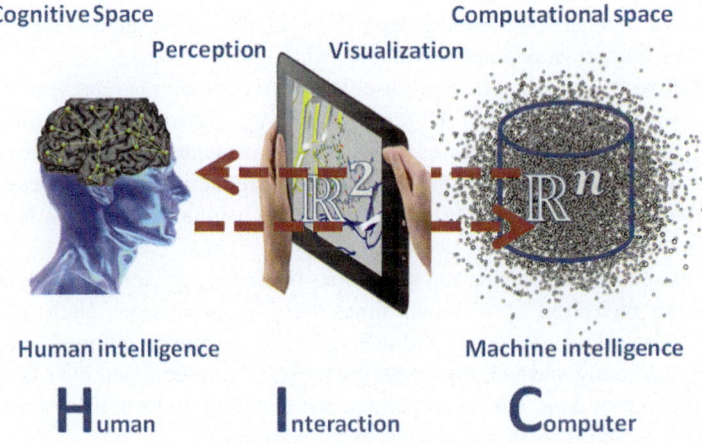

Fig. 7 See Slide 2-14

Slide 2-14: Mapping Data $\mathbb{R}^n + t \rightarrow \mathbb{R}^2 + t \text{ or } \mathbb{R}^3 + t$

Knowledge discovery from data: By getting insight into the data, the gained information can be used to build up knowledge. The grand challenge is to map higher dimensional data into lower dimensions, and hence make it interactively accessible to the end user (Holzinger 2012; Holzinger 2013).

(continued)

> (continued)
>
> This mapping from $\mathbb{R}^n \to \mathbb{R}^2$ is the core task of visualization and a major component for knowledge discovery: Enabling effective interactive human control over powerful machine algorithms to support human sense making (Holzinger 2012; Holzinger 2013).

If we have data in higher dimension, we have at first to map this data from the *computational space* into the *cognitive space* (Kaski and Peltonen 2011) of 2D or 3D representations developing in time (Note: The time *t* is an important, yet often neglected dimension in medicine):

$$\mathbb{R}^n + t \to \mathbb{R}^2 + t \quad or \quad \mathbb{R}^3 + t \qquad (2\text{-}3)$$

So far, let us define in accordance with dos Santos and Brodlie (2004):

Multivariate dataset is a dataset that has *many dependent variables* and they might be correlated to each other to varying degrees. Usually this type of dataset is associated with *discrete data models*.

Multidimensional dataset is a dataset that has *many independent variables* clearly identified, and one or more dependent variables associated to them. Usually this type of dataset is associated with *continuous data models*.

In other words, every data item (or object) in a computer is represented (stored) as a set of features. Instead of the term features we may use the term dimensions, because an object with *n*-features can also be represented as a multidimensional point in an *n*-dimensional space. Dimensionality reduction is the process of mapping an *n*-dimensional point, into a lower *k*-dimensional space—this is the main challenge in visualization see Lecture 9.

The number of dimensions can sometimes be small, e.g., simple 1D-data such as temperature measured at different times, to 3D applications such as medical imaging, where data is captured within a volume. Standard techniques—contouring in 2D; isosurfacing and volume rendering in 3D—have emerged over the years to handle this sort of data. There is no dimension reduction issue in these applications, since the data and display dimensions essentially match.

Dataset Example: A relatively recent development is the creation of the UCI "Knowledge Discovery in Databases Archive" at http://kdd.ics.uci.edu. This contains a range of large and complex datasets as a challenge to the data mining research community to scale up its algorithms as the size of stored datasets, especially commercial ones, inexorably rises (Bramer 2013).

Scale	Empirical Operation	Math. Group Structure	Transf. in $\mathbb{R}$	Basic Statistics	Mathematical Operations
NOMINAL	Determination of equality	Permutation $x' = f(x)$ x ... 1-to-1	$x \mapsto f(x)$	Mode, contingency correlation	=, ≠
ORDINAL	Determination of more/less	Isotonic $x' = f(x)$ x ... monotonic incr.	$x \mapsto f(x)$	Median, Percentiles	=, ≠, >, <
INTERVAL	Determination of equality of intervals or differences	General linear $x' = ax+b$	$x \mapsto rx+s$	Mean, Std.Dev., Rank-Order Corr., Prod.-Moment Corr.	=, ≠, >, <, -, +
RATIO	Determination of equality or ratios	Similarity $x' = ax$	$x \mapsto rx$	Coefficient of variation	=, ≠, >, <, -, +, *, ÷

Fig. 8 See Slide 2-15

Slide 2-15: Categorization of Data (Classic "Scales")

Data can be categorized into qualitative (nominal and ordinal) and quantitative (interval and ratio). Interval and ratio data are parametric, and are used with parametric tools in which distributions are predictable (and often Normal).

Nominal and ordinal data are non-parametric, and do not assume any particular distribution. They are used with non-parametric tools such as the Histogram.

The classic paper on the theory of scales of measurement is from Stevens (1946).

6 A Clinical View on Data, Information, Knowledge

We can summarize what we learned so far about data: Data can be numeric, non-numeric, or both. Non-numeric data can include anything from language data (text) to categorical, image, or video data. Data may range from completely structured, such as categorical data, to semi-structured, such as an XML File containing meta information, to unstructured, such as a narrative "free-text." Note, that term unstructured does not mean that the data are without any pattern, which would mean complete randomness and uncertainty, but rather that "unstructured data" are expressed so, that only humans can meaningfully interpret it. Structure provides information that can be interpreted to determine data organization and meaning, hence it provides a **context** for the information. The inherent structure in the data can form a basis for data representation. An important, yet often neglected issue is **temporal characteristics of data**: Data of all types may have a temporal (time) association, and this association may be either discrete or continuous (Thomas and Cook 2005).

In Medical Informatics we have a permanent interaction between data, information, and knowledge, with different definitions (Bemmel and Musen 1997), see Slide 2-16.

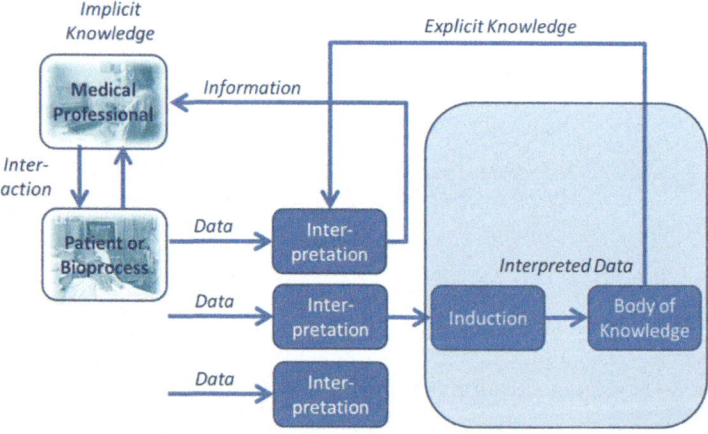

Fig. 9 See Slide 2-16

Slide 2-16: Clinical View of Data, Information, Knowledge

Data are the physical entities at the lowest abstraction level which are for example generated by a patient (e.g., patient data) or a biological process (e.g., Omics data). According to Bemmel and Musen (1997) data contain no meaning.

Information is derived by interpretation of the data by a clinician (human intelligence).

Knowledge is obtained by inductive reasoning with previously interpreted data, collected from many similar patients or processes, which is added to the so-called body of knowledge in medicine, the **explicit knowledge**. This knowledge is used for the interpretation of other data and to gain **implicit knowledge** which guides the clinician in taking further action.

For hypothesis generation and testing, four types of inferences exist (Peirce 1955): abstraction, abduction, deduction, and induction. The first two drive hypothesis generation while the latter drive hypothesis testing, see Slide 2-17.

Abstraction means that data are filtered according to their relevance for the problem solution and chunked in schemas representing an *abstract* description of the problem (e.g., abstracting that an adult male with hemoglobin concentration less than 14 g/dL is an anemic patient). Following this, hypotheses that could account for the current situation are related through a process of abduction, characterized by a "backward flow" of inferences across a chain of directed relations which identify those initial conditions from which the current abstract representation of the problem originates. This provides tentative solutions to the problem at hand by way of hypotheses. For example, knowing that disease A will cause symptom B, abduction will try to identify the explanation for B, while deduction will forecast that a patient affected by disease A will manifest symptom B: both inferences are using the same relation along two different directions (Patel and Ramoni 1997).

Abduction is characterized by a cyclical process of generating possible explanations (i.e., identification of a set of hypotheses that are able to account for the clinical case on the basis of the available data) and testing those explanations (i.e., evaluation of each generated hypothesis on the basis of its expected consequences) for the abnormal state of the patient at hand (Patel et al. 2004).

The hypothesis testing procedures can be inferred from Slide 2-17.

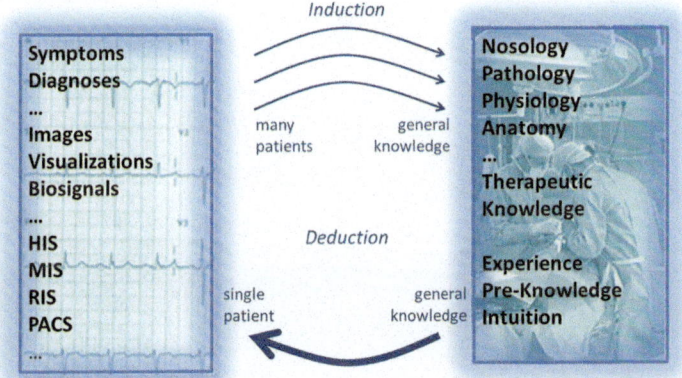

Fig. 10 See Slide 2-17

Slide 2-17: From Patient Data to Medical Knowledge

General knowledge is gained from many patients, and this general knowledge is then applied to an individual patient. We have to determine between:

Reasoning is the process by which a clinician reaches a conclusion after thinking about all the facts.

Deduction consists of deriving a particular valid conclusion from a set of *general* premises.

Induction consists of deriving a likely general conclusion from a set of particular statements.

Reasoning in the "real world" does not appear to fit neatly into any of these basic types. Therefore, a third form of reasoning has been recognized by Peirce (1955), where deduction and induction are inter-mixed.

7 A Closer Look on Information

7.1 What Is Information?

The question "What is information?" is still an open question in basic research, and any definition is depending on the view taken. For example, the definition given by Carl-Friedrich von Weizsäcker: "Information is what is understood," implies that information has both a sender and a receiver who have a common understanding of the representation and the means to convey information using some properties of the physical systems, and his addendum: "Information has no absolute meaning; it exists relatively between two semantic levels" implies the importance of context (Marinescu 2011). Without doubt information is a fundamentally important concept within our world and life is complex information, see Slide 2-14.

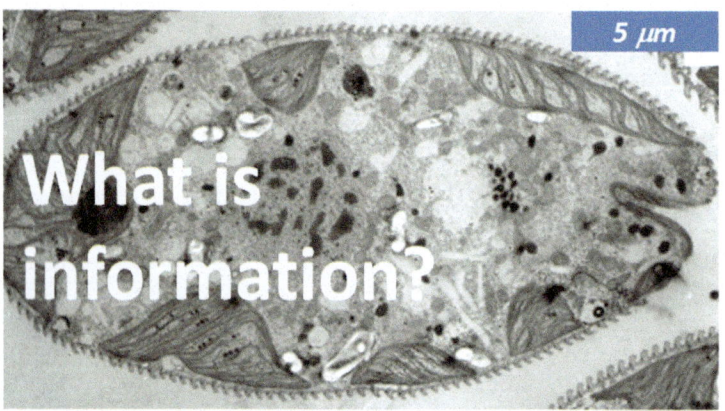

Fig. 11 See Slide 2-18

Slide 2-18: Life Is Complex Information

Many systems for example in the quantum world do not obey the classical view of information. In the quantum world and in the life sciences traditional information theory often fails to accurately describe reality ... For example in the complexity of a living cell: All complex life is composed of eukaryotic (nucleated) cells (Lane and Martin 2010). A good example of such a cell is the protist *Euglena gracilis* (in German "Augentierchen") with a length of approx. 30 μm. Life can be seen as a delicate interplay of energy, entropy and information, essential functions of living beings correspond to the generation, consumption, processing, preservation, and duplication of information.

7 A Closer Look on Information

The etymological origin of the word information can be traced back to the Greek "forma" and the Latin "information" and "informare," to bring something into a shape ("in-a-form"). Consequently, the naive definition in computer science is *"information is data in context"* and therefore different than data or knowledge.

However, we follow the notion of Boisot and Canals (2004) and define that information is an extraction from data that, by modifying the relevant probability distributions, has direct influence on an agent's knowledge base. For a better understanding of this concept, we first review the model of human information processing by Wickens (1984).

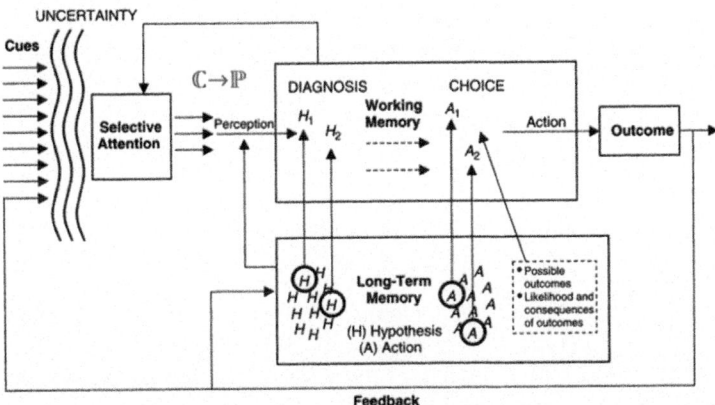

Fig. 12 See Slide 2-19

> **Slide 2-19: Human Information Processing Model**
>
> The model by Wickens (1984) beautifully emphasizes our view on data, information, and knowledge: the physical data from the real world are perceived as information through perceptual filters, controlled by selective attention and form hypotheses within the working memory. These hypotheses are the expectations depending on our previous knowledge available in our mental model, stored in the long-term memory. The subjectively best alternative hypothesis will be selected and processed further and may be taken as outcome for an action. Due to the fact that this system is a closed loop, we get feedback through new data perceived as new information and the process goes on.

By following this model, we can now understand the essential relationships between data, information, and knowledge (Slide 2-20).

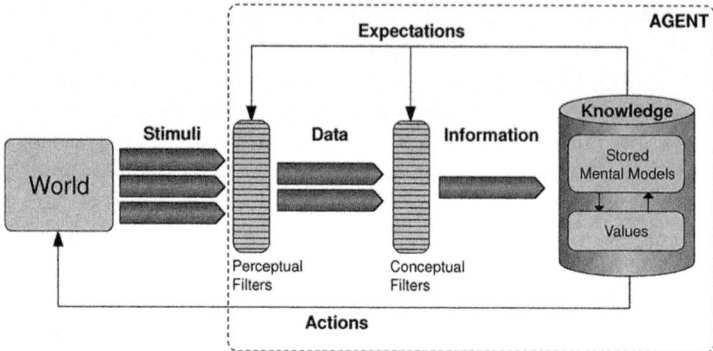

Fig. 13 See Slide 2-20

Slide 2-20: Knowledge as a Set of Expectations

Image redrawn and altered after Boisot and Canals (2004).

The incoming stimuli from the physical world must pass both a perceptual filter and a conceptual filter.

The **perceptual filter** orientates the senses (e.g., visual sense) to certain types of stimuli within a certain physical range (visual signal range, pre-knowledge, attention, etc.). Only the stimuli which pass through this filter get registered as *incoming data*—everything else is filtered out. At this point it is important to follow our physical principle of data: to differentiate between two notions that are frequently confused: an experiment's (raw, hard, measured, factual) data and its (meaningful, subjective) interpreted information results. Data are properties concerning only the instrument; it is the **expression of a fact**. The result concerns a property of the world. The following conceptual filters extract information-bearing data from what has been previously registered.

Both types of filters are influenced by the agents' cognitive and affective expectations, stored in their mental models. The enormous utility of data resides in the fact that it can carry information about the physical world. This information may modify set expectations or the state-of-knowledge. These principles allow an **agent** to act in adaptive ways in the physical world (Boisot and Canals 2004).

Confer this process with the human information processing model by Wickens (1984), seen in Slide 2-19 and discussed in Lecture 7.

7.2 Information Entropy

> **Slide 2-21: Entropy**
>
> Entropy has many different definitions and applications, originally in statistical physics and most often it can be used as a measure for disorder. In information theory, entropy can be used as a measure for the uncertainty in a dataset.
>
> The concept of entropy was first introduced in Thermodynamics (Clausius 1850), where it was used to provide a statement of the second law of thermodynamics. Later, statistical mechanics provided a connection between the macroscopic property of entropy and the microscopic state of a system by Boltzmann. Shannon was the first to define entropy and mutual information.

Shannon (1948) used a "Gedankenexperiment" (thought experiment) to propose a measure of uncertainty in a discrete distribution based on the Boltzmann entropy of classical statistical mechanics, see Slide 2-22.

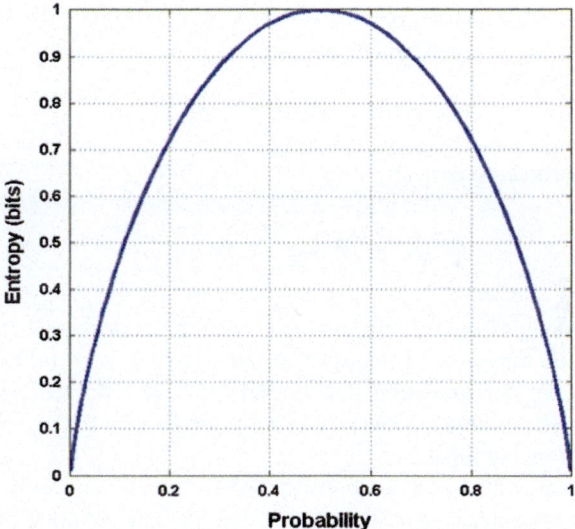

Fig. 14 See Slide 2-22

> **Slide 2-22: Entropy H as a Measure for Uncertainty (1/3)**
>
> An example shall demonstrate the usefulness of this approach:
>
> 1. Let Q be a discrete dataset with associated probabilities p_i:
>
> (continued)

(continued)
$$Q \ldots P = \{p_1, \ldots, p_n\} \tag{2-4}$$

2. Now we apply Shannon's equation (2-4):

$$H(Q) = -\sum_{i=1}^{n} p_i \log_2(p_i) \tag{2-5}$$

3. We assume that our source has two values (ball = white, ball = black)

    Let us do the famous simple "Gedankenexperiment" (thought experiment): Imagine a box which can contain two colored balls: black and white. This is our set of discrete symbols with associated probabilities. If we grab blindly into this box to get a ball, we are dealing with uncertainty, because we do not know which ball we touch. We can ask: "Is the ball black?" NO. THEN it must be white, so we need one question to surely provide the right answer. Because it is a binary decision (YES/NO) the maximum number of (binary) questions required to reduce the uncertainty is: $\log_2(N)$, where N is the number of the possible outcomes. If there are N events with equal probability p then $N = 1/p$. If you have only 1 black ball, then $\log_2(1) = 0$, which means there is no uncertainty.

$$Qb = \{a_1, a_2\} \text{ with } P = \{p, 1-p\} \tag{2-6}$$

4. Now we solve numerically (2-6):

$$H(Q_b) = p * \log \frac{1}{p} + p * \log \frac{1}{1-p} \tag{2-7}$$

Since p ranges from 0 (for impossible events) to 1 (for certain events), the entropy value ranges from infinity (for impossible events) to 0 (for certain events). So, we can summarize that the entropy is the weighted average of the surprise for all possible outcomes. For our example with the two balls we can draw the following function.

The entropy value is 1 for $p = 0.5$ and it is both 0 for either $p = 0$ or $p = 1$. This example might seem trivial, but the entropy principle has been developed a lot since Shannon and there are many different methods, which are very useful for dealing with data.

7 A Closer Look on Information

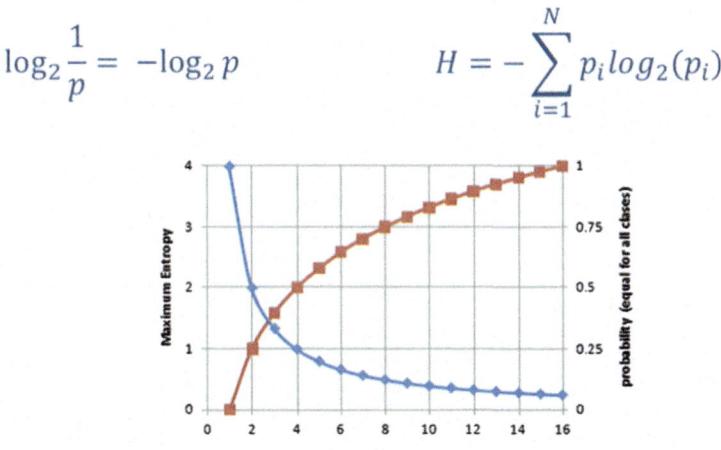

Fig. 15 See Slide 2-23

Slide 2-23: Entropy H as a Measure for Uncertainty (2/3)

Shannon called it the **information entropy** (aka Shannon entropy) and defined:

$$\log_2 \frac{1}{p} = -\log_2 p \qquad (2\text{-}8)$$

where p is the probability of the event occurring. If p is not identical for all events then the entropy H is a weighted average of all probabilities, which Shannon defined as:

$$H = -\sum_{i=1}^{N} p_i \log_2(p_i) \qquad (2\text{-}9)$$

Basically, the entropy p(x) approaches zero if we have a maximum of structure—and opposite, the entropy p(x) reaches high values if there is no structure—hence, ideally, if the entropy is a maximum, we have complete randomness, total uncertainty.

Low Entropy means differences, structure, individuality—high Entropy means no differences, no structure, no individuality. Consequently, life needs low entropy.

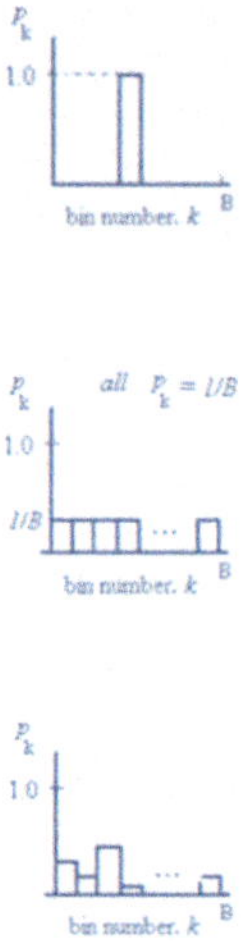

$$H_B = -\sum_{k=1} p_k \log_2 p_k = -1 * \log_2(1) = 0$$

$$H_B = -\sum_{k=1}^{B} \frac{1}{B} \log_2 \frac{1}{B} = \log_2(B)$$

$$H = H_{min} = 0 \qquad H = H_{max} = \log_2 N$$

Fig. 16 See Slide 2-24

Slide 2-24: Entropy H as a Measure for Uncertainty (3/3)

The principle what we can infer from entropy values is:

1. **Low entropy** values mean high probability, **high certainty**, and hence a high degree of structurization in the data.
2. **High entropy** values mean low probability, **low certainty** ($\cong$ high uncertainty ;-), and hence a low degree of structurization in the data.

Maximum entropy would mean complete randomness and total uncertainty.

Highly structured data contain low entropy; ideally if everything is in order and there is no surprise (no uncertainty) the entropy is low:

$$H = H_{min} = 0 \qquad (2\text{-}10)$$
$$H = H_{max} = \log_2 N. \qquad (2\text{-}11)$$

On the other hand if the data are weakly structured—as for example in biological data—and there is no ability to guess (all data is equally likely) the entropy is high:

If we follow this approach, "unstructured data" would mean complete randomness. Let us look on the history of entropy to understand what we can do in future, see Slide 2-25.

7 A Closer Look on Information

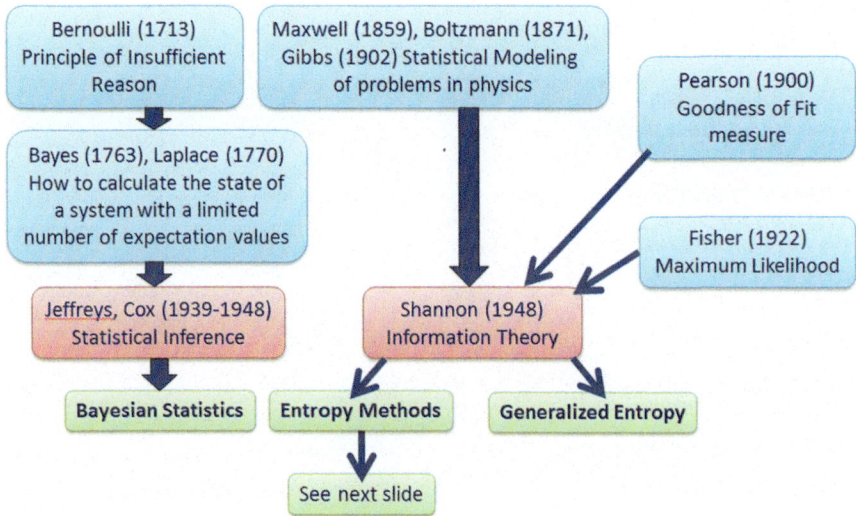

Fig. 17 See Slide 2-25

Slide 2-25: An Overview of the History of Entropy

The origin may be found in the work of Jakob Bernoulli, describing the principle of insufficient reason: we are ignorant of the ways an event can occur, the event will occur equally likely in any way. Thomas Bayes (1763) and Pierre-Simon Laplace (1774) carried on and Harold Jeffreys and David Cox solidified it in the Bayesian Statistics, aka statistical inference. The second path leading to the classical Maximum Entropy, en-route with the Shannon Entropy, can be identified with the work of James Clerk Maxwell and Ludwig Boltzmann, continued by Willard Gibbs and finally Claude Elwood Shannon. This work is geared toward developing the mathematical tools for statistical modeling of problems in information. These two independent lines of research are very similar. The objective of the first line of research is to formulate a theory/methodology that allows understanding of the *general characteristics* (distribution) of a system from partial and incomplete information. In the second route of research, the same objective is expressed as determining how to assign (initial) numerical values of probabilities when only some (theoretical) limited global quantities of the investigated system are known. Recognizing the common basic objectives of these two lines of research aided Jaynes in the development of his classical work, the Maximum Entropy formalism. This formalism is based on the first line of research and the mathematics of the second line of research. The interrelationship between information theory, statistics and inference, and the Maximum Entropy (MaxEn) principle became clear in 1950s, and many different methods arose from these principles (Golan 2008), see next slide

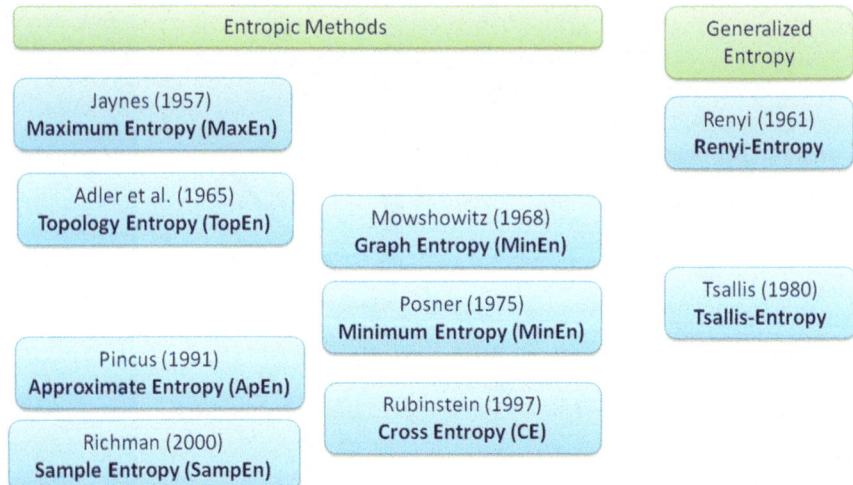

Fig. 18 See Slide 2-26

Slide 2-26: Towards a Taxonomy of Entropic Methods

Maximum Entropy (MaxEn), described by Jaynes (1957), is used to estimate unknown parameters of a multinomial discrete choice problem, whereas the Generalized Maximum Entropy (GME) model includes noise terms in the multinomial information constraints. Each noise term is modeled as the mean of a finite set of a priori known points in the interval [−1, 1] with unknown probabilities where no parametric assumptions about the error distribution are made. A GME model for the multinomial probabilities and for the distributions, associated with the noise terms is derived by maximizing the joint entropy of multinomial and noise distributions, under the assumption of independence (Jaynes 1957).

Topological Entropy (TopEn), was introduced by Adler et al. (1965) with the purpose to introduce the notion of entropy as an invariant for continuous mappings: Let (X, T) be a topological dynamical system, i.e., let X be a nonempty compact Hausdorff space and $T : X \rightarrow X$ a continuous map; the TopEn is a nonnegative number which measures the complexity of the system (Adler et al. 2008).

Graph Entropy was described by Mowshowitz (1968) to measure structural information content of graphs, and a different definition, more focused on problems in information and coding theory, was introduced by Körner (1973). Graph entropy is often used for the characterization of the structure of graph-based systems, e.g., in mathematical biochemistry. In these applications the entropy of a graph is interpreted as its structural information content and serves as a complexity measure, and such a measure is associated with an equivalence relation defined on a finite graph; by application of Shannon's

(continued)

(continued)

(2-4) with the probability distribution we get a numerical value that serves as an index of the structural feature captured by the equivalence relation (Dehmer and Mowshowitz 2011).

Minimum Entropy (MinEn), described by Posner (1975), provides us the least random, and the least uniform probability distribution of a dataset, i.e., the minimum uncertainty, which is the limit of our knowledge and of the structure of the system. Often, the classical pattern recognition is described as a quest for minimum entropy. Mathematically, it is more difficult to determine a minimum entropy probability distribution than a maximum entropy probability distribution; while the latter has a global maximum due to the concavity of the entropy, the former has to be obtained by calculating all local minima, consequently the minimum entropy probability distribution may not exist in many cases (Yuan and Kesavan 1998).

Cross Entropy (CE), discussed by Rubinstein (1997), was motivated by an adaptive algorithm for estimating probabilities of rare events in complex stochastic networks, which involves variance minimization. CE can also be used for combinatorial optimization problems (COP). This is done by translating the "deterministic" optimization problem into a related "stochastic" optimization problem and then using rare event simulation techniques (De Boer et al. 2005).

Rényi entropy is a generalization of the Shannon entropy (information theory), and **Tsallis entropy** is a generalization of the standard Boltzmann–Gibbs entropy (statistical physics).

For us more important are:

Approximate Entropy (ApEn), described by Pincus (1991), is useable to quantify regularity in data without any a priori knowledge about the system, see an example in Slide 2-20.

Sample Entropy (SampEn) was used by Richman and Moorman (2000) for a new related measure of time series regularity. SampEn was designed to reduce the bias of ApEn and is better suited for datasets with known probabilistic content.

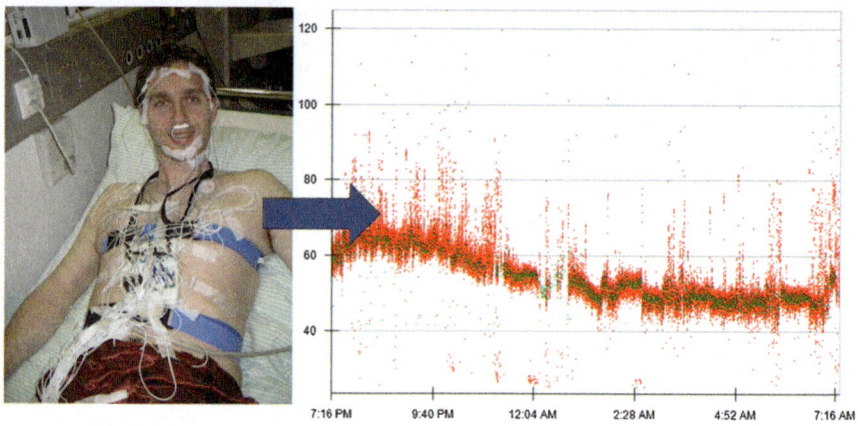

Fig. 19 See Slide 2-27

Slide 2-27: Example of the Usefulness of ApEn (1/3)

Problem: Monitoring body movements along with vital parameters during sleep provides important medical information regarding the general health, and can therefore be used to detect trends (large epidemiology studies) to discover severe illnesses including hypertension (which is enormously increasing in our society).

This seemingly simple data—only from one night period—demonstrates the complexity and the boundaries of standard methods (for example Fast Fourier Transformation) to discover knowledge (for example deviations, similarities, etc.).

Due to the complexity and uncertainty of such datasets, standard methods (such as FFT) comprise the danger of modeling artifacts. Since the knowledge of interest for medical purposes is in anomalies (alterations, differences, atypicalities, irregularities), the application of entropic methods provides benefits.

Photograph taken during the EU Project EMERGE and used with permission.

Slide 2-28: Example of the Usefulness of ApEn (2/3)

1. We have a given dataset $\langle x_n \rangle$ where capital N is the number of data points:

$$\text{Let } \langle x_n \rangle = \{x_1, x_2, \ldots, x_N\}. \qquad (2\text{-}12)$$

2. Now we form m-dimensional vectors

(continued)

(continued)

$$\vec{X}_i = \left(x_i, x_{(i+1)}, \ldots, x_{(i+m-1)}\right). \tag{2-13}$$

3. We measure the distance between every component, i.e., the maximum absolute difference between their scalar components

$$\left\|\vec{X}_i, \vec{X}_j\right\| = \max_{k=1,2,\ldots,m} \left(\left|x_{(i+k-1)} - x_{(j+k-1)}\right|\right). \tag{2-14}$$

4. We look—so to say—in which dimension is the biggest difference; as a result we get the Approximate Entropy (if there is no difference we have zero relative entropy):

$$\mathrm{ApEn}(m,r) = \lim_{N \to \infty} \left[\phi^m(r) - \phi^{m+1}(r)\right] \tag{2-15}$$

where m is the run length and r is the tolerance window r (let us assume that m is equal to r), ApEn (m,r) could also be written as $\widetilde{H}(m,r)$.

5. $\phi^m(r)$ is computed by

$$\phi^m(r) = \frac{1}{N-m+1} \sum_{t=1}^{N-m+1} \ln C_r^m(i) \tag{2-16}$$

with

$$C_r^m(i) = \frac{N^m(i)}{N-m+1}. \tag{2-17}$$

6. C_r^m measures within the tolerance r the regularity of patterns similar to a given one of window length m.
7. Finally we increase the dimension to $m + 1$ and repeat the steps before and get as a result the approximate entropy ApEn(m, r).

ApEn(m, r, N) is approximately the negative natural logarithm of the conditional probability (CP) that a dataset of length N, having repeated itself within a tolerance r for m points, will also repeat itself for $m + 1$ points. An important point to keep in mind about the parameter r is that it is commonly

(continued)

(continued)

expressed as a fraction of the Standard Deviation (SD) of the data and in this way makes ApEn a scale-invariant measure. A low value arises from a high probability of repeated template sequences in the data (Hornero et al. 2006).

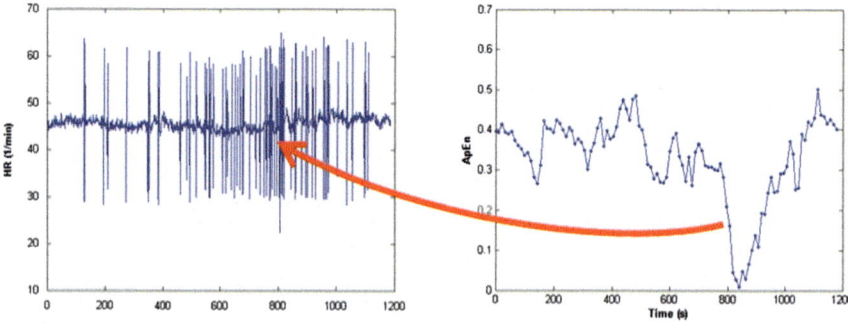

Fig. 20 See Slide 2-29

Slide 2-29: ApEn

A final example should make the advantage of such an entropy method totally clear: In the right diagram it is hard to discover irregularities for a medical professional—especially over a longer period, but an anomaly can easily be detected by displaying the measured relative ApEn.

What can we learn from this experiment? Approximate entropy is relatively unaffected by noise; it can be applied to complex time series with good reproduction; it is finite for stochastic, noisy, composite processes; the values correspond directly to irregularities; and it is applicable to many other areas—for example for the classification of large sets of texts—the ability to guess algorithmically the subject of a text collection without having to read it would permit automated classification.

8 Future Outlook

Where are the research perspectives?
What about big data?
Forrester Hype?

8 Future Outlook

Georges Grinstein: We are at a fork in the road. We have reached a fundamental display limit of about one million items with most of our current techniques. How can we break that barrier? The human perceptual system seems to be able to handle large quantities of data of few dimensions but has great difficulty as the dimensionality of the data increases. Is this another view of the curse of dimensionality? What fundamental element is holding us back? I will argue that the key challenge is to focus not simply on the computer techniques of displaying large quantities of data but on the perceptual consumption of such large amounts of data. We must focus on how the process of computer visualization can be improved to mirror the process of natural visualization, that is, the visualization of nature. Our perceptual systems were designed specifically for survival in and understanding of the surrounding external environment, not abstract objects and images. Simply put: how can we exploit human perception in the service of data visualization?

Sharon Laskowski: The thorny issue I would like to address is the difficulty of applying3D to abstract information visualization. There are several questions that must be answered to overcome this inherent difficulty. (a) When is 3D visualization useful? (b) How can 3D be applied effectively for abstract information visualization? (c) How does one know a visualization is effective? Or efficient? Just because it is good for entertainment, it does not mean it is a good choice for serious work. There are many fine examples of 2D visualizations of large databases. But why do we see so many examples of 3D that do not seem to get out of the lab and into actual practice? These difficulties with 3D I believe stem from the following problems: (1) It is difficult to design good 3D visualizations. (2) 2D often can replace 3D because the third dimension is not being used wisely. (3) It is difficult to evaluate a 3D interface and we do not have a lot of guidelines as we do for GUI design. (4) 3D is clearly suitable for scientific visualization and computer aided design applications because the dimensions map directly to physical world entities and/or the data is heavily numerical. Navigation in this context is easier as well. The artificial mapping required to represent abstract information in a 3Dworld increases the cognitive load on the user. The metaphor is once removed from the meaning and the data. For this panel discussion, I will illustrate my points by describing an effort to use 3D for the visualization of document sets and our struggles in trying to evaluate the usability of these visualizations as compared to simple 1D and 2D structures. In summary, it is time to get more serious about evaluation. For GUIs, visual design is important, of course, but there are also engineering methodologies, processes, and guidelines to test for usability. The visualization community needs this rigor as well. Alfred Inselberg: Imagine a veritable mess of old bolts, nuts and what seems like minute parts of every conceivable kind mixed in deep amorphous piles. From such data, the best analysis can reveal very little. This must be in principle true for it cannot be distinguished whether such objects derive from automobiles, airplanes, or a wide variety of machinery. If, on the other hand, these objects are assembled into components until they become distinguishable (e.g., many automobile parts can be distinguished from those of other machinery) the problem at least of identification and partial analysis is tractable. This is a simplistic but useful example. We argue that looking at a set of disorganized high dimensional data is even less informative than the situation in the "Junkyard Metaphor" above. We maintain that it is not possible in general to identify an N-dimensional object by just looking at its points (which are 0D); not because Dimensionality is Curse but rather because we are using "spectacles" with the wrong "pre-Proceedings of the IEEE Visualization 98 (VIZ'98)0-8186-9176-x/98 $10.00 © 1998 I.E. scription." Specifically, we will show how an N-dimensional object can be identified from its $(N - 1)$-dimensional components. This leads to a recursive construction where from the 0-dimensionalcomponents (points or 0-flats), lines (or 1D components—1-flats) are formed, leading to the construction of 2D components (planes or 2-flats) and so on until the $(N - 1)$-dimensional components are constructed to distinguish and identify the N-dimensional body. In turn this geometrical identification enables us to provide a rigorous description of the "structure" in the data, in terms of the geometrical properties of the N-D object which models them. Examples will be provided of such recursion not only for "precise," in the mathematical sense, objects, but also those which have been perturbed and contain errors; showing that "To Iterate is Human ... and to Recurse Divine!". All this lead us that the conclusion that Dimensionality is not a curse. Rather it seems so because of the way we have unjustly treated it so far—i.e., "pursuing," "projecting," and even mutilating (i.e., "reducing") it. To ameliorate this wrong we will propose a prize for the best composition of a rousing ecumenical blessing ... to Dimensionality. BIOGRAPHIES Georges Grinstein is a full-time Professor of Computer Science at the University of Massachusetts Lowell, Director of the Graphics Research Laboratory, and Director of the Institute for Visualization and Perception Research. He received his B.S. from the City College of N.Y. in 1967, his M.S. from the Courant Institute of Mathematical Sciences of New York University in 1969, and his Ph.D. in Mathematics from the University of Rochester in 1978. He is a member of IEEE, ACM, Eurographics, and has served on the editorial boards of Computers and Graphics, Computer Graphics Forum, and Knowledge Discovery in Databases and Data Mining Journals. He served as a member of the executive board of ANSI X3H3, as chair of X3H3.6, and as vice-chair of IFIP WG 5.10 Computer Graphics Group. He has been involved with the IEEE Visualization and AAAI KDD Conferences since 1990, and has chaired numerous workshops related to visual and analytic data mining. His areas of research include graphics, imaging, virtual environments, user interfaces and interaction, with a very strong interest in the visualization and analysis of complex systems. Alfred Inselberg received his Ph.D. in Applied Mathematics and Physics from the University of Illinois

(Champaign-Urbana). He has held several academic positions abroad and in the USA (with the University of Illinois, UCLA, and USC), and was recently elected Senior Fellow in Visualization at the San DiegoSuper Computing Center. Prior to that he was Senior Technical Corporate Staff Member at IBM doing research for about 30 years. He now has his own company, Multidimensional Graphs Ltd., and teaches at Tel Aviv University. He invented and developed Parallel Coordinates, on which he has contributed to several patents, has over 60 refereed technical papers, and is now writing a book on the subject. Sharon Laskowski is a computer scientist and group manager of the Visualization and Virtual Reality Group in the Information Technology Laboratory of the National Institute of Standards and Technology, where she is currently investigating the use of information visualization techniques for information retrieval and exploration. She is also developing a set of rapid, remote, and automated tools and methods for Web usability evaluation and testing called WebMetrics. She has also been involved in research into evaluation methodologies for collaborative systems. She is aco-founder and organizer of the NIST series of Usability Engineering in Government Systems symposia. Previously, she conducted research and development in text analysis, information fusion, and plan recognition at the Artificial Intelligence Center of the MITRE Corporation. She received her Ph.D. in Computer Science from Yale University.

9 Exam Questions

9.1 Yes/No Questions

Please check the following sentences and decide whether the sentence is true = YES; or false = NO; for each correct answer you will be awarded 2 credit points.

01	An array is a composite data type on physical level.	☐ Yes ☐ No	2 total
02	In a Von-Neumann machine "List" is a widely used data structure for applications which do not need random access.	☐ Yes ☐ No	2 total
03	The edges in a graph can be multidimensional objects, e.g. vectors containing the results of multiple Gen-expression measures.	☐ Yes ☐ No	2 total
04	Each item of data is composed of variables, and if such a data item is defined by more than one variable it is called a multivariable data item	☐ Yes ☐ No	2 total
05	A dendrogram is a tree diagram frequently used to illustrate the arrangement of the clusters produced by hierarchical clustering.	☐ Yes ☐ No	2 total
06	Nominal and ordinal data are parametric, and do assume a particular distribution.	☐ Yes ☐ No	2 total
07	Abstraction is characterized by a cyclical process of generating possible explanations and testing those explanations.	☐ Yes ☐ No	2 total
08	A metric space has an associated metric, which enables us to measure distances between points in that space and, in turn, implicitly define their neighborhoods.	☐ Yes ☐ No	2 total
09	Induction consists of deriving a likely general conclusion from a set of particular statements.	☐ Yes ☐ No	2 total
10	In the model of Boisot & Canals (2004), the perceptual filter orientates the senses (e.g. visual sense) to certain types of stimuli within a certain physical range.	☐ Yes ☐ No	2 total

Sum of Question Block A (max. 20 points)

9.2 Multiple Choice Questions (MCQ)

01	Nucleotides ... ☐ a) ... are arranged in a graph structure. ☐ b) ... are the joined molecules which form the structural units of the RNA and DNA. ☐ c) ... can be illustrated by a tree structure. ☐ d) ... are a typical example for a list data structure.	4 total
02	"Big data pools" in the health care domain include ... ☐ a) ... clinical data provided by the HIS. ☐ b) ... images stored in the PACS. ☐ c) ... laboratory data. ☐ d) ... clinical trial data.	4 total
03	Proteomics data ... ☐ a) ... are mostly stored in open data repositories. ☐ b) .... Contain a set of all RNA molecules produced in the cells. ☐ c) ... are large data sets and require specialized analytical tools, which can deal with large lists of objects. ☐ d) ... represent the collection of all metabolites in a cell, tissue, organ or organism.	4 total
04	Standardized data is ... ☐ a) ... is the majority of all data. ☐ b) ... is the basis for accurate communication. ☐ c) ... contains tags or markers to separate structure and content. ☐ d) ... ensures that information is interpreted by all users with the same understanding.	4 total
05	Ordinal data ... ☐ a) ... allow a determination of more/less. ☐ b) ... use median and percentiles as basic statistics. ☐ c) ... allow all known basic mathematical operations. ☐ d) ... are used for the determation of equality.	4 total
06	Explicit knowledge in medicine is ... ☐ ... the so called "body of knowledge" ☐ ... the result of perceptually and conceptually filtered information. ☐ ... data containing meaning. ☐ ... based on hypotheses stored in so called "mental models".	4 total
07	Knowledge in the sense of the model of Boisot & Canals (2004) ... ☐ a) ... is a set of expectations. ☐ b) ... an essential component of the human cardio-vascular system. ☐ c) ... part of the digestive system, e.g. esophagus, stomach and intestines. ☐ d) ... an ensemble of similar cells, carrying out a function as collective.	4 total
08	Entropy ... ☐ a) ... can be used as a measure for uncertainty. ☐ b) ... was first introduced in the field of thermodynamics. ☐ c) ... can be used as a measure for disorder. ☐ **d) ... can be used as a measure for the degree of structurization of data.**	4 total

09	High entropy measurement values indicate... ❏ a) ... high probability. ❏ b) ... low probability. ❏ c) ... high uncertainty. ❏ d) ... high degree of structurization in the data.	4 total
10	The concept of Approximate Entropy (ApEn) ... ❏ a) ... is useable to quantify regularity in data without any a priori knowledge about the system. ❏ b) ... provides the least random and the least uniform probability distribution. ❏ c) ... was the first entropy method used in information theory. ❏ d) ... can be used for time series data.	4 total

Sum of Question Block B (max. 40 points)

9.3 Free Recall Questions

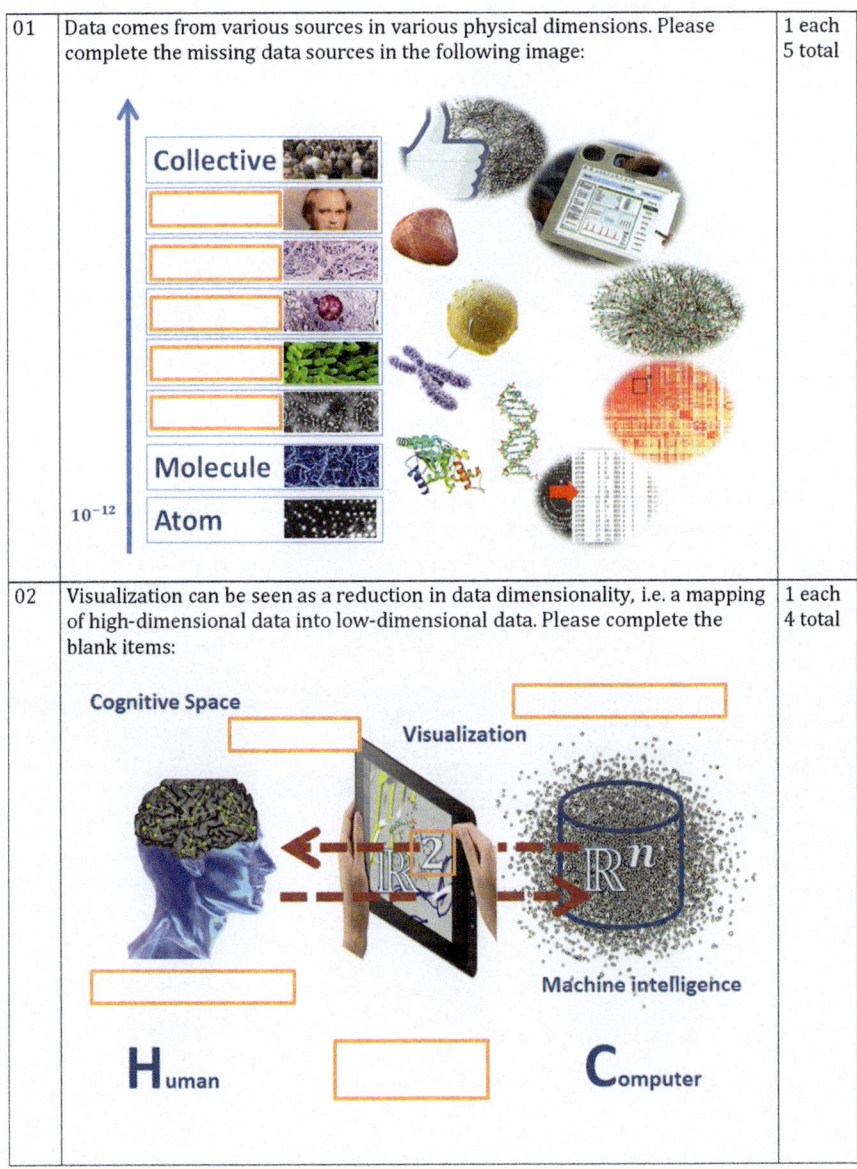

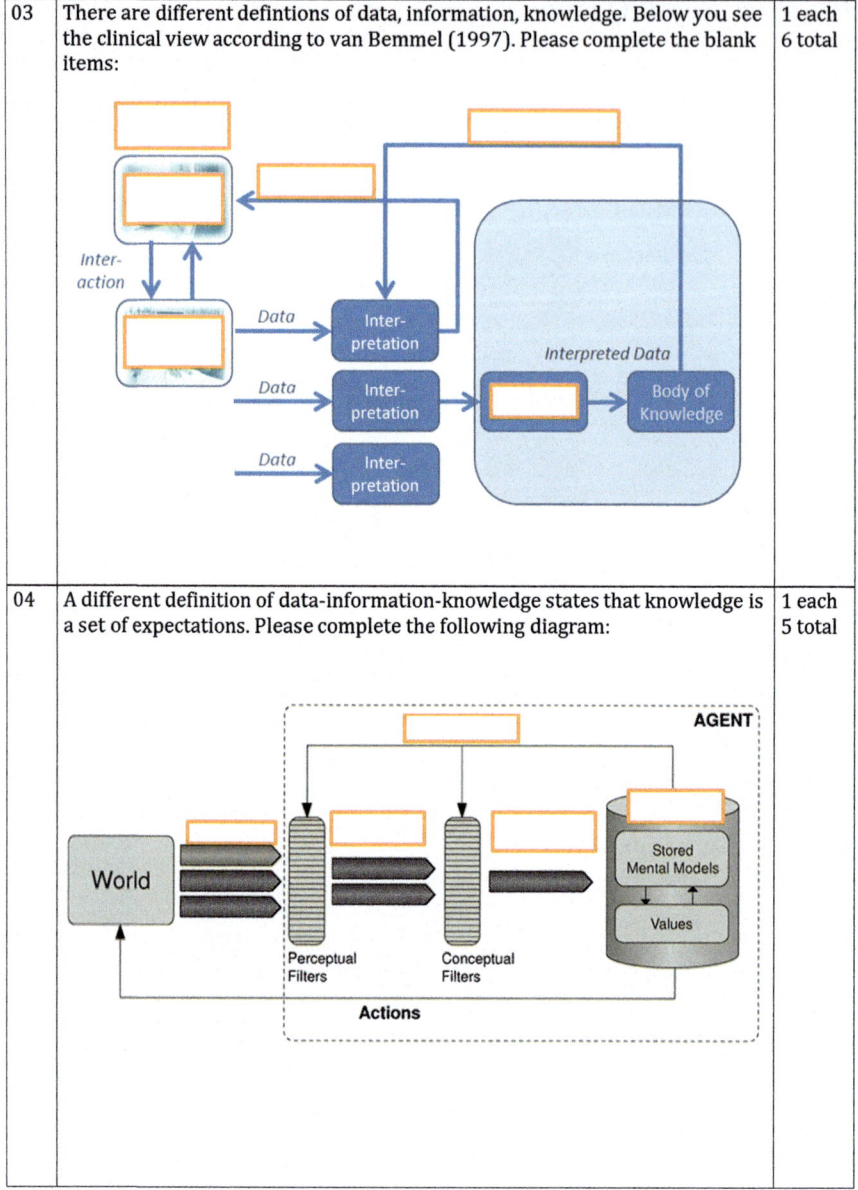

9 Exam Questions

05	Let X and Y represent random variables with associated probability distributions $p(x)$ and $p(y)$. X and Y are not independent. Their conditional probability distributions are $p(x\|y)$ and $p(y\|x)$ and their joint probability distribution is $p(x\|y)$. 1) What is the marginal entropy $H(X)$ of variable X, and what is the mutual information of X? 2) What is the mutual information $I(X;Y)$?	2 each 4 total

06	Abstract data structures are very important concepts in Informatics. Please assign the correct labels to the methods below:	2 each 6 total

(graph diagram with nodes v_1, v_2, v_3, v_4, v_5 and weights)	Tree
(linked list with head and $a_1, a_2, ..., a_n$)	List
(phylogenetic tree diagram)	Graph

Sum of Question Block C (max. 30 points)

10 Answers

10.1 Answers to the Yes/No Questions

Please check the following sentences and decide whether the sentence is true = YES; or false = NO; for each correct answer you will be awarded 2 credit points.

01	An array is a composite data type on physical level.	☐ Yes ☒ No	2 total
02	In a Von-Neumann machine "List" is a widely used data structure for applications which do not need random access.	☒ Yes ☐ No	2 total
03	The edges in a graph can be multidimensional objects, e.g. vectors containing the results of multiple Gen-expression measures.	☒ Yes ☐ No	2 total
04	Each item of data is composed of variables, and if such a data item is defined by more than one variable it is called a multivariable data item	☒ Yes ☐ No	2 total
05	A dendrogram is a tree diagram frequently used to illustrate the arrangement of the clusters produced by hierarchical clustering.	☒ Yes ☐ No	2 total
06	Nominal and ordinal data are parametric, and do assume a particular distribution.	☐ Yes ☒ No	2 total
07	Abstraction is characterized by a cyclical process of generating possible explanations and testing those explanations.	☐ Yes ☒ No	2 total
08	A metric space has an associated metric, which enables us to measure distances between points in that space and, in turn, implicitly define their neighborhoods.	☒ Yes ☐ No	2 total
09	Induction consists of deriving a likely general conclusion from a set of particular statements.	☒ Yes ☐ No	2 total
10	In the model of Boisot & Canals (2004), the perceptual filter orientates the senses (e.g. visual sense) to certain types of stimuli within a certain physical range.	☒ Yes ☐ No	2 total

Sum of Question Block A (max. 20 points)

10.2 Answers to the Multiple Choice Questions (MCQ)

01	Nucleotides ... ☐ a) ... are arranged in a graph structure. ☒ b) ... are the joined molecules which form the structural units of the RNA and DNA. ☐ c) ... can be illustrated by a tree structure. ☒ d) ... are a typical example for a list data structure.	4 total
02	"Big data pools" in the health care domain include ... ☒ a) ... clinical data provided by the HIS. ☒ b) ... images stored in the PACS. ☒ c) ... laboratory data. ☒ d) ... clinical trial data.	4 total
03	Proteomics data ... ☒ a) ... are mostly stored in open data repositories. ☐ b) .... Contain a set of all RNA molecules produced in the cells. ☒ c) ... are large data sets and require specialized analytical tools, which can deal with large lists of objects. ☐ d) ... represent the collection of all metabolites in a cell, tissue, organ or organism.	4 total
04	Standardized data is ... ☐ a) ... is the majority of all data. ☒ b) ... is the basis for accurate communication. ☐ c) ... contains tags or markers to separate structure and content. ☒ d) ... ensures that information is interpreted by all users with the same understanding.	4 total
05	Ordinal data ... ☒ a) ... allow a determination of more/less. ☒ b) ... use median and percentiles as basic statistics. ☐ c) ... allow all known basic mathematical operations. ☐ d) ... are used for the determination of equality.	4 total
06	Explicit knowledge in medicine is ... ☒ ... the so called "body of knowledge" ☒ ... the result of perceptually and conceptually filtered information. ☐ ... data containing meaning. ☒ ... based on hypotheses stored in so called "mental models".	4 total
07	Knowledge in the sense of the model of Boisot & Canals (2004) ... ☒ a) ... is a set of expectations. ☐ b) ... an essential component of the human cardio-vascular system. ☐ c) ... part of the digestive system, e.g. esophagus, stomach and intestines. ☒ d) ... an ensemble of similar cells, carrying out a function as collective.	4 total
08	Entropy ... ☒ a) ... can be used as a measure for uncertainty. ☒ b) ... was first introduced in the field of thermodynamics. ☒ c) ... can be used as a measure for disorder. ☒ d) ... can be used as a measure for the degree of structurization of data.	4 total

09	High entropy measurement values indicate... ☐ a) ... high probability. ☒ b) ... low probability. ☒ c) ... high uncertainty. ☐ d) ... high degree of structurization in the data.	4 total
10	The concept of Approximate Entropy (ApEn) ... ☒ a) ... is useable to quantify regularity in data without any a priori knowledge about the system. ☐ b) ... provides the least random and the least uniform probability distribution. ☐ c) ... was the first entropy method used in information theory. ☒ d) ... can be used for time series data.	4 total

Sum of Question Block B (max. 40 points)	

10 Answers

10.3 Answers to the Free Recall Questions

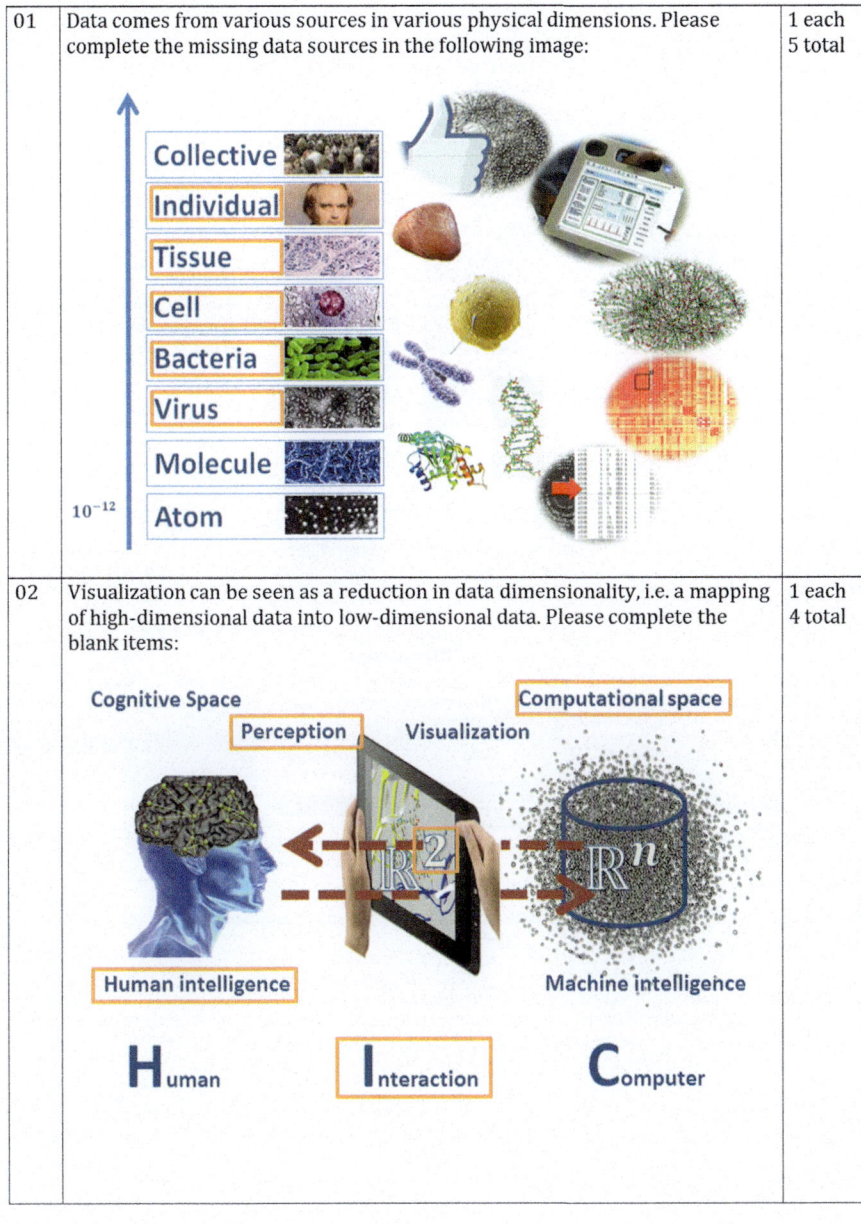

03	There are different defintions of data, information, knowledge. Below you see the clinical view according to van Bemmel (1997). Please complete the blank items:	1 each 6 total
04	A different definition of data-information-knowledge states that knowledge is a set of expectations. Please complete the following diagram:	1 each 5 total

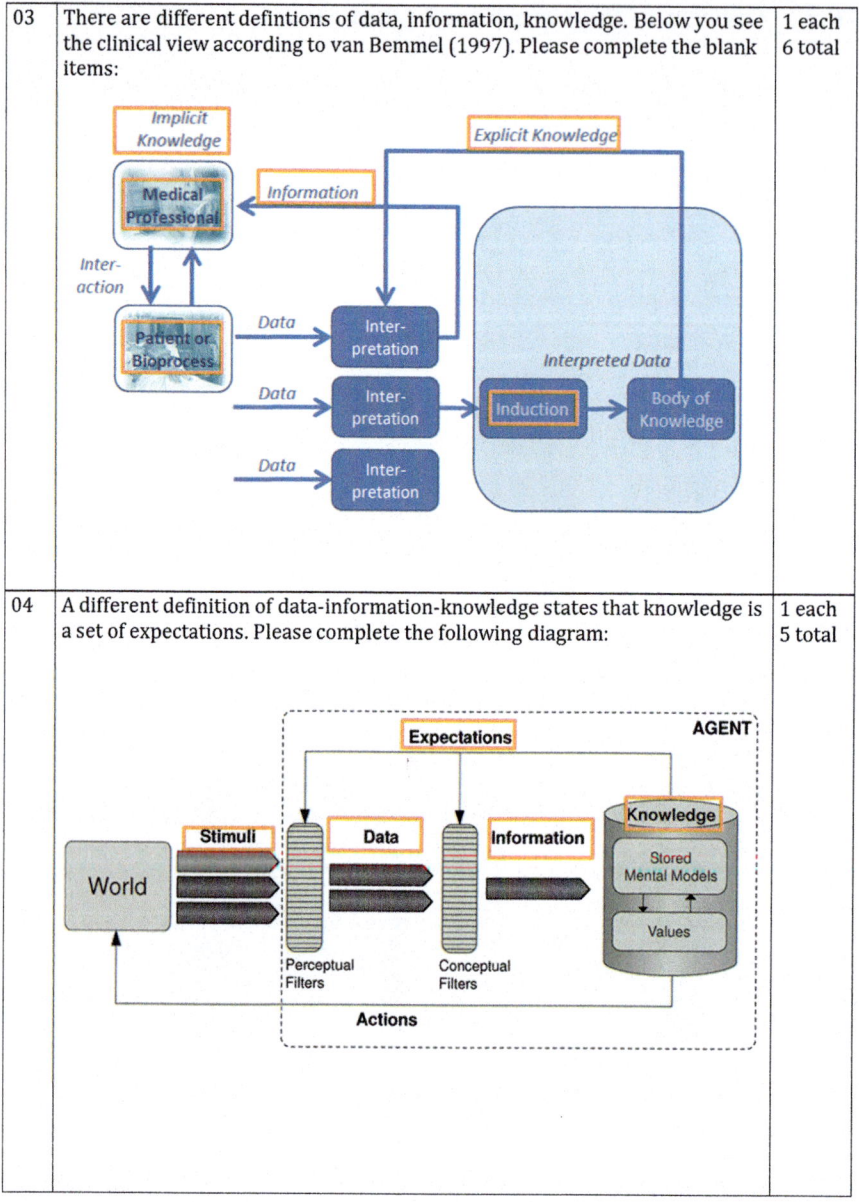

05	Let X and Y represent random variables with associated probability distributions $p(x)$ and $p(y)$. X and Y are not independent. Their conditional probability distributions are $p(x|y)$ and $p(y|x)$ and their joint probability distribution is $p(x|y)$. 1) What is the marginal entropy $H(X)$ of variable X, and what is the mutual information of X? Answer 1 : $H(X) = -\sum_x p(x)\log_2 x$ is the marginal entropy of X and its mutual information with itself. 2) What is the mutual information $I(X;Y)$? Answer 2: $I(X;Y) = H(X) - H(X|Y) = H(X) + H(Y) - H(X,Y)$	2 each 4 total
06	Abstract data structures are very important concepts in Informatics. Please assign the correct labels to the methods below: （graph diagram） — Tree （linked list diagram with head） — List （hierarchical cluster/dendrogram） — Graph	2 each 6 total

Sum of Question Block C (max. 30 points)

References

Adler R, Downarowicz T, Misiurewicz M (2008) Topological entropy (Online). http://www.scholarpedia.org/article/Topological_entropy. Accessed 10 Aug 2013

Adler RL, Konheim AG, Mcandrew MH (1965) Topological entropy. Trans Am Math Soc 114 (2):309–319

Ahmadian L, Van Engen-Verheul M, Bakhshi-Raiez F, Peek N, Cornet R, De Keizer NF (2011) The role of standardized data and terminological systems in computerized clinical decision support systems: literature review and survey. Int J Med Inform 80(2):81–93

Aho AV, Hopcroft JE, Ullman JD (1983) Data structures and algorithms. Addison-Wesley, Boston, MA

Bemmel JHV, Musen MA (1997) Handbook of medical informatics. Springer, Heidelberg

Bessarabova M, Ishkin A, Jebailey L, Nikolskaya T, Nikolsky Y (2012) Knowledge-based analysis of proteomics data. BMC Bioinform 13(Suppl 16):S13

Bleiholder J, Naumann F (2008) Data fusion. ACM Comput Surv (CSUR) 41(1):1

Boisot M, Canals A (2004) Data, information and knowledge: have we got it right? J Evol Econ 14(1):43–67

Bramer M (2013) Principles of data mining, 2nd edn. Springer, Heidelberg

Clausius R (1850) On the motive power of heat, and on the laws which can be deduced from it for the theory of heat (reprint, 1960). Dover, New York

Cormen T (2013) Algorithms unlocked. The MIT Press, Cambridge, MA

Cormen TH, Leiserson CE, Rivest RL, Stein C (2009) Introduction to algorithms, 3rd edn. The MIT Press, Cambridge, MA

Darwin C (1859) On the origin of species by means of natural selection, or the preservation of favoured races in the struggle for life. John Murray, London

De Boer P-T, Kroese DP, Mannor S, Rubinstein RY (2005) A tutorial on the cross-entropy method. Ann Oper Res 134(1):19–67

Dehmer M, Mowshowitz A (2011) A history of graph entropy measures. Inform Sci 181(1):57–78

Dos Santos S, Brodlie K (2002) Visualizing and investigating multidimensional functions. Proceedings of the symposium on data visualisation 2002. Eurographics Association. pp. 173–182

Dos Santos S, Brodlie K (2004) Gaining understanding of multivariate and multidimensional data through visualization. Comput Graph 28(3):311–325

Duda RO, Hart PE, Stork DG (2000) Pattern Classification, 2nd edn. Wiley, New York

Edelsbrunner H, Harer JL (2010) Computational topology: an introduction. American Mathematical Society, Providence, RI

Ghrist R (2008) Barcodes: the persistent topology of data. Bull Am Math Soc 45(1):61–75

Golan A (2008) Information and entropy econometric: a review and synthesis. Found Trends Econ 2(1–2):1–145

Holzinger A (2003) Basiswissen IT/Informatik. Band 2: Informatik. Vogel Buchverlag, Wuerzburg

Holzinger A (2012) On knowledge discovery and interactive intelligent visualization of biomedical data: challenges in human–computer interaction & biomedical informatics. In: Helfert M, Fancalanci C, Filipe J (eds) DATA—International conference on data technologies and applications. INSTICC, Rome, pp 5–16

Holzinger A (2013) Human–computer interaction & knowledge discovery (HCI-KDD): what is the benefit of bringing those two fields to work together? In: Alfredo Cuzzocrea CK, Simos DE, Weippl E, Xu L (eds) Multidisciplinary Research and practice for information systems, Springer lecture notes in computer science LNCS 8127. Springer, New York, pp 319–328

Hornero R, Aboy M, Abasolo D, Mcnames J, Wakeland W, Goldstein B (2006) Complex analysis of intracranial hypertension using approximate entropy. Crit Care Med 34(1):87–95

Hufford MB, Xu X, Van Heerwaarden J, Pyhajarvi T, Chia J-M, Cartwright RA, Elshire RJ, Glaubitz JC, Guill KE, Kaeppler SM, Lai J, Morrell PL, Shannon LM, Song C, Springer NM, Swanson-Wagner RA, Tiffin P, Wang J, Zhang G, Doebley J, Mcmullen MD, Ware D, Buckler ES, Yang S, Ross-Ibarra J (2012) Comparative population genomics of maize domestication and improvement. Nat Genet 44(7):808–811

Jaynes ET (1957) Information theory and statistical mechanics. Phys Rev 106(4):620

Joyce AR, Palsson BØ (2006) The model organism as a system: integrating "omics" data sets. Nat Rev Mol Cell Biol 7(3):198–210

Kaski S, Peltonen J (2011) Dimensionality reduction for data visualization (applications corner). IEEE Sig Process Mag 28(2):100–104

Körner J (1973) Coding of an information source having ambiguous alphabet and the entropy of graphs. 6th Prague conference on information theory. pp. 411–425

References

Kreuzthaler M, Bloice MD, Faulstich L, Simonic KM, Holzinger A (2011) A comparison of different retrieval strategies working on medical free texts. J Univ Comput Sci 17(7):1109–1133

Lane N, Martin W (2010) The energetics of genome complexity. Nature 467(7318):929–934

Lanzagorta M, Uhlmann J (2008) Quantum computer science. Morgan & Claypool, San Francisco

Lieberman E, Hauert C, Nowak MA (2005) Evolutionary dynamics on graphs. Nature 433 (7023):312–316

Manyika J, Chui M, Brown B, Bughin J, Dobbs R, Roxburgh C, Byers AH (2011) Big data: the next frontier for innovation, competition, and productivity. McKinsey Global Institute, Washington, DC

Marinescu DC (2011) Classical and quantum information. Academic, Burlington, MA

Mowshowitz A (1968) Entropy and the complexity of graphs: I. An index of the relative complexity of a graph. Bull Math Biol 30(1):175–204

Ottmann T, Widmayer P (2012) Algorithmen und Datenstrukturen (5. Auflage). Spektrum Akademischer Verlag, Heidelberg

Patel VL, Arocha JF, Zhang J (2004) Thinking and reasoning in medicine Key. In: Holyoak K (ed) Cambridge handbook of thinking and reasoning. Cambridge University Press, Cambridge

Patel VL, Ramoni MF (1997) Cognitive models of directional inference in expert medical reasoning. In: Feltovich PJ, Ford KM (eds) Expertise in context: human and machine. The MIT Press, Cambridge, MA, pp 67–99

Peirce CS (1955) Abduction and induction. In: Peirce CS, Buchler J (eds) Philosophical writings of Peirce. Dover Publications, New York, pp 150–156

Pincus SM (1991) Approximate Entropy as a measure of system complexity. Proc Natl Acad Sci U S A 88(6):2297–2301

Posner E (1975) Random coding strategies for minimum entropy. IEEE Trans Inform Theory 21 (4):388–391

Richman JS, Moorman JR (2000) Physiological time-series analysis using approximate entropy and sample entropy. Am J Physiol Heart Circ Physiol 278(6):H2039–H2049

Riffle M, Eng JK (2009) Proteomics data repositories. Proteomics 9(20):4653–4663

Rubinstein RY (1997) Optimization of computer simulation models with rare events. Eur J Oper Res 99(1):89–112

Shannon CE (1948) A mathematical theory of communication. Bell Syst Tech J 27:379–423

Simon HA (1973) The structure of ill structured problems. Artif Intell 4(3–4):181–201

Simonic KM, Holzinger A, Bloice M, Hermann J (2011) Optimizing long-term treatment of rheumatoid arthritis with systematic documentation. Proceedings of pervasive health—5th international conference on pervasive computing technologies for healthcare. IEEE, Dublin. pp. 550–554

Stevens SS (1946) On the theory of scales of measurement. Science 103:677–680

Thomas JJ, Cook KA (2005) Illuminating the path: the research and development agenda for visual analytics. IEEE Computer Society Press, New York

Wickens CD (1984) Engineering psychology and human performance. Charles Merrill, Columbus, OH

Yuan L, Kesavan H (1998) Minimum entropy and information measure. IEEE Trans Syst Man Cybern C Appl Rev 28(3):488–491

Zomorodian AJ (2005) Topology for computing. Cambridge University Press, Cambridge, MA

Lecture 3

Structured Data: Coding and Classification

1 Learning Goals

At the end of the third lecture, you:

- would have acquired background knowledge on some issues in standardization and structurization of data.
- would have a general understanding of modeling knowledge in medicine and biomedical informatics.
- would get some basic knowledge on medical ontologies and be aware of the limits, restrictions, and shortcomings of them.
- would know the basic ideas and the history of the International Classification of Diseases (ICD).
- would have a view on the Standardized Nomenclature of Medicine Clinical Terms (SNOMED CT).
- would have some basic knowledge on Medical Subject Headings (MeSH).
- would understand the fundamentals and principles of the Unified Medical Language System (UMLS).

2 Advance Organizer

Abstraction	Process of mapping (biological) processes onto a series of concepts (expressed in mathematical terms)
Biological system	Collection of objects ranging in size from molecules to populations of organisms, which interact in ways that display a collective function or role
Coding	Any process of transforming descriptions of medical diagnoses and procedures into standardized code numbers, i.e., to track health conditions and for

	reimbursement, e.g., based on Diagnosis-Related Groups (DRG)
Data model	Definition of entities, attributes, and their relationships within complex sets of data
DSM	Diagnostic and Statistical Manual of Mental Disorders, a multiaxial, multidimensional categorization of all (known) mental health disorders, used for clinical diagnostics
Extensible Markup Language (XML)	Set of rules for encoding documents in machine-readable form
GALEN	Generalized Architecture for Languages, Encyclopedias and Nomenclatures in Medicine, project aiming at the development of a reference model for medical concepts
ICD	International Classification of Diseases, the archetypical coding system for patient record abstraction (est. 1900)
Medical classification	Provides the terminologies of the medical domain (or parts of it), 100+ classifications in use
MeSH	Medical Subject Headings is a classification to index the world medical literature and forms the basis for UMLS
Metadata	Data that describes the data
Model	A simplified representation of a process or object, which describes its behavior under specified conditions (e.g., conceptual model)
Nosography	Science of description of diseases
Nosology	Science of classification of diseases
Ontology	Structured description of a domain and formalizes the terminology (concepts-relations, e.g., IS-A relationship provides a taxonomic skeleton), e.g., gene ontology
Ontology engineering	Subfield of knowledge engineering, which studies the methods and methodologies for building ontologies
SNOMED	Standardized Nomenclature of Medicine, est. 1975, multiaxial system with 11 axes
SNOP	Systematic Nomenclature of Pathology (on four axes: topography, morphology, etiology, function), basis for SNOMED
System features	Static or dynamic; mechanistic or phenomenological; discrete or continuous; deterministic or stochastic; single-scale or multi-scale
Terminology	Includes well-defined terms and usage
UMLS	Unified Medical Language System is a long-term project to develop resources for the support of intelligent information retrieval

3 Acronyms

ACR	American College of Radiologists
API	Application Programming Interface
DAML	DARPA Agent Markup Language
DICOM	Digital Imaging and Communications in Medicine
DL	Description Logic
ECG	Electrocardiogram
EHR	Electronic Health Record
FMA	Foundational Model of Anatomy
FOL	First-Order Logic
GO	Gene Ontology
ICD	International Classification of Diseases
IOM	Institute of Medicine
KIF	Knowledge Interchange Format
LOINC	Logical Observation Identifiers Names and Codes
MeSH	Medical Subject Headings
MRI	Magnetic Resonance Imaging
NCI	National Cancer Institute (US)
NEMA	National Electrical Manufacturer Association
OIL	Ontology Inference Layer (description logic)
OWL	Ontology Web Language
RDF	Resource Description Framework
RDF Schema	A vocabulary of properties and classes added to RDF
SCP	Standard Communications Protocol
SNOMED CT	Systematized Nomenclature of Medicine—Clinical Terms
SOP	Standard Operating Procedure
UMLS	Unified Medical Language System

4 Key Problems

- To find a trade-off between standardization and personalization
- Large amounts of non-standardized information (free text)
- Low integration of standardized terminologies in the daily clinical practice
- Low acceptance of classification codes amongst practitioners

> It is my impression that most of the ontologies that have been created for data collected in many of the fields of science have been ignored or abandoned by their intended beneficiaries. They are simply too difficult to understand and too difficult to implement (Berman 2013).

5 Standardization

5.1 The Need for Standardization in Medicine

A grand challenge in medicine and health care is complexity. Standardization is a systematic approach to create order, making selections, and formulating rules and practices. Consequently, it is indispensable for creating context (using the same terminologies, vocabularies, etc.), exchange data, provide standard operating procedures (SOP's), and enable interoperability of devices. We define:

Standard is a recognized norm that establishes criteria, methods, processes, practices, etc. which lead to interoperability, compatibility, and repeatability. Note: The existence of a published and recognized standard does not necessarily imply that it is useful or correct. For practical purposes only two generic types of standards exist: standards of quality and standards of production (aka standards of quantity).

Standards of quality are measured by the attributes or properties of a product, material, process etc., which defines the goals of a desired performance. Standards of production refer to the execution of a repeated process not necessarily characterized by product quality as much as by end-product reproducibility. Both standards have high value for a health-care system (Brown and Loweli 1972).

Standardization is the process of developing and implementing standards.

An example is the Evidence Based Medicine (EBM) approach, using techniques from science, engineering, and statistics, including systematic review of medical literature, meta-analysis, risk–benefit analyses, and randomized controlled trials (RCTs). This quality approach aims for the ideal that health-care professionals should make "conscientious, explicit, and judicious use of current best evidence" in their everyday practice.

IEEE TRANSACTIONS ON BIOMEDICAL ENGINEERING, VOL. BME-19, NO. 5, SEPTEMBER 1972 331

HEWLETT-PACKARD LIBRARY
AUG 18 1972
NON-CIRCULATING Do Not Remove From Library

Standardization and Health Care

J. H. U. BROWN, SENIOR MEMBER, IEEE, AND DEWITT JAMES LOWELL

Abstract—In order to deliver reasonable health care to all people, it is essential that standards be established. Standards vary with the type of control and with the approach desired in determining the quality of care. This paper discusses various kinds of standards and their application in the health care field. Standards may be determined as a process or as a direct regulation. It is probable that regulation of standards by process is the most satisfactory method.

INTRODUCTION

SOCIETY cannot exist without a yardstick by which its accomplishments or failures are measured. Such yardsticks are called *standards*. They are created by the need for regulation and control as an escape from anarchy or to motivate towards greater achievement. In the ultimate, society dictates these limits by the demands it places upon itself. Standards provide opportunities for security and augmentation of process and output by virtue of the goal and process structure that they provide.

arbiter may be the market place or agencies that rely on expertise from many sources to set acceptable standards of quality or performance. For these reasons, the final moderator may be found in a governmental authority, and its delegation into a system of regulation, law, and judicial action, so that an established code can become the focal point of resolution.

THE OBJECTIVES OF STANDARDIZATION

Standards have value within themselves in that they help establish quality. However, they accomplish more for society than the mere establishment of a level of quality and performance. A standard allows coordination of effort between producers so that like products can be produced. It permits the reproduction of similar units in mass quantity and permits the consumer to judge one product or service against another by performance. It establishes *freedom* of *interchange* of material and ideas, and permits the activity in one part of society

Fig. 1 See Slide 3-1

Slide 3-1: The Quest for Standardization Is as Old as Medical Informatics

Brown and Loweli (1972) describe the need for standards in order to deliver reasonable health care to all people—at a time when medical informatics was in its infancy and electronic patient records were still science fiction.

5.2 *Inaccuracy of Medical Data*

Slide 3-2: Still a big problem: Inaccuracy of medical data

Komaroff (1979) describes clinical data as being disturbingly "soft," having an obvious degree of variability and inaccuracy. Taking a medical history, the performance of a physical examination, the interpretation of laboratory tests, even the definitions of diseases … are surprisingly inexact. Data is defined, collected, and interpreted with variability and inaccuracy, which falls far short of the standards which engineers do expect from most data. Moreover,

(continued)

(continued)

standards might be interpreted variably by different medical doctors, different hospitals, different medical schools, and different medical cultures. In particular the last issue is of extreme importance: every clinic, every department, every hospital has its own established standards, and if you are a patient transferred from one to another hospital it is like changing between "different worlds." Organizational culture and communication has actually an important influence on the implementation of IT in Hospitals (Xie et al. 2013).

Slide 3-3: The Patient–Clinician Dialog (from 1979)

In order to provide information a patient must first become a patient, so the patient must perceive himself as sick, but patients have different thresholds for the definition of sickness or healthy respectively. The typical patient–doctor dialog uses two types of data: (1) expressed by the patient or the doctor; (2) directly obtained from the patient by the doctor. This is important, because as we will learn in Lecture 7, the data expressed passes a complex series of perceptive, emotional, and cognitive "filters," and thus subject to distortion. The types of medical data differentiated by Komaroff (1979) includes expressed data: Verbally expressed objective (past medical history, current illness description, statements, etc.) and verbally expressed subjective (feelings, assumptions, etc.), and nonverbally expressed (appearance, habitus, mimic, gestures, etc.). The second type is directly obtained data: Elements of physical examination, diagnostic laboratory tests, images, pathognomonic (signs, patterns, etc.).

The big difference between medicine and engineering is, that in medicine a substantial degree of uncertainty may be inevitable; it may not be possible to acquire the needed data, because the measurements cannot be made without destructive consequences for the patient, or of practical limitations, or the length of time required to take adequate measurements. In engineering, given adequate resources, the goal is to reduce uncertainty to a measurably trivial level, and to experimentally demonstrate that the predicted specifications have been met (Komaroff 1979).

5.3 Data Standardization

Slide 3-4: Standardized Data

Standardized data shall now ensure that information is interpreted by all users with the same understanding. Moreover, standardized data shall support the reusability of the data, improving the efficiency of health-care services and avoid errors by reducing duplicated efforts in data entry;

Data standardization refers to:

(a) The data content.
(b) The terminologies that are used to represent the data.
(c) How data is exchanged.
(d) How knowledge, e.g., clinical guidelines, protocols, decision support rules, checklists, standard operating procedures, is represented in the health information system (refer to IOM).

Elements for sharing require standardization of identification, record structure, terminology, messaging, privacy, etc.

The most used standardized dataset to date is the International Classification of Diseases (ICD), which was first adopted in 1900 for collecting statistics (Ahmadian et al. 2011).

Let us look first at possibly the most difficult example: linguistics in Slide 3-5 and then a manageable example from the recording of an Electrocardiogram (ECG) in Slide 3-6 to emphasize why interoperability is important.

Slide 3-5: Complex Example: Linguistic Data

Although we live in a "multimedia age" and some scientists foresee a world without text, in the hospital the major medical documentation is only available in text format and the amount of this unstructured data is immensely increasing (Holzinger et al. 2008; Holzinger et al. 2013). **Text is the written form of natural language**. Representation of natural language data presents many major challenges. It is difficult to automatically interpret even well-edited texts as well as a native speaking reader would understand it. However, there have been advances in natural language processing (NLP), e.g., the so-called "bag of words" methods: in which a document is treated as a collection of words occurring with some frequency; this works because they do not obscure this inherent meaning when presented to the analyst.

The first mechanized methods were developed by Salton (1968) for information retrieval. Salton's work on identifying salient terms in a corpus,

(continued)

(continued)

indexing, and constructing high-dimensional signature vectors that represent a corpus' topics or articles remains key to most of the current tools for analyzing big text data (Salton et al. 1975). A challenge is in mapping back the high dimensional vectors into 2D (or 3D) representations to support visualizations that end users may understand and work on. In addition to Salton's work, centuries of general linguistic study of language provide a foundation for the computer-based analysis of language. The general structure of language provides a framework for the eventual reduction of text to its meaningful logical form for computer-based analysis. While computer-based linguistic analysis is not a solved problem, current capabilities provide some reliable results that add **semantic richness** to the "bag of words" approach. Linguistics defines the levels of structure based on analysis across and within languages, and computational linguistics provides the methods for assigning structure to textual data.

As shown in Slide 3-5, the major levels of structure applicable on text are phonological, morphological, syntactic, semantic, and the pragmatic level.

Phonological level deals with the structure of the sounds that convey linguistic content in a language. However, this level of structure applies to writing and sign language as well. It is basically the lowest level containing the elements that distinguish meaning and can be defined physically as a means of linguistic production. Each language has its own set of sounds that are used in words to convey meaning at any time in its history. These elements are not usually equivalent to the graphemic elements (the smallest elements of meaning) in the writing system. Instead, phonological elements are related to graphemic elements by rules, to a greater or lesser degree. The graphemic system can influence the phonological system. **Morphological level** of a language is the level at which meaning can be assigned to parts of words and the level that describes how morphemes (the smallest meaning elements of words) are combined to produce such a word. Some written systems, such as English and Chinese, are morphological in nature. For example, the morpheme "sign" is not always pronounced the same way in English words in which it appears (sign, signal, signature, resign, resignation).

In highly agglutinative languages, words are built by affixing morphemes to one another, making word boundaries sparse within sentences. In such languages, such as Turkish and many Native American languages, an entire sentence can appear in one word. Obviously, this fact plays havoc with the "bag of words" approach because word boundaries are not easily identifiable using information within the corpus. Even in non-agglutinative languages, segmentation may be required because of the written system's lack of word delimitation. Chinese is such a language.

(continued)

(continued)

Syntactic level of structure concerns the structure of the sentence, i.e., the categories of words and the order in which they are assembled to form a grammatical sentence. The categories used in syntax are known as parts of speech. The main parts of speech are nouns and verbs. Verbs govern the roles that the nouns in the sentence can play, and the ordering and/or case marking of nouns determine their roles. Roles can be characterized at various levels. Most commonly, the syntactic roles are those like subject and object. The roles can also be viewed with degrees of semantic content, such as agent, instrument, or location. The predicate-argument structure of the sentence is used to represent the logical form of the sentence, which is a semantic representation.

Semantic level of structure of the sentence is computationally defined to be the level of representation supporting inferencing and other logical operations. Within linguistic theory, the semantics by Montague (1974) was one of the bases for this approach. Other representations important to the semantic level include, but are not limited to, the meanings of the words. Lexicology is as old as writing, but modern lexicology includes psycholinguistic knowledge concerning how the brain stores the words in our memory. WordNet is the preeminent lexicon structured along psycholinguistic principles (Miller 1998). The utility of WordNet for computational linguistics has been immeasurable. It contains an ontology of the words of English and allows the user to find synonyms, antonyms, hypernyms (more general terms), and hyponyms (more specific terms). It also distinguishes the sense of the words. Other languages have WordNets developed for them and the senses of the words have been linked cross-lingually for use in sense disambiguation within and across languages (see EuroWordNet at http://www.illc.uva.nl/EuroWordNet). The discourse structure of language is the level of structure of the exchange or presentation of information in a conversation or a written piece (Thomas and Cook 2005).

After this complex example, let us look on the recording of an Electrocardiogram (ECG) to explain why interoperability is so important.

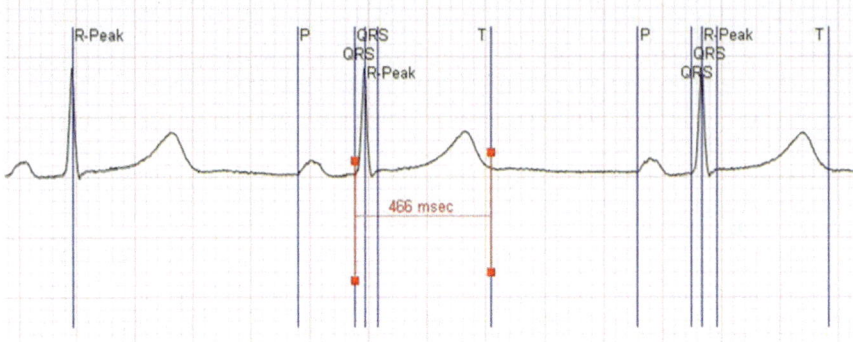

Fig. 2 See Slide 3-6

Slide 3-6: Example: Annotated ECG Signal in HL7 Standard

Electrocardiograms are used to measure the rate and regularity of heartbeats, as well as the size and position of the heart chambers. The importance of creating a standardized ECG data format is reinforced with the increasing demand for interoperability, which is concerned with the coherent exchange of clinical data within and between heterogeneous Hospital Information Systems (HIS). The aim is to facilitate the exchange of medical data, ideally on a global scale. With regard to the ECG, interoperability can only be achieved by following the creation of a standardized ECG storage format.

Slide 3-7: Standardization of a Workflow on the Example of ECG Data

The aim of the standardized data is that the interpretation and diagnosis can be done technically trans-cultural and inter-subjective.

Above we see the typical procedure in the recording and management of an ECG. The importance of creating a standardized ECG storage format is reinforced with the increasing demand for interoperability.

Interoperability is concerned with the coherent exchange of clinical documents within and between heterogeneous Hospital Information Systems. This concept is important since its ultimate aim is to facilitate the exchange of medical data on a global scale. However, it is estimated that this could take still 20 years to achieve effective interoperability in Europe.

Below we can see the rationales for creating a standard electronic ECG storage format (Bond et al. 2011), e.g., the possibility of the application of data mining algorithms on ECGs, or the easy exchange with other health providers.

Slide 3-8: Standardization of ECG Data (1/2)

A huge problem was that so many researchers had proposed their own ECG storage formats and there are many formats proclaiming to promote interoperability, with three predominant ones:

1. SCP-ECG—developed in 1993, stores in binary form, and has been the official European standard for the storage and transmission of ECGs since 2005. In July 2002, the SCP-ECG format became the promotion of a European funded consortium called **OpenECG**, which is a body of at least 464 members who are dedicated to the interoperability in digital electrocardiography. Advantage: small file sizes; Disadvantage: lacking human readability and large number of optional features.
2. DICOM-ECG—originally a image standard called ACR-NEMA in 1985—it became European standard in 1995, and although the DICOM format was originally created to store and transmit radiographic images, it can now support all diagnostic modalities. As a result NEMA has been extending the DICOM format by developing and publishing supplements. In the year 2000, DICOM-WS 30 was introduced to support the storage of raw medical waveforms, which in effect stores actual sample values as opposed to storing raster images. This supplement has enabled the DICOM format to store various waveform datasets including blood pressure, audio, and ECG. Advantage: The power of this format (can display and work as a PACS system, e.g., an ECG and an angiogram at the same time); Disadvantage: Binary based, and therefore lacks human readability; too complex.
3. HL7 aECG (annotated ECG)—In November 2001 released by the FDA as a Health Level 7 standard—the first which used XML. Advantage: XML; Disadvantage: verbosity of XML files, consequently large file sizes, uses a lot of definable metadata. For more details please refer to Bond et al. (2011).

Slide 3-9: Overview on Current ECG Storage Formats

This slide shows an overview of some important ECG storage formats, for details please refer to Bond et al. (2011).

Slide 3-10: Example of a Binary ECG File

To demonstrate the big difference of the two file formats, Slide 3-10 shows an example file in Binary, and Slide 3-11 an example file in XML.

Slide 3-11: Example of a XML ECG File

Here we see a typical example of an aECG file indicating the increment element which defines the interval in seconds between each sample. The value 0.002 s indicates that there is a 2 ms gap between each sample. This, in effect would be the frequency equivalent of 500 Hz (Bond et al. 2011).

6 Modeling Biomedical Knowledge

Slide 3-12 Modeling of Biomedical Knowledge

Medical environments have enormous complexity and poses high demands on medical professionals. Here we see an example of a traditional modeling approach for medical reasoning used as a basis for developing decision support systems. Such models may be faithful to what is known about biomedical knowledge, but they have (serious!) limitations for human problem solving, especially in unanticipated situations. This example shows the physiological factors and relations affecting mean arterial blood pressure (Hajdukiewicz et al. 2001).

Slide 3-13: Creating a Work Domain Model (WDM)

This Slide illustrates the process of generating a WDM of the patient (i.e., the human body) in an operating room (OR). The OR consists of a team of medical personnel (two nurses, two surgeons, one anesthesiologist) who interact with each other, the patient, and with medical equipment to perform a surgical procedure.

We define "work domain" as an object in this environment that is controlled and, due to its complexity and purpose, can require problem solving by the medical personnel. A work domain could be the patient himself or a complex medical device (e.g., anesthesia workstation). In the example by Hajdukiewicz et al. the work domain is the patient. The patient WDM is divided into different levels of abstraction (Abstraction Hierarchy, AH) and aggregation (part–whole hierarchy, PWH). Each "cell" (another meaning of the word "cell" ;-) in this patient WDM matrix defines a complete and different causal representation of the same patient work domain, uniquely defined by the particular levels of abstraction and aggregation.

Slide 3-14: Partial Abstraction of the Cardiovascular System

If we now "zoom-in" we see the structural means–ends links between the different levels of abstraction for parts of the patient cardiovascular system. The lower levels include the cardiac and circulatory functions necessary to support the higher-level purposes of adequate circulation and blood volume; the higher levels provide reasons for lower level functions. Here the problem solving can occur by shifting the mental focus across these levels of abstraction. Information will be required from the Abstraction Hierarchy (AH) level currently in the practitioner's mental focus, including the functional structure, state, and what needs to be controlled (i.e., the What). For example, the current task may be to control systemic circulation. Information is also required from the AH level above, which indicates the reason of the control decision (i.e., the Why). In this slide the reasons for controlling systemic circulation are to support the functions of mass transfer and balance to the organs. Finally, information is required from the AH level below, which indicates the physiological resources available for implementing the decision (i.e., the How) (Hajdukiewicz et al. 2001).

Slide 3-15: WDM of the Human Body

A further "zoom-in" into each cell, where we find a model consisting of different objects or functions connected by causal relations—further detailed in Slide 3-16.

Slide 3-16: WDM of the Cardiovascular System

Here we further "zoom-in" and see the causal arrangements for selected parts of the human body that are reasonable to illustrate, given the complexity of the cardiovascular system (i.e., levels of abstraction, balances and processes; levels of aggregation, system, subsystem, organ).

Slide 3-17: Example: Mapping OR Sensors onto the WDM

You will now have asked yourself: what is the purpose of such modeling? In Slide 3-17 you see a typical example how useful it can be: We see four types

(continued)

(continued)

of mapping between the patient WDM and operating room sensors: one-to-one, convergent, divergent, and no mapping. With a one-to-one mapping, one sensor maps onto one patient variable. For example, checking a patient's pulse provides information about heart rate. With convergent mapping (or redundancy), many sensors map onto one patient variable. Practitioners use this method to reduce the high level of uncertainty in measurements from the environment (e.g., artifact, noise, and calibration errors). For example, heart rate can be determined directly from the ECG signal as well as indirectly from other monitor signals (e.g., arterial blood pressure waveforms). With divergent mapping, some sensors provide evidence for many patient variables. The ECG waveform provides evidence for heart rate, heart rhythm, and adequate myocardial oxygenation, etc. Finally, with no mapping, some sensors do not map onto any of those patient variables, e.g., the pressure in an unused oxygen tank (Hajdukiewicz et al. 2001).

Slide 3-18: Integrated Medical Informatics Design for HCI

Our last example demonstrates the usefulness of WDM for human–computer interaction, in a way that is compatible with how medical practitioners can perform problem solving in the context of a medical environment. Note: The information requirements for surgeons are very different compared with anesthesiologists, although both need the same information from overlapping regions of the patient WDM (Hajdukiewicz et al. 2001).

7 Ontologies

When talking about standardization we immediately touch ontologies. In computer science ontology represents formal knowledge as a set of concepts within a (strictly limited) domain, and the relationships between those concepts. It is similar to what we have seen before, as it can be used for domain modeling.

The most important aspect is that an ontology provides a standardized (shareable) vocabulary, which can then be used to model such a domain. The first question is, why should we need such an approach. The answer is a typical example from semantics—seen in Slide 3-19.

Slide 3-19: Let's Start with a Very Simple Question: What Is a Jaguar?

If you put the keyword "Jaguar" into a search engine—you will get different results. The search engine—as a typical von Neumann machine—does not know what you are looking for: a sports-car, an animal, a jet plane, or a tractor? Our current computers cannot know in what **context** you use the word "Jaguar"—so additional (meta-) information is needed. Meta-information is information about information.

A categorization may help—the first known categorization was done by Aristotle.

Slide 3-20: The First "Ontology of What Exists"

Ontology is defined as a theory of our reality (in philosophy) or a conceptualization of what exists (in artificial intelligence). For our practice, an ontology consists of **categories** of individual objects organized in **taxonomies** and connected by various **relationships**. To represent such an ontology, often a graph data structure is used.

A recommendable introduction is Zhang and Bodenreider (2006).

Slide 3-21: Ontology: Classic Definition

In this slide we see the classic definition: An ontology is a formal, explicit specification of a shared conceptualization (Studer et al. 1998).

Ontology IS-A a structured description of a domain in form of concepts ↔ relations; this **IS-A relation** provides a taxonomic skeleton. Other relations reflect the domain semantics and formalize the terminology in this particular domain. The terminology contains the terms and their definitions and usage in a specific context. The knowledge base is the instance classification and concept classification; the classification itself provides the domain terminology (Holzinger 2000).

- Ontology = a structured description of a domain in form of <u>concepts</u> ↔ <u>relations</u>;
- The <u>IS-A relation</u> provides a taxonomic skeleton;
- Other relations reflect the <u>domain semantics</u>;
- Formalizes the <u>terminology</u> in the domain;
- Terminology = terms definition and usage in the specific context;
- Knowledge base = <u>instance classification</u> and <u>concept classification</u>;
- Classification provides the <u>domain terminology</u>

Fig. 3 See Slide 3-22

Slide 3-22: Ontology: Terminology

Some of the most important terms in the context of Ontology.

Slide 3-23: Additionally an ontology may satisfy:

Please note: An *ontology* is composed of at least one taxonomy and may comprise several distinct taxonomies. Concepts across taxonomies do not stand in a taxonomic relation. Concepts in an ontology represent categories of things existing in reality or abstractions generated for classification purposes. Each category or abstraction is represented exactly by one concept. Additionally, an ontology may satisfy the following (Zhang and Bodenreider 2006):

1. In addition to the IS-A relationship, partitive (meronomic) relationships may hold between concepts, denoted by PART-OF. Every part-of relationship is irreflexive, asymmetric, and transitive. IS-A and PART-OF are also called hierarchical relationships.
2. In addition to hierarchical relationships, associative relationships may hold between concepts. Some associative relationships are domain-specific (e.g., the branching relationship between arteries in anatomy and rivers in geography).
3. Relationships r and r' are inverses if, for every pair of concepts x and y, the relations x, r, y, and y, r', x hold simultaneously. A symmetric relationship is its own inverse. Inverses of hierarchical relationships are called INVERSE-IS-A and HAS-PART, respectively.

(continued)

> (continued)
> 4. Every non-taxonomic relation of x to z, x, r, z, is either inherited (y, r, z) or refined $(y, r, z'$ where z' is more specific than z) by every child y of x. In other words, every child y of x has the same properties (z) as it parent or more specific properties (z').
> 5. In addition to inter-concept relationships, concepts may have various properties, some of which are constrained by type (e.g., Boolean and integer), value range, cardinality, etc.
>
> Bottom right in Slide 3-23 you can see as an example the top-level of the anatomy taxonomy along with the classification criteria.

Ok, let us review: Ontology is defined as a theory of the reality (in philosophy) or a conceptualization of what exists (in artificial intelligence). In practice, ontologies consist of categories of entities organized in taxonomies and connected by relationships. ARISTOTLE attempted to classify the things in the world, consequently researchers adopted the term "ontology" to describe what can be computationally represented of the world within machine language (software): An ontology is a formal, explicit specification of a shared conceptualization. Explicit means that the type of concepts used, and the constraints on their use are explicitly defined (Studer et al. 1998).

The well-known "tree of Porphyry," for example, identifies animals and plants as subcategories of living things distinguished by animals being sensitive and plants being insensitive. In contrast, in computer science it is an engineering artifact, usually a so-called "conceptual ontological model" of the world; it introduces vocabulary describing various aspects of the domain being modeled and provides an explicit specification of the intended meaning of the vocabulary by describing the relationships between different vocabulary terms. In a logic-based (=formal) ontology, the vocabulary can be thought of as predicates and constants. For example, Animal and Plant can be thought of as (unary) predicates and Socrates, Plato, and Aristotle as constants.

Such a logical theory enables the possibility of using **automated reasoning** to check, for example, for internal consistency (the theory does not entail "false") and other (non-) entailments. In the life sciences, several large biomedical ontologies are in use, including the Biological Pathways Exchange (BioPAX) ontology, the GALEN ontology, or the Foundational Model of Anatomy (FMA) (Horrocks 2013).

> **Slide 3-24: Ontologies: Taxonomy**
>
> In Slide 3-24 we see a hierarchy of ontologies regarding formalization on the x-axis and expressivity on the y-axis.

(continued)

(continued)

Whereas typical dictionaries are on the left-down corner, first-order logic is on the right-up corner.

Note: Logic programming is a well-known declarative method of knowledge representation based on first-order logic. Logic programming was developed in the early 1970s based on work in automated theorem proving. A logic program consists of a set of rules (Horn clauses), where each rule has the form head body, where head is a logical atom and body is a conjunction of logical atoms. The logical semantics of such a rule is given by the implication body head. The semantics of a pure logic program is completely independent of the order in which its clauses are given, and of the order of the single atoms in each rule body. In PROLOG, the paradigm of logic programming is practically usable. The clause matching and backtracking algorithms at the core of PROLOG are sensitive to the order of the clauses in a program and of the atoms in a rule body. In application areas such as knowledge representation and databases there is a predominant need for full declarativeness, and hence for pure logic programming. In knowledge representation, declarative extensions of pure logic programming, such as negation in rule bodies and disjunction in rule heads, are used to formalize common sense reasoning. In the database context, the query language DATALOG was designed (Dantsin et al. 2001; Eiter et al. 2006).

Slide 3-25: Example of a Conceptual Structure from CogSci

Here we see an example from cognitive science: The knowledge about the brain domain (aka anatomy–functional ontology) is expressed through semantic relationships between the concepts of the three ontologies. This ontology has been used to support the discovery of relationships between the cognitive function and the anatomical regions of the brain (Simonet et al. 2009).

Slide 3-26: Examples of Biomedical Ontologies

In this slide we see some biomedical ontologies, including scope, number of entities (concepts), distribution of the number of terms per entity (minimum, maximum, median, and average), and existence of a sub-sumption hierarchy,

(continued)

(continued)

based on information present in the UMLS version of 2007 (Bodenreider 2008).

Ontologies generally serve as a source of vocabulary, i.e., a list of names for the entities represented in these ontologies, however, collecting names is the function of terminology, not ontology, and ontology languages such as OWL, the Web Ontology Language, treat names as labels or annotations. In practice, however, most biomedical ontologies (with the notable exception of LOINC) provide lists of names for the entities they accommodate, in addition to properties and relations for these entities. The terminological component of biomedical ontologies is an important resource for natural language processing systems and supports knowledge management tasks such as annotation (or indexing) of resources, information retrieval, access to information, and mapping across resources. However, the corpus of entity names present in biomedical ontologies covers only in part the lexicon of the domain (especially for languages other than English) and only forms the basis for managing term variation (Bodenreider 2008).

7.1 Ontology Languages

Slide 3-27: Taxonomy of Ontology Languages

Ontology languages are formal languages used to construct ontologies, allow the knowledge representation within specific domains, and include reasoning rules, which support knowledge processing. Ontology languages are usually declarative languages, are almost always generalizations of frame languages, and are commonly based on either first-order logic or on description logic.

A coarse taxonomy is to determine between three concepts:

1. Graphical notations (semantic networks, topic maps, UML, RDF, ...).
2. Logical based languages (e.g., description logics, OIL, DAML + OIL, OWL; rules, RuleML, LP/PROLOG; first order logics, KIF; conceptual graphs, syntactically higher order logics, Flogic, Non-Mon, modalities).
3. Probabilistic/fuzzy approaches.

Note: KIF = Knowledge Interchange Format (e.g., Ontolingua), See: http://www.ksl.stanford.edu/knowledge-sharing/kif; OWL=Web Ontology Language

Slide 3-28: Example for (1) Graphical Notation: RDF

In this slide we can see the conversion of Table I into triples contained in a named graph (the source data for this figure is available at: www.nature.com/msb).

The Table I is an example of a properties table (its canonical table counterpart has the same structure) and was obtained from a study to test whether the yeast gene, MDM20, is necessary for mitochondrial inheritance and organization of the actin cytoskeleton (Hermann et al. 1997). It lists the different yeast strains in three columns (name, genotype, and source). Each table row corresponds to a specific yeast strain. We can apply the following rules to convert this table into RDF triples:

1. Each row is mapped to a subject
2. Each column header is mapped to a property
3. Each column value (cell) is mapped to a property value

The figure in the left depicts the mapping process and some of the mapping results. For the subject of each triple, we may check to see if it is an instance of an existing ontology class (represented using OWL or RDFS). For example, each subject (e.g., "FY10") derived from Table I is an instance of (represented by a dotted line) the class "yeast strain" in some organism ontology. Although the column name can be used to name the property, we may want to map it to some standard property name, if available. The generated triples represent a RDF graph. To this end, we use the named graph technique to identify the RDF graph generated from the table and to store the provenance information including the title, description (e.g., the table caption), creator, source (e.g., the paper), and so on. The properties (e.g., title, description, creator, and source) are derived from the Dublin Core metadata standard http:// dublincore.org (Cheung et al. 2010).

7.2 OWL

Slide 3-29: Example for (2) OWL Axioms

The Web Ontology Language (OWL) is the most widely used ontology language, was developed by the W3C and thus designed specifically for use on the semantic Web; it exploits existing Web standards (XML and RDF), adding the familiar ontological primitives of object and frame based systems, and the formal rigor of a very expressive description logic (DL) that emerges from research in the field of Artificial Intelligence.

(continued)

(continued)

As we can see in Slides 3-29 and 3-30 the OWL consists of a rich set of knowledge representation constructs that can be used to formally specify medical-domain knowledge, which in turn can be exploited by description logic reasoners for purposes of inferencing, i.e., deductively inferring new facts from knowledge that is explicitly available.

The knowledge base (KB) of a typical DL based system comprises of two components, the TBox and the ABox; The TBox introduces the terminology, i.e., the vocabulary of an application domain (e.g., "Neoplastic Process is-a Biological Function"), while the ABox contains assertions about named individuals in terms of this vocabulary ("Cancer is-a-instance of a Neoplastic Process"). The logical basis of the language means that reasoning services can be provided in order to make OWL described resources more accessible to automated processes. Formally, OWL is similar to a very expressive DL, with the OWL ontology corresponding to a DL terminology (TBox) and instance data pertaining to the ontology making up the assertions (ABox), and therefore it is widely used in the medical domain (Bhatt et al. 2009).

Slide 3-30: OWL Class Constructors

A Primer on OWL 2 as W3C recommendation from 11 December 2012 can be found here: http://www.w3.org/TR/owl2-primer/

8 Medical Classifications

Slide 3-31: Medical Classifications: Rough Overview

Medical classification, called coding by the professionals, is the process of transforming descriptions of medical diagnoses and procedures into a universal medical classification scheme.

A classification is a hierarchy of objects that conforms to the following principles (Berman 2012):

1. The classes of the hierarchy have a set of properties that extend to every member of the class and to all of the subclasses of the class, to the exclusion of all other classes. A subclass is itself a type of class wherein the members have the defining class properties of the parent class plus some additional properties specific for the subclass.

(continued)

(continued)
2. In a hierarchical classification, each subclass may have no more than one parent class. The root class has no parent class.
3. The members of classes may be highly similar to each other, but their similarities result from their membership in the same class (i.e., conforming to class properties), and not the other way around (i.e., similarity alone cannot define class inclusion).

The father of classification was Carl von Linne (1707–1778) who began in 1735 with a classification of species. Today, more than 100 various biomedical classifications are in use, for example:

International Statistical Classification of Diseases (ICD)
Systematized Nomenclature of Medicine—Clinical Terms (SNOMED CT)
Medical Subject Headings (MeSH)
Foundational Model of Anatomy (FMA)
Gene Ontology (GO)
Unified Medical Language System (UMLS)
Logical Observation Identifiers Names and Codes (LOINC)
National Cancer Institute Thesaurus (NCI Thesaurus)

Medical classification systems are used for a variety of applications in medicine, public health, and medical informatics, including the reimbursement, e.g., based on diagnosis-related groups (DRG), but also for statistical analysis, therapeutic actions, and knowledge engineering and decision support systems.

Meanwhile, taxonomy is a science of classifying the elements of a knowledge domain, and assigning names to the classes and the elements. In the case of terrestrial life forms, taxonomy involves assigning a name and a class to every species of life—on earth approx. 50 million species—a huge task. The central rules include Berman (2012):

1. All living organisms on earth contain DNA, which is transcribed into a less-stable, single-stranded molecule called RNA, which is translated into proteins. All living organisms replicate their DNA and produce more organisms of the same genotype.
2. All living organisms on earth can be divided into two broad classes: the prokaryotes (organisms with a simple string of DNA and without a membrane-delimited nucleus, or any other membrane-delimited organelles), the class that includes all bacteria; and eukaryotes (organisms with a membrane-delimited nucleus).
3. The prokaryotes preceded the emergence of the eukaryotes, and the first eukaryotes were built from the union of two or more prokaryotes.

(continued)

(continued)
4. Every eukaryotic organism that lives today is a descendant of a single eukaryotic ancestor.
5. Every organism belongs to a species that has a set of features that characterizes every member of the species and that distinguishes the members of the species from organisms belonging to any other species.

For more information please refer to Berman (2012) and Scamardella (2010).

Slide 3-32: International Classification of Diseases (ICD)

The International Classification of Diseases (ICD) is the standard diagnostic tool for epidemiology, health management and clinical purposes and includes the analysis of the general health situation of population groups. It is used to monitor the incidence and prevalence of diseases and other health problems as well as to classify diseases and other health problems recorded on many types of health and vital records including death certificates and health records. In addition to enabling the storage and retrieval of diagnostic information for clinical, epidemiological, and quality purposes, these records also provide the basis for the compilation of national mortality and morbidity statistics by WHO Member States. It is used for reimbursement and resource allocation decision making by countries. ICD-10 was endorsed by the 43rd World Health Assembly in May 1990 and came into use in WHO Member States as from 1994. The 11th revision of the classification has already started and will continue until 2015, for more details see: http://www.who.int/classifications/icd/en/.

Slide 3-33: The International Classification of Diseases ICD

The oldest classification is the ICD, the roots can be traced back to:
  1629 London Bills of Mortality
  1855 William Farr (London, one of the founders of medical statistics): List of causes of death, list of diseases
  1893 Jacques Bertillot: List of causes of death
  1900 International Statistical Institute (ISI) accepts the Bertillot list
  1938 5th Edition
  1948 WHO

(continued)

(continued)

1965 ICD-8
1989 ICD-10
2015 ICD-11 due
1965 SNOP, 1974 SNOMED, 1979 SNOMED II
1997 Logical Observation Identifiers Names and Codes (LOINC) integrated into SNOMED
2000 SNOMED RT
2002 SNOMED CT

Jacques Bertillon, actually, introduced the Bertillon Classification of Causes of Death at a congress of the International Statistical Institute in Chicago in 1893 and thereof a number of countries and cities adopted his system, which was based on the principle of distinguishing between general diseases and those localized to a particular organ or anatomical site. Subsequent revisions represented a synthesis of English, German and Swiss classifications, expanding from the original 44 titles to 161 titles. In 1898, the American Public Health Association (APHA) recommended that the registrars of Canada, Mexico, and the USA also adopt it. The APHA also recommended revising the system every 10 years to ensure the system remained current with medical practice advances. As a result, the first international conference to revise the International Classification of Causes of Death took place in 1900.

Slide 3-34: Systematized Nomenclature of Medicine SNOMED

SNOMED CT is the Systematized Nomenclature of Medicine Clinical Terms and covers diseases, clinical findings, and procedures. Originally developed by the College of American Pathologists, the ownership of SNOMED CT was transferred to a new public body called the International Health Terminology Standards Development Organization (IHTSDO) in 2006. Presently, IHTSDO has 15 charter member countries with the common goal to develop, maintain, and promote this terminology standard. The July 2009 version of SNOMED CT contains over 388,000 concepts, 1.14 million descriptions, and 1.38 million relationships. There is a new release every 6 months through the National Release Centers of the respective charter member countries. With each release, there are changes that can affect the use of SNOMED CT within an organization's electronic patient record (EPR) systems. These include the fully specified name/preferred term, concept status, primitive/fully defined status, defining attributes, normal forms, and position within the "is a"

(continued)

(continued)

hierarchy. Some of these changes may lead to unexpected consequences in subsequent encoding, equivalency and subsumption testing, and querying of a SNOMED CT (Lee et al. 2010).

Slide 3-35: SNOMED Example Hypertension

Here we see two examples: (A) SNOMED Representation for increased blood pressure. (B) SNOMED Representation for decreased blood pressure.

A big issue in clinical information systems is the distinction between observables and findings. Although there exists no universal consensus on the distinction, the term "observable" generally refers to an aspect of the patient that can be quantified or qualified, e.g., blood pressure, skin color, body mass index.

A "finding," on the other hand, usually refers to something which is either present or absent, possibly with additional qualification (diabetes, fractures, ...), or to the state of some observable such as "increased blood pressure" which likewise may be present or absent. In SNOMED, distinctions are made between the classes "finding" and "observable entity." Figure A in Slide 3-35 makes this clear: the finding of increased blood pressure implies a finding of "abnormal blood pressure" that interprets the observable entity "blood pressure." The fact that a finding of an "increased blood pressure" qualifies the blood pressure as abnormally high as opposed to abnormally low is not reflected at all in this expression! This is a common phenomenon. In many cases, most of the intended meaning behind concepts such as finding of increased blood pressure remains in the term name and is not reflected in a definition. This is even more obvious when comparing SNOMED's (primitive) definition of a decreased blood pressure as shown in Figure B below (Rector and Brandt 2008).

Slide 3-36: Medical Subject Headings (MeSH)

The MeSH thesaurus is produced by the National Library of Medicine (NLM) since 1960 and is used for cataloging documents and as an index to search these documents in a database, as part of the metathesaurus of the Unified Medical Language System (UMLS). This thesaurus originates from keyword

(continued)

(continued)

lists of the Index Medicus (today Medline); MeSH is polyhierarchic, i.e., every concept can occur multiple times. It consists of the three parts:

1. MeSH Tree Structures (see the Example in Slide 3-37)
2. MeSH Annotated Alphabetic List
3. Permuted MeSH

Slide 3-37: The 16 Trees in MeSH

The 16 trees in MeSH are as follows:

1. Anatomy [A]
2. Organisms [B]
3. Diseases [C]
4. Chemicals and Drugs [D]
5. Analytical, Diagnostic, and Therapeutic Techniques and Equipment [E]
6. Psychiatry and Psychology [F]
7. Biological Sciences [G]
8. Natural Sciences [H]
9. Anthropology, Education, Sociology, Social Phenomena [I]
10. Technology, Industry, Agriculture [J]
11. Humanities [K]
12. Information Science [L]
13. Named Groups [M]
14. Health Care [N]
15. Publication Characteristics [V]
16. Geographicals [Z]

Slide 3-38: MeSH Hierarchy on the Example of the Heading Hypertension—1/2

This is an example for the MeSH Hierarchy for the heading Hypertension (Hersh 2010)—the same example can be seen in the next slide as it looks originally in the Mesh Descriptor Database of the NLM.

Slide 3-39: MeSH Hierarchy on the Example of the Heading Hypertension—2/2

The example of Slide 3-38 as seen in the MeSH Database.

MeSH descriptors are arranged in both an alphabetic and a hierarchical structure. At the most general level of the hierarchical structure, there are very broad headings such as "Anatomy." More specific headings are found at more narrow levels of the 12-level hierarchy, such as "Ankle." In the 2013 version there are 26,853 descriptors and over 213,000 entry terms that assist in finding the most appropriate MeSH Heading, for example, "Vitamin C" is an entry term to "Ascorbic Acid." In addition to these headings, there are more than 214,000 headings called Supplementary Concept Records (formerly Supplementary Chemical Records) within a separate thesaurus. http://www.nlm.nih.gov/mesh

Slide 3-40: MeSH Interactive Tree-Map Visualization (See Lecture 9)

This is a very nice example of a possibility of visualization of such structures. We will discuss this in detail in Lecture 9. The idea of such an approach is that the end user has an idea of the overall structure (of the thesaurus) or selected parts of it. This example is a tree-map (Shneiderman 1992): arbitrary trees are shown with a 2D space-filling representation. With such a treemap, two additional aspects can be displayed beside the thesaurus structure: One is represented by the size of the partitions, the other by its color. The hierarchy is visualized through the nesting of areas. The color of the different areas is used to represent the result of the different measures introduced above, for more details consult:

http://www.ieee-tcdl.org/Bulletin/v4n2/eckert/eckert.html

Slide 3-41: UMLS—Unified Medical Language System

UMLS is a set of files and software that brings together many health and biomedical vocabularies and standards to enable interoperability between computer systems (refer also to Slide 3-43). UMLS can be used to enhance or develop applications, such as electronic health records, classification tools, dictionaries and language translators.

> **Slide 3-42: UMLS Homepage**
>
> http://www.nlm.nih.gov/research/umls
> The Metathesaurus forms the base of the UMLS and comprises over one million biomedical concepts and five million concept names (!), all of which stem from the over 100 incorporated controlled vocabularies and classification systems. Some examples of the incorporated controlled vocabularies are ICD-10, MeSH, SNOMED CT, DSM-IV, LOINC, WHO Adverse Drug Reaction Terminology, UK Clinical Terms, RxNorm, Gene Ontology, and OMIM (to mention only a few).

> **Slide 3-43: UMLS Metathesaurus Integrates subdomains**
>
> In this slide we see the UMLS metathesaurus, integrating various other terminologies and serving as link between them and the subdomains they represent:
> SNOMED—as link to clinical repositories
> OMIM—Online Mendelian Inheritance - as link to genetic knowledge bases
> MeSH—as link to biomedical literature (MEDLINE)
> GO—as link used for the annotation of gene products across various model organisms
> UWDA University of Washington Digital Anatomist—as link to the Digital Anatomist Symbolic Knowledge Base
> NCBI—taxonomy used for identifying organisms
> Although the UMLS was not specifically developed for the needs of bioinformaticists, it includes terminologies used in bioinformatics. Integrated terminologies include the NCBI taxonomy, used for identifying organisms, and Gene Ontology, used for the annotation of gene products across various model organisms. The Metathesaurus also covers the biomedical literature with the MeSH, the controlled vocabulary used to index MEDLINE. Core subdomains such as anatomy, used across the spectrum of biomedical applications, are also represented in the Metathesaurus with the Digital Anatomist Symbolic Knowledge Base. Finally, the subdomain represented best is probably the clinical component of biomedicine, with general terminologies such as SNOMED International (and SNOMED-CT). Clinical genetics resources include the Online Mendelian Inheritance in Men OMIM represented in part, and the Online Multiple Congenital Anomaly/Mental Retardation (MCA/MR) Syndromes. Other categories of terminologies in the Metathesaurus include specialized disciplines (e.g., nursing, psychiatry) and components of the

(continued)

(continued)

clinical information system (e.g., diseases, drugs, procedures, adverse effects). The figure illustrates how the UMLS Metathesaurus, by integrating these various terminologies, can serve as a link between not only the vocabularies but also the subdomains they represent (Bodenreider 2004).

Slide 3-44: Example of Proteins and Diseases in the UMLS

For example, Neurofibromatosis 2 is an autosomal dominant disease characterized by tumors called schwannomas involving the acoustic nerve, as well as other features, where the disorder is caused by mutations of the NF2 gene resulting in the absence or inactivation of the protein product. The protein product of NF2 is commonly called merlin and functions as a tumor suppressor. Neurofibromatosis 2, NF2 and Merlin are concepts in the UMLS, for which the Metathesaurus provides many synonyms, including those listed above. In the slide we can see that these three concepts are linked by associative relationships: Each concept is part of a hierarchy of concepts. Neurofibromatosis 2 inherits from ancestors such as "Benign neoplasms of cranial nerves," which reflects the nonmalignant behavior of schwannomas. Similarly, the function of NF2 is expressed through its direct parent "Tumor suppressor genes." Semantic types from the UMLS semantic network provide a direct categorization to Metathesaurus concepts, making it easy to distinguish between the disease Neurofibromatosis 2 (Neoplastic Process) and the gene NF2 (Bodenreider 2004).

An example for the mix of different data for solving a medical problem can be seen in Slide 2-8.

9 Future Outlook

Slide 3-45: Future Challenges

A grand challenge is in data integration and data fusion in the life sciences and to make relevant data accessible to the clinical workplace. While there is much research on the integration of heterogeneous information systems, a shortcoming is in the integration of available data. **Data fusion** is the process

(continued)

(continued)

of merging multiple records representing the same real-world object into a single, consistent, accurate, and useful representation (Bleiholder and Naumann 2008; McCray and Lee 2013; Horrocks 2013).

Knowledge representation is an emerging field of artificial intelligence and stimulated ontologies in particular in the Web and its recent evolution, the so-called Semantic Web. The idea of the Semantic Web is consistent with some of the basic goals of knowledge representation. The vision is to enable semantic interoperability and machine interpretability of datasets from various sources and to provide the mechanisms that enable such data to be used to support the user in an automated and intelligent way. In order to establish a completely automated knowledge acquisition in the future, advances must be made both in the fields of natural language understanding and techniques of machine learning. The next generation of semantic applications will thus be characterized by the acquisition of knowledge from several sources instead of acquiring it from merely one source covering all the needs of target applications. Similar trends can also be expected in the use of knowledge available in existing ontologies. As it is not likely for a single ontology to satisfy all the needs of a certain application, the trends nowadays move towards ontology integration (also known as ontology alignment, matching, or mapping). Integrating ontologies is one of the most complex and at the same time most important issues related to the practical implementation of Semantic Web. Consequently, the trend of integrating ontologies has lately gained substantial attention also in the research spheres and has actually become one of the most active fields of research. Although the results are very encouraging, so far integrated ontologies have not been used in practice in most cases.

Due to the integration of knowledge from different sources, one of the challenges is ensuring a homogenous conceptualization of domains, as the contents of individual ontologies are very diverse and their vocabularies inhomogeneous, not to mention the differences in the quality of the presented knowledge.

Knowledge representation holds one of the key roles in the development of context awareness. Challenges in this field comprise of the formal presentation of the context, the determination of the formal relationships between different contexts of ontology use, the development of mechanisms for the selection of the appropriate context in a given situation, and reasoning based on context. The development of reasoning based on context is especially important for user profiling, application personalization, and mobility support. The examples of applications including the aforementioned areas are nowadays very popular social networks. To summarize, the results achieved

(continued)

(continued)

in the domain of knowledge representation so far seem tentative and incomplete. Much work remains to be done. It is expected that under the auspices of Semantic Web and other accompanying concepts and visions, such as intelligent and personalized content retrieval, cloud computing, ubiquitous computing, and, last but not least, artificial intelligence, the development of the field will continue (Jakus et al. 2013).

10 Exam Questions

10.1 Yes/No Questions

Please check the following sentences and decide whether the sentence is true = YES; or false = NO; for each correct answer you will be awarded 2 credit points.

01	In the daily clinical practice we observe a high integration of standardized terminologies, e.g. SNOMED, MeSH, UMLS etc.	☐ Yes ☐ No	2 total
02	Standardization is the process of developing and implementing standards, e.g. the Evidence Based Medicine (EBM) approach.	☐ Yes ☐ No	2 total
03	Komaroff (1979) describes clinical data as being disturbingly "soft", having an obvious degree of variability and inaccuracy.	☐ Yes ☐ No	2 total
04	In computer science an ontology represents formal knowledge as a set of concepts within an unlimited domain, and the relationships between those concepts.	☐ Yes ☐ No	2 total
05	Standardized data shall now ensure that information is interpreted by all users with the same understanding.	☐ Yes ☐ No	2 total
06	In practice, ontologies consist of categories of entities organized in taxonomies and connected by relationships.	☐ Yes ☐ No	2 total
07	A logic program consists of a set of rules (Horn clauses), where each rule has the form head-body, where head is a logical atom and body is a conjunction of logical atoms.	☐ Yes ☐ No	2 total
08	The International Classification of Diseases (ICD) is the standard diagnostic tool for epidemiology, health management and clinical purposes and includes the analysis of the general health situation.	☐ Yes ☐ No	2 total
09	Data integration and data fusion in the life sciences is mostly solved and of no further research interest.	☐ Yes ☐ No	2 total
10	SNOMED CT is the Systematized Nomenclature of Medicine Clinical Terms and covers diseases, clinical findings and procedures.	☐ Yes ☐ No	2 total
Sum of Question Block A (max. 20 points)			

10.2 Multiple Choice Questions (MCQ)

01	Knowledge representation ... ☐ a) ... uses imperative extensions of pure logic programming ☐ b) ... is an emerging field of artificial intelligence. ☐ c) ... is primarily used for dealing with imprecise, vague and noisy data. ☐ d) ... hold one of the key roles in the development of context awareness.	4 total
02	The famous Forrester reports speak of ... ☐ a) ... the challenge that most of our data is unstructured. ☐ b) ... the increasing complexity of data. ☐ c) ... the rapidly growing amount of data. ☐ d) ... the difficulty of handling these data.	4 total
03	Medical data is characterized by .... ☐ a) ... high variability. ☐ b) ... high inaccuracy. ☐ c) ... high noise. ☐ d) ... weak structures.	4 total
04	The most used standardized data sets in the clinical practice to date is ☐ a) ... UMLS. ☐ b) ... SNOMED-CT. ☐ c) ... LOINC. ☐ d) ... ICD.	4 total
05	In standardization of ECG data ... ☐ a) ... HL7 aECG is using XML. ☐ b) ... DICOM-WS 30 is using binary. ☐ c) ... SCP-ECG is using XML. ☐ d) ... MFER is using binary.	4 total
06	A high degree of formalization is provided by... ☐ ... First-order logic. ☐ ... data dictionaries. ☐ ... glossaries. ☐ ... PROLOG.	4 total
07	In taxonomy of ontology languages, graphical notations are represented by ... ☐ a) ... Topic Maps. ☐ b) ... OWL ☐ c) ... RDF. ☐ d) ... UML.	4 total
08	$P1 \sqsubseteq P2$ means ☐ a) ... sub property. ☐ b) ... sub class. ☐ c) ... functional property. ☐ d) ... transitive property.	4 total
09	In the set of OWL class constructors, the formula for intersection is... ☐ a) ... $C1 \sqcup C2$ ☐ b) ... $C1 \wedge C2$ ☐ c) ... $C1 \sqcap C2$ ☐ d) ... $C1 \geqslant C2$	4 total

10	Medical Classifications ... ☐ a) ... has been established since Linne's classificiation of species in 1735. ☐ b) ... are called codes by practioners. ☐ c) ... are hierarchies of objects that conforms to certain principles. ☐ d) ... are used for example for reimbursement in the clinical routine (DRG).	4 total

Sum of Question Block B (max. 40 points)	

10 Exam Questions

10.3 Free Recall Questions

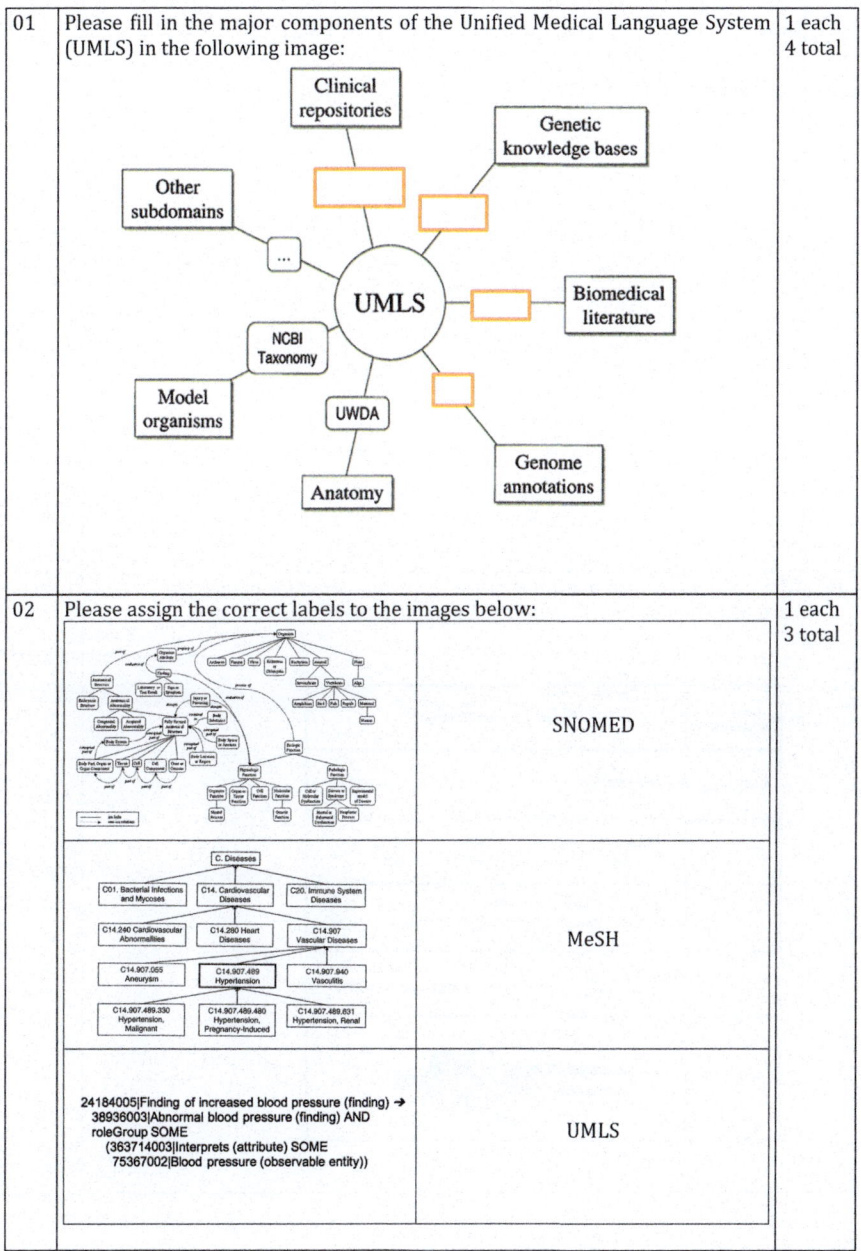

03	The Web Ontology Language (OWL) is the most widely used ontology language. Please assign the correct axioms to the examples below.	1 each 3 total
	Alga ⊑ Plant ⊑ Organism — Disjoint with	
	Sea Horse ⊑ ¬Horse — Sub class	
	Vertebrate ⊑ ¬Invertebrate — Different from	
04	There are three main questions in the abstraction example of the cardiovascular system below. Please complete the following image:	1 each 3 total

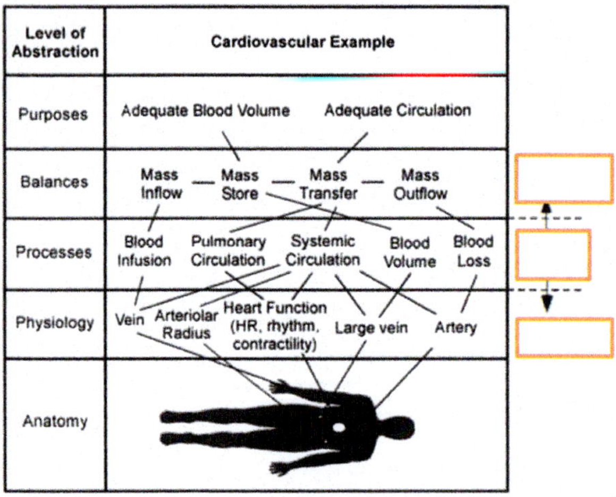

10 Exam Questions

05	Please assign the correct numbers to the biomedical ontologies		1 each 3 total
	ICD-10	310,314 concepts	
	UMLS	12,318 concepts	
	SNOMED CT	1,400,000 concepts	

Sum of Question Block C (max. 40 points)

11 Answers

11.1 Answers to the Yes/No Questions

Please check the following sentences and decide whether the sentence is true = YES; or false = NO; for each correct answer you will be awarded 2 credit points.

01	In the daily clinical practice we observe a high integration of standardized terminologies, e.g. SNOMED, MeSH, UMLS etc.	☐ Yes ☒ No	2 total
02	Standardization is the process of developing and implementing standards, e.g. the Evidence Based Medicine (EBM) approach.	☒ Yes ☐ No	2 total
03	Komaroff (1979) describes clinical data as being disturbingly "soft", having an obvious degree of variability and inaccuracy.	☒ Yes ☐ No	2 total
04	In computer science an ontology represents formal knowledge as a set of concepts within an unlimited domain, and the relationships between those concepts.	☐ Yes ☒ No	2 total
05	Standardized data shall now ensure that information is interpreted by all users with the same understanding.	☒ Yes ☐ No	2 total
06	In practice, ontologies consist of categories of entities organized in taxonomies and connected by relationships.	☒ Yes ☐ No	2 total
07	A logic program consists of a set of rules (Horn clauses), where each rule has the form head-body, where head is a logical atom and body is a conjunction of logical atoms.	☒ Yes ☐ No	2 total
08	The International Classification of Diseases (ICD) is the standard diagnostic tool for epidemiology, health management and clinical purposes and includes the analysis of the general health situation.	☒ Yes ☐ No	2 total
09	Data integration and data fusion in the life sciences is mostly solved and of no further research interest.	☐ Yes ☒ No	2 total
10	SNOMED CT is the Systematized Nomenclature of Medicine Clinical Terms and covers diseases, clinical findings and procedures.	☒ Yes ☐ No	2 total
Sum of Question Block A (max. 20 points)			

11 Answers

11.2 Answers to the Multiple Choice Questions (MCQ)

01	Knowledge representation ... ☐ a) ... uses imperative extensions of pure logic programming ☒ b) ... is an emerging field of artificial intelligence. ☐ c) ... is primarily used for dealing with imprecise, vague and noisy data. ☒ d) ... hold one of the key roles in the development of context awareness.	4 total
02	The famous Forrester reports speak of ... ☒ a) ... the challenge that most of our data is unstructured. ☒ b) ... the increasing complexity of data. ☒ c) ... the rapidly growing amount of data. ☒ d) ... the difficulty of handling these data.	4 total
03	Medical data is characterized by .... ☒ a) ... high variability. ☒ b) ... high inaccuracy. ☒ c) ... high noise. ☒ d) ... weak structures.	4 total
04	The most used standardized data sets in the clinical practice to date is ☐ a) ... UMLS. ☐ b) ... SNOMED-CT. ☐ c) ... LOINC. ☒ d) ... ICD.	4 total
05	In standardization of ECG data ... ☒ a) ... HL7 aECG is using XML. ☒ b) ... DICOM-WS 30 is using binary. ☐ c) ... SCP-ECG is using XML. ☒ d) ... MFER is using binary.	4 total
06	A high degree of formalization is provided by... ☒ ... First-order logic. ☐ ... data dictionaries. ☐ ... glossaries. ☒ ... PROLOG.	4 total
07	In taxonomy of ontology languages, graphical notations are represented by ... ☒ a) ... Topic Maps. ☐ b) ... OWL ☒ c) ... RDF. ☒ d) ... UML.	4 total
08	$P1 \sqsubseteq P2$ means ☒ a) ... sub property. ☐ b) ... sub class. ☐ c) ... functional property. ☐ d) ... transitive property.	4 total
09	In the set of OWL class constructors, the formula for intersection is... ☐ a) ... $C1 \sqcup C2$ ☐ b) ... $C1 \wedge C2$ ☒ c) ... $C1 \sqcap C2$ ☐ d) ... $C1 \geqslant C2$	4 total

10	Medical Classifications ... ☒ a) ... has been established since Linne's classificiation of species in 1735. ☒ b) ... are called codes by practioners. ☒ c) ... are hierarchies of objects that conforms to certain principles. ☒ d) ... are used for example for reimbursement in the clinical routine (DRG).	4 total

Sum of Question Block B (max. 40 points)

11 Answers

11.3 Answers to the Free Recall Questions

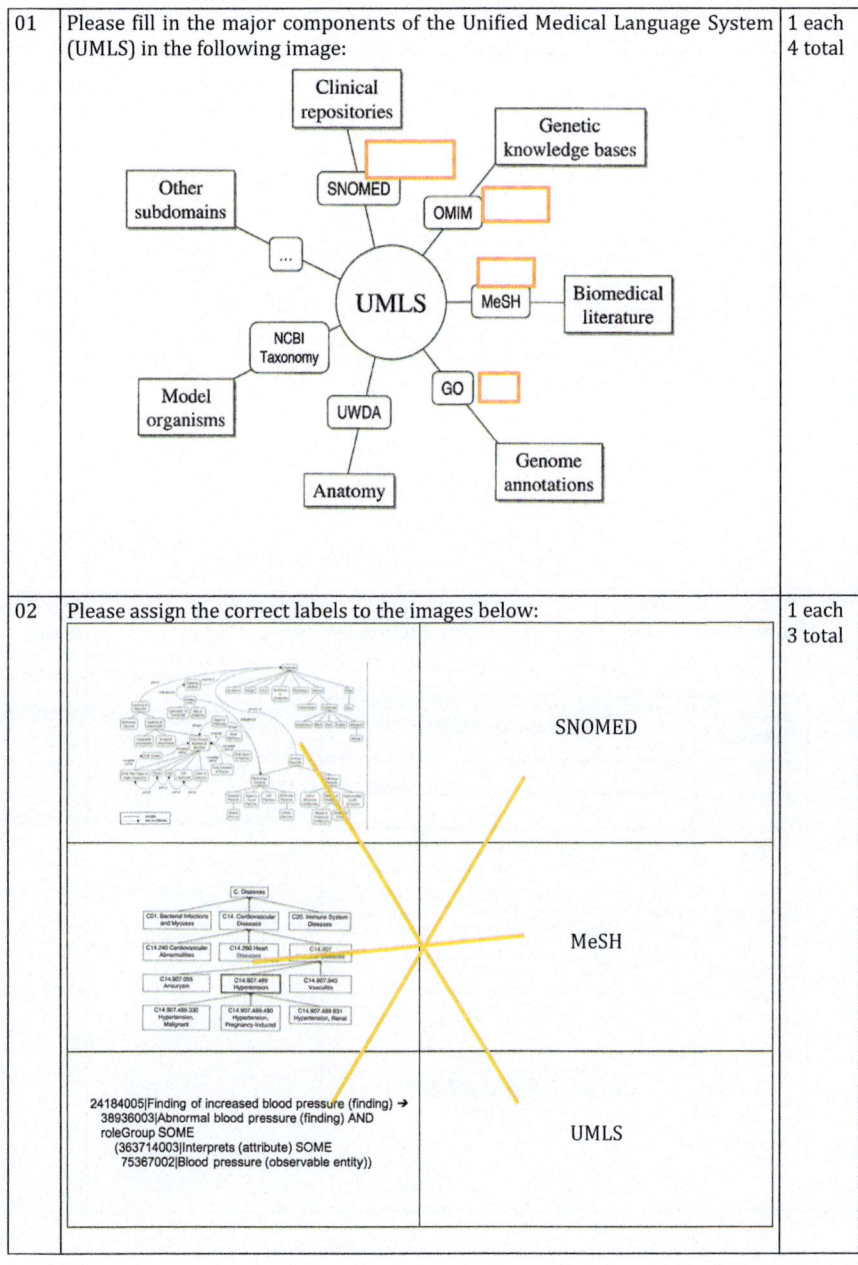

03	The Web Ontology Language (OWL) is the most widely used ontology language. Please assign the correct axioms to the examples below.	1 each 3 total

Alga ⊑ Plant ⊑ Organism	Disjoint with
Sea Horse ⊑ ¬Horse	Sub class
Vertebrate ⊑ ¬Invertebrate	Different from

04	There are three main questions in the abstraction example of the cardiovascular system below. Please complete the following image:	1 each 3 total

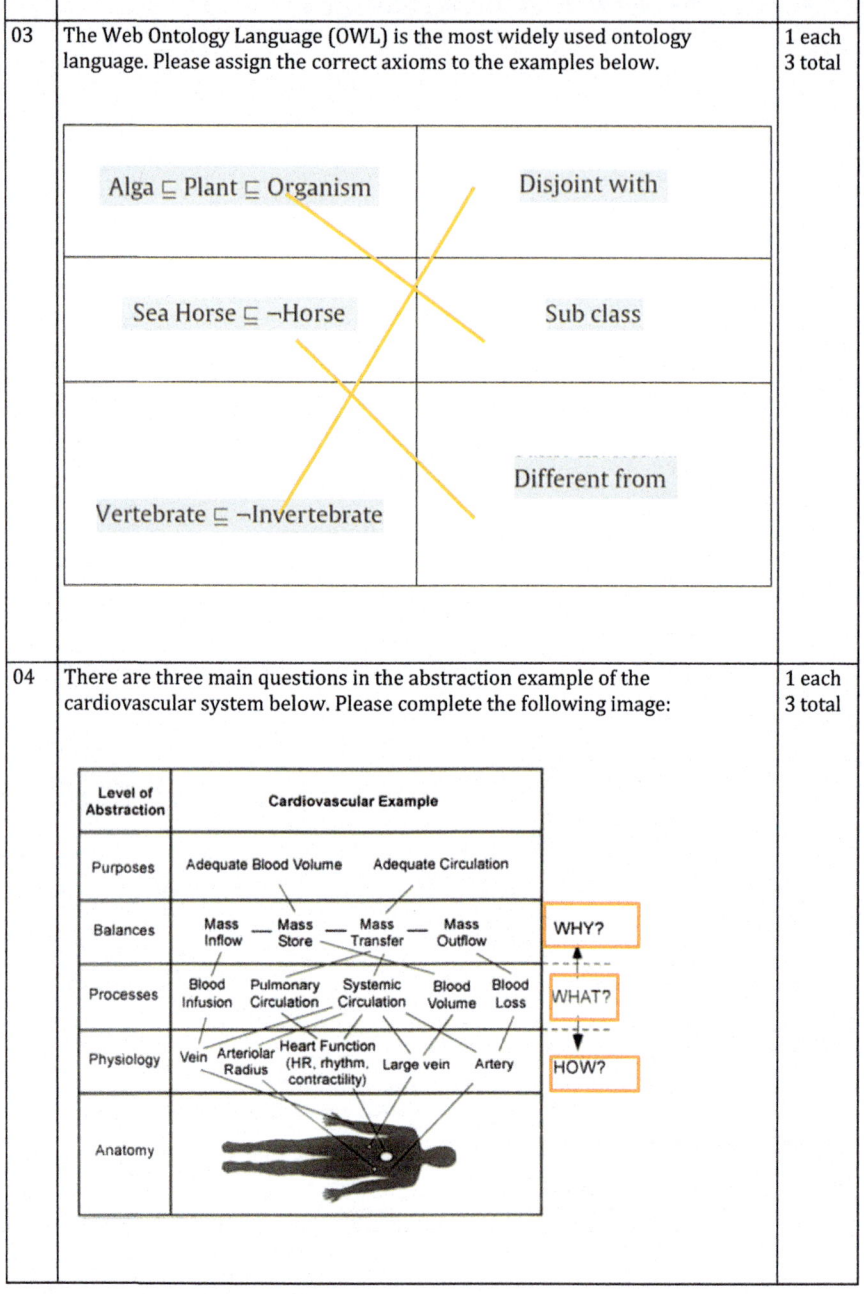

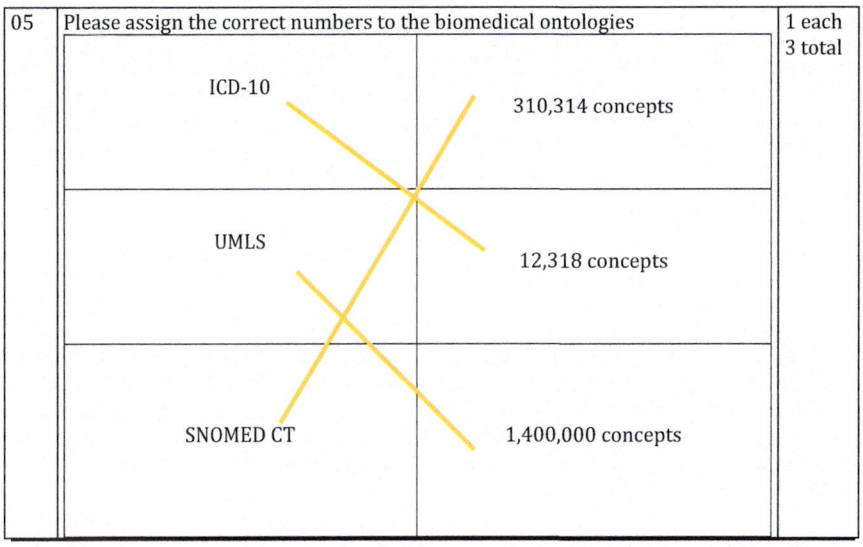

References

Ahmadian L, Van Engen-Verheul M, Bakhshi-Raiez F, Peek N, Cornet R, De Keizer NF (2011) The role of standardized data and terminological systems in computerized clinical decision support systems: literature review and survey. Int J Med Inform 80(2):81–93

Berman JJ (2012) Taxonomic guide to infectious diseases: understanding the biologic classes of pathogenic organisms. Academic, Amsterdam

Berman JJ (2013) Chapter 3—Ontologies and semantics. Principles of big data. Morgan Kaufmann, Boston, MA, pp 35–48

Bhatt M, Rahayu W, Soni SP, Wouters C (2009) Ontology driven semantic profiling and retrieval in medical information systems. Web Semantics Sci Serv Agents World Wide Web 7(4):317–331

Bleiholder J, Naumann F (2008) Data fusion. ACM Comput Surv (CSUR) 41(1):1

Bodenreider O (2004) The Unified Medical Language System (UMLS): integrating biomedical terminology. Nucleic Acids Res 32:D267–D270

Bodenreider O (2008) Biomedical ontologies in action: role in knowledge management, data integration and decision support. Methods Inform Med 47(1):67–79

Bond RR, Finlay DD, Nugent CD, Moore G (2011) A review of ECG storage formats. Int J Med Inform 80(10):681–697

Brown JHU, Loweli DJ (1972) Standardization and health care. IEEE Trans Biomed Eng BME-19 (5):331–334

Cheung K-H, Samwald M, Auerbach RK, Gerstein MB (2010) Structured digital tables on the Semantic Web: toward a structured digital literature. Mol Syst Biol 6:403

Dantsin E, Eiter T, Gottlob G, Voronkov A (2001) Complexity and expressive power of logic programming. ACM Comput Surv (CSUR) 33(3):374–425

Eiter T, Ianni G, Polleres A, Schindlauer R, Tompits H (2006) Reasoning with rules and ontologies. Reasoning Web. Springer, New York, pp 93–127

Hajdukiewicz JR, Vicente KJ, Doyle DJ, Milgram P, Burns CM (2001) Modeling a medical environment: an ontology for integrated medical informatics design. Int J Med Inform 62(1):79–99

Hermann GJ, King EJ, Shaw JM (1997) The yeast gene, MDM20, is necessary for mitochondrial inheritance and organization of the actin cytoskeleton. J Cell Biol 137(1):141–153

Hersh W (2010) Information retrieval: a health and biomedical perspective. Springer, New York

Holzinger A (2000) Modul 2: Wissensorganisation und Informatik. In: Holzinger A (ed) Basiswissen Multimedia, Band 2 Lernen Kognitive Grundlagen multimedialer Informationssysteme. Vogel-Buchverlag, Würzburg, pp 51–102

Holzinger A, Geierhofer R, Modritscher F, Tatzl R (2008) Semantic information in medical information systems: utilization of text mining techniques to analyze medical diagnoses. J Univ Comput Sci 14(22):3781–3795

Holzinger A, Stocker C, Ofner B, Prohaska G, Brabenetz A, Hofmann-Wellenhof R (2013) Combining HCI, natural language processing, and knowledge discovery—potential of IBM content analytics as an assistive technology in the biomedical domain. Springer lecture notes in computer science LNCS 7947. Springer, Heidelberg, pp 13–24

Horrocks I (2013) What are ontologies good for? In: Küppers B-O, Hahn U, Artmann S (eds) Evolution of semantic systems. Springer, Berlin, pp 175–188

Jakus G, Milutinović V, Omerović S, Tomažič S (2013) Concepts, ontologies, and knowledge representation. Springer, Berlin

Komaroff AL (1979) The variability and inaccuracy of medical data. Proc IEEE 67(9):1196–1207

Lee D, Lau F, Quan H (2010) A method for encoding clinical datasets with SNOMED CT. BMC Med Inform Decis Mak 10(1):53

Mccray A, Lee K (2013) Taxonomic change as a reflection of progress in a scientific discipline. In: Küppers B-O, Hahn U, Artmann S (eds) Evolution of semantic systems. Springer, Berlin, pp 189–208

Miller GA (1998) Nouns in wordnet. In: Felbaum C (ed) WordNet: an electronic lexical database. MIT, Cambridge, MA, pp 24–25

Montague R (1974) Formal philosophy; selected papers of Richard Montague. In: Thomason R (ed) Yale University Press, New Haven, CT

Rector AL, Brandt S (2008) Why do it the hard way? The case for an expressive description logic for SNOMED. J Am Med Inform Assoc 15(6):744–751

Salton G (1968) Automatic information organization and retrieval. McGraw-Hill, New York, NY

Salton G, Wong A, Yang CS (1975) Vector-space model for automatic indexing. Commun ACM 18(11):613–620

Scamardella JM (2010) Not plants or animals: a brief history of the origin of Kingdoms Protozoa, Protista and Protoctista. Int Microbiol 2(4):207–216

Shneiderman B (1992) Tree visualization with tree-maps: 2-d space-filling approach. ACM Trans Graph (TOG) 11(1):92–99

Simonet M, Messai R, Diallo G, Simonet A (2009) Ontologies in the health field. In: Berka P, Rauch J, Zighed DA (eds) Data mining and medical knowledge management: cases and applications. Medical Information Science Reference, New York, pp 37–56

Studer R, Benjamins VR, Fensel D (1998) Knowledge engineering: principles and methods. Data Knowledge Eng 25(1–2):161–197

Thomas JJ, Cook KA (2005) Illuminating the path: the research and development agenda for visual analytics. IEEE Comput Soc Press, New York

Xie S, Helfert M, Lugmayr A, Heimgärtner R, Holzinger A (2013) Influence of organizational culture and communication on the successful implementation of information technology in hospitals. In: Rau PLP (ed) Cross-cultural design. Cultural differences in everyday life. Lecture notes in computer science, LNCS 8024. Springer, Berlin, pp 165–174

Zhang S, Bodenreider O (2006) Law and order: assessing and enforcing compliance with ontological modeling principles in the Foundational Model of Anatomy. Comput Biol Med 36(7–8):674–693

Lecture 4

Biomedical Databases: Acquisition, Storage, Information Retrieval, and Use

1 Learning Goals

At the end of this lecture, you:

- would have an overview about the general architecture of an Hospital Information System (details in Lecture 10: Medical Information Systems and Biomedical Knowledge Management).
- would know some principles of hospital databases.
- would have an overview on some biomedical databases.
- would be familiar with some basics of information retrieval.

2 Advance Organizer

AceDB	Object oriented database architecture, developed by the Sanger Institute, http://www.acedb.org
Business intelligence (BI)	A type of application software designed to report, analyze, and present information on real-time management dashboards, i.e., integrated displays of metrics that measure the performance of a system
Cassandra	An open source database management system designed to handle huge amounts of data on a distributed system. This system was developed at Facebook and is now managed as a project of the Apache Software foundation
Cladogram	A phylogenetic tree to show evolutionary relationships with species represented by nodes and lines of descent represented by links (unrooted or rooted)

Classification system	Arbitrary in nature, there is no standard measure of difference that defines a species, genus, family, or order
Cloud computing	A computing paradigm in which highly scalable computing resources, often configured as a distributed system, are provided as a service
CPOE (Computerized physician order entry)	A process of electronic entry of medical practitioner instructions for the treatment of patients (particularly hospitalized patients)
Data Mart (DM)	Access layer of a data warehouse environment that is used to get data to the users. DM is a subset of DW (data warehouse), oriented to a specific business team to provide data to users usually through business intelligence tools
DBGET	A data retrieval tool (simpler than ENTREZ) from the Kyoto University, which covers more than 20 databases, related to the Kyoto Encyclopedia of Genes and Genomes
Distance matrix method	Work by two most closely related taxa in a distance matrix and clustering them
EnsEMBL	Database format
ENTREZ	A dedicated data retrieval tool
Extract, transform, and load (ETL)	Software tools used to extract data from outside sources, transform them to fit operational needs, and load them into a database or data warehouse
Federated database system	Type of meta-database management system, which integrates multiple autonomous database systems into a single federated database
Genetic algorithm	A technique used for optimization inspired by the process of natural evolution or "survival of the fittest." Often described as a type of "evolutionary algorithm," these algorithms are well suited for solving nonlinear problems
GOLD	Genomes OnLine Databases, genomics gateway
Hadoop	An open source (free) software framework for processing huge datasets on certain kinds of problems on a distributed system. Its development was inspired by Google's MapReduce and Google File System
Hbase	An open source (free), distributed, non-relational database modeled on Google's Big Table. It was originally developed by Powerset and is now managed as a project of the Apache Software foundation as part of the Hadoop.

2 Advance Organizer

Information Extraction (IE)	Automatic assignment of meaning to elementary textual entities and complex structured information objects
Information Retrieval (IR)	Indexing and retrieval of information in documents
KEGG	Kyoto Encyclopedia of Genes and Genomes, a combined database which stores information on particular types of proteins (receptors, signal transduction components, and enzymes)
MapReduce	A software framework introduced by Google for processing huge datasets on certain kinds of problems on a distributed system. Also implemented in Hadoop.
Mashup	Uses and combines data presentation or functionality from two or more sources to create new services
MEDLINE	Literature data bank
Metadata	Data that describes the content and context of data files, e.g., means of creation, purpose, time and date of creation, and author.
MMMDB	Molecular Modeling Database can be accessed at the NCBI (National Center for Biotechnology Information) using ENTREZ
Natural Language Processing (NLP)	A set of machine learning techniques from computer science and linguistics that uses computer algorithms to analyze human (natural) language
Neural networks	Computational models, inspired by the structure and workings of biological neural networks (i.e., the cells and connections within a brain), that find non linear patterns in data
Non-relational database	A database that does not store data in tables (rows and columns) (In contrast to relational database).
Online Mendelian Inheritance in Man (OMIM)	A database as a resource for the study of human genetics and human molecular medicine
PDB	Protein Data Bank contains data derived from X-ray crystallography and NMR (nuclear magnetic resonance) studies
Phylogenetics	Similarities and differences among species can be used to infer evolutionary relationships (=phylogenies); Examples for phylogenetic software: PAUP, PHYLIP
PROSITE	Database containing sequence patterns associated with protein family membership, specific protein functions, and posttranslational modifications

R	An open source programming language and software environment for statistical computing and graphics
Relational database	A database made up of a collection of tables (relations), i.e., data are stored in rows and columns. Relational database management systems (RDBMS) store a type of structured data. SQL is the most widely used language for managing relational databases
Semi-structured data	Data that do not conform to fixed fields but contain tags and markers to separate data elements. Examples include XML or HTML-tagged text. Contrast with structured data, weakly structured, and unstructured data.
Similarity table	Distance table
SQL	Structured query language is a language designed for managing data in relational databases. This technique includes the ability to insert, query, update, and delete data, as well as manage data schema (database structures) and control access to data in the database
SRS	Sequence Retrieval System, a data retrieval tool based on open source software
SWISS-PROT	is a databank containing a collection of confirmed protein sequences with annotations relating to structure, function, and protein family assignment. Database interrogation can be searched by sequence and annotations, e.g., ENTREZ
UniGene	Experimental facility for the clustering of GenBank sequences, related to EST (expressed sequence tag) data

3 Acronyms

ACeDB	A *C elegans* Database
ADE	Adverse drug events
CDSS	Clinical decision support system
CPOE	Computerized physician order entry
DBMS	Database Management System
EMAC	Electronic medication administration chart
EO	Electronic order

ERT	Error registration table
GFR	Glomerular filtration rate
HIS	Hospital Information System (DE: KIS = Krankenhaus Informations System)
HWO	Handwritten order
NICU	Neonatal intensive care unit
NOE	Nurse order entry (followed by physician's verification and countersignature)
PBMAC	Paper-based medication administration chart
POE	Physician order entry
RR	Rate ratio
UniProt	Universal Protein Resource

4 Key Problems

- Increasingly large and complex datasets "Big Data" due to data intensive research.
- Increasing amounts of non-standardized and unstructured information (e.g., free text).
- Data quality, data integration, universal access.
- Privacy, security, safety, and data protection issues (see Lecture 11).
- Time aspects in databases (Gschwandtner et al. 2012; Johnston and Weis 2010).

> Big Data resources are all a waste of time and money if data analysts cannot find, or fail to comprehend, the basic information that describes the data held in the resources (Berman 2013b)

Data identification is certainly the most underappreciated and least understood Big Data issue. Measurements, annotations, properties, and classes of information have no informational meaning unless they are attached to an identifier that distinguishes one data object from all other data objects and that links together all of the information associated with the identified data object (Berman 2013a).

Communication of data between application systems must ensure security to avoid improper access, because trust or the lack thereof, is the most essential factor blocking the adoption of rapidly evolving Web technology paradigm such as software as a service (SaaS) or data distribution services such as Cloud computing (Sreenivasaiah 2010).

5 A First View on Hospital Information Systems

5.1 Goals and Challenges of Hospital Information Systems

Slide 4-1: Hospital Information System: Typical Scenario

Let us start with a look into the hospital: In this slide we see a typical hospital scenario: medical professionals are surrounded by information technology. An old dream of hospital managers was always to have an "all digital hospital" to digitalize all workflows and to store all data in an electronic way—towards a paperless hospital. Although much effort has been spent towards a paperless hospital, most hospitals worldwide are still far away from being a "all-digital hospital" (Waterson et al. 2012). An interesting study: All hospitals in the province of Styria (Austria) are well equipped with sophisticated Information Technology, which provides all-encompassing on-screen patient information. Previous research made on the theoretical properties, advantages and disadvantages, of reading from paper vs. reading from a screen has resulted in the assumption that reading from a screen is slower, less accurate and more tiring. However, recent flat screen technology, especially on the basis of LCD, is of such high quality that obviously this assumption should now be challenged. As the electronic storage and presentation of information has many advantages in addition to a faster transfer and processing of the information, the usage of electronic screens in clinics should outperform the traditional hardcopy in both execution and preference ratings. In a study in the County hospital Styria, Austria, with 111 medical professionals, working in a real-life setting, they were each asked to read original and authentic diagnosis reports, a gynecological report and an internal medical document, on both screen and paper in a randomly assigned order. Reading comprehension was measured by the Chunked Reading Test, and speed and accuracy of reading performance was quantified. In order to get a full understanding of the clinicians' preferences, subjective ratings were also collected. Wilcoxon Signed Rank Tests showed **no significant differences on reading performance between paper vs. screen**. However, medical professionals showed a significant (90 %) preference for reading from paper. Despite the high quality and the benefits of electronic media, paper still has some qualities which cannot be provided electronically do date (Holzinger et al. 2011).

Slide 4-2: HIS: Typical View on the Clinical Workplace

Mega issues related with hospital information systems include: data integration, data fusion, standardization issues, clinical process analysis, modeling, compliance issues, evidence-based treatment and decision support, privacy, security, safety and data protection, and knowledge discovery and data mining—all connected with the central topic of this lecture: databases.

Slide 4-3: Much of the Hospital Work Is Teamwork

The teamwork in the hospital requires a lot of communication and information exchange. The vision of a business enterprise hospital information system is to cover all workflows, organizational processes, and information flows electronically.

Note: The quality of the work of physicians is heavily influenced by the usability of their available equipment. In the slide you see a typical work meeting of medical professionals, where they discuss the patient cases jointly. It is important to study and understand the workflows of the end users and to involve them into the development of information systems as early as possible by a user-centered design process (Holzinger 2003). Experiments showed that by studying the workflows the engineers get deep insights into how to develop an appropriate application for a specified target end user group (Holzinger et al. 2005).

Slide 4-4: The Medical Report Is Still the Most Important Output

The aforementioned goal of an "all-digital hospital" produces "big data" and remarkably much of the data is unstructured text. Interestingly, the main and most important output is the medical report (in German: Arztbrief): In the example it is the report of a medical image—not the image itself is the relevant issue—it is the report (Holzinger et al. 2007). The handling with unstructured data is a mega challenge and brings along a lot of challenges for computers.

Slide 4-5: Example: Chess Game vs. Natural Language

Let us briefly compare human intelligence with machine intelligence.

A good example on the complexity which we are facing in hospital information processing is the differences between chess and human natural language processing:

(continued)

(continued)

Whereas chess is a finite, mathematically well-defined search space, and hence we have a **well defined computational space**, with limited numbers of moves and states and grounded in explicit, unambiguous mathematical rules, human language is exactly the opposite: Ambiguous, contextual, and implicit; grounded in the **human cognitive space**, with a seemingly infinite number of ways to express one and the same meaning.

Note: IBM Deep Blue defeated the World Chess Champion Garry Kasparov in a six-game match in 1997. There were a number of factors that contributed to this success, including: a single-chip chess search engine, a massively parallel system with multiple levels of parallelism, a strong emphasis on search extensions, a complex evaluation function, and effective use of a Grandmaster game database. Technically, Deep Blue was a massively parallel system designed for carrying out chess game tree searches. The system was composed of a 30-node IBM RS/6000 SP computer and 480 single-chip chess search engines, with 16 chess chips per SP processor. The SP system consists of 28 nodes with 120 MHz P2SC processors, and 2 nodes with 135 MHz P2SC processors. The nodes communicated with each other via a high speed switch and all nodes had 1 GB of RAM, and 4 GB of disk. During the 1997 match with Kasparov, the system ran the AIX 4.2 operating system. The chess chips in Deep Blue were each capable of searching up to 2.5 million chess positions per second, and communicate with their host node via a fast micro channel bus (Campbell et al. 2002).

5.2 *Workflows*

Slide 4-6: Typical Hospital Workflow: Order Entry and Result Reporting

Health-care processes require the cooperation of different organizational units and various medical disciplines. In such an environment optimal process support becomes crucial. In this slide we see a typical organizational process for medical order entry and result reporting, which is used to coordinate the interdepartmental communication between a ward (ambulatory setting) and the radiology unit. The depicted process is not tailored to a specific clinical pathway, but shows an example for a characteristic organizational procedure of the hospital: An order (in German: Anweisung, Verschreibung) is placed by a physician at the ward or at an ambulatory setting. The indication is checked in the radiology department and depending on the result the order placer is informed whether the request has been rejected or scheduled. The

(continued)

(continued)

actual radiological examination and corresponding documentation is done in the examination room. The radiology report is generated afterwards, which has to be validated by the physician with his signature. The report is sent back to the order placer. This is an example for a fundamental process of clinical practice and captures the organizational knowledge necessary to coordinate the health-care process among different people and organizational units, i.e., the focus is on the support of core organizational processes (Lenz and Reichert 2007).

Slide 4-7: Remember the Diagnostic-Therapeutic Cycle

The medical treatment process is often described as diagnostic–therapeutic cycle (Bemmel and Musen 1997) including: observation, reasoning, and action. Please remember that in medicine we deal with **uncertain information** (Holzinger and Simonic 2011) and each pass of the diagnostic-therapeutic cycle can be seen as a step in decreasing the uncertainty about the patient's disease. Consequently, the observation process always starts with the patient history ("looking into the past") and proceeds with diagnostic procedures which are selected based on available information. The aim of the HIS is to assist health-care personnel in making informed decisions. Maybe the most important question to be answered is **how to determine what is relevant**. Availability of relevant information is a precondition for (good) medical decisions—and the medical knowledge guides these decisions (Lenz and Reichert 2007).

Following the principles of **Evidence Based Medicine (EBM)** physicians are required to formulate questions based on patients' problems, search the literature for answers, evaluate the evidence for its validity and usefulness, and finally apply the (new) information to patients treatment (Hawkins 2005). The limiting factor is the short time a clinician has to make a decision (Gigerenzer and Gaissmaier 2011).

5.3 Architecture of HIS

Slide 4-8 HIS: Classic Conceptual Model (in Use Since 1984)

This slide shows a classical conceptual model: The heart is a central data and communication structure. The patients "enter" (logically) the system through the admission on the left side, transfer and discharge functions of the core and leaves the system, at least partially, through the right side. In the main focus is

(continued)

(continued)

a central database, although alternative solutions have opted for a more distributed construction of databases; nonetheless, central ordering principles have to be kept to achieve the necessary integration of information and the distribution to the various points where it is needed, be it in the area of hospital management or in the field of care provision. This central database is serving the central operational purposes of the hospital in the context of its dual goals (Haux et al. 1998; Reichertz 2006; Haux 2006).

Slide 4-9: Modern Enterprise HIS: Sample Architecture

There are many different application architectures in use, and we will come back to it later, in Lecture 10, so here just ONE example for an enterprise business hospital information system as it is called professionally. However, we now want to concentrate on some technical issues of databases.

6 Databases

Slide 4-10: HIS Central Components: Databases

In this classical image by Shortliffe et al. (2001) it becomes very obvious that databases are central components for a hospital information system. A very interesting slide is the next, where we see an historical example from the "stone-age" of computer science.

Slide 4-11: Historical Example: The HELP System (1967)

This picture by Gardner et al. (1999) is insofar interesting as it shows us clearly a mega issue up to the present: to integrate and fusion different data and to make it accessible to the clinician. While there is much research on the integration of heterogeneous information systems, a shortcoming is in the integration of available data. Just to clarify the differences between data *integration* and data *fusion*:

Data integration involves combining data residing in different distributed sources and providing users with a unified view of and access to these data. It has become the focus of extensive theoretical and practical work, and numerous open problems remain unsolved (Lenzerini 2002).

(continued)

(continued)

Data fusion is the process of merging multiple records representing the same real-world object into a single, consistent, accurate, and useful representation (Bleiholder and Naumann 2008).

The trend towards P4 medicine (Predictive, Preventive, Participatory, Personalized) has resulted in a sheer mass of the generated (-omics) data, and hence, a main challenge is in the integration and fusion of heterogeneous data sources, especially in the integration of data from the clinical domain with sources from the biological domain.

Slide 4-12: Databases: Fundamental Terms and Definitions

Database (DB) is the organized collection of data through a certain data structure (e.g., hash-table, adjacency matrix, graph structure).

Database management system (DBMS) is software which operates the DB. Well-known DBMSs include Oracle, IBM DB2, Microsoft SQL Server, Microsoft Access, MySQL, and SQLite. Examples for Graph Databases include InfoGrid, Neo4j, or BrightstarDB.

The used DB is not generally portable, but different DBMSs can interoperate by using standards such as SQL and ODBC.

Database system (DBS) = DB + DBMS. The term database system emphasizes that data is managed in terms of accuracy, availability, resilience, and usability.

Data warehouse (DWH) is an integrated repository used for reporting and long term storage of analysis data.

Data Marts (DM) are access layers of a DWH and are used as temporary repositories for data analysis.

Recommendable Reading include (Plattner 2013; Robinson et al. 2013):

Robinson, I., Webber, J. & Eifrem, E. 2013. Graph Databases, O'Reilly Media.

Plattner, H. 2013. A Course in In-Memory Data Management: The Inner Mechanics of In-Memory Databases, Heidelberg New York Dordrecht London, Springer.

One of the standard textbooks is the 6th edition of "Database System Concepts" by Silberschatz et al. (2010).

6.1 Data Warehouse

Slide 4-13: Example Hospital Data Warehouse

A DWH is an integrated system, specifically designed for enterprise business decision support and can be used in hospitals and in biomedical applications. In Slide 4-13 we see an example of a hospital data warehouse: On the left there are the (heterogeneous) data sources, such as PACS (Picture Archiving and Communication System) and RIS (Radiological Information System), and apart from the core HIS, some special databases which can also include proprietary and legacy systems. For the data staging and area servers the Common Object Request Broker Architecture (CORBA) is used, a standard defined by the Object Management Group (OMG) that supports multiple platform interoperability (Zhang et al. 2004).

This is a standard hospital information architecture and—typically—with no integration of laboratory data sources and most of all no Omics-data integration, as for example from the pathology or a biobank.

6.2 Data Marts

Slide 4-14: HIS Central Components: Databases

A DWH can be subdivided into so-called data marts (DM), which can be seen as specific access layer of a DWH, oriented to a specific team. Slide 4-14 shows the architecture of the Mayo clinic DWH, which is incrementally instantiating each component of the architecture on demand. Data integration proceeds from left to right (leftmost you see the primary data sources; moving right, the data are integrated into staging and replication services, with further refinement). The layers are as follows:

1. Subjects = the highest level areas that define the activities of the enterprise (e.g., Individual).
2. Concepts = the collections of data that are contained in one or more subject areas (e.g., Patient, Provider, Referrer).
3. Business Information Models = the organization of the data that support the processes and workflows of the enterprise's defined Concepts (Chute et al. 2010).

6.3 Biomedical Databanks and Cloud Computing

> **Slide 4-15: Traditional Genome Information System**
>
> A standard environment for production and processing of genomic data can be seen in Slide 4-15: Sequencing labs submit their data to large databases, e.g., GenBank; National Center of Biotechnology Information (NCBI); European Bioinformatics Institute (EMBL) database; DNA Data Bank of Japan (DDBJ); Short Read Archive (SRA); Gene Expression Omnibus (GEO); or Microarray database Array Express. These maintain, organize and distribute the sequencing data. Most users access the information either through Web-based applications or through integrators, such as Ensembl, the University of California at Santa Cruz (UCSC) Genome Browser, or Galaxy. The end users have to download genomic data from these primary and secondary sources (Stein 2010).
>
> Remember: Sequencing is the process of determining the precise order of nucleotides within a DNA molecule to determine the order of the four bases—adenine, guanine, cytosine, and thymine—in a strand of DNA. The advent of rapid DNA sequencing methods has greatly accelerated biological and medical research and discovery and produces large datasets. Sequencing has become indispensable for basic biological research, and in numerous applied fields such as diagnostics and biotechnology.
>
> Note: A biobank is a physical place which stores biological specimens—and in some cases also data (Roden et al. 2008).

6.4 Cloud Computing

> **Slide 4-16: Genome Information Ecosystem: Cloud Computing**
>
> Here we see a cloud-based genome informatics system. Instead of separate genome datasets stored at various locations, the datasets are stored in the cloud as virtual databases. Web services run on top of these datasets, including the primary archives and the integrators, running as virtual machines within the cloud. Casual users, who are accustomed to accessing the data via the NCBI, DDBJ, Ensembl, or UCSC, work as usual; the fact that these servers are located inside the cloud is invisible to them. Power users can continue to download the data, but have an attractive alternative. Instead of moving the data to the computational cluster, they move the computational cluster to the data (Stein 2010).
>
> Note: Cloud computing is based on sharing of resources to achieve coherence and economies of scale over a network (similar to the electricity

(continued)

(continued)

grid). The foundation of cloud computing is the broader concept of converged infrastructure and shared services. Cloud providers offer their services according to several fundamental models

1. Infrastructure as a service (IaaS).
2. Platform as a service (PaaS).
3. Software as a service (SaaS).

Slide 4-17: Example for Clinical Cloud Computing: PACS Cloud

Just an example for a cloud-based service: The Master Index is the PACS Cloud core entity and contains information about other modules, including Gateways and Cloud Slaves (repository and database). It also provides authentication services to institutional gateways and all identifiable information related with patients are stored in a master index database, fundamental to ensure solutions for confidentiality and privacy. The Cloud Slaves provide, on the one hand, storage of sightless data (objects repositories) and, on the other hand, a database containing all no identifiable metadata extracted from DICOM studies, i.e., the most demanding task concerning computational power (Bastiao-Silva et al. 2011).

Slide 4-18: Federated Data vs. Warehoused Data

We have to determine between federated data and warehoused data. A federated database system is a meta-database management system, which transparently maps multiple heterogeneous and autonomous database systems into a single federated database and this can be a "virtual database"—without data integration as it is in data warehouses. In the slide we can see on the y-axis the data integration architecture and on the x-axis the knowledge representation methodologies and where current data integration systems lie along this continuum. The essence of this image is that there is no "best-solution": A system designed to have full control of data and fast queries can have difficulty expressing complex biological concepts and integrating them. Systems that employ highly expressive knowledge representation methodologies such as Ontologies are more able to represent and integrate complex biological concepts but have much less tractable queries (Louie et al. 2007).

6.5 Biomedical Databases

Slide 4-19: Biomedical Databases

Whereas databases for the use in HIS are process centered and central for the electronic patient record, biomedical databases are libraries of all sorts of life science data, collected from scientific experiments and computational analyses. Such databases contain experimental biological data from clinical work, genomics, proteomics, metabolomics, microarray gene expression, phylogenetics, pharmacogenomics, etc.

Examples:

Text: e.g., PubMed, OMIM (Online Mendelian Inheritance in Man).

Sequence data: e.g., Entrez, GenBank (DNA), UniProt (protein).

Protein structures: e.g., PDB, Structural Classification of Proteins (SCOP), CATH (Protein Structure Classification).

An overview can be found here: (Masic and Milinovic 2012)

Online open access via: http://www.ncbi.nlm.nih.gov/pmc/articles/PMC3544328

Note: Pharmacogenomics is the technology for the analytics of how genetic makeup affects an individual's response to drugs—so it deals with the influence of genetic variation on drug response in patients by correlating gene expression or single-nucleotide polymorphisms with efficacy and toxicity. The central aim is to optimize drug therapy to ensure maximum effectiveness with minimal adverse effects and is a core towards personalized medicine.

Slide 4-20: Example Database: PDB

Remember: Proteins are the molecules used by the cell for performing and controlling cellular processes, including degradation and biosynthesis of molecules, physiological signaling, energy storage and conversion, formation of cellular structures, etc.

Protein structures are determined with crystallographic X-ray methods or by nuclear magnetic resonance spectroscopy. Once the atomic coordinates of the protein structure have been determined, a table of these coordinates is deposited into the protein database (PDB), an international repository for 3D structure files: http://www.rcsb.org/pdb/

This database is handled by the RCSB (Research Collaboratory for Structural Biology) at the Rutgers University and UC San Diego. PDB is the most important source for protein structures. Before a new structure of a protein is added, a careful examination of the data must be carried out to guarantee the quality of the structure. The PDB data file contains, among others, the coordinates of all the atoms of the protein (Wiltgen and Holzinger 2005; Wiltgen et al. 2007).

Slide 4-21: Databases: From Molecules to Systems

Remember the structural dimensions which we discussed in Lectures 1 and 2. This slide by Kampen (2013) is a very nice overview of various databases addressing the different microscopic dimensions. Additionally, the data on the level of the hospital information systems are added—so that you have a good summary of the aforementioned. If we take aside literature databases and ontologies (in the upper right corner of this slide) we start with:

Genome databases: Ensembl http://www.ensembl.org/index.html
Nucleotide sequence EMBL-Bank http://www.ebi.ac.uk/ena/
Gene expression: ArrayExpress http://www.ebi.ac.uk/arrayexpress
Proteomes: UniProt http://www.uniprot.org/
Proteins: InterPro http://www.ebi.ac.uk/interpro/
Protein structure: PDB http://www.rcsb.org/pdb/home/home.do
Protein Interactions: IntAct http://www.ebi.ac.uk/intact/
Chemical entities: ChEMBL https://www.ebi.ac.uk/chembl/
Pathways: Reactome http://www.reactome.org/
Systems: BioModels http://www.ebi.ac.uk/biomodels-main/

Slide 4-22: Example Genome Database: Ensembl

Ensembl (not to mix up with Ensemble ;-) is a good example for a Genome database and is a joint project between the European Bioinformatics Institute and the Wellcome Trust Sanger Institute, which was launched in 1999 in response to the imminent completion of the Human Genome Project (Flicek et al. 2011). Its aim remains to provide a centralized resource for geneticists, molecular biologists studying the genomes of our own species and other vertebrates and model organisms. Ensembl provides one of several well-known genome browsers for the retrieval of genomic information.

Slide 4-23: Example Gene Expression Database: ArrayExpress

ArrayExpress is a database of functional genomics experiments that can be queried and the data downloaded. It includes gene expression data from microarray and high throughput sequencing studies. Data is collected to MIAME and MINSEQE standards. Experiments are submitted directly to ArrayExpress or are imported from the NCBI GEO database.

MIAME = Minimum Information About a Microarray Experiment. This is the data that is needed to enable the interpretation of the results of the experiment unambiguously and potentially to reproduce the experiment (Brazma et al. 2001).

(continued)

(continued)

The six most critical elements contributing towards MIAME are:

(1) The raw data for each hybridization (e.g., CEL or GPR files),
(2) The final processed (normalized) data for the set of hybridisations in the experiment;
(3) The essential sample annotation including experimental factors and their values,
(4) the experimental design including sample data relationships;
(5) Annotation of the array (e.g., gene identifiers, genomic coordinates, probe oligonucleotide sequences, or reference commercial array catalog number), and
(6) Laboratory and data processing protocols (e.g., what normalization method has been used to obtain the final processed data); see: http://www.mged.org/Workgroups/MIAME/miame.html

Slide 4-24: Example for Protein Interactions Database: IntAct

IntAct is an open source database for protein–protein interactions (PPI). The Web interface provides both textual and graphical representations of such protein interactions, and allows for exploring interaction networks in the context of the GO annotations of the interacting proteins. Moreover, a Web service allows direct computational access to retrieve interaction networks in XML format. IntAct (http://www.ebi.ac.uk/intact) contains binary and complex interactions imported from the literature and curated in collaboration with the Swiss-Prot team, making intensive use of controlled vocabularies to ensure data consistency (Hermjakob et al. 2004).

Slide 4-25: Example for Systems Database: BioModels

The BioModels Database is a freely accessible online resource for storing, viewing, retrieving, and analyzing published, peer-reviewed quantitative models of biochemical and cellular systems. The structure and behavior of each simulation model are thoroughly checked; in addition, model elements are annotated with terms from controlled vocabularies as well as linked to relevant data resources. Models can be examined online or downloaded in various formats and reaction network diagrams can be generated from the models in several formats. BioModels Database (http://www.ebi.ac.uk/biomodels) also provides features such as online simulation and the extraction of components from large-scale models into smaller sub-models. The system provides a range of Web services that external software systems can use to access up-to-date data from the database (Li et al. 2010).

(continued)

> (continued)
>
> Note: Quantitative models of biochemical and cellular systems are used to answer research questions in the biological sciences and digital modeling is of growing interest in molecular and systems biology. A well-known example is the Virtual Human (Kell 2007).

7 Information Retrieval

Before we start we should answer the question: What is the difference between data mining/knowledge discovery and information retrieval? Slide 4-26 gives the answer:

> **Slide 4-26: Data Mining/KDD vs. Information Retrieval**
>
> Maimon and Rokach (2010) define Knowledge Discovery in Databases (KDD) as an automatic, exploratory analysis and modeling of large data repositories and the organized process of identifying valid, novel, useful and understandable patterns from large and complex datasets. Data Mining (DM) is the core of the KDD process (Witten et al. 2011).
>
> The term KDD actually goes back to the machine learning and Artificial Intelligence (AI) community (Piatetsky-Shapiro 2000). Interestingly, the first application in this area was again in medical informatics: The program Rx was the first that analyzed data from about 50,000 Stanford patients and looked for unexpected side-effects of drugs (Blum and Wiederhold 1985). The term really became popular with the paper by Fayyad et al. (1996), who described the KDD process consisting of nine subsequent steps:
>
> 1. **Learning from the application domain**: includes understanding relevant previous knowledge, the goals of the application and a certain amount of domain expertise.
> 2. **Creating a target dataset**: includes selecting a dataset or focusing on a subset of variables or data samples on which discovery shall be performed.
> 3. **Data cleansing (and preprocessing)**: includes removing noise or outliers, strategies for handling missing data, etc.
> 4. **Data reduction and projection**: includes finding useful features to represent the data, dimensionality reduction, etc.
> 5. **Choosing the function of data mining**: includes deciding the purpose and principle of the model for mining algorithms (e.g., summarization, classification, regression and clustering).
> 6. **Choosing the data mining algorithm**: includes selecting method(s) to be used for searching for patterns in the data, such as deciding which models and parameters may be appropriate (e.g., models for categorical data are different from models on vectors over reals) and matching a particular data mining method with the criteria of the KDD process.
>
> (continued)

(continued)
7. **Data mining**: searching for patterns of interest in a representational form or a set of such representations, including classification rules or trees, regression, clustering, sequence modeling, dependency, and line analysis.
8. **Interpretation**: includes interpreting the discovered patterns and possibly returning to any of the previous steps, as well as possible visualization of the extracted patterns, removing redundant or irrelevant patterns and translating the useful ones into terms understandable by users.
9. **Using discovered knowledge**: includes incorporating this knowledge into the performance of the system, taking actions based on the knowledge or documenting it and reporting it to interested parties, as well as checking for, and resolving, potential conflicts with previously believed knowledge (Holzinger 2013).

In Information retrieval a query q is defined as a formulation (N,L) = q and the matches with an index I Matching (q,I) retrieves relevant data to satisfy the search query (Baeza-Yates and Ribeiro-Neto 2011).

7.1 Data Retrieval vs. Information Retrieval

Factor	Data Retrieval (DR)	Information Retrieval (IR)
Model	Deterministic	Probabilistic
Matching	Exact match	Partial (best match)
Inference	Deduction	Induction
Classification	Monothetic*	Polythetic**
Query language	Artificial (abstract)	Natural
Query specification	Must be complete	Can be incomplete
Items wanted	matching	relevant
Error response	sensitive	insensitive

Fig. 1 See Slide 4-27

Slide 4-27: Data Retrieval (DR) vs. Information Retrieval (IR)

An excellent start in the determination between DR and IR is the work of Van Rijsbergen (1979): The most important difference is that the data model in DR is **deterministic**, whereas we speak about probable information in the IR Model, and hence, information retrieval is **probabilistic** (Simonic and Holzinger 2010).

*Monothetic = type in which all members are identical on all characteristics; **Polythetic = type in which all members are similar, but not identical.

7.2 Text Retrieval

Slide 4-28 Information Retrieval on the Example of Text

IR can be defined as a recall of already existing information, not aiming at the discovery of new structures as it is the goal in Knowledge Discovery and Data Mining (see Lecture 6). As we have already heard several times, in hospital information systems most of the data consists of medical documents, which consist mostly of unstructured information: text. But: What is text?

From a computational perspective, text consists of sequences of character strings, the syntax (Hotho et al. 2005), and hence, it is an abstract representation of natural language, and the challenges are in semantics (meaning). Text processing belongs to the field of Natural language processing (NLP) which is highly interdisciplinary, dealing with the interaction between the cognitive space (natural languages) and the computational space (formal languages). As such, NLP is closely related to HCI. Text mining is a subfield of data mining.

The original goal of IR was to find documents which contain answers to questions and not the finding of answers itself (Hearst 1999). For this purpose statistical measures and methods are used, and we need a formal description first.

7.3 IR Process

Slide 4-29: IR Process Principle

This is the general principle: The end user formulates his query via the user interface, in form of a Text Operations ("user need"). The next step is the representation (logical document view D in the formal model in Slide 4-30) of the documents and the representation of the reasoning strategy, query logical view Q (compare with Slides 4-30 and 4-31). The result is a ranking of the retrieved documents, which will be displayed via the user interface.

7.4 Formal Notation

Slide 4:30: Formal Description of IR Models

Modeling the IR-process is complex, because we are dealing with imprecise, vague, and uncertain elements, and thus, it is difficult to formalize due to high influences of human factors, i.e., relevance and information needs, which are highly subjective and context specific. However, in the definition of any IR-model we can identify some common aspects (Canfora and Cerulo 2004). The first step is the representation of documents and information needs. From

(continued)

(continued)

these representations a reasoning strategy can be defined, which solves a representation similarity problem to compute the relevance of documents with respect to the queries. Various strategies have been introduced with the aim of improving the IR-process. We classify these methodologies under two main aspects: Representation (query and document, see Slide 4-33) and Reasoning (application of diverse methods, see Slide 4-34).

Let the IR Model be a quadruple

$$IR = D, Q, \mathcal{F}, R(q_i, d_j)\} \qquad (4\text{-}1)$$

D is a set composed of logical views (representation component) of the documents within a collection;

Q is a set of logical views (representation component) of the user information needs (these are called queries);

$\mathcal{F}$ is a framework for modeling document representations, queries and their relationships (reasoning component); This includes sets and Boolean relations, vectors and linear algebra operations, sample spaces and probability distributions;

R (q_i, d_j) is a ranking function (Slide 4-31) that associates a real number with a query representation $q_i \in Q$ and a document representation $d_j \in D$. Such ranking defines an ordering among the docs with regard to the query q_i.

The end user in Slide 4-29 formulates his query in form of a text operation, the next step is the representation (logical view D) of the documents and the representation of the reasoning strategy, both logical views D and Q (compare with Slide 4-31) result in a ranking of the retrieved documents.

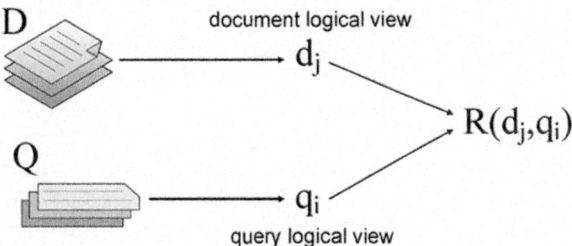

Fig. 2 See Slide 4-31

Slide 4-31 Logical View (Query Q and Document D)

The logical views D and Q result in the ranking function R (q_i, d_j) according to Baeza-Yates and Ribeiro-Neto (2011).

Slide 4-32: A Famous Example

Let $L_{ij} = 1$ if page j is pointing to page i, and zero otherwise. Let $c_j = \sum_{i=1}^{N} L_{ij}$ equal the number of pages pointed to by page j (this is the number of outgoing links). The so-called Google *PageRanks* p_i are defined by the recursive relationship:

$$p_i = (1-d) + d \sum_{j=1}^{N} \left(\frac{L_{ij}}{c_j}\right) p_j$$

where d is a positive constant (and apparently set to 0.85). The idea is now that the importance of page i equals to the sum of the importance of all pages that point to that page. The sums are weighted by $1/c_j$, i.e., each page distributes a total vote of 1 to other pages. The constant d ensures that each page gets a *PageRank* of at least $1 - d$. In matrix notation this can be written as:

$$p = (1-d)e + d \cdot \mathbf{LD}_c^{-1} p$$

where $\mathbf{e}$ is a vector out of N and $\mathbf{D}_c = diag(\mathbf{c})$ is a diagonal matrix with diagonal elements c_j. Introducing the normalization $eTp = N$ (i.e., the average *PageRank* is 1) it follows:

$$p = \left[(1-d)ee^T/N + d\mathbf{LD}_c^{-1}\right]p = \mathbf{A}p$$

matrix $\mathbf{A}$ is the expression within the square braces.

Exploiting a connection with Markov chains, it can be shown that the matrix $\mathbf{A}$ has a real eigenvalue equal to one, and one is its largest eigenvalue, i.e., we can find $\hat{\mathbf{p}}$ by the power method, starting with $\mathbf{p} = \mathbf{p}_0$ we can iterate

$$p_k \leftarrow Ap_{k-1}; p_k \leftarrow N \frac{p_k}{e^T p_k}$$

The fixed points p are the desired *PageRanks*. Interestingly, in the original paper of Page et al. (1999), the authors considered *PageRank* as a model of user behavior, where a random Web surfer clicks on links at random, without regard to content. This means that the surfer does a **random walk** on the Web, choosing among available outgoing links at random, for more details refer to Hastie et al. (2009).

7.5 Taxonomy of IR Models

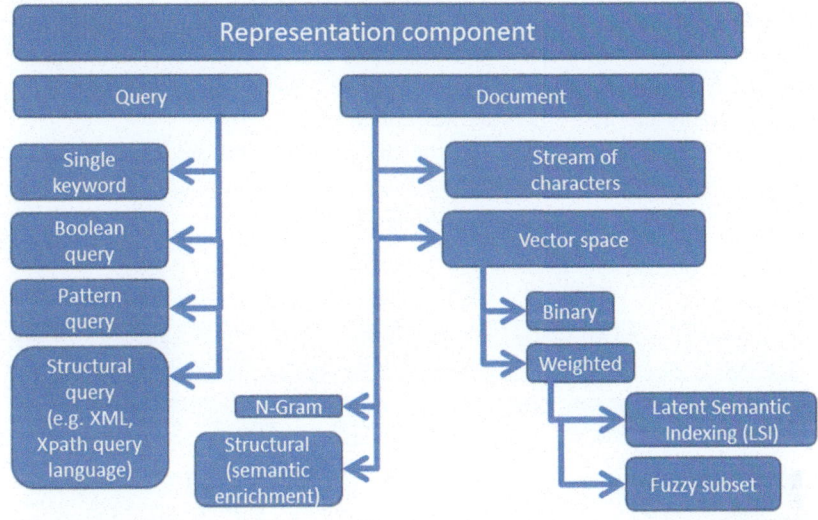

Fig. 3 See Slide 4-33

Slide 4-33: Taxonomy of Information Retrieval Models 1/3

For both the representation (this slide) and the reasoning component (Slide 4-34) we can develop a tree of possible approaches and solutions:

The representation component is an essential part of every IR system, as it is the representation of the information itself (visible to the user): information can be processed if it is represented in an appropriate way. Queries are the representation of information needs of a user.

Note: A text can be characterized by using four attributes: syntax, structure, semantics, and style. A text has a given syntax and a structure, which are usually dictated by the application or by the person who created it. Text also has semantics, specified by the author of the document. Additionally, a document may have a presentation style associated with it, which specifies how it should be displayed or printed. In many approaches to text representation the style is coupled with the document syntax and structure (LaTeX). XML separates the representation of syntax and structures, defined by either a DTD or an XSD, and style, which is captured by XSL (Canfora and Cerulo 2004).

An n-gram is a subsequence of n items from a given sequence. The items in question can be phonemes, syllables, letters, words or base pairs according to the application.

(continued)

(continued)

An n-gram of size 1 is referred to as a "unigram"; size 2 is a "bi-gram" (or, less commonly, a "di-gram"); size 3 is a "tri-gram"; size 4 is a "four-gram" and size 5 or more is simply called an "n-gram." Some language models built from n-grams are "(n − 1)-order Markov models."

An n-gram model is a type of probabilistic model for predicting the next item in such a sequence. n-gram models are used in various areas of statistical natural language processing and genetic sequence analysis.

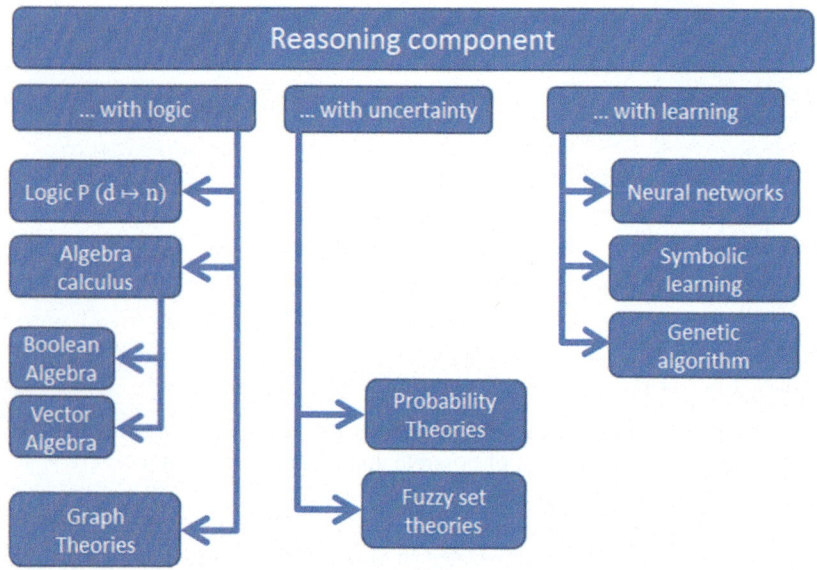

Fig. 4 See Slide 4-34

Slide 4-34: Taxonomy of Information Retrieval Models 2/3

The second part is the reasoning component:

Reasoning refers to the set of methods, models, and technologies used to match document and query representations in the retrieval task. Strictly related with the reasoning component is the concept of relevance. The primary goal of an IR system is to retrieve the documents **relevant to a query**. The reasoning component defines the framework to measure the relevance between documents and queries using their representations (Canfora and Cerulo 2004).

Google, for example, uses a keyword-based vector space model (see Slide 4-38) along with graph-based probability theories and Fuzzy set theories. Slide 4-35 shows a concise overview of some selected methods, according to various document properties.

7 Information Retrieval

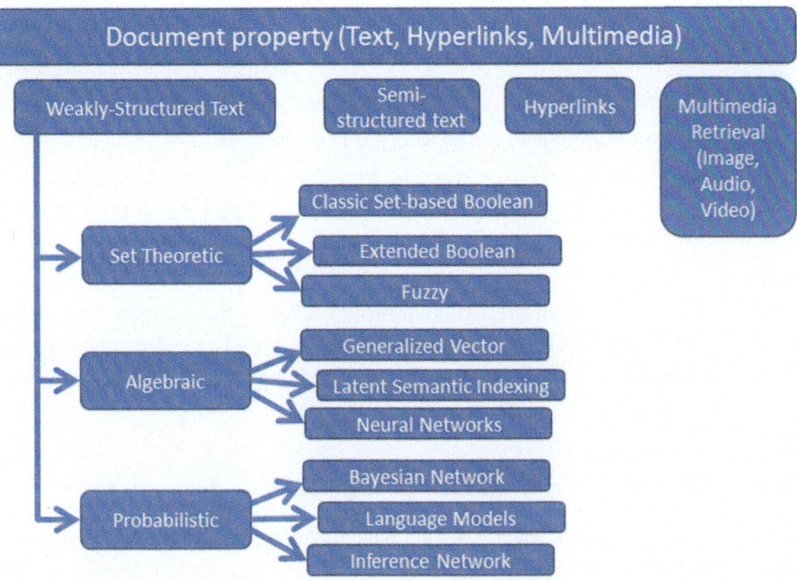

Fig. 5 See Slide 4-35

Slide 4-35: Taxonomy of Information Retrieval Models 3/3

There are many methods of IR, for details consult a standard reference, e.g., Baeza-Yates and Ribeiro-Neto (2011). Set theoretic approaches include the Classic Set-based Boolean, the Extended Boolean and the Fuzzy Approach; Algebraic approaches include the Generalized Vector Model, Latent Semantic Indexing (LSI), Neural Networks; and the Probabilistic approach includes Bayesian Networks, Language Models and Inference Networks. We will discuss only a few and these very briefly, so that you have a quick overview: The set theoretic approach: Boolean Model in Slides 4-36 and 4-37; the Vector Space Model in Slide 4-38 to Slide 4-42; and the Probabilistic Model in Slide 4-43 to Slide 4-44.

7.5.1 Set Theoretic Example: Boolean Model

Slide 4-36: Set Theoretic Example: Boolean Model

Documents/queries are represented as a set of index terms; queries are Boolean expressions (AND, OR, NOT); For the Boolean model, the index term weight variables are binary, i.e., $w_{i,j} \in \{0|1\}$. A query q is a conventional Boolean expression. Let $\vec{q}_{dnf}$ be the disjunctive normal form of the

(continued)

(continued)

query q. Further, let $\vec{q}_{cc}$ be any of the conjunctive components of $\vec{q}_{dnf}$. The similarity of a document d_j to the query q is defined as

$$\sim(d_j, q) = \begin{cases} 1 & \text{if } \exists \ \vec{q}_{cc}\left(\vec{q}_{cc} \in \vec{q}_{dnf}\right) \wedge \left(\forall_{k_i}, g_i(\vec{d}_j) = g_i(\vec{q}_{cc})\right) \\ 0 & \text{otherwise} \end{cases}$$

If $sim(d_j, q) = 1$ then the Boolean model predicts that the document d_j is relevant to query q. Otherwise the prediction is that the document is not relevant. For details please refer to Baeza-Yates and Ribeiro-Neto (2011).

Slide 4-37: Set Theoretic Model: Boolean Model—Pros and Cons

The Boolean Model has several advantages, including easy to understand, exact formalism and the query language is expressive; however, serious disadvantages, e.g., no partial matches, the "bag-of-words" representation does not accurately consider the semantics of documents (Vallet et al. 2005), and the query language is complicated, finally the retrieved documents cannot be ranked.

The Extended Boolean Model (EBM) by Salton et al. (1983) overcomes some disadvantages by making use of partial matching and term weights, similar as in the vector space model. Moreover, as the vector-processing system suffers from one major disadvantage: the structure inherent in the standard Boolean query formulation is absent, the EBM combines the characteristics of the Vector Space Model with the properties of Boolean algebra. Hence, the EBM can also be applied, when the initial query statements are available as natural language formulations of user needs, rather than as conventional Boolean formulations.

7.5.2 Example Algebraic Model: Vector Space Model

Slide 4-38 Example Algebraic Model: Vector Space Model

The vector space model (VSM) represents documents as vectors in the m-dimensional space (Salton et al. 1975).

Let D be a collection of (medical) documents:

$$D = \langle d_1, d_2, \ldots d_n \rangle \qquad (4\text{-}2)$$

(continued)

(continued)

And every document d_i consists of terms:

$$d_i = t_1, t_2, \ldots t_k \ldots \quad (4\text{-}3)$$

Now we carry out a document transformation:

$$w_{i,j} = \begin{cases} 1, t_i \in d_j \\ 0, t_i \notin d_j \end{cases} \to d_j = (0, 1, 1, 0, 1, \ldots, 1)^T \quad (4\text{-}4)$$

Now we count the frequency of the terms and get:

$$w_{i,j} = \begin{cases} \left(1 + \log f_{i,j}\right) * \log \dfrac{N}{n_i}, & \text{if } f_{i,j} > 0 \\ 0 & \text{otherwise} \end{cases} \quad (4\text{-}5)$$

As a result we get a matrix:

$$D_{m \times n} = \begin{pmatrix} w_{1,1} & w_{1,2} & \cdots & w_{1,n-1} & w_{1,n} \\ w_{2,1} & w_{2,2} & & w_{2,n-1} & w_{2,n} \\ \vdots & & \ddots & & \vdots \\ w_{m-1,1} & w_{m-1,2} & & w_{m-1,n-1} & w_{m-1,n} \\ w_{m,1} & w_{m,2} & \cdots & w_{m,n-1} & w_{m,n} \end{pmatrix} \quad (4\text{-}6)$$

Thus, documents can be compared by vector operations and queries can be performed by encoding the query terms similar to the documents in a query vector. This query vector can be compared to each document, which returns a result list by ordering the documents according to the computed similarity. The main task of the vector space representation of documents is to find an appropriate encoding of the feature vector. Each element of a vector usually represents a word (see Slide 4-40) of the document collection. The size of the vector is defined by the number of words of the complete document collection. The easiest way of document encoding is to use binary term vectors, which means a vector element is set to 1 if the corresponding word is used in the document and to 0 if the word is not (4-4). This encoding results in a simple Boolean comparison. To improve the performance usually term weighting schemes are used, where the weights reflect the importance of a word in a specific document of the considered collection. Large weights are assigned to terms that are used frequently in relevant documents but rarely in the whole document collection (Salton and Buckley 1988). Thus, a weight w (d; t) for a term t in document d is computed by term frequency tf (d; t) times inverse document frequency idf(t), which describes the term specificity within the document collection.

(continued)

(continued)

The ranking can be made by using the Cosines Similarity (see Slide 4-41). The cosine of the angle between two vectors is a measure of how "similar" they are, which in turn, is a measure of the similarity of these strings. If the vectors are of unit length, the cosine of the angle between them is simply the dot product of the vectors (Tata and Patel 2007).

$$D_{m \times n} = \begin{Bmatrix} w_{1.1} & w_{1.2} & \cdots & w_{1.n-1} & w_{1.n} \\ w_{2.1} & w_{2.2} & & w_{2.n-1} & w_{2.n} \\ \vdots & & \ddots & & \vdots \\ w_{m-1.1} & w_{m-1.2} & & w_{m-1.n} & w_{m-1.n} \\ w_{m.1} & w_{m.2} & \cdots & w_{m.n-1} & w_{m.n} \end{Bmatrix}$$

Fig. 6 See Slide 4-39

Slide 4-39: As a Result We Get a Matrix ...

As a result we get a matrix representation, and now we can apply vector algebra, or particular linear algebra—here still in R^3. Mathematically, we can work in arbitrarily high dimensional spaces. The major problem involved is the mapping back into R^2.

One very positive aspect is that we can look for getting sparse matrices, i.e., we save a lot of computational power.

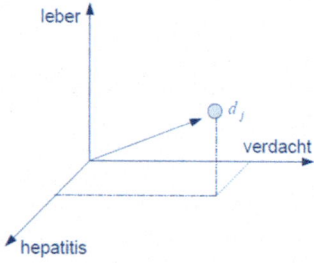

Fig. 7 See Slide 4-40

Slide 4-40: dj Can Thus Be Seen as a Point in n-dim Space

Remember that, computationally mathematically we can produce arbitrarily high dimensional spaces.

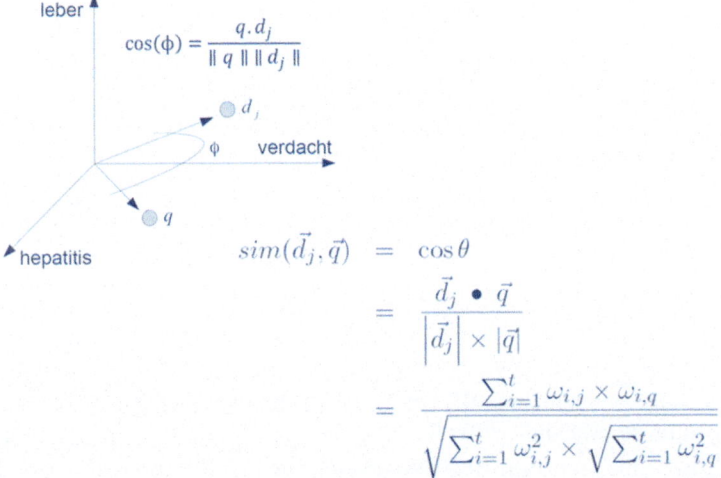

Fig. 8 See Slide 4-41

Slide 4-41: Use the cos-Similarity for Ranking Similar Docs

The ranking can be made by using the Cosines Similarity (Fig. 8). The cosine of the angle between two vectors is a measure of how "similar" they are, which in turn, is a measure of the similarity of these strings. If the vectors are of unit length, the cosine of the angle between them is simply the dot product of the vectors (Tata and Patel 2007).

Slide 4-42: Algebraic Vector Space Model: Pros and Cons

The advantages of the algebraic VSM include that it is easy to understand, partial matches are possible, documents can be sorted by rank, and it uses term-weighting schemes; on the other side there is a higher computational effort to calculate similarity, and the "bag-of-words" representation does not accurately consider the semantics of documents (Vallet et al. 2005).

7.5.3 Example: Probabilistic Model (Bayes' Rule)

$$sim(d_j, q) = \frac{P(R|\vec{d_j})}{P(\overline{R}|\vec{d_j})}$$

$$sim(d_j, q) = \frac{P(\vec{d_j}|R) \times P(R)}{P(\vec{d_j}|\overline{R}) \times P(\overline{R})}$$

Rev. Thomas Bayes (1702-1761)

$$sim(d_j, q) \sim \frac{(\prod_{g_i(\vec{d_j})=1} P(k_i|R)) \times (\prod_{g_i(\vec{d_j})=0} P(\overline{k_i}|R))}{(\prod_{g_i(\vec{d_j})=1} P(k_i|\overline{R})) \times (\prod_{g_i(\vec{d_j})=0} P(\overline{k_i}|\overline{R}))}.$$

Fig. 9 See Slide 4-43

Slide 4-43: Example: Probabilistic Model (Bayes' Rule)

For the probabilistic model, the index weight variables are all binary, i.e., $\omega ij \in [0, 1]$, $\omega iq \in [0, 1]$. A query q is a subset of index terms. Let R be the set of documents known (or initially guessed) to be relevant. Let $\overline{R}$ be the complement of R (this is the set of nonrelevant documents). Let P(R/dj) be de probability that the document dj is relevant to the query q and $P(\overline{R}/dj)$ be the probability that dj is non relevant to q. The similarity sim(dj,q) of the document dj to the query q is defined as the ratio.

Slide 4-44: Probabilistic Model: Pros and Cons

As in all models we have certain pros and cons, the probabilistic model has a big advantage: the documents can be ranked by relevance; however, on the disadvantageous side it is a binary model (binary weights), the index terms are assumed to be independent and lack of document normalization and there is a need to guess the initial separation of documents into relevant and nonrelevant sets.

7 Information Retrieval

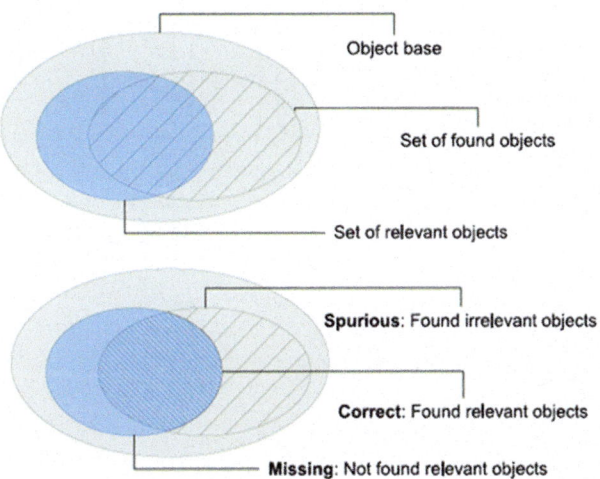

$$Recall = \frac{Correct}{Correct + Missing} \qquad Precision = \frac{Correct}{Correct + Spurious}$$

Fig. 10 See Slide 4-45

Slide 4-45: Measuring the Quality of Information Retrieval

Precision P is the fraction of retrieved documents that are relevant to the search:

$$P = \frac{|\{set\ of\ relevant\ docs\} \cap \{set\ of\ found\ docs\}|}{\{set\ of\ found\ docs\}} \qquad (4\text{-}7)$$

Recall R is the fraction of the documents that are relevant to the query that are successfully retrieved:

$$R = \frac{|\{set\ of\ relevant\ docs\} \cap \{set\ of\ found\ docs\}|}{\{set\ of\ relevant\ docs\}} \qquad (4\text{-}8)$$

A combination of precision and recall is the harmonic mean of both, which is called F-measure:

$$F = 2 \cdot \frac{P \cdot R}{P + R} \qquad (4\text{-}9)$$

In classification five terms are used: true positives (=correct); true negatives (=correct); false positives (=spurious); false negatives (=spurious); not detected (=missing).

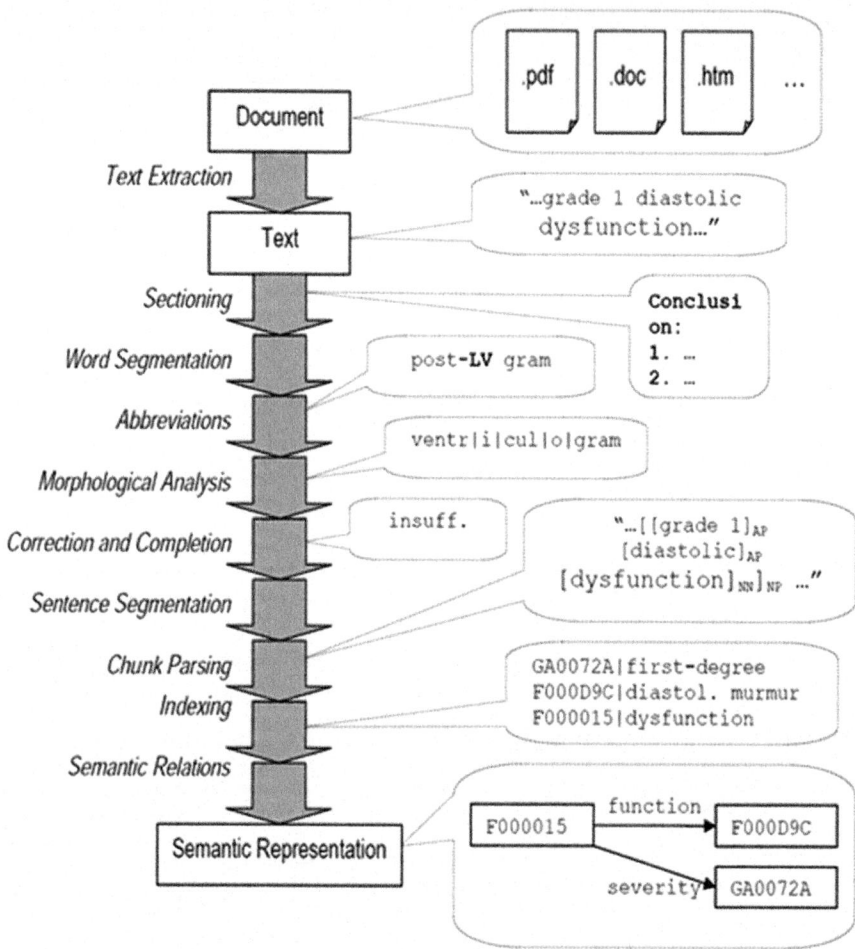

Fig. 11 See Slide 4-46

Slide 4-46: Example: Linguistic Processing Pipeline

In this slide we see an overview of the linguistic processing pipeline that describes the steps that are performed from the document to its semantic representation.

The domain knowledge used in the semantic retrieval system is modeled in the form of the medical semantic network ID MACSR (MSN). It uses the Wingert Nomenclature (WNC) as its medical terminology. The WNC is

(continued)

(continued)

based on the German version of SNOMED developed by Friedrich Wingert. Although its main focus is on German, it, to a lesser extent, supports several other languages including English and French. The MSN forms a simple ontology whose concepts are organized in a taxonomy (isA-hierarchy) and a merology (anatomical partOfhierarchy). Further relations between concepts are modeled by labeled edges. The MSN is divided into several subdomains, including:

- Topography (i.e., anatomical concepts)
- Morphology (e.g., fracture, fever)
- Function (e.g., respiration)
- Diseases (e.g., glaucoma)
- Agents (e.g., pathogens, pharmaceutical substances)

Currently, the MSN contains more than 90,000 terms and 300,000 unique relations.

The query language follows a simple grammar, namely:
Query::= Disjunction
Disjunction::= Conjunction | Conjunction ";" Disjunction
Conjunction::= Atom | Atom ","Conjunction
Atom::= Term | "!" Term

Thus, a query forms a Boolean expression in disjunctive form over search terms. Semantic query expansion has been discussed in several previous work (Kingsland et al. 1993; Aronson et al. 1994; Efthimiadis 1996). The approach is as follows: each search term is indexed (using the linguistic processing methods described above) and replaced by the identifier of the WNC concept matching the term. These concept identifiers are called WNC indices. If the search term refers to a combination of several concepts in the WNC (e.g., Gastroparesis = Stomach + Paresis), the search term is replaced by a conjunction of the WNC (Kreuzthaler et al. 2011).

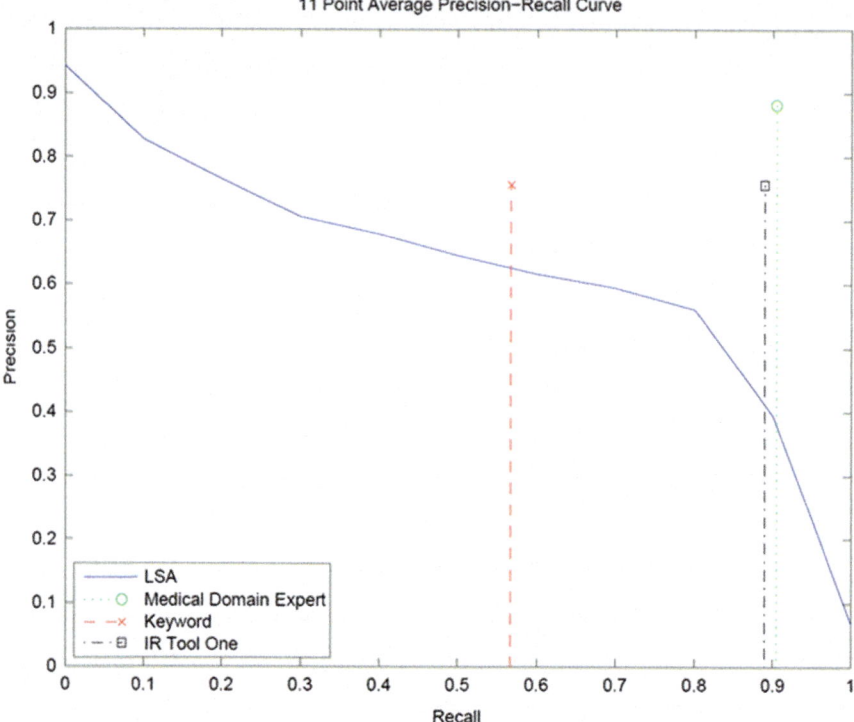

Fig. 12 See Slide 4-47

Slide 4-47: Example: Point Average Precision-Recall Graph

As can be seen from this slide the medical domain expert outperforms the other retrieval methods, achieving high precision at a high recall level. Interestingly, the semantic-based information retrieval tool achieves approximately the same recall level as the medical domain expert while having a lower precision value. This performance result is good, remembering the fact what effort the medical domain expert has to make to translate the information need into a query string. In contrast to this, the input for the information retrieval tool is short and clear so therefore less effort has to be made to transform the information need to the query language understood by the information retrieval tool. Keyword search has a high precision value but a lower recall value. This result is clear when considering the fact that information needs that can be described by using these keyword(s) will achieve a high precision value. So if documents are found, they will be relevant, but the recall level will generally suffer. Looking at the Slide 4-47, keyword search achieves approximately the same precision as IR Tool One but a far worse

(continued)

(continued)

recall. It is also possible that no search results are found at all when using the keyword search methodology. In contrast to this, for this information need, IR Tool One has about the same precision recall levels as the medical domain expert, reflecting the semantic processing chain of the tool. The LSA statistical retrieval method has, when compared to the other methods, a lower precision for all measured recall levels. This result gives the impression that LSA is applicable for getting high precision values for a particular amount of search results but hard to use to achieve both high precision and high recall values, which is needed for example in clinical studies (Kreuzthaler et al. 2011).

8 Future Outlook

Slide 4-48: Big Data: A Growing Torrent in the Future

"Prediction is very difficult, especially if it's about the future."—Niels Bohr

Future aspects include most of all aspects of data integration, data fusion, and preprocessing of heterogeneous datasets. Information Retrieval (retrieving objects from a database) is important and Knowledge Discovery (finding patterns and novel insights into n data which were previously unknown) will become more and more important to deal with the flood of data. In the health area the use of secondary data for research issues are becoming important and the production of open datasets to support international collaboration, e.g., to fight cancer.

The ultimate vision of the data integration field is to be able to integrate large numbers of sources with large amounts of data. Ultimately, the ability to scale up search resources, to spend more processing power to get more accurate matches and more extensive searches, will require the redesign of many algorithms to exploit the power of large clusters. Problems including schema matching, entity resolution, data cleaning, indexing, etc. will need to be tackled in a much more parallelizable and scalable way (Doan et al. 2012).

9 Exam Questions

9.1 Yes/No Decision Questions

Please check the following sentences and decide whether the sentence is true = YES; or false = NO; for each correct answer you will be awarded 2 credit points.

01	Regular, recurring arrangements in space of adjacent amino acid residues in a polypeptide chain are described by the secondary structure	☐ Yes ☐ No	2 total
02	Ubiquitous computing is a post-desktop model of Human–Computer Interaction in which the information processing is integrated into everyday objects.	☐ Yes ☐ No	2 total
03	The central dogma of molecular biology states that DNA is transcribed into RNA and translated into protein: DNA → RNA → Protein → Cellular Phenotype.	☐ Yes ☐ No	2 total
04	Proteins are molecules consisting of one or more chains of chromosomes and they vary from one to another mainly in their sequence of these chromosomes.	☐ Yes ☐ No	2 total
05	X-ray crystallography is a standard method to analyze the arrangement of objects (atoms, molecules) within a crystal structure.	☐ Yes ☐ No	2 total
06	The Church–Turing thesis states that Turing machines capture the informal notion of effective methods in logic and mathematics, and provide a precise definition of an algorithm	☐ Yes ☐ No	2 total
07	Within a Von Neumann Structure, the internal memory executes the instructions with help of the logic controller, the "heart" of the computer.	☐ Yes ☐ No	2 total
08	Biomarkers are measured molecules which indicate the presence of an abnormal condition within a patient, and can be a gene (e.g., SNP), protein (e.g., prostate-specific antigen), or metabolite.	☐ Yes ☐ No	2 total
09	Eukaryotic cells are usually single-celled organisms, while prokaryotic cells can be either single-celled or part of multicellular organisms..	☐ Yes ☐ No	2 total
10	Total Quality Management (TQM) provides a useful, simple yet important definition of quality: "consistently meeting customer's expectations.	☐ Yes ☐ No	2 total

Sum of Question Block A (max. 20 points)	

9 Exam Questions

9.2 Multiple Choice Questions (MCQ)

The following questions are composed of two parts: the stem, which identifies the question or problem and a set of alternatives which can contain 0, 1, 2, 3, or 4 correct answers, along with a number of distractors that might be plausible—but are incorrect. Please **select the correct answers** by ticking ☒—and do not forget that it can be none. Each question will be awarded 4 points *only if everything is correct*.

01	Blood as biomarker has the advantage ... ☐ a) ... of having a wide range of protein concentrations. ☐ b) ... that it is available in large quantities. ☐ c) ... that it can be collected non-invasive. ☐ d) ... provides a good reflection of the physiologic state of the body.	4 total
02	Living "things" are able ... ☐ a) ... to reproduce. ☐ b) ... to grow ☐ c) ... to process information. ☐ d) ... to self-replicate.	4 total
03	At the beginning of medical informatics in the 1970ies .... ☐ a) ... the focus was mostly on data acquisition, storage and accounting. ☐ b) ... there was an emerging trend to use Web based health applications. ☐ c) ... personalized medicine was already part of some health strategies. ☐ d) ... the first end-user programmable Mash-ups were introduced.	4 total
04	The "Quality Era" of biomedical informatics is mainly characterized by ☐ a) ... focus on data processing and storage. ☐ b) ... health care networks, telemedicine and CPOE-Systems. ☐ c) ... pervasive and ubiquitous computing technologies. ☐ d) ... patient empowerment and individual molecular medicine.	4 total
05	The problems in Biomedical Informatics are rooted mostly in the ... ☐ a) ... high dimensionality and complexity of the data. ☐ b) ... non-standardization of most of the data available. ☐ c) ... lacking memory capacity and computational power. ☐ d) ... heterogeneity and weak structurization of the available data.	4 total
06	Part of the definition of Biomedical Informatics is the ... ☐ ... effective use of biomedical data. ☐ ... motivation to improve computational capacities. ☐ ... effort to expand the technological capabilities. ☐ ... motivation to improve human health.	4 total
07	Tissue is defined as ... ☐ a) ... cellular organization between cells and a complete organism. ☐ b) ... an essential component of the human cardio-vascular system. ☐ c) ... part of the digestive system, e.g. esophagus, stomach and intestines. ☐ d) ... an ensemble of similar cells, carrying out a function as collective.	4 total
08	A Von Neumann Machine ... ☐ a) ... is theoretically equivalent to a Universal Turing Machine. ☐ b) ... is today present in nearly all of our "computers". ☐ c) ... uses a central processing unit and a single separate memory. ☐ d) ... can process continuous variables and multiple instructions.	4 total

09	Epigenetics ... ☐ a) ... is the study beyond protein folding, hence called epi-genetics. ☐ b) ... is the study of changes in gene expression or cellular phenotypes,. ☐ c) ... concerns mechanisms other than changes in the underlying DNA ☐ d) ... provides a genetical snapshot of the cell.	4 total
10	Chromatography ... ☐ a) ... is the collective term for the separation of mixtures. ☐ b) ... is the primary method for determining the atomic structure. ☐ c) ... separates molecules according to their charge and their mass. ☐ d) ... typically includes Gas-Chromatography and Liquid-Chromatography.	4 total

Sum of Question Block B (max. 40 points)

9 Exam Questions

9.3 Free Recall Block

Please follow the instructions below. At each question you will be assigned the credit points indicated if your option is correct (partial points may be given).

01	Please sketch the principle tertiary structure of a protein and indicate the α-helix and the β-sheet:	1-15 1 each 3 total
02	The Von-Neumann Architecture is the fundamental computer organization structure of nearly all of our todays computing systems (e.g. in your PC, smartphone, microwave oven, car, etc.), please roughly sketch the Von Neumann Architecture and indicate the main parts:	1-28 1 each 6 total

03	Please draw the overlapping circles of the three fields: Medicine – Genomics – Informatics, and indicate the areas of Bioinformatics – Medical Informatics – Biomedical Informatics and Clinical Genomics:	1/42 1 each 7 total
04	It is important to recognize the dimensions where the data comes from and the respective fields dealing with it. Please complete the following image: 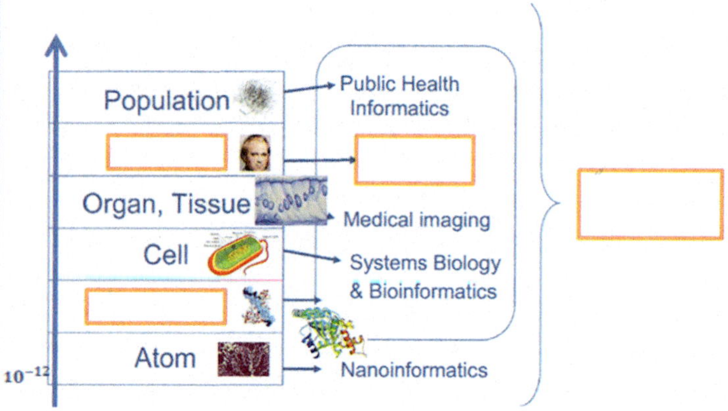 Population → Public Health Informatics Organ, Tissue → Medical imaging Cell → Systems Biology & Bioinformatics Atom → Nanoinformatics 10^{-12}	1/47 1 each 4 total

| 05 | Protein analytics is a primary source of data for research. Please assign the correct labels to the methods below (just by drawing a connecting line between the correct items): | 1/16
1 each
3 total |

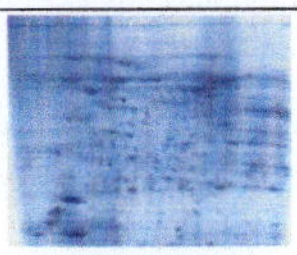

	[gel image]	Liquid Chromatography
	[diffraction pattern]	Gel Electrophoresis
	[chromatogram]	X-Ray Crystallography

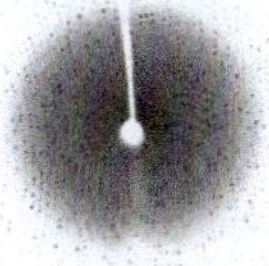

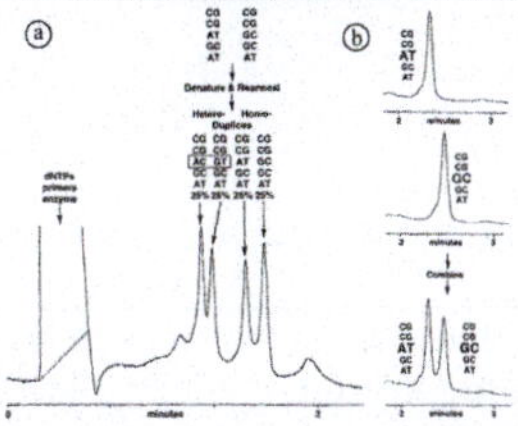

Sum of Question Block C (max. 40 points)

10 Answers

10.1 Answers to the Yes/No Questions

Please check the following sentences and decide whether the sentence is true = YES; or false = NO; for each correct answer you will be awarded 2 credit points.

01	Regular, recurring arrangements in space of adjacent amino acid residues in a polypeptide chain are described by the secondary structure	☒ Yes ☐ No	2 total
02	Ubiquitous computing is a post-desktop model of Human–Computer Interaction in which the information processing is integrated into everyday objects.	☒ Yes ☐ No	2 total
03	The central dogma of molecular biology states that DNA is transcribed into RNA and translated into protein: DNA → RNA → Protein → Cellular Phenotype.	☒ Yes ☐ No	2 total
04	Proteins are molecules consisting of one or more chains of chromosomes and they vary from one to another mainly in their sequence of these chromosomes.	☐ Yes ☒ No	2 total
05	X-ray crystallography is a standard method to analyze the arrangement of objects (atoms, molecules) within a crystal structure.	☒ Yes ☐ No	2 total
06	The Church–Turing thesis states that Turing machines capture the informal notion of effective methods in logic and mathematics, and provide a precise definition of an algorithm	☒ Yes ☐ No	2 total
07	Within a Von Neumann Structure, the internal memory executes the instructions with help of the logic controller, the "heart" of the computer.	☐ Yes ☒ No	2 total
08	Biomarkers are measured molecules which indicate the presence of an abnormal condition within a patient, and can be a gene (e.g., SNP), protein (e.g., prostate-specific antigen), or metabolite.	☒ Yes ☐ No	2 total
09	Eukaryotic cells are usually single-celled organisms, while prokaryotic cells can be either single-celled or part of multicellular organisms..	☐ Yes ☒ No	2 total
10	Total Quality Management (TQM) provides a useful, simple yet important definition of quality: "consistently meeting customer's expectations.	☒ Yes ☐ No	2 total

Sum of Question Block A (max. 20 points)	

10.2 Answers to the Multiple Choice Questions (MCQ)

01	Mega issues related with hospital information systems include ... ☒ a) ... data mining. ☒ b) ... data integration. ☒ c) ... data security. ☒ d) ... data fusion.	4 total
02	The diagnostic-therapeutic cycle encompasses ... ☒ a) ... decreasing of uncertainty. ☒ b) ... looking into the past. ☒ c) ... reasoning. ☒ d) ... action.	4 total
03	Warehoused data is produced mainly by .... ☒ a) ... Biobanks. ☐ b) ... Microarrays. ☐ c) ... Clinical practice genomics. ☐ d) ... relational drug design.	4 total
04	The "Quality Era" of biomedical informatics is mainly characterized by ☐ a) ... focus on data processing and storage. ☐ b) ... health care networks, telemedicine and CPOE-Systems. ☐ c) ... pervasive and ubiquitous computing technologies. ☒ d) ... patient empowerment and individual molecular medicine.	4 total
05	Ensembl ... ☐ a) ... is the collection of biological data in an organized way. ☐ b) ... is the name for a famous protein database. ☒ c) ... provides a genome browser for the retrieval of genomic information. ☒ d) ... is the name of genome database.	4 total
06	Information Retrieval is defined as ... ☒ ... satisfy the search query by an end user. ☐ ... searching for patterns of interest. ☐ ... finding something which was previously unknown. ☒ ... bringing back an already known object.	4 total
07	Data retrieval ... ☒ a) ... is deterministic. ☐ b) ... can be incomplete. ☐ c) ... is probabilistic. ☒ d) ... is deductive.	4 total
08	In the formal description of IR models ... ☒ a) ... D is a set composed of logical views. ☒ b) ... Q is a set of logical views of the user information needs. ☐ c) ... F is the ranking function that associates a real number with a query. ☐ d) ... R (qi, dj) is a framework for modeling document representations.	4 total
09	Set Theoretic approaches include ... ☐ a) ... Language Models. ☒ b) ... Extended Boolean. ☒ c) ... Fuzzy approaches ☐ d) ... Neural Networks.	4 total

10	Advantages of the VSM include ... ☒ a) ... that partial matches are possible. ☐ b) ... that the query language is expressive. ☐ c) ... documents can be ranked by relevance. ☒ d) ... documents can be sorted by rank.	4 total

Sum of Question Block B (max. 40 points)

10 Answers 197

10.3 Answers to the Free Recall Questions

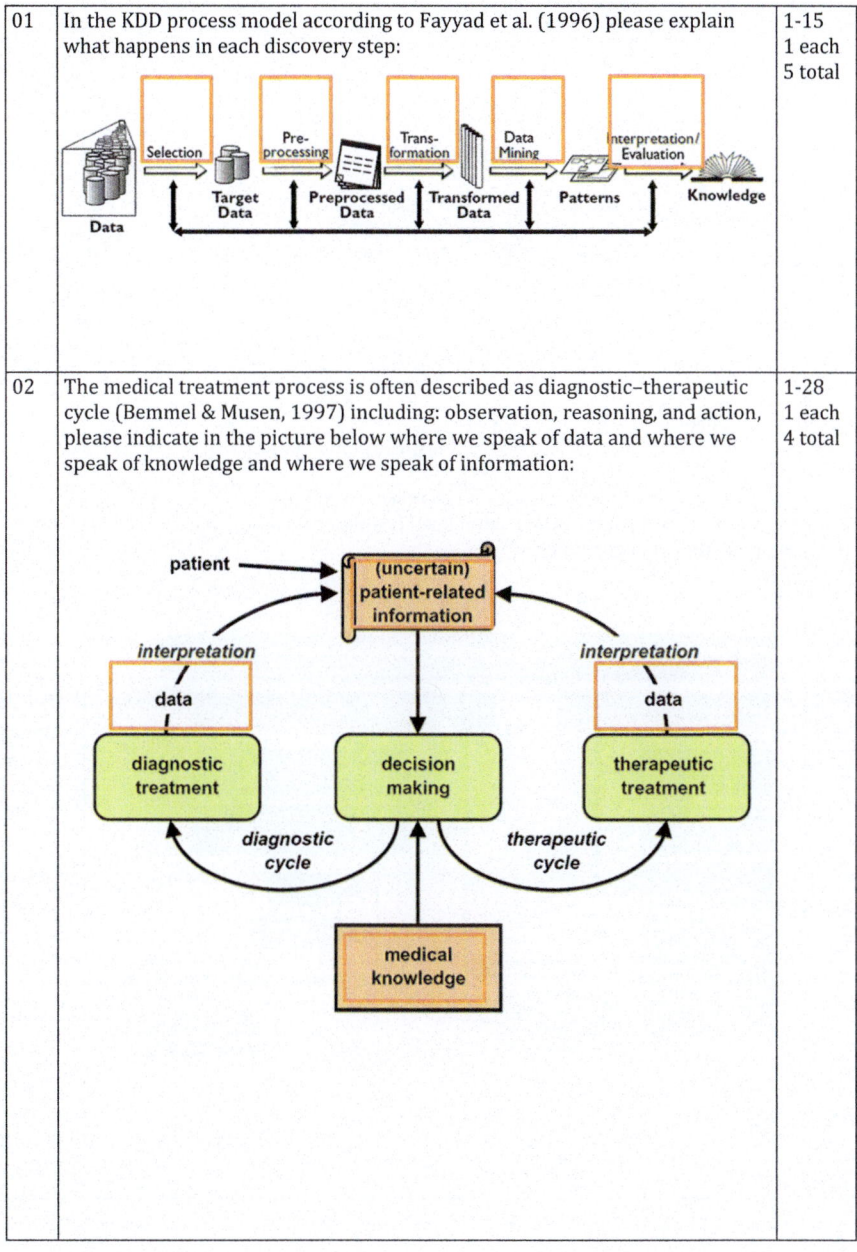

01	In the KDD process model according to Fayyad et al. (1996) please explain what happens in each discovery step:	1-15 1 each 5 total
02	The medical treatment process is often described as diagnostic–therapeutic cycle (Bemmel & Musen, 1997) including: observation, reasoning, and action, please indicate in the picture below where we speak of data and where we speak of knowledge and where we speak of information:	1-28 1 each 4 total

03	In the IR Process principle below please fill in the missing terms: 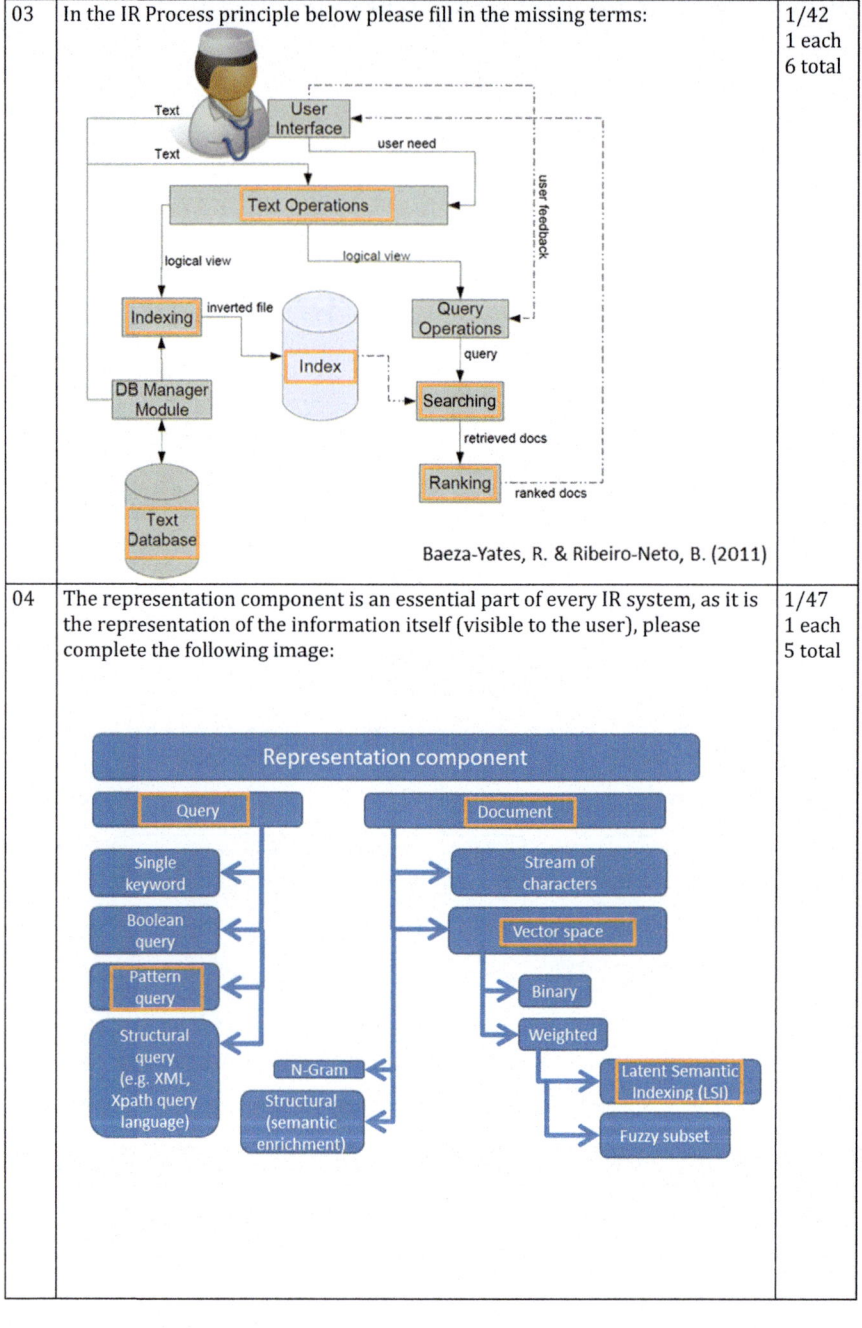 Baeza-Yates, R. & Ribeiro-Neto, B. (2011)	1/42 1 each 6 total
04	The representation component is an essential part of every IR system, as it is the representation of the information itself (visible to the user), please complete the following image:	1/47 1 each 5 total

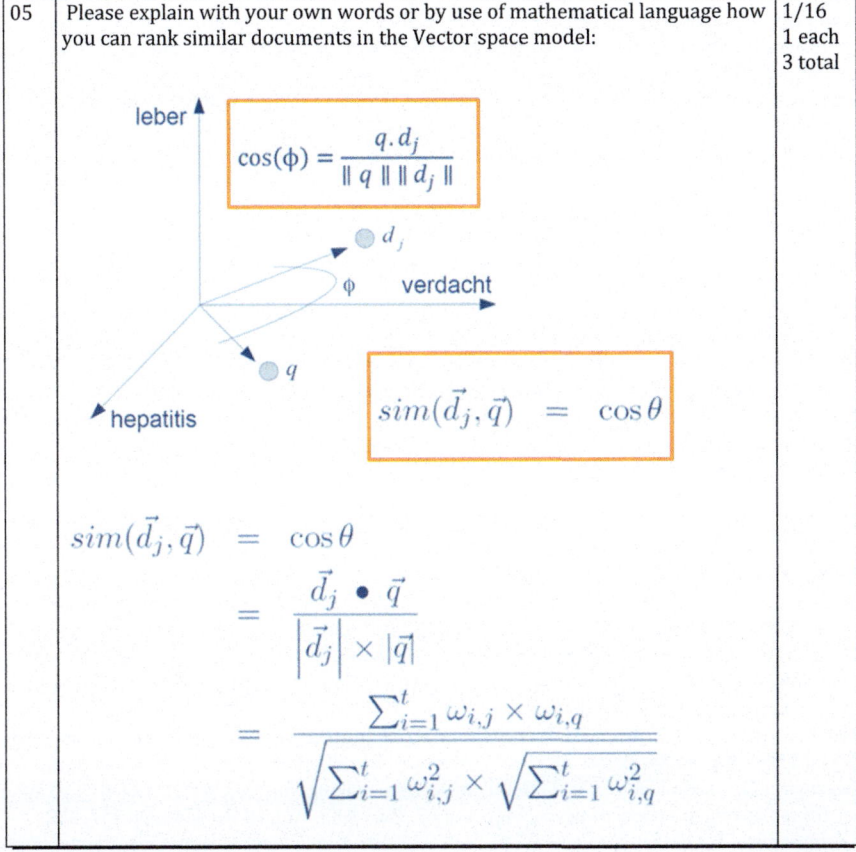

References

Aronson A, Rindflesch T, Browne A (1994) Exploiting a large thesaurus for information retrieval RIAO Recherche d'Information assistee par Ordinateur—Computer-aided information retrieval, pp 197–216

Baeza-Yates R, Ribeiro-Neto B (2011) Modern information retrieval: the concepts and technology behind search. Pearson, Harlow

Bastiao-Silva LA, Costa C, Silva A, Oliveira JL (2011) A PACS gateway to the cloud. 6th Iberian conference on information systems and technologies (CISTI), pp 1–6

Bemmel JHV, Musen MA (1997) Handbook of medical informatics. Springer, Heidelberg

Berman JJ (2013a) Chapter 2—Identification, deidentification, and reidentification. In: Berman JJ (ed) Principles of big data. Morgan Kaufmann, Boston, pp 15–33

Berman JJ (2013b) Chapter 4—Introspection. In: Berman JJ (ed) Principles of big data. Morgan Kaufmann, Boston, pp 49–61

Bleiholder J, Naumann F (2008) Data fusion. ACM Comput Surv 41(1):1
Blum RL, Wiederhold GC (1985) Studying hypotheses on a time-oriented clinical database: an overview of the RX project. Computer-assisted medical decision making. Springer, New York, NY, pp 242–253
Brazma A, Hingamp P, Quackenbush J, Sherlock G, Spellman P, Stoeckert C, Aach J, Ansorge W, Ball CA, Causton HC, Gaasterland T, Glenisson P, Holstege FCP, Kim IF, Markowitz V, Matese JC, Parkinson H, Robinson A, Sarkans U, Schulze-Kremer S, Stewart J, Taylor R, Vilo J, Vingron M (2001) Minimum information about a microarray experiment (MIAME) [mdash]toward standards for microarray data. Nat Genet 29(4):365–371
Campbell M, Hoane AJ, Hsu F-H (2002) Deep blue. Artif Intell 134(1):57–83
Canfora G, Cerulo L (2004) A taxonomy of information retrieval models and tools. J Comput Inf Technol 12(3):175–194
Chute CG, Beck SA, Fisk TB, Mohr DN (2010) The Enterprise Data Trust at Mayo Clinic: a semantically integrated warehouse of biomedical data. J Am Med Inform Assoc 17(2):131–135
Doan A, Halevy A, Ives Z (2012) 19—The future of data integration. In: Doan A, Halevy A, Ives Z (eds) Principles of data integration. Morgan Kaufmann, Boston, pp 453–457
Efthimiadis E (1996) Query expansion. In: Williams ME (ed) Annual review of information systems and technology (ARIST), Vol 31, pp 121–187
Fayyad U, Piatetsky-Shapiro G, Smyth P (1996) The KDD process for extracting useful knowledge from volumes of data. Commun ACM 39(11):27–34
Flicek P, Amode MR, Barrell D, Beal K, Brent S, Chen Y, Clapham P, Coates G, Fairley S, Fitzgerald S, Gordon L, Hendrix M, Hourlier T, Johnson N, Kahari A, Keefe D, Keenan S, Kinsella R, Kokocinski F, Kulesha E, Larsson P, Longden I, Mclaren W, Overduin B, Pritchard B, Riat HS, Rios D, Ritchie GR, Ruffier M, Schuster M, Sobral D, Spudich G, Tang YA, Trevanion S, Vandrovcova J, Vilella AJ, White S, Wilder SP, Zadissa A, Zamora J, Aken BL, Birney E, Cunningham F, Dunham I, Durbin R, Fernandez-Suarez XM, Herrero J, Hubbard TJ, Parker A, Proctor G, Vogel J, Searle SM (2011) Ensembl 2011. Nucleic Acids Res 39(Database issue):D800–D806
Gardner RM, Pryor TA, Warner HR (1999) The HELP hospital information system: update 1998. Int J Med Inform 54(3):169–182
Gigerenzer G, Gaissmaier W (2011) Heuristic decision making. In: Fiske ST, Schacter DL, Taylor SE (eds) Annual review of psychology, vol 62. Palo Alto, Annual Reviews, pp 451–482
Gschwandtner T, Gärtner J, Aigner W, Miksch S (2012) A taxonomy of dirty time-oriented data. In: Quirchmayr G, Basl J, You I, Xu L, Weippl E (eds) Multidisciplinary research and practice for information systems. Springer, Berlin, pp 58–72
Hastie T, Tibshirani R, Friedman J (2009) The elements of statistical learning: data mining, inference, and prediction, 2nd edn. Springer, New York
Haux R (2006) Health information systems-past, present, future. Int J Med Inform 75(3–4):268–281
Haux R, Lagemann A, Knaup P, Schmücker P, Winter A (1998) Management von Informationssystemen: Analyse, Bewertung, Auswahl, Bereitstellung und Einführung von Informationssystemkomponenten am Beispiel von Krankenhausinformationssystemen. Teubner, Stuttgart
Hawkins RC (2005) The evidence based medicine approach to diagnostic testing: practicalities and limitations. Clin Biochem Rev 26(2):7
Hearst MA (1999) Untangling text data mining. Proceedings of the 37th annual meeting of the Association for Computational Linguistics on computational linguistics. Association for Computational Linguistics, pp 3–10
Hermjakob H, Palazzi L, Lewington C, Mudali S, Kerrien S, Orchard S, Vingron M, Roechert B, Roepstorff P, Valencia A (2004) IntAct: an open source molecular interaction database. Nucleic Acids Res 32(Suppl 1):D452–D455
Holzinger A (2003) Experiences with User Centered Development (UCD) for the Front End of the Virtual Medical Campus Graz. In: Jacko JA, Stephanidis C (eds) Human-computer interaction, theory and practice. Lawrence Erlbaum, Mahwah, NJ, pp 123–127

Holzinger A (2013) Human–computer interaction & knowledge discovery (HCI-KDD): what is the benefit of bringing those two fields to work together? In: Alfredo Cuzzocrea C, Simos DE, Weippl E, Xu L (eds) Multidisciplinary research and practice for information systems, vol 8127, Lecture notes in computer science (LNCS). Springer, Heidelberg, pp 319–328

Holzinger A, Baernthaler M, Pammer W, Katz H, Bjelic-Radisic V, Ziefle M (2011) Investigating paper vs. screen in real-life hospital workflows: performance contradicts perceived superiority of paper in the user experience. Int J Hum Comput Stud 69(9):563–570

Holzinger A, Geierhofer R, Ackerl S, Searle G (2005) CARDIAC@VIEW: the user centered development of a new medical image viewer. In: Zara J, Sloup J (eds) Central European multimedia and virtual reality conference (available in Eurographics Library). Czech Technical University (CTU), Prague, pp 63–68

Holzinger A, Geierhofer R, Errath M (2007) Semantische Informationsextraktion in medizinischen Informationssystemen. Informatik Spektrum 30(2):69–78

Holzinger A, Simonic K-M (eds) (2011) Information quality in e-Health, vol 7058, Lecture notes in computer science (LNCS). Springer, Heidelberg

Hotho A, Nürnberger A, Paaß G (2005) A brief survey of text mining. GLDV J Comput Linguist Lang Technol 20(1):19–62

Johnston T, Weis R (2010) Managing time in relational databases: how to design, update and query temporal data. Morgan Kaufmann, San Francisco

Kampen AV (2013) Medical bioinformatics and e-bioscience [Online]. http://www.bioinformaticslaboratory.nl/twiki/bin/view/BioLab/EducationMIK1-2. Accessed 22 Apr 2013

Kell DB (2007) The virtual human: towards a global systems biology of multiscale, distributed biochemical network models. IUBMB Life 59(11):689–695

Kingsland LC, Harbourt AM, Syed EJ, Schuyler PL (1993) Coach: applying UMLS knowledge sources in an expert searcher environment. Bull Med Libr Assoc 81:178–183

Kreuzthaler M, Bloice MD, Faulstich L, Simonic KM, Holzinger A (2011) A comparison of different retrieval strategies working on medical free texts. J Univ Computer Sci 17(7):1109–1133

Lenz R, Reichert M (2007) IT support for healthcare processes-premises, challenges, perspectives. Data Knowl Eng 61(1):39–58

Lenzerini M (2002) Data integration: a theoretical perspective. Proceedings of the twenty-first ACM SIGMOD-SIGACT-SIGART symposium on principles of database systems. ACM, pp 233–246

Li C, Donizelli M, Rodriguez N, Dharuri H, Endler L, Chelliah V, Li L, He E, Henry A, Stefan MI, Snoep JL, Hucka M, Le Novere N, Laibe C (2010) BioModels database: an enhanced, curated and annotated resource for published quantitative kinetic models. BMC Syst Biol 4:92

Louie B, Mork P, Martin-Sanchez F, Halevy A, Tarczy-Hornoch P (2007) Data integration and genomic medicine. J Biomed Inform 40(1):5–16

Maimon O, Rokach L (eds) (2010) Data mining and knowledge discovery handbook, 2nd edn. Springer, New York

Masic I, Milinovic K (2012) On-line biomedical databases–the best source for quick search of the scientific information in the biomedicine. Acta Inform Med 20(2):72

Page L, Brin S, Motwani R, Winograd T (1999) The PageRank citation ranking: bringing order to the web. Technical Report Stanford InfoLab

Piatetsky-Shapiro G (2000) Knowledge discovery in databases: 10 years after. ACM SIGKDD Explor Newslett 1(2):59–61

Plattner H (2013) A course in in-memory data management: the inner mechanics of in-memory databases. Springer, Heidelberg

Reichertz PL (2006) Hospital information systems—past, present, future. Int J Med Inform 75(3–4):282–299

Robinson I, Webber J, Eifrem E (2013) Graph databases. O'Reilly Media, Sebastopol

Roden DM, Pulley JM, Basford MA, Bernard GR, Clayton EW, Balser JR, Masys DR (2008) Development of a large-scale de-identified DNA biobank to enable personalized medicine. Clin Pharmacol Therapeut 84(3):362–369

Salton G, Buckley C (1988) Term-weighting approaches in automatic text retrieval. Inf Process Manag 24(5):513–523

Salton G, Fox EA, Wu H (1983) Extended Boolean information retrieval. Commun ACM 26 (11):1022–1036

Salton G, Wong A, Yang CS (1975) A vector space model for automatic indexing. Commun ACM 18(11):613–620

Shortliffe EH, Perrault LE, Wiederhold G, Fagan LM (2001) Medical informatics: computer applications in health care and biomedicine, 2nd edn. Springer, New York

Silberschatz A, Korth HF, Sudarshan S (2010) Database system concepts, 6th edn. McGraw-Hill Hightstown, New York

Simonic K-M, Holzinger A (2010) Zur Bedeutung von Information in der Medizin. OCG J 35(1):8

Sreenivasaiah PK (2010) Current trends and new challenges of databases and web applications for systems driven biological research. Front Physiol 1:147

Stein LD (2010) The case for cloud computing in genome informatics. Genome Biol 11(5)

Tata S, Patel JM (2007) Estimating the selectivity of tf-idf based cosine similarity predicates. ACM SIGMOD Rec 36(2):7–12

Vallet D, Fernández M, Castells P (2005) An ontology-based information retrieval model. In: Gómez-Pérez A, Euzenat J (eds) The semantic web: research and applications. Springer, Berlin, pp 103–110

Van Rijsbergen CJ (1979) Information retrieval, 2nd edn. Butterworths, London

Waterson P, Glenn Y, Eason K (2012) Preparing the ground for the 'paperless hospital': a case study of medical records management in a UK outpatient services department. Int J Med Inform 81(2):114–129

Wiltgen M, Holzinger A (2005) Visualization in bioinformatics: protein structures with physico-chemical and biological annotations. In: Zara J, Sloup J (eds) Central European multimedia and virtual reality conference (available in EG Eurographics Library). Czech Technical University (CTU), Prague, pp 69–74

Wiltgen M, Holzinger A, Tilz GP (2007) Interactive analysis and visualization of macromolecular interfaces between proteins. In: Holzinger A (ed) HCI and usability for medicine and health care, vol 4799, Lecture notes in computer science (LNCS). Springer, Berlin, pp 199–212

Witten IH, Frank E, Hall MA (2011) Data mining: practical machine learning tools and techniques. Morgan Kaufmann, San Francisco

Zhang M, Zhang H, Tjandra D, Wong STC (2004) DBMap: a space-conscious data visualization and knowledge discovery framework for biomedical data warehouse. IEEE Trans Inf Technol Biomed 8(3):343–353

Lecture 5

Semi-structured, Weakly Structured, and Unstructured Data

1 Learning Goals

At the end of this fourth lecture, you:

- would have acquired background knowledge on some issues in standardization and structurization of data;
- would have a general understanding of modeling knowledge in medicine and biomedical informatics;
- would get some basic knowledge on medical Ontologies and be aware of the limits, restrictions, and shortcomings of them;
- would know the basic ideas and the history of the International Classification of Diseases (ICD);
- would have a view on the Standardized Nomenclature of Medicine Clinical Terms (SNOMED CT);
- would have some basic knowledge on Medical Subject Headings (MeSH);
- would be able to understand the fundamentals and principles of the Unified Medical Language System (UMLS).

2 Advance Organizer

Abstraction	Process of mapping (biological) processes onto a series of concepts (expressed in mathematical terms)
Biological system	Collection of objects ranging in size from molecules to populations of organisms, which interact in ways that display a collective function or role
Coding	Any process of transforming descriptions of medical diagnoses and procedures into standardized code numbers, i.e., to track health conditions and for

	reimbursement, e.g., based on Diagnosis Related Groups (DRG)
Data model	Definition of entities, attributes, and their relationships within complex sets of data
DSM	Diagnostic and Statistical Manual for Mental Disorders, a multiaxial, multidimensional categorization of all (known) mental health disorders, used for clinical diagnostics
Extensible Markup Language (XML)	Set of rules for encoding documents in machine-readable form
GALEN	Generalized Architecture for Languages, Encyclopedias and Nomenclatures in Medicine, project aiming at the development of a reference model for medical concepts
ICD	International Classification of Diseases, the archetypical coding system for patient record abstraction (est. 1900)
Medical Classification	Provides the terminologies of the medical domain (or parts of it), 100+ classifications in use
MeSH	Medical Subject Headings is a classification to index the world medical literature and forms the basis for UMLS
Metadata	Data that describes the data
Model	A simplified representation of a process or object, which describes its behavior under specified conditions (e.g., conceptual model)
Nosography	Science of description of diseases
Nosology	Science of classification of diseases
Ontology	Structured description of a domain and formalizes the terminology (concepts-relations, e.g., IS-A relationship provides a taxonomic skeleton), e.g., gene ontology
Ontology engineering	Subfield of knowledge engineering, which studies the methods and methodologies for building ontologies
SNOMED	Standardized Nomenclature of Medicine, est. 1975, multiaxial system with 11 axes
SNOP	Systematic Nomenclature of Pathology (on four axes: topography, morphology, etiology, function), basis for SNOMED
System features	Static or dynamic; mechanistic or phenomenological; discrete or continuous; deterministic or stochastic; single-scale or multi-scale
Terminology	Includes well-defined terms and usage
UMLS	Unified Medical Language System is a long-term project to develop resources for the support of intelligent information retrieval

3 Acronyms

ACR	American College of Radiologists
API	Application Programming Interface
DAML	DARPA Agent Markup Language
DICOM	Digital Imaging and Communications in Medicine
DL	Description Logic
ECG	Electrocardiogram
EHR	Electronic Health Record
FMA	Foundational Model of Anatomy
FOL	First-order logic
GO	Gene Ontology
ICD	International Classification of Diseases
IOM	Institute of Medicine
KIF	Knowledge Interchange Format
LOINC	Logical Observation Identifiers Names and Codes
MeSH	Medical Subject Headings
MRI	Magnetic Resonance Imaging
NCI	National Cancer Institute (US)
NEMA	National Electrical Manufacturer Association
OIL	Ontology Inference Layer (description logic)
OWL	Ontology Web Language
RDF	Resource Description Framework
RDF Schema	A vocabulary of properties and classes added to RDF
SCP	Standard Communications Protocol
SNOMED CT	Systematized Nomenclature of Medicine—Clinical Terms
SOP	Standard Operating Procedure
UMLS	Unified Medical Language System

4 Key Problems

> **Slide 5-1: Mathematically Seen Our World Is Complex and High Dimensional**
>
> Key problems in dealing with data in the life sciences include:
>
> - Complexity of our world
> - High-dimensionality (curse of dimensionality (Catchpoole et al. 2010))
> - Most of the data is weakly structured and unstructured

(continued)

(continued)

A grand challenge in health care is the complexity of data, implicating two issues: structurization and standardization. As we have learned in Lecture 2, very little of the data is structured. Most of our data is weakly structured (Holzinger 2012). In the language of business there is often the use of the word "unstructured," but we have to use this word with care; unstructured would mean—in a strict mathematical sense—that we are talking about total randomness and complete uncertainty, which would mean noise, where standard methods fail or lead to the modeling of artifacts, and only statistical approaches may help. The correct term would be unmodeled data—or we shall speak about **unstructured information**. Please mind the differences.

To the image in Slide 5-1: Advances in genetics and genomics have accelerated the discovery-based (=hypotheses generating) research that provides a powerful complement to the direct hypothesis-driven molecular, cellular, and systems sciences.

For example, genetic and functional genomic studies have yielded important insights into neuronal function and disease. One of the most exciting and challenging frontiers in neuroscience involves harnessing the power of large-scale genetic, genomic, and phenotypic datasets, and the development of tools for data integration and data mining (Geschwind and Konopka 2009).

5 Review on Data

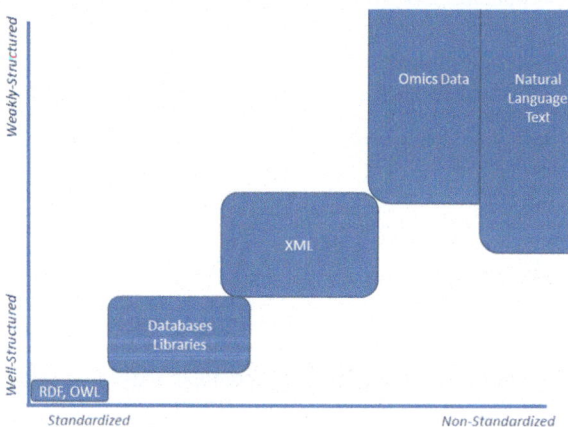

Fig. 1 See Slide 5-2

Slide 5-2: Remember: Standardization Versus Structurization

Before we proceed, please review the four different definitions of data in terms of structurization and standardization: well-structured and standardized data, semi-structured data (e.g., XML), weakly structured data (e.g., Omics data), and "unmodelled data"—unstructured information (text).

Do not confuse structure with standardization (see Slide 2-9). Data can be standardized (e.g., numerical entries in laboratory reports) and non-standardized. A typical example is non-standardized text—imprecisely called "Free-Text" or "unstructured data" in an electronic patient record (Kreuzthaler et al. 2011; Holzinger 2011; Holzinger et al. 2013c).

Well-structured data is the minority of data and an idealistic case when each data element has an associated defined structure, relational tables, or the resource description framework RDF, or the Web Ontology Language OWL (see Lecture 3).

Note: Ill-structured is a term often used for the opposite of well-structured, although this term originally was used in the context of problem solving (Simon 1973).

Semi-structured is a form of structured data that does not conform with the strict formal structure of tables and data models associated with relational databases but contains tags or markers to separate structure and content, i.e., they are schema-less or self-describing; a typical example is a markup language such as XML (see Lectures 3 and 4).

Weakly structured data is the most of our data in the whole universe, whether it is in macroscopic (astronomy) or microscopic structures (biology)—see Lecture 5.

Non-structured data or unstructured data is an imprecise definition used for information expressed in natural language, when no specific structure has been defined. This is an issue for debate: Text has also some structure: words, sentences, paragraphs. If we are very precise, unstructured data would meant that the data is complete randomized—which is usually called noise and is defined by (Duda et al. 2000) as any property of data which is not due to the underlying model but instead to randomness (either in the real world, from the sensors or the measurement procedure).

5.1 Well-Structured Data

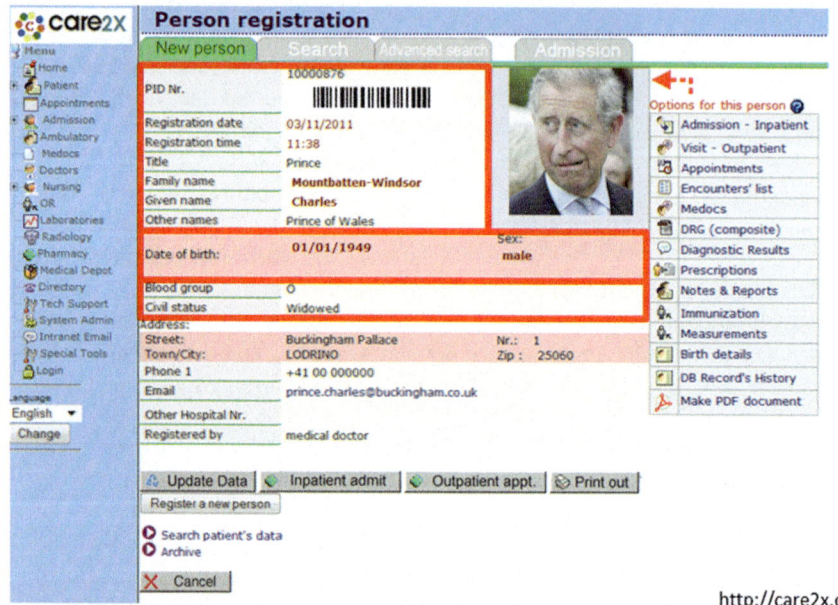

Fig. 2 See Slide 5-3

Slide 5-3: Example: Well-Structured Data

A look on the typical view of a hospital information system shows us the organization of well-structured data: Standardized and well-structured data is the basis for accurate communication. In the medical domain, many different people work at different times in various locations. Data standards can ensure that information is interpreted by all users with the same understanding. Moreover, standardized data facilitate comparability of data and interoperability of systems. It supports the reusability of the data, improves the efficiency of health-care services and avoids errors by reducing duplicated efforts in data entry. Remember: Data standardization refers to (a) the data content; (b) the terminologies that are used to represent the data; (c) how data is exchanged; and (d) how knowledge, e.g., clinical guidelines, protocols, decision support rules, checklists, standard operating procedures are represented in the health information system.

Note: The opposite, i.e., non-standardized data is the majority of data and inhibit data quality, data exchange, and interoperability.

Remark: Care2x is an Open Source Information System, see: http://care2x.org

See Lecture 10 for more details.

5.2 Semi-structured Data

Fig. 3 See Slide 5-4

```
<?xml version="1.0"?>
<patient>
        <patient-id>11111</patient-id>
        <Name>Chen</Name>
        <Date of Birth>1.1.1900</Date of Birth>
        <diagnosis>
                <code>123</code>
                        <diagnosistext>Myocardinfarct</diagnosistext>
        </diagnosis>
</patient>
```

Slide 5-4: Example: Semi-structured Data: XML

This is a Medical example for semi-structured data in XML (Holzinger 2003). The eXtensible Markup Language (XML) is a flexible text format recommended by the W3C for data exchange and derived from SGML (ISO 8879), (Usdin and Graham 1998).

XML is often classified as semi-structured; however, this is in some way misleading, as the data itself is still **structured**, but in a flexible rather than a static way (Forster and Vossen 2012). Such data does not conform to the formal structure of tables and data models as for example in relational databases, but at least contains tags/markers to separate semantic elements and enforce hierarchies of records and fields within these data.

Slide 5-5: Example: Generic XML Template for a Medical Report

This example by Rassinoux et al. (2003) shows how XML can be used in the hospital information system: The structure of any new document edited in the Patient Record (here: DPI) is based on a template defined in XML format (left). These templates play the role of DTDs or XML schemas as they precisely define the structure and content type of each paragraph, thus validating the document at the application level. Such a structure embeds a <HEADER> and a <BODY>. The header encapsulates the properties that are inherent to the new document and that will be useful to further classify it, according to various criteria, including the patient identification, the document type, the identifier of its redactors and of the hospitalization stay, or ambulatory consultation to which the document will be attached in the patient

(continued)

(continued)

trajectory, etc. The body encapsulates the content, and is divided into two parts: The <STRUCDOC> part describes the semantic entities that compose the document. The <FULLDOC> part embeds the document itself with its page layout information, which can be stored either as a draft, a temporary text or as a definitive text. This format guarantees the storage of dynamic and controlled fields for data input, thus allowing the combination of free text and structured data entry in the document. Once the document is no longer editable, it is definitively saved into the RTF format. A CDATA section is utilized for storing the rough document whatever its format, as it permits to disregard blocks of text containing characters that would otherwise be regarded as markup (Rassinoux et al. 2003).

Slide 5-6: Comparison of XML: RDF/OWL in Bioinformatics

On top in this slide you can see a sample XML describing genes from Drosophila melanogaster involved in long-term memory. Nested within the gene elements, are sub-elements related to the parent. The first gene includes two nucleic acid sequences, a protein product, and a functional annotation. Additional information is provided by attributes, such as the organism. This example illustrates the difficulty of modeling many-to-many relationships, such as the relationship between genes and functions. Information about functions must be repeated under each gene with that function. If we invert the nesting, then we must repeat information about genes with more than a single function. Below the XML we see the information about genes using both RDF and OWL. Both genes are instances of the class Fly Gene, which has been defined as the set of all Genes for the organism D. melanogaster. The functional information is represented using a hierarchical taxonomy, in which Long-Term Memory is a subclass of Memory (Louie et al. 2007).

Remark: Drosophila melanogaster is a model organism and shares many genes with humans. Although Drosophila is an insect whose genome has only about 14,000 genes (half of humans), a remarkable number of these have very close counterparts in humans; some even occur in the same order in the fly's DNA as in our own. This, plus the organism's more than 100-year history in the lab, makes it one of the most important models for studying basic biology and disease (see for example http://www.lbl.gov/Science-Articles/Archive/sabl/2007/Feb/drosophila.html).

Note: The relational data model requires preciseness: The data must be regular, complete and structured. However, in Biology the relationships are

(continued)

(continued)

mostly un-precise. Genomic medicine is extremely data intensive and there is an increasing diversity in the type of data: DNA sequence, mutation, expression arrays, haplotype, proteomic, etc. In bioinformatics many heterogeneous data sources are used to model complex biological systems (Rassinoux et al. 2003; Achard et al. 2001). The challenge in genomic medicine is to integrate and analyze these diverse and huge data sources to elucidate physiology and in particular disease physiology. XML is suited for describing semi-structured data, including a kind of natural modeling of biological entities, because it allows features as for example nesting (see Slide 5-6 on top). Still a key limitation of XML is that it is difficult to model complex relationships; for example, there is no obvious way to represent many-to-many relationships, which are needed to model complex pathways. On top in Fig. 5-9 we can see a sample XML, describing genes involved in the long-term memory of a sample specimen d. melanogaster. Nested within the gene elements, are sub-elements related to the parent. The first gene includes two nucleic acid sequences, a protein product, and a functional annotation. Additional information is provided by attributes, such as the organism. This example illustrates the difficulty of modeling many-to-many relationships, such as the relationship between genes and functions. Information about functions must be repeated under each gene with that function. If we invert the nesting (i.e., nesting genes inside function elements), then we must repeat information about genes with more than a single function. At the bottom in Slide 5-6 we see the same information about genes, but using RDF and OWL. Both genes are instances of the class Fly Gene, which has been defined as the set of all Genes for the organism D. melanogaster. The functional information is represented using a hierarchical taxonomy, in which Long-Term Memory is a subclass of Memory (Louie et al. 2007).

5.3 Weakly Structured Data

Slide 5-7: Example: Weakly Structured Data—Protein–Protein Interactions

Here we see a human protein interaction network and its connections: Proteins likely to be under positive selection are colored in shades of red (light red, low likelihood of positive selection; dark red, high likelihood). Proteins estimated not to be under positive selection are in yellow, and proteins for which the likelihood of positive selection was not estimated are in white (Kim et al. 2007).

5.4 On the Topology of Data

Data has shape!

> **Slide 5-8: On the Topology of Data: Data Has Shape!**
>
> Such data does not conform to the formal structure of tables and data models as for example in relational databases, but at least contains tags/markers to separate semantic elements and enforce hierarchies of records and fields within these data.

> **Slide 5-9: Again: What Is a Mathematical, What Is a Physical Space?**
>
> Such data does not conform to the formal structure of tables and data models as for example in relational databases, but at least contains tags/markers to separate semantic elements and enforce hierarchies of records and fields within these data.

There are many different types of topology:

Point-set topology, aka general topology, studies properties of spaces and the structures defined on them, where the spaces may be very general, and do not have to be similar to manifolds (a manifold of dimension n is a topological space that near each point resembles an n-dimensional Euclidean space). General topology provides the most general framework where fundamental concepts of topology such as open/closed sets, continuity, interior/exterior/boundary points, and limit points can be defined (Gaal 1966).

Algebraic topology uses tools from abstract algebra to study topological spaces. The basic goal is to find algebraic invariants that classify topological spaces up to homeomorphism (Hatcher 2002). A function f: $X \rightarrow Y$ between two topological spaces (X, T_X) and (Y, T_Y) is called a **homeomorphism** if f is bijective, continuous, and the inverse function f^{-1} is continuous.

Before we go into examples, let us answer the question: "What is a space?"

6 Networks = Graphs + Data

6.1 Networks in Biological Systems

> **Slide 5-10: Complex Biological Systems: Key Concepts**
>
> The concept of network structures is fascinating, compelling, and powerful and applicable in nearly any domain at any scale.
>
> Network theory can be traced back to **graph theory**, developed by Leonhard Euler in 1736 (see Slide 5-11). However, stimulated by works for example from Barabási et al. (1999), research on complex networks has only recently been applied to biomedical informatics. As an extension of classical graph theory, see for example Diestel (2010), complex network research focuses on the characterization, analysis, modeling and simulation of complex systems involving many elements and connections, examples including the internet, gene regulatory networks, protein–protein networks, social relationships and the Web, and many more. Attention is given not only to try to identify special patterns of connectivity, such as the shortest average path between pairs of nodes (Newman 2003), but also to consider the evolution of connectivity and the growth of networks, an example from biology being the evolution of protein–protein interaction (PPI) networks in different species (Slide 5-11). In order to understand complex biological systems, the three following key concepts need to be considered:
>
> (i) **emergence**, the discovery of links between elements of a system because the study of individual elements such as genes, proteins, and metabolites is insufficient to explain the behavior of whole systems;
> (ii) **robustness**, biological systems maintain their main functions even under perturbations imposed by the environment; and
> (iii) **modularity**, vertices sharing similar functions are highly connected. Network theory can largely be applied for biomedical informatics, because many tools are already available (Costa et al. 2008).

6.2 Network Theory

6.2.1 Basic Concepts of Networks

> **Slide 5-11: Networks on the Example of Bioinformatics**
>
> A graph $G(V, E)$ describes a structure which consists of nodes aka vertices V, connected by a set of pairs of distinct nodes (links), called edges $E\{a, b\}$ with $a, b \in V; a \neq b$.
>
> Graphs containing cycles and/or alternative paths are referred to as networks. The vertexes and edges can have a range of properties defined as colors, which also may have quantitative values, referred to as weights. In this slide we see the basic building block symbols of a biological network as used in bioinformatics. The blue dots are serving as network hubs, the red block is a critical node (on a critical link), the white balls are bottle necks, the stars second order hubs etc. (Hodgman et al. 2010).

6.2.2 Computational Graph Representation

Adjacency (ə-'jā-s^ən(t)-sē) Matrix $A = (a_{jk})$ $\qquad a_{jk} = \begin{cases} 1, & if \{j, k\} \in E \\ 0, & otherwise \end{cases}$

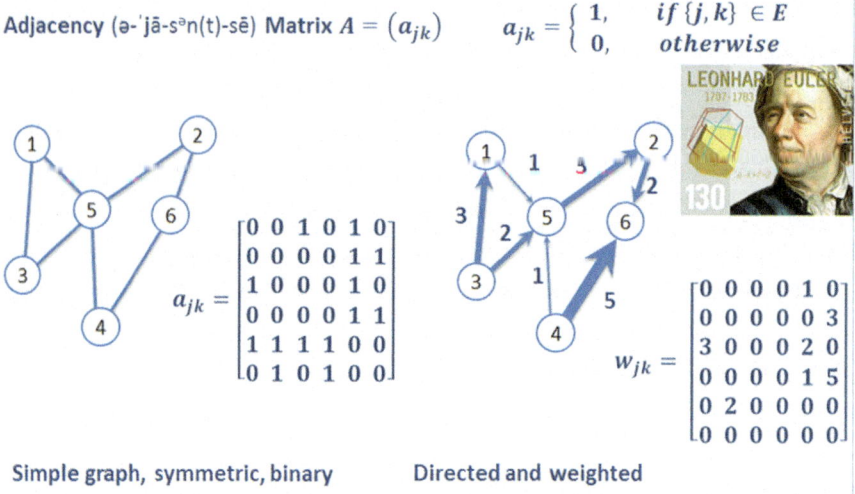

Simple graph, symmetric, binary   Directed and weighted

Fig. 4 See Slide 5-12

> **Slide 5-12: Computational Graph Representation**
>
> In order to represent network data in computers it is not comfortable to use sets; more practical are matrices. The simplest form of a graph representation

(continued)

6 Networks = Graphs + Data   215

(continued)

is the so-called adjacency matrix. In this Slide we see an undirected (left) and a directed graph and their respective adjacency matrices. If the graph is undirected, the adjacency matrix is symmetric, i.e., the elements $a_{ij} = a_{ji}$ for any i and j.

Left: a simple undirected binary graph and its mapping in a square adjacency matrix (symmetric), right: a directed and weighted graph (nonsymmetric); there is full correspondence between the network, the graph, and the adjacency matrix.

Slide 5-13: Example: Tool for Node-Link Visualization

This Tool is a nice example on the usefulness of adjacency matrices: The InfoVis Toolkit is an interactive graphics toolkit developed by Jean-Daniel Fekete at INRIA (The French National Institute for Computer Science and Control). The toolkit implements nine types of visualization: Scatter Plots, Time Series, Parallel Coordinates, and Matrices for tables; Node-Link diagrams, Icicle trees, and Tree maps for trees; Adjacency Matrices and Node-Link diagrams for graphs. Node-Link visualizations provide several variants (eight for graphs and four for trees). There are also a number of interactive controls and information displays, including dynamic query sliders, fisheye lenses, and excentric labels. Information about the InfoVis toolkit can be found at http://ivtk.sourceforge.net

The InfoVis Toolkit provides interactive components such as range sliders and tailored control panels required to configure the visualizations. These components are integrated into a coherent framework that simplifies the management of rich data structures and the design and extension of visualizations. Supported data structures include tables, trees, and graphs. All visualizations can use fisheye lenses and dynamic labeling (Fekete 2004).

6.2.3 Network Metrics

Slide 5-14: Some Network Metrics (1/2)

The truly multidisciplinary network science has led to a wide variety of quantitative measurements of their topological characteristics (Costa et al. 2007). The identification between a graph and an adjacency matrix

(continued)

(continued)

makes all the powerful methods of linear algebra, graph theory, and statistical mechanics available to us for investigating specific network characteristics:

Order (a in Figure Slide 5-14) = total number of nodes n
Size = total number of links:

$$\sum_i \sum_j a_{ij}$$

Clustering Coefficient (b in Slide 5-14) = the degree of concentration of the connections of the node's neighbors in a graph and gives a measure of local inhomogeneity of the link density, i.e., the level of connectedness of the graph. It is calculated as the ratio between the actual number t_i of links connecting the neighborhood (the nodes immediately connected to a chosen node) of a node and the maximum possible number of links in that neighborhood:

$$C_i = \frac{2t_i}{k(k_i - 1)}$$

For the whole network, the clustering coefficient is the arithmetic mean:

$$C = \frac{1}{n}\sum_i C_i$$

Path length (c in Slide 5-14) = is the arithmetical mean of all the distances; The characteristic path length of node i provides information about how close node i is connected to all other nodes in the network and is given by the distance $d(i,j)$ between node i and all other nodes j in the network.

The Path length l provides important information about the level of global communication efficiency of a network:

$$l = \frac{1}{n(n-1)}\sum_{i \neq j} d_{ij}$$

Note: *Numerical methods, e.g., the Dijkstra's algorithm (1959), are used to calculate all the possible paths between any two nodes in a network.*

Slide 5-15: Some Network Metrics (2/2)

Centrality (d in Slide 5-15) = the level of "betweenness-centrality" of a node i; it indicates how many of the shortest paths between the nodes of the network pass through node i. A high "betweenness-centrality" indicates that this node is important in interconnecting the nodes of the network, marking a potential hub role (refer to Slide 5-11) of this node in the overall network.

Nodal degree (e in Slide 5-15) = number of links connecting i to its neighbors. The degree of node i is defined as its total number of connections.

$$k_i = \sum_i a_{ij}$$

The degree probability distribution P(k) describes the p(x) that a node is connected to k other nodes in the network.

Modularity (f in Slide 5-15) = describes the possible formation of communities in the network, indicating how strong groups of nodes form relative isolated subnetworks within the full network (refer also to Slide 5-11).

Two further metrics include:

Density = the ratio between m and the maximum possible number of links that a graph may have:

$$\delta = \frac{2m}{n(n-1)}$$

Path = a series of consecutive links connecting any two nodes in the network, the distance between two vertices is the length of the shortest path connecting them, the diameter of a graph is the longest distance (the maximum shortest path) existing between any two vertices in the graph:

$$D = \max d_{ij}$$

Slide 5-16: Some Network Topologies

Regular network (a in Slide 5-16) has a local character, characterized by a high clustering-coefficient (c in Slide 5-16) and a high path length (L, Slide

(continued)

(continued)

5-16). It takes a large number of steps to travel from a specific node to a node on the other end of the graph. A special case of a regular network is the:

Random network, where *all* connections are distributed randomly across the network; the result is a graph with a random organization (outer right in Slide 5-16). In contrast to the local character of the regular network, a random network has a more global character, with a low C and a much shorter path length L than the regular network. A particular case is the:

Small-world network (center of Slide 5-16) which are very robust and combine a high level of local and global efficiency. Watts and Strogatz (1998b) showed that with a low probability p of randomly reconnecting a connection in the regular network, a so-called small-world organization arises. It has both a high C and a low L, combining a high level of local clustering with still a short average travel distance. Many networks in nature are small-world (e.g., Internet, protein networks, social networks, functional and structural brain network), combining a high level of segregation with a high level of global information integration. In addition, such networks can have a heavy tailed connectivity distribution, in contrast to random networks in which the nodes roughly all have the same number of connections.

Scale-free networks (B in Slide 5-16) are characterized by a degree probability distribution that follows a power-law function, indicating that on average a node has only a few connections, but with the exception of a small number of nodes that are heavily connected. These nodes are often referred to as hub nodes (see Slide 5-11) and they play a central role in the level of efficiency of the network, as they are responsible for keeping the overall travel distance in the network to a minimum. As these hub nodes play a key role in the organization of the network, scale-free networks tend to be vulnerable to specialized attack on the hub nodes.

Modular networks (c in Slide 5-16) show the formation of so-called communities, consisting of a subset of nodes that are mostly connected to their direct neighbors in their community and to a lesser extend to the other nodes in the network. Such networks are characterized by a high level of modularity of the nodes.

Slide 5-17: Small-World Networks

Taking a connected graph with a high graph diameter and adding a very small number of edges randomly, results in the small world phenomenon: the diameter drops drastically. It is also known as "six degrees of separation"

(continued)

(continued)

since, in the social network of the world, any person turns out to be linked to any other person by roughly six connections (Milgram 1967). The human short-term memory uses small world networks between the neurons.

In the Slide we see the random rewiring procedure for interpolating between a regular ring lattice (rightmost) and a random network (leftmost), without altering the number of vertices or edges in the graph. We start with a ring of n vertices, each connected to its k nearest neighbors by undirected edges. We choose a vertex and the edge that connects it to its nearest neighbor in a clockwise sense. With probability p, we reconnect this edge to a vertex chosen uniformly at random over the entire ring, with duplicate edges forbidden; otherwise we leave the edge in place. We repeat this process by moving clockwise around the ring, considering each vertex in turn until one lap is completed. Then we consider the edges that connect vertices to their second-nearest neighbors clockwise. We randomly rewire each of these edges with probability p, and continue this process, circulating around the ring and proceeding outward to more distant neighbors after each lap, until each edge in the original lattice has been considered once. For intermediate values of p, the graph is a small-world network: highly clustered like a regular graph, yet with small characteristic path length, like a random graph (Watts and Strogatz 1998a).

6.2.4 Graphs from Point Cloud Datasets

Slide 5-18: Graphs from Point Cloud Datasets

There are many ways to construct a proximity graph representation from a set of data points that are embedded in $\mathbb{R}^d$.

Let us consider a set of data points $\{x_1, \ldots, x_n\} \in \mathbb{R}^d$.

To each data point we associate a vertex of a proximity graph $\mathcal{G}$ to define a set of vertices $\mathcal{V} = \{v1, v2, \ldots, vn\}$. Determining the edge set $\mathcal{E}$ of the proximity graph $\mathcal{G}$ requires defining the neighbors of each vertex vi according to its embedding xi.

Consequently, a **proximity graph** is a graph in which two vertices are connected by an edge *if* the data points associated to the vertices satisfy particular geometric requirements. Such particular geometric requirements are usually based on a metric measuring the distance between two data points. A usual choice of metric is the Euclidean metric. Look at the slide:

(a) is our initial set of points in the plane $\mathbb{R}^2$

(continued)

(continued)
(b) ε-ball graph $vi \sim vj$ if $xj \in \mathcal{B}(vi; \varepsilon)$
(c) k-nearest-neighbor graph (k-NNG): $vi \sim vj$ if the distance between xi and xj is among the k-th smallest distances from xi to other data points. The k-NNG is a directed graph since one can have xi among the k-nearest neighbors of xj but not vice versa.
(d) Euclidean Minimum Spanning Tree (EMST) graph is a connected tree subgraph that contains all the vertices and has a minimum sum of edge weights. The weight of the edge between two vertices is the Euclidean distance between the corresponding data points.
(e) Symmetric k-nearest-neighbor graph (Sk-NNG): $vi \sim vj$ if xi is among the k-nearest neighbors of y or vice versa.
(f) Mutual k-nearest-neighbor graph (Mk-NNG): $vi \sim vj$ if xi is among the k-nearest neighbors of y and vice versa. All vertices in a mutual k-NN graph have a degree upper-bounded by k, which is not usually the case with standard k-NN graphs.
(g) Relative Neighborhood Graph (RNG): $vi \sim vj$ if there is no vertex in $B(vi; D(vi, vj)) \cap B(vj; D(vi, vj))$.
(h) Gabriel Graph (GG)
(i) The β-Skeleton Graph (β-SG):

For details please refer to Lézoray and Grady (2012), or to a classical graph theory book, e.g., Harary (1969), Bondy and Murty (1976), Golumbic (2004), Diestel (2010).

Slide 5-19: Graphs from Images

In this slide we see the examples of:

(a) a real image with the quadtree tessellation,
(b) the region adjacency graph associated to the quadtree partition,
(c) irregular tessellation using image-dependent superpixel Watershed Segmentation (Vincent and Soille 1991)
(d) irregular tessellation using image-dependent SLIC superpixels (Lucchi et al. 2010)

SLIC = Simple Linear Iterative Clustering

Slide 5-20: Example: Watershed Algorithm

A straightforward implementation of the original Vincent–Soille algorithm is difficult if plateaus occur. Therefore, an alternative approach was proposed by Meijster and Roerdink (1995), in which the image is first transformed to a directed valued graph with distinct neighbor values, called the components graph of f. On this graph the watershed transform can be computed by a simplified version of the Vincent–Soille algorithm, where fifo queues are no longer necessary, since there are no plateaus in the graph (Roerdink and Meijster 2000).

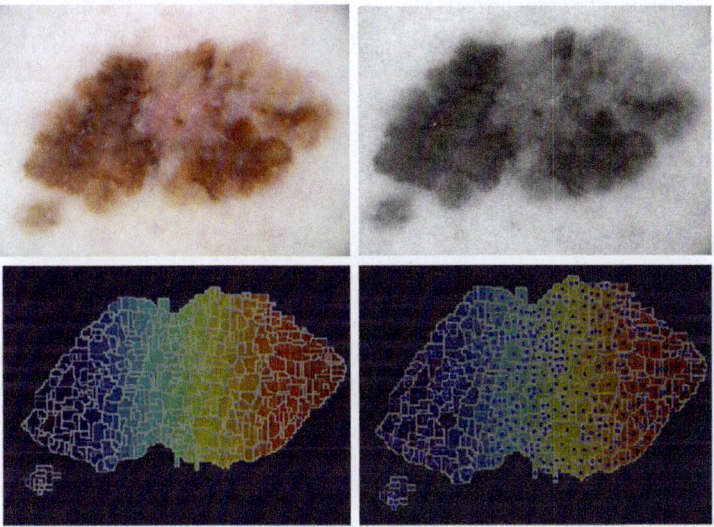

Fig. 5 See Slide 5-21

Slide 5-21: From Graphs to Images: Watershed + Centroid

The original natural digital image is first transformed into grey scale, then the Watershed algorithm is applied and then the centroid function calculated, the results are representative point sets in the plane.

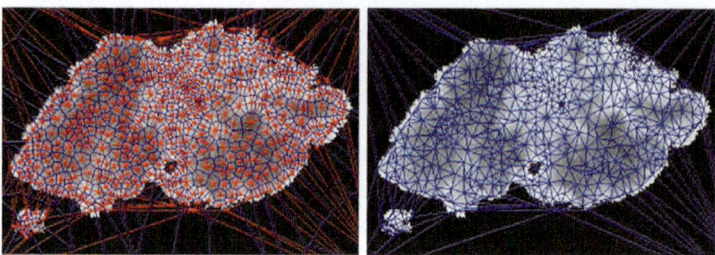

Fig. 6 See Slide 5-22

Slide 5-22: Graphs from Images: Voronoi ↔ Delauney

The Delaunay Triangulation (DT): $vi \sim vj$ if there is a closed ball $\mathcal{B}(\cdot; r)$ with vi and vj on its boundary and no other vertex vk contained in it. The dual to the DT is the Voronoi irregular tessellation where each Voronoi cell is defined by the set $\{x \in Rn \mid D(x, vk) \leq D(x, vj) \text{ for all } vj = vk\}$. In such a graph, $\forall\, vi$, $deg\,(vi) = 3$ (Lézoray and Grady 2012).

Slide 5-23: Points → Voronoi → Delauney

This animation shows the construction of a Delaunay graph: First the red points on the plane are drawn, then we insert the blue edges and the blue vertices on the Voronoi graph, finally the red edges drawn build the Delaunay graph (Kropatsch et al. 2001).

Slide 5-24: Example: Graph Entropy Measures

In this Slide we see the evaluated information-theoretic network measures on publication networks. Here from the excellence network of RWTH Aachen University. Those measures can be understood as graph complexity measures which evaluate the structural complexity based on the corresponding concept. A possible useful interpretation of these measures helps to understand the differences in subgraphs of a cluster. For example one could apply community detection algorithms and compare entropy measures of such detected communities. Relating these data to social measures (e.g., balanced score card data) of subcommunities could be used as indicators of collaboration success or lack thereof. The node size shows the node degree, whereas the node color shows the betweenness centrality, and darker color means higher centrality (Holzinger et al. 2013a).

Fig. 7 See Slide 5-25

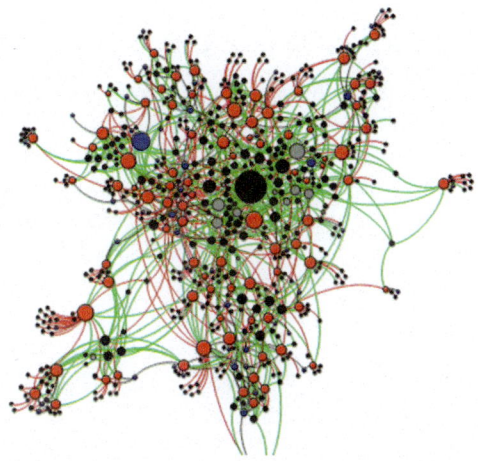

> **Slide 5-25: Example for a Medical Knowledge Space**
>
> A further example shall demonstrate the usefulness of graph theory and network analysis: This graph shows the medical knowledge space of a standard quick reference guide for emergency doctors and paramedics in the German speaking area. It has been subsequently developed, tested in the medical real world and constantly improved for 20 years by Dr. med. Ralf Müller, emergency doctor at Graz-LKH University Hospital and is practically in the pocket of every emergency and family doctor and paramedics in the German speaking area (Holzinger et al. 2013b).
>
> Up to know we know that Graphs and Graph-Theory are powerful tools to map data structures and to find novel connections between single data objects (Strogatz 2001; Dorogovtsev and Mendes 2003). The inferred graphs can be further analyzed by using graph-theoretical and statistical and machine learning techniques (Dehmer et al. 2011). A mapping of the already existing and in the medical practice approved "knowledge space" as a conceptual graph and the subsequent visual and graph-theoretical analysis may provide novel insights on hidden patterns in the data. Another benefit of the graph-based data structure is in the applicability of methods from network topology and network analysis and data mining, e.g., small-world phenomenon (Barabasi and Albert 1999; Kleinberg 2000), and cluster analysis (Koontz et al. 1976; Wittkop et al. 2011).
>
> The graph-theoretic data of the graph seen in this slide include:
>
> Number of nodes = 641, number of edges = 1,250, red are agents, black are conditions, blue are pharmacological groups, grey are other documents. The average degree of this graph = 3.888, the average path length = 4.683, the network diameter = 9.

Slide 5-26: Medical Details of the Graph

The nodes of the sample graph represent: drugs, clinical guidelines, patient conditions (indication, contraindication), pharmacological groups, tables and calculations of medical scores, algorithms, and other medical documents; and the edges represent three crucial types of relations inducing medical relevance between two active substances, i.e., pharmacological groups, indications, and contraindications. The following example will demonstrate the usefulness of this approach.

Fig. 8 See Slide 5-27

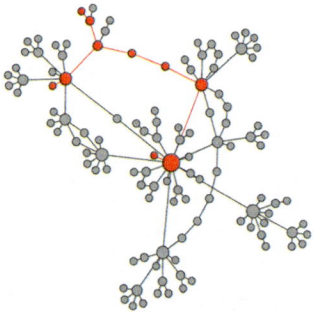

Slide 5-27: Example for the Shortest Path

This example shows us how convenient we can find which path between two nodes is the shortest as well as the navigation way between these nodes. Computing shortest paths is a fundamental and ubiquitous problem in network analysis. We can for example apply the Dijkstra-algorithm, solves the shortest path problem for a graph with non-negative edge path costs, producing a shortest path tree. This algorithm is often used in routing and as a subroutine in other graph algorithms: For a given node, the algorithm finds the path with lowest cost (i.e., the shortest path) between that node and every other node(Henzinger et al. 1997).

Slide 5-28: Example for Finding Related Structures

Here we see the relationship between Adrenaline (center black node) and Dobutamine (top left black node), Blue: Pharmacological Group,

(continued)

(continued)

Dark red: Contraindication; Light red: Condition, the Green nodes (from dark to light) are:

1. Application (one or more indications + corresponding dosages)
2. Single indication with additional details (e.g., "VF after 3rd Shock")
3. Condition (e.g., VF, Ventricular Fibrillation)

6.3 Network Examples

6.3.1 The Human Brain as Network

Slide 5-29: Example: The Brain Is a Complex Network

Our brain forms one integrative complex network, linking all brain regions and subnetworks together (Van Den Heuvel and Hulshoff Pol 2010). Examining the organization of this network provides insights in how our brain works. Graph theory provides a framework in which the topology of complex networks can be examined, and thus can reveal novelties about both the local and global organization of functional brain networks. In the slide we can see how the modeling of the functional brain by a graph works: edges are the connections between regions that are functionally linked. First, the collection of nodes is defined (A), second the existence of functional connections between the nodes in the network needs to be defined, resulting in a connectivity matrix (B). Finally, the existence of a connection between two points can be defined as whether their level of functional connectivity exceeds a certain predefined threshold (C) (Van Den Heuvel and Hulshoff Pol 2010).

6.3.2 Systems Biology and Human Diseases

Slide 5-30: Representative Examples of Disease Complexes

Insight into the biology of molecular networks is an important field, as anomalies in these systems underlie a wide spectrum of polygenetic human disorders, ranging from schizophrenia to congenital heart disease (CHD). Understanding the functional architecture of networks that organize the development of organs, see for example Chien et al. (2008), lays the foundation of novel approaches in *regenerative medicine*, since manipulation of

(continued)

(continued)

such systems is necessary for success of *tissue engineering* technologies and *stem cell therapy*.

Lage et al. (2010) developed a framework for gaining new insights into the systems biology of the protein networks driving organ development and related polygenic human disease phenotypes, exemplified with heart development and CHD. In the slide we see examples of four functional networks driving the development of different anatomical structures in the human heart. These four networks are constructed by analyzing the interaction patterns of four different sets of cardiac development (CD): proteins corresponding to the morphological groups "atrial septal defects," "abnormal atrioventricular valve morphology," "abnormal myocardial trabeculae morphology," and "abnormal outflow tract development," CD proteins from the relevant groups are shown in orange and their interaction partners are shown in grey. Functional modules annotated by literature curation are indicated with a colored background. Centrally in the Figure is a hematoxylin–eosin stained frontal section of the heart from a 37-day human embryo, where tissues affected by the four networks are marked; AS (developing atrial septum), EC (endocardial cushions, which are anatomical precursors to the atrioventricular valves), VT (developing ventricular trabeculae), and OFT (developing outflow tract).

Slide 5-31: Example: Cell-Based Therapy

In this slide we see an overview of the modular organization of heart development: (A) Protein interaction networks are plotted at the resolution of functional modules. Each module is color coded according to functional assignment as determined by literature curation. The amount of proteins in each module is proportional to the area of its corresponding node. Edges indicate direct (lines) or indirect (dotted lines) interactions between proteins from the relevant modules. (B) Recycling of functional modules during heart development. The bars represent functional modules and recycling is indicated by arrows. The bars follow the color code of (A) and the height of the bars represent the number of proteins in each module, as shown left on the y axis (Lage et al. 2010).

Note: **Phenotype** = an organism's observable characteristics (traits), e.g., morphology, biochemical/physiological properties, behavior. Phenotypes result from the expression of an organism's genes as well as the influence of environmental factors and the interactions between them. **Genotype** = inherited instructions within its genetic code.

6.3.3 Gene Networks

> **Slide 5-32: Identifying Networks in Disease Research**
>
> Diseases (e.g., obesity, diabetes, and atherosclerosis) result from multiple genetic and environmental factors, and importantly, interactions between genetic and environmental factors. This slide shows the vast networks of molecular interactions. It can be seen that the gastrointestinal (GI) tract, vasculature, immune system, heart, and brain are all potentially involved in either the onset of diseases such as atherosclerosis or in comorbidities such as myocardial infarction and stroke brought on by such diseases. Further, the risks of comorbidities for diseases such as atherosclerosis are increased by other diseases, such as hypertension, which may, in turn, involve other organs, such as kidney. The role that each organ and tissue type plays in a given disease is largely determined by genetic background and environment, where different perturbations to the genetic background (perturbations corresponding to DNA variations that affect gene function, which, in turn, leads to disease) and/or environment (changes in diet, levels of stress, level of activity, and so on) define the subtypes of disease manifested in any given individual. Although the physiology of diseases such as atherosclerosis is beginning to be better understood, what have not been fully exploited to data are the vast networks of molecular interactions within the cells.
>
> We see clearly in the Slide that there is a diversity of molecular networks functioning in any given tissue, including genomics networks, networks of coding and noncoding RNA, protein interaction networks, protein state networks, signaling networks, and networks of metabolites. Further, these networks are not acting in isolation within each cell, but instead interact with one another to form complex, giant molecular networks within and between cells that drive all activity in the different tissues, as well as signaling between tissues. Variations in DNA and environment lead to changes in these molecular networks, which, in turn, induce complicated physiological processes that can manifest as disease. Despite this vast complexity, the classic approach to elucidating genes that drive disease has focused on single genes or single linearly ordered pathways of genes thought to be associated with disease. This narrow approach is a natural consequence of the limited set of tools that were available for querying biological systems; such tools were not capable of enabling a more holistic approach, resulting in the adoption of a reductionist approach to teasing apart pathways associated with complex disease phenotypes. Although the emerging view that complex biological systems are best modeled as highly modular, fluid systems exhibiting a plasticity that allows them to adapt to a vast array of conditions, the history of science demonstrates that this view, although long the ideal, was never

(continued)

(continued)

within reach, given the unavailability of tools adequate to carrying out this type of research. The explosion of large-scale, high-throughput technologies in the biological sciences over the past 15–20 years has motivated a rapid paradigm shift away from reductionism in favor of a systems-level view of biology (Schadt and Lum 2006).

6.4 The Essence: Three Types of Biomedical Networks

Slide 5-33: Three Main Types of Biomedical Networks

In this Slide we see the three main types of biological networks: (1) a transcriptional regulatory network has two components: transcription factor (TF) and target genes (TG), where TF regulates the transcription of TGs; (2) PPI networks: two proteins are connected if there is a docking between them; (3) a metabolic network is constructed considering the reactants, chemical reactions, and enzymes (Costa et al. 2008).

6.4.1 Transcriptional Regulatory Networks

Slide 5-34: Example Transcriptional Regulatory Network

The extreme complexity of the E. coli transcriptional regulatory network. In this graphical representation, nodes are genes, and edges represent regulatory interactions. The network was reconstructed using data from the RegulonDB. This figure highlights the extreme complexity in regulatory networks. To obtain a deeper understanding of regulatory complexity, scientists must first discover biologically relevant organizational principles to unravel the hidden architecture governing these networks (Salgado et al. 2006).

The complexity of organisms arises rather as a consequence of elaborated regulations of gene expression than from differences in genetic content in terms of the number of genes. The **transcription network** is a critical system that regulates gene expression in a cell. Transcription factors (TFs) respond to changes in the cellular environment, regulating the transcription of target genes (TGs) and connecting functional protein interactions to the genetic information encoded in inherited genomic DNA in order to control the timing and sites of gene expression during biological development. The interactions between TFs and TGs can be represented as a directed graph: The two types

(continued)

(continued)

of nodes (TF and TG) are connected by arcs (see Slide 5-33, arrows) when regulatory interaction occurs between regulators and targets. Transcriptional regulatory networks display interesting properties that can be interpreted in a biological context to better understand the complex behavior of gene regulatory networks. At a local network level, these networks are organized in substructures such as motifs and modules. **Motifs** represent the simplest units of a network architecture required to create specific patterns of inter-regulation between TFs and TGs. Three most common types of motifs can be found in gene regulatory networks:

(1) single input,
(2) multiple input and
(3) feed-forward loop

Target genes belonging to the same single and multiple input motifs tend to be co-expressed, and the level of co-expression is higher when multiple transcription factors are involved.

Modularity in the regulatory networks arises from groups of highly connected motifs that are hierarchically organized, in which modules are divided into smaller ones. The evolution of gene regulatory networks mainly occurs through extensive duplication of transcription factors and target genes with inheritance of regulatory interactions from ancestral genes while the evolution of motifs does not show common ancestry but is a result of convergent evolution (Costa et al. 2008).

6.4.2 Protein–Protein Networks

Slide 5-35: Network Representations of Protein Complexes

The interactions between proteins are essential to keep the molecular systems of living cells working properly. PPI is important for various biological processes such as cell–cell communication, the perception of environmental changes, protein transport and modification. Complex network theory is suitable to study PPI maps because of its universality and integration in representing complex systems. In complex network analysis each protein is represented as a node and the physical interactions between proteins are indicated by the edges in the network.

Many complex networks are naturally divided into communities or modules, where links within modules are much denser than those across modules (e.g., human individuals belonging to the same ethnic groups interact more

(continued)

(continued)

than those from different ethnic groups). Cellular functions are also organized in a highly modular manner, where each module is a discrete object composed of a group of tightly linked components and performs a relatively independent task. It is interesting to ask whether this modularity in cellular function arises from modularity in molecular interaction networks such as the transcriptional regulatory network and PPI network.

The slide shows a hypothetical protein complex (A). Binary PPI are depicted by direct contacts between proteins. Although five proteins (A, B, C, D, and E) are identified through the use of a bait protein (red), only A and D directly bind to the bait. (B) shows the true PPI network topology of the protein complex is shown in. (C) depicts the PPI network topology of the protein complex inferred by the "matrix" model, where all proteins in a complex are assumed to interact with each other. Finally (D) demonstrates the PPI network topology of the protein complex inferred by the "spoke" model, where all proteins in a complex are assumed to interact with the bait; but no other interactions are allowed (Wang and Zhang 2007).

Slide 5-36: Correlated Motif Mining (CMM)

Correlated motif mining (CMM) is the challenge to find overrepresented pairs of patterns (motifs), in sequences of interacting proteins. Algorithmic solutions for CMM thereby provide a computational method for predicting binding sites for protein interaction. The task is basically to represent motifs X and Y (Fig. 119) to truly represent an overrepresented consensus pattern in the sequences of the proteins in VX, respectively VY, in order to increase the likelihood that they correspond or overlap with a so-called binding site—a site on the surface of the molecule that makes interactions between proteins from VX and VY possible through a molecular lock-and-key mechanism.

We call $\{X, Y\}$ a $(k_x\ k_y\ k_{xy})$-motif pair of a PPI network.

$G = (V, E, \lambda)$ if $|V_x| = k_x$, $|V_y| = k_y$ and $|V_x \cap V_y| = k_{xy}$.

It is called complete if all vertices from V_x are connected with all vertices from V_y (Boyen et al. 2011).

Slide 5-37: Steepest Ascent Algorithm Applied to CMM

Since the decision problem associated with CMM is in NP,[1] we can efficiently check if a motif pair has higher support than another which makes it possible to tackle CMM as a search problem in the space of all possible (l,d)-motif pairs. If we add the assumption that similar motifs can be expected to get similar support, it has the typical form of a combinatorial optimization problem. In combinatorial optimization, the objective is to find a point in a discrete search space which maximizes a user-provided function f. A number of heuristic algorithms called metaheuristics are known to yield stable results, e.g., the steepest ascent algorithm (Aarts and Lenstra 1997), illustrated as pseudocode in the slide.

6.4.3 Metabolic Networks

Slide 5-38: Metabolic Networks

Metabolism is primarily determined by genes, environment and nutrition. It consists of chemical reactions catalyzed by enzymes to produce essential components such as amino acids, sugars and lipids, and also the energy necessary to synthesize and use them in constructing cellular components. Since the chemical reactions are organized into metabolic pathways, in which one chemical is transformed into another by enzymes and cofactors, such a structure can be naturally modeled as a complex network. In this way, metabolic networks are directed and weighted graphs, whose vertices can be metabolites, reactions and enzymes, and two types of edges that represent mass flow and catalytic reactions. One widely considered catalogue of metabolic pathways available on-line is the Kyoto Encyclopedia of Genes and Genomes (KEGG). In the slide we see a simple metabolic network involving five metabolites M1–M5 and three enzymes E1–E3, of which the latter catalyzes an irreversible reaction (Hodgman et al. 2010).

---

[1] NP = nondeterministic polynomial time; in computational complexity theory NP is one of the fundamental complexity classes.

Slide 5-39: Metabolic Networks are Usually Big … Big Data

Such metabolic structures can be very large, as can be seen in this slide. The enzyme-coding genes under TrmB (this is the thermococcus regulator of maltose binding) acts as a repressor for genes encoding glycolytic enzymes and as activator for genes encoding gluconeogenic enzymes control included in the metabolic pathways shown in the slide (13 are unique to archaea and 35 are conserved across species from all three domains of life). Integrated analysis of the metabolic and gene regulatory network architecture reveals various interesting scenarios (Schmid et al. 2009).

Slide 5-40: Using EPRs to Discover Disease Correlations

Electronic patient records (EPR remain an unexplored, but rich data source for discovering for example correlations between diseases). Roque et al. (2011) describe a general approach for gathering phenotypic descriptions of patients from medical records in a systematic and non-cohort dependent manner: By extracting phenotype information from the "free-text" (=unstructured information) in such records they demonstrated that they can extend the information contained in the structured record data, and use it for producing fine-grained patient stratification and disease co-occurrence statistics. Their approach uses a dictionary based on the International Classification of Disease (ICD-10) ontology and is therefore in principle language independent. As a use case they show how records from a Danish psychiatric hospital lead to the identification of disease correlations, which subsequently can be mapped to systems biology frameworks.

Slide 5-41: Heatmap of Disease-Disease Correlations (ICD)

Roque et al. (2011) have used text mining to automatically extract clinically relevant terms from 5,543 psychiatric patient records and mapped these to disease codes in the ICD10. They clustered patients together based on the similarity of their profiles. The result is a patient stratification, based on more complete profiles than the primary diagnosis, which is typically used. Figure 124 illustrates the general approach to capture correlations between different disorders. Several clusters of ICD10 codes relating to the same anatomical area or type of disorder can be identified along the diagonal of the heatmap, ranging from trivial correlations (e.g., different arthritis

(continued)

(continued)

disorders), to correlations of cause and effect codes (e.g., stroke and mental/behavioral disorders), to social and habitual correlations (e.g., drug abuse, liver diseases, and HIV).

6.5 Structural Homologies

Slide 5-42: Example: ὁμολογέω (Homologeo)

Homology (plural: homologies) origins from Greek ὁμολογέω (homologeo) and means "to conform" (in German: übereinstimmen) and has its origins in Biology and Anthropology, where the word is used for a correspondence of structures in two life forms with a common evolutionary origin (Darwin 1859).

In chemistry it is used for the relationship between the elements in the same group of the periodic table, or between organic compounds in a homologous series.

In mathematics homology is a formalism for talking in a quantitative and unambiguous manner about how a space is connected (Edelsbrunner and Harer 2010).

Basically, homology is a concept that is used in many branches of algebra and topology. Historically, the term was first used in a topological sense by Henry Poincaré (1854–1912).

In Bioinformatics, homology modelling is a mature technique that can be used to address many problems in molecular medicine. Homology modelling is one of the most efficient methods to predict protein structures. With the increase in the number of medically relevant protein sequences, resulting from automated sequencing in the laboratory, and in the fraction of all known structural folds, homology modelling will be even more important to personalized and molecular medicine in the future. Homology modelling is a knowledge-based prediction of protein structures. In homology modelling a protein sequence with an unknown structure (the target) is aligned with one or more protein sequences with known structures (the templates).

The method of homology modelling is based on the principle that homologue proteins have similar structures. The prerequisite for successful homology modelling is a detectable similarity between the target sequence and the template sequences (more than 30 %) allowing the construction of a correct alignment. Homology modelling is a knowledge-based structure prediction relying on observed features in known homologous protein structures. By

(continued)

(continued)

exploiting this information from template structures the structural model of the target protein can be constructed (Wiltgen and Tilz 2009).

Two well-known homology modelling programs, which are free for academic research, are

MODELLER (http://salilab.org/modeller) and
SWISSMODEL (http://swissmodel.expasy.org).

The slide shows the comparison of two proteins: The sequences of both proteins are 95 % (53 of 56) identical (only residues 20, 30, and 45 differ), yet the structures are totally different.

Slide 5-43: Towards Personalized Medicine

Homology modeling is a knowledge-based prediction of protein structures.

In homology modeling a protein sequence with an unknown structure (the target) is aligned with one or more protein sequences with known structures (the templates).

The method is based on the principle that homologue proteins have similar structures.

Homology modeling will be extremely important to personalized and molecular medicine in the future.

7 Future Outlook

Slide 5-: Future Outlook

All these approaches are producing gigantic amounts of highly complex datasets, and the amounts are rising. In particular the amount of unstructured data (or information respectively) rises. Predictive modeling and machine learning are increasingly central to the business models of data-driven businesses (Dhar 2013).

8 Exam Questions

8.1 Yes/No Decision Questions

Please check the following sentences and decide whether the sentence is true = YES; or false = NO; for each correct answer you will be awarded 2 credit points.

01	One of the most exciting and challenging frontiers in neuroscience involves harnessing the power of large-scale genetic, genomic and phenotypic data sets.	❏ Yes ❏ No	2 total
02	In the medical domain, many different people work at different times in various locations, therefore standardized data is the basis for accurate communication.	❏ Yes ❏ No	2 total
03	XML is often classified as semi-structured, however this is in some way misleading, as the data itself is still structured, but in a flexible rather than a static way.	❏ Yes ❏ No	2 total
04	Non-standardized data is an realistic case and is the minority of data but support data quality, data exchange and interoperability in information systems.	❏ Yes ❏ No	2 total
05	In order to understand complex biological systems, the three following key concepts need to be considered: emergence, robustness, and standardization.	❏ Yes ❏ No	2 total
06	A transcriptional regulatory network has two components: transcription factor (TF) and target genes (TG), where TF regulates the transcription of TGs.	❏ Yes ❏ No	2 total
07	The complexity of organisms arises rather as a consequence of elaborated regulations of gene expression than from differences in genetic content in terms of the number of genes.	❏ Yes ❏ No	2 total
08	In genetics, a sequence motif is a nucleotide or amino-acid sequence pattern that is widespread and has, or is conjectured to have, a biological significance.	❏ Yes ❏ No	2 total
09	The decision problem associated with Correlated Motif Mining (CMM) is solvable in P.	❏ Yes ❏ No	2 total
10	Our brain forms one integrative complex network, linking all brain regions and sub-networks together.	❏ Yes ❏	2 total

Sum of Question Block A (max. 20 points)	

8.2 Multiple Choice Questions (MCQ)

The following questions are composed of two parts: the stem, which identifies the question or problem and a set of alternatives which can contain 0, 1, 2, 3, or 4 correct answers, along with a number of distractors that might be plausible—but are incorrect. Please **select the correct answers** by ticking ☒—and do not forget that it can be none. Each question will be awarded 4 points *only if everything is correct*.

01	Homology … ☐ a) … In mathematics homology is a formalism for talking in a quantitative and unambiguous manner about how a space is connected. ☐ b) … origins from Greek ὁμολογέω (homologeo) and means "to conform". ☐ c) … is used for a correspondence of structures in two life forms with a common evolutionary origin. ☐ d) … has its origins in Darwinian Biology.	4 total
02	The four network representations of protein networks include … ☐ a) … protein complex structure. ☐ b) … true PPI structure. ☐ c) … Spoke model. ☐ d) … Matrix model with bait in the center.	4 total
03	Homology modelling … ☐ a) … is extremely important for personalized and molecular medicine. ☐ b) … is based on the principle that homologue proteins are very different. ☐ c) … uses a protein sequence with known structures (targets) to align it with a protein structure with unknown structures (templates). ☐ d) … is a knowledge-based prediction of protein structures.	4 total
04	Drosophila melanogaster … ☐ a) … is an insect which has only some 140 genes. ☐ b) … is a very recently found laboratory animal and very important for research in personalized medicine. ☐ c) … has been used for many years and is of no more use in genomics. ☐ d) … is a model organism and shares many genes with humans.	4 total
05	The centrality of a network … ☐ a) … measures the level of "betweeness" of a node (the "importance"). ☐ b) … indicates how many of the shortest paths between the nodes of the network pass through node i. ☐ c) … describes the possible formation of communities in the network. ☐ d) … indicates how strong groups of nodes form relative isolated sub-networks within the full network.	4 total
06	Scale-free Topology … ☐ … ensures that there are short paths between pairs of nodes, allowing rapid communication between otherwise distant parts of the network. ☐ … is a set of techniques, applied from statistics, which analyze the topological structure of a network. ☐ … is used as a model to predict future values of a topological structure in networks. ☐ … is a measure of similarity between two protein structures.	4 total

07	Semi-structured data ... ❏ a) ... does not conform with the formal structure of tables/data models associated with relational databases. ❏ b) ... means randomness, noise and uncertainty. ❏ c) ... enforces hierarchies of records and fields within the data. ❏ d) ... contains tags/markers to separate semantic elements.	4 total
08	Data standardization refers to ... ❏ a) ... the data content. ❏ b) ... the terminologies that are used to represent this data. ❏ c) ... how data is exchanged. ❏ d) ... how knowledge, e.g. clinical guidelines, protocols, decision support rules, checklists, standard operating procedures are represented in the health information system.	4 total
09	Metabolism ... ❏ a) ... is the study of DNA-sequencing methods and produces a lot of complex data. ❏ b) ... is primarily determined by genes, environment and nutrition. ❏ c) ... consists of chemical reactions catalyzed by enzymes to produce essential components such as amino acids, sugars and lipids. ❏ d) ... is a process within genetics where regulatory metabolic elements have influence on nucleotide sequences.	4 total
10	Phenotype ... ❏ a) ... an organism's observable characteristics (traits). ❏ b) ... result from the expression of an organism's genes as well as the influence of environmental factors and the interactions between them. ❏ c) ... are inherited instructions within its genetic code. ❏ d) ... includes morphology, biochemical/physiological properties, behaviour, etc.	4 total

Sum of Question Block B (max. 40 points)

8.3 Free Recall Block

Please follow the instructions below. At each question you will be assigned the credit points indicated if your option is correct (partial points may be given).

01	Identifying networks in disease research is an important aspect of systems biology, where there is a high diversity of molecular networks within and between cells. Please identify in the following picture the networks and write the name of the network in the appropriate space!	1 each 4 total
	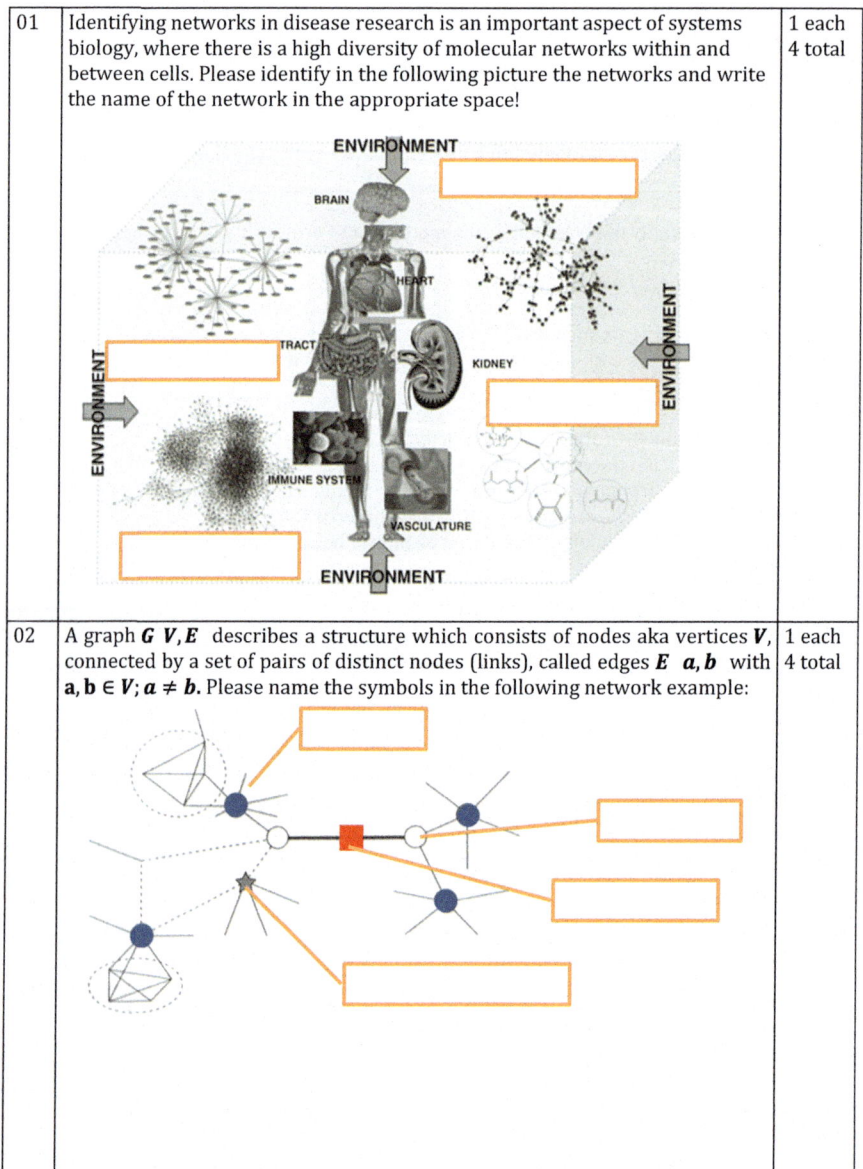	
02	A graph $G\ V, E$ describes a structure which consists of nodes aka vertices V, connected by a set of pairs of distinct nodes (links), called edges $E\ a, b$ with $a, b \in V; a \neq b$. Please name the symbols in the following network example:	1 each 4 total

| 03 | In order to represent network data in computers it is not comfortable to use sets; more practical are matrices. The simplest form of a graph representation is the so called adjacency matrix. Please set up the adjacency matrices for the following graphs: | 1 each 6 total |

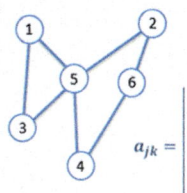

$a_{jk} =$

$w_{jk} =$

| 04 | In Biomedicine networks of all kind play an extremely important role. Please assign the correct labels to the network metrics below: | 1 each 6 total |

$\sum_i \sum_j a_{ij}$	Path length
$k_i = \sum_i a_{ij}$	Network size
$C = \dfrac{1}{n} \sum_i C_i$	Complete number of nodes
n	Clustering coefficient
$l = \dfrac{1}{n(n-1)} \sum_{i \neq j} d_{ij}$	Nodal degree

05	Please draw a metabolic network, constructed considering the reactants, chemical reactions and enzymes, consisting of A, B, C, D, E1 and E2	1 each 6 total
06	Please provide a simple example of a patient record in XML format:	1 each 4 total

Sum of Question Block C (max. 40 points)

9 Answers

9.1 Answers to the Yes/No Questions

Please check the following sentences and decide whether the sentence is true = YES; or false = NO; for each correct answer you will be awarded 2 credit points.

01	One of the most exciting and challenging frontiers in neuroscience involves harnessing the power of large-scale genetic, genomic and phenotypic data sets.	☒ Yes ☐ No	2 total
02	In the medical domain, many different people work at different times in various locations, therefore standardized data is the basis for accurate communication.	☒ Yes ☐ No	2 total
03	XML is often classified as semi-structured, however this is in some way misleading, as the data itself is still structured, but in a flexible rather than a static way.	☒ Yes ☐ No	2 total
04	Non-standardized data is an realistic case and is the minority of data but support data quality, data exchange and interoperability in information systems.	☐ Yes ☒ No	2 total
05	In order to understand complex biological systems, the three following key concepts need to be considered: emergence, robustness, and standardization.	☐ Yes ☒ No	2 total
06	A transcriptional regulatory network has two components: transcription factor (TF) and target genes (TG), where TF regulates the transcription of TGs.	☒ Yes ☐ No	2 total
07	The complexity of organisms arises rather as a consequence of elaborated regulations of gene expression than from differences in genetic content in terms of the number of genes.	☒ Yes ☐ No	2 total
08	In genetics, a sequence motif is a nucleotide or amino-acid sequence pattern that is widespread and has, or is conjectured to have, a biological significance.	☒ Yes ☐ No	2 total
09	The decision problem associated with Correlated Motif Mining (CMM) is solvable in P.	☐ Yes ☒ No	2 total
10	Our brain forms one integrative complex network, linking all brain regions and sub-networks together.	☒ Yes ☐ No	2 total

Sum of Question Block A (max. 20 points)	

9.2 Answers to the Multiple Choice Questions (MCQ)

01	Homology ... ☒ a) ... In mathematics homology is a formalism for talking in a quantitative and unambiguous manner about how a space is connected. ☒ b) ... origins from Greek ὁμολογέω (homologeo) and means "to conform". ☒ c) ... is used for a correspondence of structures in two life forms with a common evolutionary origin. ☒ d) ... has its origins in Darwinian Biology.	4 total
02	The four network representations of protein networks include ... ☒ a) ... protein complex structure. ☒ b) ... true PPI structure. ☒ c) ... Spoke model. ☒ d) ... Matrix model with bait in the center.	4 total
03	Homology modelling ... ☒ a) ... is extremely important for personalized and molecular medicine. ☐ b) ... is based on the principle that homologue proteins are very different. ☐ c) ... uses a protein sequence with known structures (targets) to align it with a protein structure with unknown structures (templates). ☒ d) ... is a knowledge-based prediction of protein structures.	4 total
04	Drosophila melanogaster ... ☐ a) ... is an insect which has only some 140 genes. ☐ b) ... is a very recently found laboratory animal and very important for research in personalized medicine. ☐ c) ... has been used for many years and is of no more use in genomics. ☒ d) ... is a model organism and shares many genes with humans.	4 total
05	The centrality of a network ... ☒ a) ... measures the level of "betweeness" of a node (the "importance"). ☒ b) ... indicates how many of the shortest paths between the nodes of the network pass through node i. ☐ c) ... describes the possible formation of communities in the network. ☐ d) ... indicates how strong groups of nodes form relative isolated sub-networks within the full network.	4 total
06	Scale-free Topology ... ☒ ... ensures that there are short paths between pairs of nodes, allowing rapid communication between otherwise distant parts of the network. ☐ ... is a set of techniques, applied from statistics, which analyze the topological structure of a network. ☐ ... is used as a model to predict future values of a topological structure in networks. ☐ ... is a measure of similarity between two protein structures.	4 total
07	Semi-structured data ... ☒ a) ... does not conform with the formal structure of tables/data models associated with relational databases. ☐ b) ... means randomness, noise and uncertainty. ☒ c) ... enforces hierarchies of records and fields within the data. ☒ d) ... contains tags/markers to separate semantic elements.	4 total

08	Data standardization refers to ... ☒ a) ... the data content. ☒ b) ... the terminologies that are used to represent this data. ☒ c) ... how data is exchanged. ☒ d) ... how knowledge, e.g. clinical guidelines, protocols, decision support rules, checklists, standard operating procedures are represented in the health information system.	4 total
09	Metabolism ... ☐ a) ... is the study of DNA-sequencing methods and produces a lot of complex data. ☒ b) ... is primarily determined by genes, environment and nutrition. ☒ c) ... consists of chemical reactions catalyzed by enzymes to produce essential components such as amino acids, sugars and lipids. ☐ d) ... is a process within genetics where regulatory metabolic elements have influence on nucleotide sequences.	4 total
10	Phenotype ... ☒ a) ... an organism's observable characteristics (traits). ☒ b) ... result from the expression of an organism's genes as well as the influence of environmental factors and the interactions between them. ☐ c) ... are inherited instructions within its genetic code. ☒ d) ... includes morphology, biochemical/physiological properties, behaviour, etc.	4 total

Sum of Question Block B (max. 40 points)

9.3 Answers to the Free Recall Questions

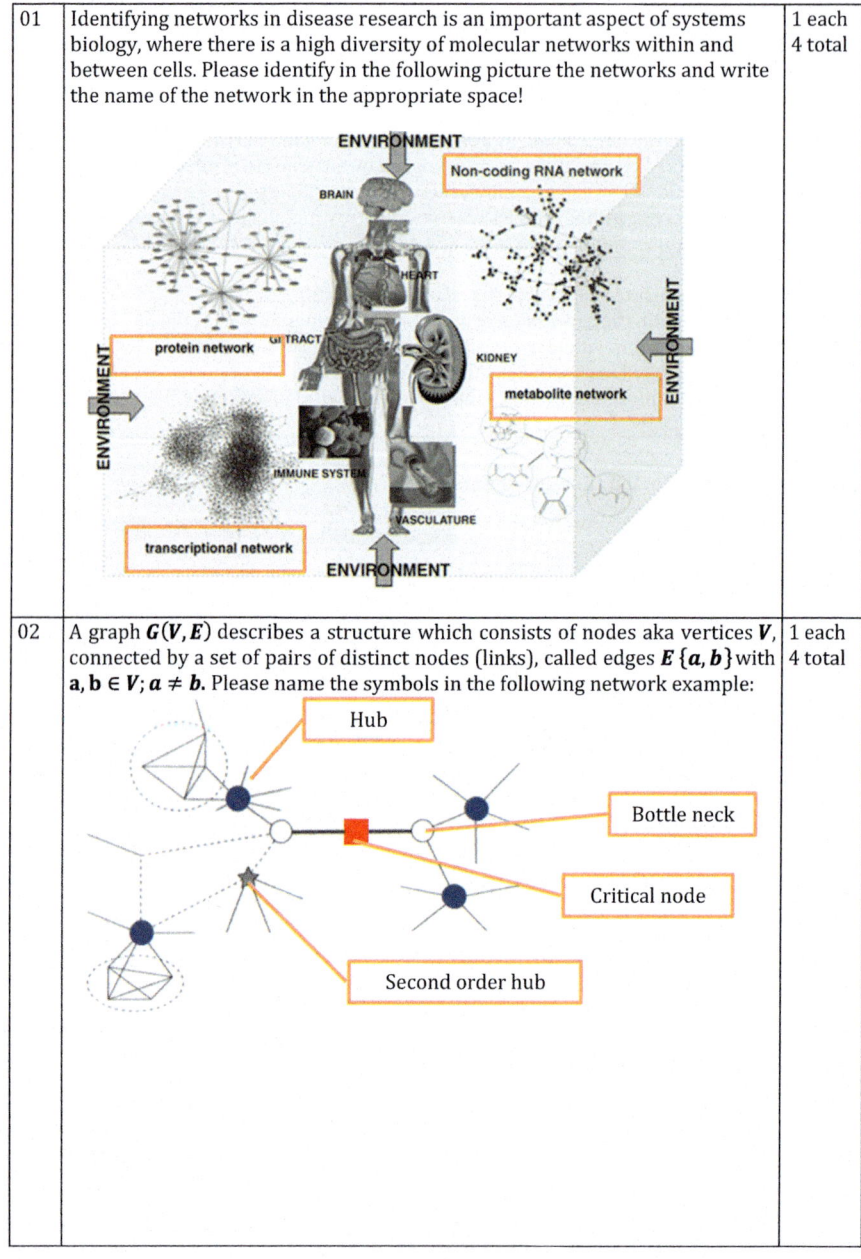

9 Answers

03	In order to represent network data in computers it is not comfortable to use sets; more practical are matrices. The simplest form of a graph representation is the so called adjacency matrix. Please set up the adjacency matrices for the following graphs:	1 each 6 total
	$$a_{jk} = \begin{bmatrix} 0 & 0 & 1 & 0 & 1 & 0 \\ 0 & 0 & 0 & 0 & 1 & 1 \\ 1 & 0 & 0 & 0 & 1 & 0 \\ 0 & 0 & 0 & 0 & 1 & 1 \\ 1 & 1 & 1 & 1 & 0 & 0 \\ 0 & 1 & 0 & 1 & 0 & 0 \end{bmatrix}$$ Simple graph, symmetric, binary $$w_{jk} = \begin{bmatrix} 0 & 0 & -3 & 0 & 1 & 0 \\ 0 & 0 & 0 & 0 & 0 & 3 \\ 3 & 0 & 0 & 0 & 2 & 0 \\ 0 & 0 & 0 & 0 & 1 & -5 \\ 0 & -2 & 0 & 0 & 0 & -10 \\ 0 & 0 & 0 & 0 & 5 & 0 \end{bmatrix}$$ Directed and weighted	
04	In Biomedicine networks of all kind play an extremely important role. Please assign the correct labels to the network metrics below:	1 each 6 total

$\sum_i \sum_j a_{ij}$	Path length
$k_i = \sum_i a_{ij}$	Network size
$C = \frac{1}{n} \sum_i C_i$	Complete number of nodes
n	Clustering coefficient
$l = \frac{1}{n(n-1)} \sum_{i \neq j} d_{ij}$	Nodal degree

05	Please draw a metabolic network, constructed considering the reactants, chemical reactions and enzymes, consisting of A, B, C, D, E1 and E2	1 each 6 total
	[Diagram: metabolic network with nodes A, B, C, D and enzymes E1, E2. E1 catalyzes reaction A+B→C; E2 catalyzes reaction B+C→D.]	

06	Please provide a simple example of a patient record in XML format:
	
	``xml
	<?xml version="1.0"?>
	<patient>
	    <patient-id>11111</patient-id>
	    <Name>Chen</Name>
	    <Date of Birth>1.1.1900</Date of Birth>
	        <diagnosis>
	            <code>123</code>
	            <diagnosistext>Myocardinfarct</diagnosistext>
	        </diagnosis>
	</patient>
	``
		1 each 4 total

Sum of Question Block C (max. 40 points)

References

Aarts E, Lenstra J (1997) Local search in combinatorial optimization. Wiley, New York, NY

Achard F, Vaysseix G, Barillot E (2001) XML, bioinformatics and data integration. Bioinformatics 17(2):115

Barabási A-L, Albert R, Jeong H (1999) Mean-field theory for scale-free random networks. Phys A Stat Mech Its Appl 272(1–2):173–187

Barabasi AL, Albert R (1999) Emergence of scaling in random networks. Science 286(5439):509–512

Bondy JA, Murty USR (1976) Graph theory with applications. Macmillan, London

Boyen P, Van Dyck D, Neven F, Van Ham RCHJ, Van Dijk ADJ (2011) SLIDER: a generic metaheuristic for the discovery of correlated motifs in protein-protein interaction networks. IEEE/ACM Trans Comput Biol Bioinform 8(5):1344–1357

Catchpoole DR, Kennedy P, Skillicorn DB, Simoff S (2010) The curse of dimensionality: a blessing to personalized medicine. J Clin Oncol 28(34):E723–E724

Chien KR, Domian IJ, Parker KK (2008) Cardiogenesis and the complex biology of regenerative cardiovascular medicine. Science 322(5907):1494

Costa LF, Rodrigues FA, Cristino AS (2008) Complex networks: the key to systems biology. Genet Mol Biol 31(3):591–601

Costa LF, Rodrigues FA, Travieso G, Boas PRV (2007) Characterization of complex networks: a survey of measurements. Adv Phys 56(1):167–242

Darwin C (1859) On the origin of species by means of natural selection, or the preservation of favoured races in the struggle for life. John Murray, London

Dehmer M, Emmert-Streib F, Mehler A (2011) Towards an information theory of complex networks: statistical methods and applications. Birkhäuser, Boston, MA

Dhar V (2013) Data science and prediction. Commun ACM 56(12):64–73

Diestel R (2010) Graph theory, 4th edn. Springer, Berlin

Dorogovtsev SN, Mendes JFF (2003) Evolution of networks: from biological nets to the internet and WWW. Oxford University Press, New York, NY

Duda RO, Hart PE, Stork DG (2000) Pattern classification, 2nd edn. Wiley, New York, NY

Edelsbrunner H, Harer JL (2010) Computational topology: an introduction. American Mathematical Society, Providence, RI

Fekete J-D (2004) The infovis toolkit. Information visualization, INFOVIS 2004. IEEE, Washington, DC, pp 167–174

Forster C, Vossen G (2012) Exploiting XML technologies in medical information systems. In: Proceedings of the 25th Bled eConference eDependability: reliable and trustworthy eStructures, eProcesses, eOperations and eServices for the future, Bled, Slovenia, pp 70–83

Gaal SA (1966) Point set topology, 2nd edn. Academic, New York, NY

Geschwind DH, Konopka G (2009) Neuroscience in the era of functional genomics and systems biology. Nature 461(7266):908–915

Golumbic MC (2004) Algorithmic graph theory and perfect graphs. Elsevier, Amsterdam

Harary F (1969) Graph theory. Addison-Wesley, Reading, MA

Hatcher A (2002) Algebraic topology. Cambridge University Press, Cambridge

Henzinger MR, Klein P, Rao S, Subramanian S (1997) Faster shortest-path algorithms for planar graphs. J Comput Syst Sci 55(1):3–23

Hodgman CT, French A, Westhead DR (2010) Bioinformatics, 2nd edn. Taylor & Francis, New York, NY

Holzinger A (2003) Basiswissen IT/Informatik. Band 2: Informatik. Das Basiswissen für die Informationsgesellschaft des 21. Jahrhunrets, Vogel Buchverlag, Wuerzburg.

Holzinger A (2011) Weakly structured data in health-informatics: the challenge for human-computer interaction. In: Baghaei N, Baxter G, Dow L, Kimani S (eds) Proceedings of INTERACT 2011 workshop: promoting and supporting healthy living by design. IFIP, Lisbon, Portugal, pp 5–7

Holzinger A (2012) On knowledge discovery and interactive intelligent visualization of biomedical data: challenges in human–computer interaction & biomedical informatics. In: Helfert M, Fancalanci C, Filipe J (eds) DATA - international conference on data technologies and applications. INSTICC, Rome, pp 5–16

Holzinger A, Ofner B, Stocker C, Valdez AC, Schaar AK, Ziefle M, Dehmer M (2013a) On graph entropy measures for knowledge discovery from publication network data. In: Cuzzocrea A, Kittl C, Simos DE, Weippl E, Xu L (eds) Multidisciplinary research and practice for information systems, vol LNCS 8127, Springer lecture notes in computer science. Springer, Heidelberg, pp 354–362

Holzinger A, Simonic KM, Geier M, Ofner B, Müller R, Heschl S, Prause G (2013) Constraints of list-based knowledge interaction on an android app for emergency medicine. Medicine 2.0 London, Oral Presentation on 23 Sept 2013 (Online). http://www.medicine20congress.com/ocs/index.php/med/med2013/paper/view/1479

Holzinger A, Stocker C, Ofner B, Prohaska G, Brabenetz A, Hofmann-Wellenhof R (2013c) Combining HCI, natural language processing, and knowledge discovery - potential of IBM content analytics as an assistive technology in the biomedical domain, vol LNCS 7947, Springer lecture notes in computer science. Springer, Heidelberg, pp 13–24

Kim PM, Korbel JO, Gerstein MB (2007) Positive selection at the protein network periphery: evaluation in terms of structural constraints and cellular context. Proc Natl Acad Sci U S A 104 (51):20274–20279

Kleinberg JM (2000) Navigation in a small world. Nature 406(6798):845

Koontz WLG, Narendra PM, Fukunaga K (1976) A graph-theoretic approach to nonparametric cluster analysis. IEEE Trans Comput 100(9):936–944

Kreuzthaler M, Bloice MD, Faulstich L, Simonic KM, Holzinger A (2011) A comparison of different retrieval strategies working on medical free texts. J Univ Comput Sci 17(7):1109–1133

Kropatsch W, Burge M, Glantz R (2001) Graphs in image analysis. In: Kropatsch W, Bischof H (eds) Digital image analysis. Springer, New York, NY, pp 179–197

Lage K, Møllgård K, Greenway S, Wakimoto H, Gorham JM, Workman CT, Bendsen E, Hansen NT, Rigina O, Roque FS (2010) Dissecting spatio-temporal protein networks driving human heart development and related disorders. Mol Syst Biol 6(1):1–9

Lézoray O, Grady L (2012) Graph theory concepts and definitions used in image processing and analysis. In: Lezoray O, Grady L (eds) Image processing and analysing with graphs: theory and practice. CRC Press, Boca Raton, FL, pp 1–24

Louie B, Mork P, Martin-Sanchez F, Halevy A, Tarczy-Hornoch P (2007) Data integration and genomic medicine. J Biomed Inform 40(1):5–16

Lucchi A, Smith K, Achanta R, Lepetit V, Fua P (2010) A fully automated approach to segmentation of irregularly shaped cellular structures in EM images. Medical image computing and computer-assisted intervention—MICCAI 2010. Springer, Berlin, pp 463–471

Meijster A, Roerdink JB (1995) A proposal for the implementation of a parallel watershed algorithm. Computer analysis of images and patterns. Springer, Berlin, pp 790–795

Milgram S (1967) The small world problem. Psychol Today 2(1):60–67

Newman MEJ (2003) The structure and function of complex networks. SIAM Rev 45:167–256

Rassinoux A-M, Lovis C, Baud R, Geissbuhler A (2003) XML as standard for communicating in a document-based electronic patient record: a 3 years experiment. Int J Med Inform 70(2–3):109–115

Roerdink JB, Meijster A (2000) The watershed transform: definitions, algorithms and parallelization strategies. Fundamenta Informaticae 41(1):187–228

Roque FS, Jensen PB, Schmock H, Dalgaard M, Andreatta M, Hansen T, Søeby K, Bredkjær S, Juul A, Werge T, Jensen LJ, Brunak S (2011) Using electronic patient records to discover disease correlations and stratify patient cohorts. PLoS Comput Biol 7(8):e1002141

Salgado H, Santos-Zavaleta A, Gama-Castro S, Peralta-Gil M, Peñaloza-Spínola MI, Martínez-Antonio A, Karp PD, Collado-Vides J (2006) The comprehensive updated regulatory network of Escherichia coli K-12. BMC Bioinform 7(1):5

Schadt EE, Lum PY (2006) Reverse engineering gene networks to identify key drivers of complex disease phenotypes. J Lipid Res 47(12):2601–2613

Schmid AK, Reiss DJ, Pan M, Koide T, Baliga NS (2009) A single transcription factor regulates evolutionarily diverse but functionally linked metabolic pathways in response to nutrient availability. Mol Syst Biol 5(1):1–9

Simon HA (1973) The structure of ill structured problems. Artif Intell 4(3–4):181–201

Strogatz SH (2001) Exploring complex networks. Nature 410(6825):268–276

Usdin T, Graham T (1998) XML: not a silver bullet, but a great pipe wrench. ACM Stand View 6(3):125–132

Van Den Heuvel MP, Hulshoff Pol HE (2010) Exploring the brain network: a review on resting-state fMRI functional connectivity. Eur Neuropsychopharmacol 20(8):519–534

Vincent L, Soille P (1991) Watersheds in digital spaces: an efficient algorithm based on immersion simulations. IEEE Trans Pattern Anal Machine Intell 13(6):583–598

Wang Z, Zhang JZ (2007) In search of the biological significance of modular structures in protein networks. PLoS Comput Biol 3(6):1011–1021

Watts DJ, Strogatz SH (1998) Collective dynamics of 'small-world' networks. Nature 393(6684):440–442

Wiltgen M, Tilz GP (2009) Homology modelling: a review about the method on hand of the diabetic antigen GAD 65 structure prediction. Wiener Medizinische Wochenschrift 159(5):112–125

Wittkop T, Emig D, Truss A, Albrecht M, Böcker S, Baumbach J (2011) Comprehensive cluster analysis with transitivity clustering. Nat Protoc 6(3):285–295

Lecture 6

Multimedia Data Mining and Knowledge Discovery

1 Learning Goals

At the end of this sixth lecture you:

- would be aware of the importance of gaining knowledge from (big) data.
- would know the differences between data mining, knowledge discovery, and information retrieval (Lecture 4).
- would have some knowledge about the various types of multimedia data in biomedical informatics.
- would be able understand the basic process of knowledge discovery from data (bases) (KDD).
- would get an overview on some data mining algorithms used in biomedical informatics.
- would have seen an example of the application of support vector machines.

2 Advance Organizer

Artificial neural network (ANN)	A computational adaptive model (inspired by biological neural networks), consisting of interconnected groups of artificial neurons; processes information using a connectionist approach.
Association rule learning	A set of techniques for discovering interesting relationships, i.e., "association rules," among variables in large databases used for data mining
Classification	A set of techniques to identify the categories in which new data points belong, based on a training set containing data points that have already been categorized; these techniques are often described as

	supervised learning because of the existence of a training set; they stand in contrast to cluster analysis, a type of unsupervised learning; used for example for data mining
Cluster analysis	Statistical method for classifying objects that splits a diverse group into smaller groups of similar objects, whose characteristics of similarity are not known in advance; a type of unsupervised learning because training data are not used—in contrast to classification; used for data mining.
Data mining	A set of techniques to extract patterns from large data by combining methods from statistics and machine learning with database management (association rule learning, cluster analysis, classification, regression, etc.)
Knowledge Discovery (KD)	Process of identifying valid, novel, useful and understandable patterns out of large volumes of data
Knowledge Extraction	Is the creation of knowledge from structured (relational databases, XML) and unstructured (text, documents, images) sources
Multimedia	Several data of different modalities are processed at the same time, i.e., encompassing audio data (sound, speech), image data (b/w and color images), video data (time-aligned sequences of images), electronic ink (sequences of time aligned 2D and 3D coordinates of a stylus, pen, data gloves, etc.)
Principal Component Analysis (PCA)	Statistical technique for finding patterns in high-dimensional data
Supervised learning	Inferring a function from supervised training data on the basis of training data which consist of a set of training examples, the input objects (typically vectors) and a desired output value (also called the supervisory signal).
Supervised learning algorithm	Analyzes the training data and produces an inferred function, called a classifier (if the output is discrete) or a regression function (if the output is continuous); the algorithm generalizes from the training data to unseen situations.
Support vector machine (SVM)	Concept for a set of related supervised learning methods to analyze data and recognize patterns, used for classification and regression analysis.
Unsupervised learning	Establishes clusters in data, where the class labels of training data is unknown.

3 Acronyms

ANN	Artificial Neural Network
ANOVA	Analysis of Variance
AUC	Area under the curve
CDT	Clinical Decision Tree
DM	Data Mining
KDD	Knowledge Discovery from Data(bases)
MDM	Multimedia Data Mining
MELD	Model for end-stage liver disease
MM	Multimedia
NLP	Natural Language Processing
ROC	Receiver-operating characteristic
SVM	Support Vector Machine

4 Key Problems

Slide 6-1: Key Challenges

Summarized, we have the following challenges:

(1) Cross-disciplinary cooperation with domain experts
(2) Data-driven challenges including (a) Massive datasets; (b) Heterogeneous data; (c) Streaming data (e.g., from large sensor networks, Multimedia); (d) Graph data (e.g., protein network data); (e) Data restrictions (accessibility, privacy, legal restrictions, etc.);
(3) Context (data mining occurs always in a particular context)
(4) Interpretability
(5) Computational resources
(6) Benchmarking against Gold-Standards
(7) Embedded data mining

(1) Cross-disciplinary cooperation with domain experts
   A big issue is the so-called knowledge acquisition bottleneck. Many researchers from the data mining community expected that machine learning techniques would automate the knowledge acquisition process and thus would exclude experts form the process of building models—but the opposite is true, we need the knowledge of the domain expert, and

(continued)

(continued)

  hence data mining approaches must put the human intelligence into the loop (Holzinger 2013).
(2) Data-driven challenges including

  (a) Massive datasets
  (b) Heterogeneous data
  (c) Streaming data (e.g., from large sensor networks)
  (d) Graph data
      Many data mining methods are designed for collections of objects well represented in rigid tabular formats. However, we are increasingly finding ourselves confronting collections of interrelated objects whose natural representation is in graphs, in particular biomedical structures.
  (e) Data restrictions
      Including data accessibility, temporal limits, legal restrictions, privacy, data provenance (such as in situations where copyright or patents may be relevant). We face a range of research challenges in developing data mining methods capable of honoring various restrictions that are not directly about the data but rather about the use of the data.

(3) Context
  In such complex domains as in biomedicine we need experts who understand the domain, the problem, and the datasets, and hence the context (Berka et al. 2007).

(4) Interpretability
  Much data mining research focuses on well-defined metrics such as classifier accuracy, which does not always match the goals of particular data mining tasks. One broad example of this is a case where there is a need for interpretability in the results of data mining. In biomedicine, customer relationship management, or domains with appropriate regulatory characteristics (where you may need to explain the results of a decision), data mining is often irrelevant if it does not produce results that can be explained to others. In considering the current state of data mining methods, especially classification methods, we appear to have a tension between effectiveness and interpretability. How can we measure interpretability if we want to investigate it scientifically? What techniques can we develop for enabling interpretability of complex models? Interpretability is in the eye of the beholder (consider what is required to achieve interpretability for an expert as opposed to a novice), so can we develop methods that allow suitably varying levels of interpretability?

(continued)

(continued)

Can we develop methods that allow us to trade off between interpretability and accuracy? We face many challenges in advancing our ability to perform data mining in suitably interpretable ways.

(5) Computing Resources

In recent years, the computing platforms on which we perform data mining have begun changing dramatically. The advent of multi core architectures begs the question of how we can best implement our algorithms on the next generation of computers. Perhaps even more importantly, data are often of such enormity that moving data to computing no longer is credible, and we now must move computing to the data. We are beginning to see the development of large scale data-intensive computing clusters by major Internet companies, and such architectures can change the nature of how data mining will be performed. One important question raised by the development of computing resources tailored to massive data is whether there is value in intentionally developing simple mining algorithms that, through the use of data-intensive computing resources, allow us to exploit more data than would otherwise be the case, in the hope that they may yield better results than would occur with less data even in the presence of greater algorithmic sophistication.

(6) Benchmarking against Gold-Standards

Some areas of data mining rely heavily on benchmark datasets (see for example Kreuzthaler et al. 2010; 2011), which allow us to compare results across competing methods. However, benchmarks are a means to an end, not the end in itself. These benchmark problems are intended to be representative of the sorts of problems our algorithms will see in practice, but what we will see in practice will change over time. We thus make sure that our datasets stay timely as technological and scientific advances allow our ambitions to grow. For example, we do not have widely available benchmark "massive" datasets. There is value in enabling widespread access to benchmark massive datasets, such as snapshots of the World Wide Web, social networks, etc. Even so, we must remain vigilant, realizing that even these benchmarks will outlive their value as research tools as our capabilities and ambitions grow.

(7) Reproducibility, both to verify the results of others and to build off of them. Unfortunately, we often do not see results documented with the rigor found in other non computing experimental sciences. How can we replicate results generated on proprietary or otherwise restricted data? How can results be replicated if they require the use of specialized equipment, such as massive data clusters found in only a handful of

(continued)

(continued)

commercial enterprises? How do we make sure academic research addresses problems that are important in practice when most academic institutions cannot match the resources available in the most heavily data-driven enterprises? More fundamentally, even when a researcher has access to identical resources, including both data as well as computing and software platforms, it is often difficult to replicate results identically in the absence of details about data cleaning, algorithm parameter selections, and the like. As a scholarly community, we need to be able to document experimental results in sufficient detail to allow reproducibility.

(8) Embedded data mining

A future goal would be to have KDD tools not separately but integrated in the workbenches, e.g., integrated into the clinical workplace in a hospital. Advanced biomedical technologies have made systems biology a promising area in which data mining faces new challenges and has an increasing importance (Hirsh 2008).

5 Knowledge Discovery

5.1 What Is Knowledge?

Slide 6-2: What Is Knowledge?

Please remember what you have learned in Lecture 2. Look at the slide 2–20: The incoming stimuli from the physical world must pass both a perceptual filter and a conceptual filter. The **perceptual filter** orientates the senses (e.g., visual sense) to certain types of stimuli (e.g., visual stimuli) within a certain physical range (visual signal range, pre-knowledge, attention, etc.). Only the particular stimuli which pass through this filter get registered as *incoming data*—everything else is filtered out. At this point it is important to follow our physical principle of data: to differentiate between two notions that are frequently confused: an experiment's (raw, hard, measured, factual) data and its (meaningful, subjective) **interpreted information** results. Data is property concerning only the physical instrument; it is the **expression of a fact**. The result concerns a property of the world and the subsequent conceptual filters extract information-bearing data from what has been previously registered. It is important to emphasize that both types of filters are influenced by the agents' cognitive and affective expectations, stored in their mental

(continued)

(continued)

models. The enormous utility of data resides in the fact that it can carry information about the physical world. This information may modify **set expectations or the state-of-knowledge**. These principles allow an **agent** (a human agent or an computer agent) to act in adaptive ways in this physical world (Boisot and Canals 2004). Confer this process with the human information processing model by Wickens (1984), seen in Slide 2-19 and discussed in Lecture 7.

Note: This is one possible theory amongst various theories about human information processing.

Remark: It takes energy for a stimulus to register as data, the amount of energy being a function of the sensitivity of the agent's sensory apparatus (attention). Subsequently, Information constitutes significant regularities in this data. What constitutes a *significant* regularity, hence information, can only be established with respect to the individual dispositions of the receiving agent. Information, in effect, sets up a relation between in-coming data and the previous knowledge of the agent.

Consequently, knowledge is a set of expectations held by agents and modified by the arrival of information.

These expectations embody the prior situated interactions between agents and the world—in short, the agent's prior learning. Such learning need not be limited—as required by the theory of rational expectations (Muth 1961)—to models specifically relevant to the expectations to which they give rise.

To summarize, we might say that information is an extraction from data that, by modifying the relevant probability distributions, has a capacity to perform useful work on an agent's knowledge base. The essential relationships between data, information and knowledge are depicted in the slide. The diagram indicates that agents operate two kinds of filters in converting incoming stimuli into information. Perceptual filters first orient the senses to certain types of stimuli that operate within a given physical range. Only stimuli passing through this initial filter get registered as data.

Conceptual filters then extract information-bearing data from what has been so registered. Both types of filters get "tuned" by the agents' cognitive and affective expectations, shaped as these are by prior knowledge, to act selectively on both stimuli and data. The schema depicted in the slide allows us to view data, information and knowledge as distinct kinds of economic goods, each possessing a specific type of utility. The utility of data resides in the fact that it can carry information about the physical world; that of information, in the fact that it can modify an expectation or a state of knowledge; finally, that of knowledge in the fact that it allows an agent to act in adaptive ways in and upon the physical world (Boisot and Canals 2004).

5.2 Implicit vs. Explicit Knowledge

> **Slide 6-3: Implicit vs. Explicit Knowledge**
>
> Remember what we have learned in Lecture 2:
>
> **Explicit Knowledge** is obtained by inductive reasoning with previously interpreted data, collected from many similar patients or processes, which is added to the so-called "body of knowledge."
>
> **Implicit Knowledge** is gained by the interpretation of other data on the basis of explicit knowledge and guides the clinician in making decisions and taking further action.

5.3 Differences Between KDD and DM

> **Slide 6-4: The Classic Differentiation Between DM and KDD**
>
> We can understand the difference between DM and KDD best, when looking at the classic KDD process chain by Fayyad et al. (1996): DM is a subset of KDD.
>
> Explanation of the steps:
>
> 1. **Learning from the application domain**: includes understanding relevant previous knowledge, the goals of the application and a certain amount of domain expertise;
> 2. **Creating a target dataset**: includes selecting a dataset or focusing on a subset of variables or data samples on which discovery shall be performed;
> 3. **Data cleansing** (this is not a typo, it is called cleansing, not cleaning;-)) and **preprocessing**: includes removing noise or outliers, strategies for handling missing data etc.;
> 4. **Data reduction and projection**: includes finding useful features to represent the data, dimensionality reduction etc.
> 5. **Choosing the function of data mining**: includes deciding the purpose and principle of the model for mining algorithms (e.g., summarization, classification, regression, and clustering);
> 6. **Choosing the data mining algorithm**: includes selecting method(s) to be used for searching for patterns in the data, such as deciding which models and parameters may be appropriate (e.g., models for categorical data are different from models on vectors over reals) and matching a particular data mining method with the criteria of the KDD process;
> 7. **Data mining**: searching for patterns of interest in a representational form or a set of such representations, including classification rules or trees, regression, clustering, sequence modelling, dependency, and line analysis;
>
> (continued)

(continued)
8. **Interpretation**: includes interpreting the discovered patterns and possibly returning to any of the previous steps, as well as possible visualization of the extracted patterns, removing redundant or irrelevant patterns, and translating the useful ones into terms understandable by users;
9. **Using discovered knowledge**: includes incorporating this knowledge into the performance of the system, taking actions based on the knowledge, or documenting it and reporting it to interested parties, as well as checking for and resolving potential conflicts with previously believed knowledge.

Slide 6-5: Interactive Knowledge Discovery and Data Mining Overview

Interaction, communication, and sensemaking are still missing, which are core topics in HCI (Blandford and Attfield 2010). Consequently, a novel approach is to combine HCI & KDD (Holzinger 2013) in order to enhance human intelligence by computational intelligence. The main contribution of HCI-KDD is to *enable* end users to *find and recognize* previously unknown and potentially useful and usable information. It may be defined as the process of identifying novel valid, and potentially useful data patterns, with the goal to understand these data patterns (Funk and Xiong 2006). The domain expert in Slide 6-5 possesses explicit domain knowledge and by enabling him to interactively look at datasets, he may be able to identify, extract, and understand useful information, to gain new, previously unknown knowledge (Holzinger et al. 2012).

Slide 6-6: Research Areas Involved

KDD historically builds on three fields: computational statistics and machine learning; databases and artificial intelligence to construct tools that let the end users gain insight into the nature of massive datasets. Pazzani (2000) emphasizes that it is necessary to take human cognitive processes into account in order to increase the usefulness of KDD systems. End user's perceptions of novelty, utility, and understandability ultimately determine the acceptance of data mining. Figure 6-4 illustrates how these four fields in combination form the field of KDD. People have been learning representations of the environment for a long time and have been using these learned models to guide decision-making. Psychological investigation has revealed factors that simplify the learning, understanding, and communication of category information.

6 Data Mining Methodologies

6.1 Definitions

Slide 6-7: Definitions

Originally, data mining was applied to laboratory research, clinical trials, actuarial[1] studies, and risk analysis. It is now applied from the infinitely small microscopic area (omics) to the infinitely large macroscopic area (astrophysics), from the most general (customer relationship management, CRM) to the most specialized (assistance to surgeries in operations), from the most open (e-commerce) to the most secret (fraud detection in mobile telephony and bank card applications), from the most practical (quality control, production management) to the most theoretical (human sciences, biology, medicine, and pharmacology), and from the most basic (life science) to the most entertaining (audience prediction for television).

The most relevant fields are those where large volumes of data ("big data") have to be analyzed, mostly with the aim of (rapid) decision making, which is in biomedicine still the key topic (Tufféry 2011).

Data mining is defined as the set of methods and techniques for exploring and analyzing (big) datasets, in an automatic or semi-automatic way, in order to find among these data certain *unknown or hidden rules, associations, or tendencies.*

Data mining is *the art* of extracting information from data to gain insight into the data (sensemaking, knowledge, wisdom)—as a core element of the whole knowledge discovery process.

Data mining is therefore both descriptive and predictive:

(a) <u>Descriptive (exploratory)</u> techniques discover information that is present but hidden in the mass of data (as in the case of automatic clustering of individuals and searches for associations between products or medicines), whilst

(b) <u>Predictive (or explanatory)</u> techniques are designed to extrapolate new information based on present information. In the concept map in the next slide we can see the main applications in the field of biomedical informatics and engineering.

---

[1] Actuarial science is the discipline that applies mathematical and statistical methods to assess risk, e.g., for insurance and finance.

6.2 Example Tasks

Slide 6-8: Data Mining in Biomedical Engineering

In this slide we see a concept map of typical applications of data mining in biomedicine (Suh et al. 2011): In biomedicine data mining is applied to four main fields: molecular biology, biosurveillance, epidemiology, and clinical field. In molecular biology researchers search for related genes and protein interactions, and they are using datasets derived from proteomics and genomics, e.g., from gene expression data or sequencing (e.g., high-throughput sequencing).

On the clinical side clinical data mining aims for discovery of patient subgroups, image and signal processing, data artifact detection and event detection (for biosurveillance and epidemiology) and decision support which can be diagnostic or prognostic.

Slide 6-9: Typical Data Mining Tasks

Data mining tasks include:

1. Clustering (assigning a set of objects into groups); classification (predicting an item class, i.e., identifying which set of categories a new observation belongs);
2. Associations (finding that A and B and C occur frequently together);
3. Visualization (to facilitate human cognition);
4. Deviation detection (e.g., finding changes out of a given limit);
5. Anomaly detection (to find anomalies, irregularities, or inconsistencies);
6. Estimation (e.g., predicting a continuous value);
7. Link analysis (finding relationships);
8. Forecasting (e.g., predicting a trend)

6.3 Taxonomy of Methods

Slide 6-10: Taxonomy of Data Mining Methods

There is a large variety of methods used for very different purposes (Maimon and Rokach 2010). This slide provides a rough overview on a possible taxonomy:

(I) **Verification Methods** deal with the evaluation of a hypothesis proposed by an external source, e.g., a human expert. Methods include traditional

(continued)

(continued)

statistics (e.g., Goodness of fit test), classic hypothesis testing (e.g., *t*-test) and most of all the Analysis of Variance (ANOVA).

(II) **Discovery Methods** are for automatic identification of patterns in the data, mostly based on inductive learning, where a model is constructed (explicitly or implicitly), by generalization from a number of training samples. The assumption is to use the trained model and apply it for future (unforeseen) examples. We have to distinguish between two different types of such discovery methods:

(a) **Description methods** aim to understand the process on *how* the underlying data is related to their parts. Typical methods include: clustering (aka *unsupervised learning*), summarization, linguistic summary, and visualization. The comparison in Slide 6-11 and an example from gene expression in Slide 6-12 shall make the differences of unsupervised vs. supervised learning clear.

(b) **Prediction methods** (aka *supervised learning*) aim for automatic building of a behavioral model, which captures new and previously unseen samples and is able to predict values of one or more variables related to this sample. Prediction methods can be further categorized into:

(1) *Classification*, e.g., by application of neural networks, bayesian networks, decision trees, support vector machines, and instance-based methods and

(2) *Regression*, e.g., in the simplest form (linear regression, or multiple regression);

The application of supervised learning to a real-world medical problem is described in Slide 6-13.

Slide 6-11: Differences Between Unsupervised and Supervised Learning

Unsupervised learning (e.g., clustering) means that the class labels of training data is unknown, i.e., given a set of measurements, observations, etc. we are aiming to establish the existence of clusters in the data;

Supervised learning (classification)

Supervision = the training data (observations, measurements, etc.) are accompanied by labels indicating the class of the observations;

New data is classified based on the training set.

Slide 6-12: Example Unsupervised vs. Supervised Learning

(i) **Unsupervised learning (clustering)**: multiple tumor samples are clustered into groups based on overall similarity of their gene expression profiles. This approach is useful for discovering previously unappreciated relationships.

(ii) **Supervised learning (classification)**: multiple tumor samples from different known classes are used to train a model capable of classifying unknown samples. This model is then applied to a test set for class label assignment.

Slide 6-13: Unsupervised → Supervised → Semi-supervised

(a) Unsupervised Learning: An clustering algorithm (latent semantic indexing, k-means, principal component analysis, etc.) is applied on the raw data and produces automated clusters, which must then be manually reviewed before the data can be used. Algorithms to unsupervised learning include: k-means, mixture models, hierarchical clustering, hidden Markov models, blind signal separation using feature extraction techniques for dimensionality reduction (e.g., principal component analysis (PCA), independent component analysis (ICA)) (Bandyopadhyay and Saha 2013).

(b) Supervised learning: Sample datasets are manually selected and used as a model for training the algorithm. The product of the trained algorithm must then be manually verified before the data can be used.

(c) Semi-supervised learning: This begins in the same manner as supervised learning, but the manual review of the data happens before the algorithm application.

6.4 Supervised Learning

Slide 6-14: Supervised Learning Principle

This slide shows a more detailed view on the supervised learning principle in three steps according to Kotsiantis (2007):

Step 1 Identification of required data: The first step is collecting the dataset by help of a medical expert. If no expert is available, then the simplest method is "brute-force," i.e., measuring everything, in the hope that the right (relevant!) features can be isolated. However, a dataset collected by the "brute-force" method is not directly suitable for induction. It contains in most cases noise, artifacts and missing feature values, hence requires much preprocessing effort.

(continued)

(continued)

Step 2 Data preprocessing: The second step is the data preparation to handle missing data. Instance selection is used not only to handle noise but also to cope with the infeasibility of learning from very large datasets. Instance selection in these datasets is an optimization problem that attempts to maintain the mining quality while minimizing the sample size (Liu et al. 2004). It reduces data and enables a data mining algorithm to function and work effectively with very large datasets. There is a variety of procedures for sampling instances from a large dataset and feature subset selection is the process of identifying and removing as many irrelevant and redundant features as possible.

Step 3 Algorithm selection: Once preliminary testing is judged to be satisfactory, the classifier (mapping from unlabeled instances to classes) is available for routine use. The classifier's evaluation is most often based on prediction accuracy (the percentage of correct prediction divided by the total number of predictions). There are only two techniques described which are used to calculate a classifier's accuracy. One technique is to split the training set by using two-thirds for training and the other third for estimating performance. In another technique, known as *cross-validation*, the training set is divided into mutually exclusive and equal-sized subsets and for each subset the classifier is trained on the union of all the other subsets. The average of the error rate of each subset is therefore an estimate of the error rate of the classifier.

6.4.1 Artificial Neural Networks

Slide 6-15: Example for Supervised Learning: ANN

In this slide we see a neural network on the example of *Macaca fascicularis* (monkey) and we can see the structure of the nerve cells, see brainmap.org for more examples; Mikula et al. (2007).

Artificial neural networks (ANN) mimic the human neural network and literally "learn" from examples to find patterns in data or to classify data. The advantage is that it is not necessary to have any specific model in mind when running the analysis. ANN finds interaction effects (e.g., effects from the combination of age and gender) which must be explicitly specified in regression. The disadvantage is that it is harder to interpret the resultant model with its layers of weights and arcane transformations. Neural networks are therefore useful in predicting a target variable when the data are highly nonlinear

(continued)

(continued)

with interactions, but they are not very useful when these relationships in the data need to be explained (Vikram and Upadhayaya 2011).

Note: The human brain can be described as a **highly parallelized** biological neural network, an interconnected network of neurons transmitting patterns of electrical signals, where the dendrites receive input signals and, based on those inputs, fire an output signal via an axon (see Slide 6-16).

Remark: When it comes to tasks other than number crunching, the human brain possesses numerous advantages over a digital (von Neumann) computer. For example, humans can quickly recognize a face, even when seen from the side in bad lighting in a room full of other objects. We can easily understand natural speech, even that of an unknown person in a noisy room. Despite years of focused research, computers are far from performing well at this level. Moreover, the brain is also remarkably robust; it does not stop working just because a few cells die. The computations of the brain are done by a highly interconnected network of neurons, which communicate by sending electric pulses through the neural wiring consisting of axons, synapses, and dendrites (see Slide 6-16). McCulloch and Pitts (1943) modelled a neuron as a switch which receives input from other neurons and, depending on the total **weighted input**, is either activated or remains inactive.

Slide 6-16: Neuron: Information Flow Through

Here we see the information flow within a neuron: the dendrites collect electrical signals, the cell body integrates the incoming signals and generates outgoing signals to the axon, the axon passes electrical signals to dendrites of another cell or to an effector cell (Freeman 2008).

The weight, by which an input from another cell is multiplied, corresponds to the strength of a synapse (the neural contacts between nerve cells). These weights can be both positive (excitatory) and negative (inhibitory).

Slide 6-17: Artificial Neural Network: ANN 1/4

In the 1960s, it was shown that networks of such model neurons have properties similar to the brain: they can perform sophisticated pattern recognition, and they can function even if some of the neurons are destroyed. Simple networks of such model neurons are called a **perceptron** (Rosenblatt 1958).

However, such models can solve only a very limited class of linearly separable problems. Nonetheless, the error back-propagation method, which can make fairly complex networks of simple neurons learn from examples, showed that these networks could solve problems that were not linearly

(continued)

(continued)

separable. In this slide we see such a model neuron perceptron: It receives input from a number of other units (or external sources), weighs each input and adds them up. If the total input is above a threshold, the output of the unit is one; otherwise it is zero. Therefore, the output changes from 0 to 1 if the total weighted sum of inputs is equal to the threshold. The points in the input space satisfying this condition define a so-called hyperplane. In 2D such a hyperplane is a line, whereas in 3D it is a normal plane. Data points on one side of the hyperplane are classified as 0 and those on the other side as 1.

So, we have a typical classification problem, which can be solved by a threshold unit if the two classes can be separated by a hyperplane (Krogh 2008)—as can be seen in Slide 6-18.

Slide 6-18: Classification Problem in Hyperplane: ANN 2/4

Such problems are called **linearly separable** and can be seen in this slide on the example of a classification problem in $\mathbb{R}^3$.

If this classification problem is separable, we still need a way to set the weights and the threshold, such that the threshold unit correctly solves the classification problem. This can be done in an iterative manner by presenting examples with known classifications: This process is called learning (or training), which can be implemented by various algorithms. During this learning process, the hyperplane moves around until it finds its correct position in space.

This is nicely illustrated by a set of Java Applets which can be found here: http://lcn.epfl.ch/tutorial/english/index.html

Medical Example: Let us think of two classes of cancer, only one responds to a certain treatment. As there is no simple biomarker to discriminate the two, we can use gene expression measurements of tumor samples to classify them. Assuming to measure gene expression values for 20 different genes in 50 tumors of class 0 (nonresponsive) and 50 of class 1 (responsive). On the basis of these data, we train a threshold unit that takes an array of 20 gene expression values as input and gives 0 or 1 as output for the two classes, respectively. If the data are linearly separable, the threshold unit will classify the training data correctly. But many classification problems are not linearly separable, so we must introduce more hyperplanes, that is, by introducing more than one threshold unit. This is usually done by adding an extra (hidden) layer of threshold units each of which does a partial classification of the input and sends its output to a final layer, which assembles the partial classifications to the final classification, the so-called **multilayer perceptron**, see Slide 6-19.

Slide 6-19: Multilayer Perceptron: ANN 3/4

In this slide we see a **multilayer perceptron** aka **feed-forward network**. These can be used for regression problems, which require continuous outputs, as opposed to binary outputs (0 and 1). By replacing the step function with a continuous function, output is a real number. A so-called **sigmoid** function is used, i.e., a soft version of the threshold function. The sigmoid function can also be used for classification problems by interpreting an output below 0.5 as class 0 and an output above 0.5 as class 1; often it makes sense to interpret the output as the probability of class 1.

In our example we can have a situation where class 1 is characterized by either a highly expressed gene 1 and a silent gene 2 or a silent gene 1 and a highly expressed gene 2; if neither or both of the genes are expressed, it is a class 0 tumor. This corresponds to the logical exclusive or function and it is the canonical example of a nonlinearly separable function. In this case, it would be necessary to use a multilayer network to classify the tumors (Krogh 2008).

Slide 6-20: Danger of Over-Fitting: ANN 4/4

Over-fitting occurs when the network has too many parameters to be learned from the number of examples available, i.e., when a few points are fitted with a function with too many free parameters. In the slide we clearly see that the original eight data points shown by the "+ symbols" are located on a parabola (apart from some noise). They were used to train three different neural networks. The networks all take an x value as input (one input) and are trained with a y value as desired output. As expected, a network with just one hidden unit (green) does not do a very good job. A network with ten hidden units (blue) approximates the underlying function remarkably well. The last network with 20 hidden units (purple) over-fits the data: the training data points are learned perfectly, but for some of the intermediate regions the network is overly "creative" (Krogh 2008).

Slide 6-21: Neural Networks in Biomedical Engineering

Neural networks have been applied to many interesting problems in medicine and biology and there are many other types of neural networks, for example:

Hopfield networks,
Boltzmann machines,
Kohonen nets, and
Unsupervised networks.

(continued)

(continued)

Hopfield networks are recurrent ANNs serving as content-addressable memory systems with binary threshold nodes (Hopfield 1982, 1987). Recurrent networks mean that the graph may contain (directed) cycles. Actually, Hopfield is the simplest form, which originated as physical models to describe magnetism.

Boltzmann machines are the stochastic, generative counterpart of Hopfield nets. They were one of the first examples of a neural network capable of learning internal representations, and are able to represent and to solve difficult combinatory problems (Barra et al. 2012; Leitner et al. 2006).

Kohonen networks are self-organizing maps and a computational method for the visualization and analysis of high-dimensional data. The basic idea of Kohonen (1982) was, that topologically correct maps of structured distributions of signals can be formed in a one or 2D array of processing units which did not have this structure initially. This principle is a generalization of the formation of direct topographic projections between two laminar structures known as retinotectal mapping.

6.4.2 Clinical Example: Model for End-Stage Liver Disease

Slide 6-22: Risky Medical Example: Liver Transplantation

Liver transplantation is an accepted treatment for patients with end-stage liver disease. Risking the imbalance between the potential number of recipients and the donor shortage, it is **of top importance to accurately predict the prognosis of patients with liver cirrhosis to establish a correct timing** of referral to a liver transplant program on one hand and prioritize the allocation of organs to the most ill patient already on the waiting list on the other (Cucchetti et al. 2007).

Case: A 51-year-old woman with polycystic liver and kidney disease had undergone renal transplantation 21 years before. She had no evidence of cerebrovascular malformations. Both her father and aunt also had polycystic kidney disease. After the renal transplantation, her liver had become progressively diseased and enlarged through cystic changes. Early satiety, malnutrition, and abdominal pain necessitated a liver transplantation. The recipient's weight at transplantation was 59 kg. A 9.1-kg liver (white arrow) was removed and replaced with a whole graft that was one tenth the weight of the diseased liver (black arrowhead). A large cyst at the dome of the native liver had to be decompressed (white arrowhead) to allow for access to the recipient's suprahepatic vena cava. She made an excellent recovery and had

(continued)

(continued)

normal kidney and liver function at 4 years of follow-up; case by William J. Wall, MD (Wall 2007).

Slide 6-23: Model for End-Stage Liver Disease (MELD)

The so-called MELD score (see also Slide 6-25) is a modification of the risk score used in the original Transjugular Intrahepatic Portosystemic Shunt (TIPS) model. For ease of use, the score was multiplied by 10 and then rounded to the nearest integer. Here we can clearly see the 3-month mortality of patients with cirrhosis awaiting liver transplantation (Cucchetti et al. 2007).

An on-line worksheet is available over the Internet at www.mayo.edu/intmed/gi/model/mayomodl.htm

Slide 6-24: ANN Application Example: Liver Transplantation

In this slide we see the schematic representation of the artificial neural network model developed to predict the 3-month mortality of patients with cirrhosis awaiting liver transplantation (Cucchetti et al. 2007).

Abbreviations: ALP = alkaline phosphatase; AST = aspartate aminotransferase; GGT = cglutamyl-transpeptidase; INR = international normalized ratio; WBC = white cell count;

Slide 6-25: Diagnostic Accuracy of the ANN

Here we see the diagnostic accuracy at different cutoff values of the ANN output in the internal validation group and external cohort compared to the MELD score (Cucchetti et al. 2007).

Abbreviations: ALP = alkaline phosphatase, AST = aspartate aminotransferase, GGT = gamma-glutamyl transpeptidase, INR = international normalized ratio, WBC = white blood cell).

Slide 6-26: Another Clinical Case Example

This slide shows the MRI of the right breast of a 49-year-old woman presented with a 12-cm mass, fixed to the chest wall. The overlying skin was erythematous and edematous. A core-biopsy specimen showed infiltrating ductal carcinoma.

(continued)

(continued)

Of all breast cancers, only 1–6 % are classified as inflammatory breast cancer. Approximately 20–40 % of patients with inflammatory breast cancer have evidence of distant metastases at the time of presentation (Overmoyer et al. 2011).

Slide 6-27: Important in Clinical Practice → Prognosis!

An important task in clinical patient management is to determine a prognosis for a patient that suffers from a disease. We define prognosis as the **prediction of the future course** of a disease process conditional on the patient's history and a projected treatment strategy. This is nontrivial since the physician often has incomplete information and treatment itself can have many **uncertain effects** (Knaus et al. 1991).

Note: Decision-trees, neural networks, and support vector machines are not always useful, because they require the availability of large amounts of high-quality data. These data are not always available and the only source of knowledge may be that of domain experts which is often condensed in the form of medical textbooks, guidelines, or procedures.

Slide 6-28: Model-Based Clinical Decision Making Strategy

This slide depicts the representation of clinical decision making and the interaction between a patient and a physician in terms of the model of how the physician makes decisions, and a model of how the patient responds to these decisions. With respect to prognosis, the medical professional wishes to predict (part of) the state of the patient for a future time period. They therefore need to take into account **how the state of the patient changes** under the influence of choices made by the physicians. This implies that we must predict how the physician will respond in the future. That is, we need to know the **treatment strategy** under which the physician operates (Always ask your medical doctor: What is your strategy?). Most traditional prognostic models sidestep the explicit representation of time by mapping the current state of the patient directly into the future time period of interest. However, the temporal nature of a problem is often essential to clinical decision-making as well. Clinical data, such as patient history, laboratory analysis, ultrasound parameters, which are the basis of day-to-day clinical decision support, are often underused to guide the clinical management of cancer in the presence of microarray data (van Gerven et al. 2008).

Note: The current patient state influences the physician's decisions, which in turn influences the next patient state.

6.4.3 Bayesian Network

Slide 6-29: Bayesian Network (BN): Definition

Gevaert et al. (2006) proposed a strategy based on Bayesian networks (BN) to treat clinical and microarray data equally. The main advantage of this probabilistic model is that it allows integrating these data in several ways and that it allows investigating and understanding the model structure and parameters. By use of the concept of a **Markov Blanket** (see Slide 6-34) we can identify all the variables that shield off the class variable from the influence of the remaining network. BN's automatically perform feature selection by identifying the (in)dependency relationships with the class variable.

A BN is a **probabilistic model** that consists of two parts: (1) a dependency structure and (2) local probability models.

Dependency structure specifies how the variables are related to each other by drawing directed edges between the variables without creating directed cycles. Each variable depends on a possibly empty set of other variables which are called the parents:

$$p(x_1,\ldots,x_n) = \prod_{i=1}^{n} p(x_i | Pa(x_i))$$

where $Pa(x_i)$ are the parents of x_i.

Usually the number of parents for each variable is small, therefore a Bayesian network is a sparse way of writing down a joint probability distribution (Pearl 1988).

Slide 6-30: Example: Breast Cancer: Probability Table

Local Probability Models specifies how the variables depend on their parents. Gevaert et al. (2006) used discrete-valued Bayesian networks which means that these local probability models can be represented with a Conditional Probability Table (CPT). In this slide we see that the features include (Wang et al. 1999):

Five factors from the patient history (drinking and smoking, taking hormones, menopause, previous pregnancies, family breast cancer cases)

Four factors from the physical examination (nipple discharge, skin thickening, breast pain, lumps)

Four factors from the mammographic findings (distortion, mass, microcalcification, asymmetry).

Slide 6-31: Breast Cancer: Big Picture—State of 1999

This slide summarizes the aforementioned and shows the "big picture" of a BN to diagnose breast cancer. The first layer includes five features related to patients' clinical history, the second layer contains one diagnostic feature, breast cancer, and the third layer involves eight features from both mammographic findings and physical examinations (Wang et al. 1999).

Slide 6-32: 10 Years Later: Integration of Microarray Data

Prognostic markers of breast cancer metastases to ensure that newly diagnosed patients receive appropriate therapy are of vital importance. Within the last years studies have demonstrated the potential value of **gene expression signatures** in assessing the risk of developing distant metastases. However, due to the small sample sizes of individual studies, the overlap among signatures is almost zero and their predictive power is often limited. Integrating microarray data from multiple studies in order to increase sample size is therefore a promising approach to the development of more robust prognostic tests.

Xu et al. (2008) have performed a study by using a highly stable data aggregation procedure based on expression comparisons, and have integrated three independent microarray gene expression datasets for breast cancer and identified a structured prognostic signature consisting of 112 genes organized into 80 pair-wise expression comparisons. Such prognostic signatures and tests could provide powerful tools to guide adjuvant systemic treatment that could greatly reduce the cost of breast cancer treatment, both in terms of toxic side effects and health-care expenditures (Xu et al. 2008).

Slide 6-33: Example: BN with Four Binary Variables

The Conditional Probability Tables (CPTs) specify the probability that a variable takes a certain value given the value of its parents. In this slide we see an example of a Bayesian network with four binary variables. The prognosis variable in this example has two parents: gene 2 and gene 3. The CPTs for each variable are shown alongside each node (Gevaert et al. 2006).

Slide 6-34: Concept Markov-Blanket

An important concept of BN is the **Markov blanket** of a variable, which is the set of variables that completely shields off this variable from the other variables. This set consists of the variable's parents, children and its children's other parents. A variable in a Bayesian network is conditionally independent of the other variables given its Markov blanket. Conditional independency means that when the Markov blanket of a certain variable x is known, adding knowledge of other variables leaves the probability of x unchanged.

Note: This is an important concept because the Markov blanket is the only knowledge that is needed to predict the behavior of that variable. For classification purposes we must focus on the Markov blanket of the outcome variable (Gevaert et al. 2006).

Slide 6-35: Dependency Structure → First Step (1/2)

Previously we mentioned that a discrete valued BN consists of two parts. Consequently, there are two steps to be performed during model building: (1) structure learning and (2) learning the parameters of the CPTs. First the structure is learned using a **search strategy**. Since the number of possible structures *increases* super exponentially with the number of variables, the well-known **greedy search algorithm K2** can be used in combination with the Bayesian Dirichlet (BD) scoring metric:

$$p(S|D) \propto p(S) \prod_{i=1}^{n} \prod_{j=1}^{q_i} \left[\frac{\Gamma(N'_{ij})}{\Gamma(N'_{ij}+N_{ij})} \prod_{k=1}^{r_i} \frac{\Gamma(N'_{ijk}+N_{ijk})}{\Gamma(N'_{ijk})} \right]$$

N_{ijk} ... number of cases in the dataset D having variable i in state k associated with the j-th instantiation of its parents in current structure S. n is the total number of variables.

Slide 6-36: Dependency Structure → First Step (2/2)

Next, N_{ij} is calculated by summing over all states of a variable:
$N_{ij} = \sum_{k=1}^{r_i} N_{ijk} \cdot N'_{ijk}$ and N'_{ij} have similar meanings but refer to prior knowledge for the parameters.

When no knowledge is available they are estimated using $N_{ijk} = N/(r_i q_i)$

(continued)

(continued)

with N the equivalent sample size, r_i the number of states of variable i and q_i the number of instantiations of the parents of variable i.

$\Gamma(.)$ corresponds to the gamma distribution.

Finally $p(S)$ is the prior probability of the structure. $p(S)$ is calculated by:

$$p(S) = \prod_{i=1}^{n} \prod_{l_i=1}^{p_i} p(l_i \rightarrow x_i) \prod_{m_i=1}^{o_i} p(m_i x_i)$$

with p_i the number of parents of variable x_i and o_i all the variables that are not a parent of x_i.

Next, $p(a \rightarrow b)$ is the probability that there is an edge from a to b while $p(ab)$ is the inverse, i.e., the probability that there is no edge from a to b.

Slide 6-37: Parameter Learning → Second Step

Estimating the parameters of the local probability models corresponding with the dependency structure. CPTs are used to model these local probability models. For each variable and instantiation of its parents there exists a CPT that consists of a set of parameters. Each set of parameters was given a uniform Dirichlet prior:

$$p(\theta_{ij}|S) = Dir\left(\theta_{ij}|N'_{ij1}, \ldots, N'_{ijk}, \ldots, N'_{ijr_i}\right)$$

Note: With θ_{ij} a parameter set where i refers to the variable and j to the j-th instantiation of the parents in the current structure. θ_{ij} contains a probability for every value of the variable x_i given the current instantiation of the parents. Dir corresponds to the Dirichlet distribution with $\left(N'_{ij1}, \ldots, N'_{ijr_i}\right)$ as parameters of this Dirichlet distribution. Parameter learning then consists of updating these Dirichlet priors with data. This is straightforward because the multinomial distribution that is used to model the data, and the Dirichlet distribution that models the prior, are conjugate distributions. This results in a Dirichlet posterior over the parameter set:

$$p(\theta_{ij}|D,S) = Dir\left(\theta_{ij}|N'_{ij1} + N_{ij1}, \ldots, N'_{ijk} + N_{ijk}, \ldots, N'_{ijr_i} + N_{ijr_i}\right)$$

with N_{ijk} defined as before.

6.4.4 Alterative Approach: Support Vector Machines

Slide 6-38: Support Vector Machine SVM: (Boser et al. 1992)

A support vector machine (SVM)—the name machine comes from machine learning—is a concept for a set of related supervised learning methods that analyze data and recognize patterns. The standard SVM takes a set of input data and predicts, for each given input, which of two possible classes comprises the input, making the SVM a non-probabilistic binary linear classifier. Given a set of training examples, each marked as belonging to one of two categories, an SVM training algorithm builds a model that assigns new examples into one category or the other. An SVM model is a representation of the examples as points in space, mapped so that the examples of the separate categories are divided by a clear gap that is as wide as possible. New examples are mapped into the same space and predicted to belong to a category based on which side of the gap they fall on. Beneficial of SVM is, that they are well suited for high dimensional data, because the complexity of trained classifier is characterized by the number of support vectors—*not the dimensionality of the data*. The support vectors are the essential or critical training examples — they lie closest to the decision boundary. If all other training examples are removed and the training is repeated, the same separating hyperplane would be found. The number of support vectors found can be used to compute an (upper) bound on the expected error rate of the SVM classifier, which is independent of the data dimensionality. Thus, an SVM with a small number of support vectors can have good generalization, even when the dimensionality of the data is high (Han et al. 2011).

Slide 6-39: SVM vs. ANN

The differences between SVM and ANN in a nutshell:

SVM is a deterministic algorithm, whereas ANN is non-deterministic;
SVM has good generalization properties, whereas ANN generalizes well, but does not have a strong mathematical foundation;
SVM is difficult to learn—thus learned in batch mode using quadratic programming techniques, whereas ANN can easily be learned in incremental fashion;
SVM by using kernels can learn very complex functions, whereas for ANN to learn complex functions would need multilayer perceptron (complex);

Slide 6-40: Clinical Use: SVM Are More Accurate than ANN

In this slide we see the receiver operating characteristic (ROC) curve analysis of clinical decision support systems using support vector machine (SVM) and artificial neural network (ANN) models. It is clearly to see that the area under the ROC curve (AUC) value of SVM was superior to ANN (Kim et al. 2011).

Slide 6-41: The Ten Top Data Mining Algorithms

C4.5 = for generation of decision trees used for classification, (statistical classifier, Quinlan 1993);

k-means = a simple iterative method for partition of a given dataset into a user-specified number of clusters, k (Lloyd 1982);

Apriori = for finding frequent item sets using candidate generation (Agrawal and Srikant 1994);

EM = Expectation–Maximization algorithm for finding maximum likelihood estimates of parameters in models (Dempster et al. 1977);

PageRank = a search ranking algorithm using hyperlinks on the Web (Brin and Page 1998);

Adaptive Boost = one of the most important ensemble methods (Freund and Schapire 1995);

k-Nearest Neighbor = a method for classifying objects based on closest training examples in the feature space (Fix and Hodges 1951);

Naive Bayes = can be trained very efficiently in a supervised learning setting (Domingos and Pazzani 1997);

CART = Classification And Regression Trees as predictive model mapping observations about items to conclusions about the goal (Breiman et al. 1984);

SVM = support vector machines offer one of the most robust and accurate methods among all well-known algorithms (Cortes and Vapnik 1995);

6.5 Text Mining and Semantic Methods

Slide 6-42: Selection of Semantic Methods

Patient information in health-care systems mostly consists of textual data, and so-called "free text" (refer to Lectures 2 and 3 for the correct definitions) in particular makes up a significant amount of it. There are many different methods to deal with text, e.g., Latent Semantic Analysis (LSA), Probabilistic latent semantic analysis (PLSA), Latent Dirichlet allocation (LDA),

(continued)

(continued)

Hierarchical Latent Dirichlet Allocation (hLDA), Semantic Vector Space Model (SVSM), Latent semantic mapping (LSM), and Principal component analysis (PCA) to name only a few.

6.5.1 Latent Semantic Analysis

Slide 6-43: Latent Semantic Analysis (LSA)

LSA is both: a theory and a method for both: extracting and representing the meaning of words in their contextual environment by application of statistical analysis to a large amount of text. LSA is a general theory of acquired similarities and knowledge representations, originally developed to explain learning of words and psycholinguistic problems (Landauer et al. 1998). The general idea was to induce global knowledge indirectly from local co-occurrences in the representative text. Originally, LSA was used for explanation of textual learning of the English language at a comparable rate amongst schoolchildren. LSA does not use any prior linguistic or perceptual similarity knowledge, i.e., it is based on a general mathematical learning method that achieves powerful inductive effects by extracting the right number of dimensions to represent both objects and contexts. The fundamental suggestion is that the aggregate of all words in contexts in which a given word does and does not appear provides a set of mutual constraints that largely determines the similarity of meaning of words and sets of words to each other. For the combination of Informatics and Psychology it is interesting to note that the adequacy of LSA's reflection of human knowledge has been established in a variety of ways (Foltz et al. 1998). For example, the scores overlap closely to those of humans on standard vocabulary and subject matter tests and interestingly it emulates human word sorting behavior and category judgments (Landauer and Dumais 1997). Consequently, as a practical outcome, it can estimate passage coherence and the learnability of passages, and both the quality and quantity of knowledge contained in a textual passage.

LSA is primarily used for measuring the coherence of texts. By comparing the vectors for two adjoining segments of text in a high-dimensional semantic space, the method provides a characterization of the degree of semantic relatedness between the segments. LSA can be applied as an automated method that produces coherence predictions similar to propositional modelling, thus having potential as a psychological model of coherence effects in text comprehension (Foltz et al. 1998). Having t terms and d documents one

(continued)

(continued)

can build a $t \times d$ matrix X. Often the terms within this matrix are weighted according to the term frequency—inverse document frequency (fd-idf)[x]. The main method now is to apply the Singular Value Decomposition (SVD) on X[y]. Therefore X can be disjointed into three components $X = TSD^T$. T and D^T are orthonormal matrices with the eigenvectors of XX^T and X^TX respectively. S contains the roots of the eigenvalues of XX^T and X^TX. Reducing the dimensionality can now be achieved via step-by-step eliminating the lowest eigenvalue with the corresponding eigenvectors to a certain value k. A given Query q can now be projected into this space by applying the equation:

$$Q = \frac{q^T U_k}{diag(S_k)}$$

Having Q and the documents in the same semantic space a similarity measure can now be done. Often used is the so-called cosine similarity between a document in the semantic space and a query Q. Having two vectors A and B in the n-dimensional space, the cosine similarity is defined as:

$$\cos(\theta) = \frac{A \cdot B}{|A||B|}$$

Advantages include: LSA tends to solve the synonym problem;
LSA reflects text semantic, e.g., similar concepts can be found.
Disadvantages include: High mathematical complexity (SVD);
Recalculation of the singular value decomposition when adding new documents or terms;
Difficult estimation of k, which means how many eigenvalues to keep getting good results;

6.5.2 Latent Dirichlet Allocation

Slide 6-44: Latent Dirichlet Allocation (LDA)

Latent Dirichlet allocation (LDA) is a generative probabilistic model for collections of discrete data such as text corpora, based on a three-level hierarchical Bayesian model, in which each item of a collection is modelled as a finite mixture over an underlying set of topics (Blei et al. 2003). Each topic is, in turn, modelled as an infinite mixture over an underlying set of

(continued)

(continued)

topic probabilities. In the context of text modelling, the topic probabilities provide an explicit representation of a document, consequently LDA can be seen as a "bag-of-words" type of language modelling and dimension reduction method (Kakkonen et al. 2006). One interesting application of LDA is fraud detection, in order to build user profile signatures and assumes that any significant unexplainable deviations from the normal activity of an individual end user is strongly correlated with fraudulent activity; thereby, the end user activity is represented as a probability distribution over call features which surmises the end user's calling behavior (Xing and Girolami 2007). LDA is often assumed to be better performing than LSA or PLSA (Girolami and Kaban 2005).

6.5.3 Principal Components Analysis

Slide 6-45: Principal Component Analysis (PCA)

Principal component analysis (PCA) is a technique used to reduce multidimensional datasets to lower dimensions for analysis. Depending on the field of application, it is also named the discrete Karhunen–Loève transform, the Hotelling transform, or proper orthogonal decomposition (POD).

PCA was invented in 1901 by Karl Pearson. Now it is mostly used as a tool in exploratory data analysis and for making predictive models. PCA involves the calculation of the eigenvalue decomposition of a data covariance matrix or singular value decomposition of a data matrix, usually after mean centering the data for each attribute. The results of a PCA are usually discussed in terms of component scores and loadings. Assuming mathematical knowledge about standard deviation, covariance, eigenvectors, and eigenvalues, let us assume to have M observations, each observation having N features. So at any observation point we can collect a N-dimensional feature vector Γ. All feature vectors together form the $N \times M$ observation matrix $\theta(\Gamma_1, \Gamma_2, \ldots, \Gamma_M)$. Furthermore the average feature vector is given by:

$$\psi = \frac{1}{M}\sum_{n=1}^{M}\Gamma_n$$

By adjusting every feature vector Γ_i by $\phi_i = \Gamma_i - \psi$ we can form the covariance matrix of the mean adjusted data:

(continued)

(continued)

$$C = \frac{1}{M}\sum_{n=1}^{M} \phi_n \phi_n^T$$

Now, the PCA is prepared be the following steps:

(1) Calculate the eigenvalues and the corresponding eigenvectors of C, resulting in $\lambda_1, \lambda_2, \ldots, \lambda_M$ eigenvalues with the corresponding eigenvectors $u_1, u_2, \ldots, u_M$;
(2) Keep the highest eigenvalues $M' < M$ forming a matrix $P = [u_1, u_2, \ldots, u_{M'}]^T$ with the corresponding eigenvectors where $\lambda_1 > \lambda_2 > \ldots \lambda_{M'}$.
(3) Transform the data into the reduced space applying $Y = PC$.

Note that $cov(Y) = \frac{1}{M}YY^T$ is a diagonal matrix. That means we found a representation of the data with minimum redundancy and noise (off-diagonal elements of the covariance matrix are zero). One diagonal element of the covariance matrix represents the variance of one typical feature measured. The eigenvectors point towards the direction of greatest variance of the data.

Let us collect n-dimensional i observations in the Euclidean vector space $\mathbb{R}^n$ and we get:

$$x_i = [x_{i1}, \ldots, x_{in}] \qquad \text{(Eq. 2-1)}$$

7 Future Outlook

Slide 6-46: Understanding Natural Language ...

... is *the* grand challenge in future computing.

First of all we must acknowledge the fact that most of the data (Gardner and IBM speak about 80 % of all data—whatever "80 % of all data" really is) is composed of unstructured information.

Note: Please make a difference between unstructured data (which would mean noise, randomness) and unmodelled data, which is called unstructured information (which is a synonym for natural language, aka "free text").

These masses of information are not easily understandable by humans—hence we need "machine intelligence" for dealing with this flood of data. However, legacy approaches have failed, "searching" is not the right

(continued)

(continued)

approach—search is a way to gather information—but not to answer questions. Consequently, a new approach is needed, leveraging content analysis and natural language processing. One step towards reaching this goal is IBM Watson (see next slide).

Remark: Since its beginning, AI was based on the assumption that a physical symbol system has the necessary and sufficient means for general intelligent action (Newell and Simon 1976), i.e., a computer with a program operating on data is able to have a general intelligent behavior. The main objective of AI is then to create programs that solve NP-hard problems, which some natural cognitive systems (including humans) can solve, but for which no satisfactory algorithmic description is known (Sabah 2011).

Slide 6-47: Watson: A Workload Optimized System

The computer system who won the Jeopardy Quiz in 2011 is made of a cluster of 90 IBM Power 750 servers (plus additional I/O, network, and cluster controller nodes). Watson has ten racks, nine for the servers and one for the associated I/O, network, and cluster controller nodes. The Power 750 server is commercially available and not specially designed for Watson. The Power 750 servers use 3.5 GHz POWER7 eight-core processors, with four threads per core—in total Watson has control of 2,880 POWER7 cores—enabling a massively parallel processing capability. Watson runs the DeepQA software, which executes multiple threads in parallel. The cluster has a total of 16 TB RAM. The file system containing all of Watson's data is on a 4 TB disk. When Watson starts up, much of the data on the disk is loaded into memory and replicated across many servers in the cluster.

Watson can operate more than 80 Teraflops and uses Apache the Unstructured Information Management Architecture (UIMA), which was originally developed at IBM, but is now in open source. UIMA does not answer questions, but supports building and scaling applications and provides standards for interoperability and scale-out of deep text and multimodal (speech, image, video) content analytics. Watson's Power 750 servers are clustered over a 10 Gb Ethernet network.

Watson receives a question in text form at the same time the human contestants receive them in voice/text. Interestingly, Watson is not connected to the Internet or any outside source of information and moreover Watson can neither "see nor hear," but it implements a buzzer and it provides spoken

(continued)

(continued)

output to answer the questions. These limitations provide a lot of further research opportunities in artificial intelligence (Ferrucci et al. 2010, 2013).

The research unit HCI4MED is the first Austrian IBM Watson Think Tank in close cooperation with the Software Group of IBM Austria, http://www.hci4all.at/about-us/

Slide 6-48: Example: IBM Watson for Health care

In this slide we see a top-level view of the DeepQA architecture (Ferrucci et al. 2010).

DeepQA has a massively parallel, component-based pipeline architecture (Ferrucci and Lally 2004), which uses an extensible set of both structured and unstructured content sources and a range of pluggable search and scoring components that allow the integration of different analytic techniques.

Machine learning is used to learn the weights for combining scores from different scorers. Each answer is linked to its supporting evidence. DeepQA is informed by extensive research in question answering systems: Such systems analyze an input question and generate and evaluate answer candidates. DeepQA analyzes an input question to determine precisely what it is asking for and generates possible candidate answers through a broad search of large volumes of content. For each of these candidate answers, a hypothesis is formed based on considering the candidate in the context of the original question and topic. For each hypothesis, DeepQA spawns an independent thread that attempts to prove it, and then it searches the content sources for evidence that supports or refutes each hypothesis. For each evidence–hypothesis pair, DeepQA applies hundreds of algorithms that dissect and analyze the evidence along different dimensions of evidence such as type classification, time, geography, popularity, passage support, source reliability, and semantic relatedness. This analysis produces hundreds of features. These features are then combined based on their learned potential for predicting the right answer. The final result of this process is a ranked list of candidate answers, each with a confidence score indicating the degree to which the answer is believed correct, along with links back to the evidence (Ferrucci et al. 2013).

Slide 6-49:

In recent years, multimedia data like pictures, audio, videos, text, graphics, animations, and other multimodal sensory data have grown at a phenomenal rate and are almost ubiquitous. As a result, not only the methods and tools to organize, manage, and search such data have gained widespread attention but the methods and tools to discover hidden knowledge from such data have also

(continued)

(continued)

become extremely important. The task of developing such methods and tools is facing the big challenge of overcoming the semantic gap of multimedia data. But in certain sense data mining techniques are attempting to bridge this semantic gap in analytical tools. This is because such tools can facilitate decision making in many situations. Data mining refers to the process of finding interesting patterns in data that are not ordinarily accessible by basic queries and associated results with the objective of using discovered patterns to improve decision making. For example, it might not be possible to easily detect suspicious events using simple surveillance systems. But MDM tools that perform mining on captured trajectories from surveillance videos can potentially help find suspicious behavior, suspects and other useful information. MDM brings in strengths from both multimedia and data mining fields along with challenging problems in these fields. In terms of strength, we can say image, audio, video, etc. are more information rich than the simple text data alone in most of the domains. The knowledge available from such multimedia data can be universally understandable. Also, there can be certain situations where there is no other efficient way to represent the information other than the multimodal representation of the scenario. Mining of multimedia data is more involved than that of traditional business data because multimedia data are unstructured by nature. There are no well-defined fields of data with precise and nonambiguous meaning, and the data must be processed to arrive at fields that can provide content information. Such processing often leads to nonunique results with several possible interpretations. In fact, multimedia data are often subject to varied interpretations even by human beings. For example, it is not uncommon to have different interpretation of an image by different people. Another difficulty in mining of multimedia data is its heterogeneous nature. The data are often the result of outputs from various kinds of sensor modalities with each modality needing sophisticated preprocessing, synchronization and transformation procedures. Yet another distinguishing aspect of multimedia data is its sheer volume. The high dimensionality of the feature spaces and the size of the multimedia datasets make feature extraction a difficult problem. MDM works focus their effort to handle these issues while following the typical data mining process. The typical data mining process consists of several stages and the overall process is inherently interactive and iterative. The main stages of the data mining process are (1) domain understanding; (2) data selection; (3) data preprocessing, cleaning, and transformation; (4) discovering patterns; (5) interpretation; and (6) reporting and using discovered knowledge. The domain understanding stage requires learning how the results of data-mining will be used so as to gather all relevant prior knowledge before mining. For example, while mining sports video for a particular sport like tennis, it is important to have a good knowledge and understanding of the game to detect

(continued)

(continued)

interesting strokes used by players. The data selection stage requires the user to target a database or select a subset of fields or data records to be used for data mining. A proper understanding of the domain at this stage helps in the identification of useful data. The quality and quantity of raw data determines the overall achievable performance. The goal of the preprocessing stage is to discover important features from raw data. The preprocessing step involves integrating data from different sources and/or making choices about representing or coding certain data fields that serve as inputs to the pattern discovery stage. Such representation choices are needed because certain fields may contain data at levels of details not considered suitable for the pattern discovery stage. This stage is of considerable importance in multimedia data mining, given the unstructured and heterogeneous nature and sheer volume of multimedia data. The preprocessing stage includes data cleaning, normalization, transformation, and feature selection. Cleaning removes the noise from data. Normalization is beneficial as there is often large difference between maximum and minimum values of data. Constructing a new feature may be of higher semantic value to enable semantically more meaningful knowledge. Selecting subset of features reduces the dimensionality and makes learning faster and more effective. Computation in this stage depends on modalities used and application's requirements. The pattern discovery stage is the heart of the entire data mining process. It is the stage where the hidden patterns, relationships and trends in the data are actually uncovered. There are several approaches to the pattern discovery stage. These include association, classification, clustering, regression, time-series analysis, and visualization. Each of these approaches can be implemented through one of several competing methodologies, such as statistical data analysis, machine learning, neural networks, fuzzy logic, and pattern recognition. It is because of the use of methodologies from several disciplines that data mining is often viewed as a multidisciplinary field. The interpretation stage of the data mining process is used to evaluate the quality of discovery and its value to determine whether the previous stages should be revisited or not. Proper domain understanding is crucial at this stage to put a value to the discovered patterns. The final stage of the data mining process consists of reporting and putting to use the discovered knowledge to generate new actions or products and services or marketing strategies as the case may be. This stage is application dependent. Among the above-mentioned stages of data mining process Data preprocessing, cleaning, and transformation; Discovering patterns; Interpretation; and Reporting and using discovered knowledge contains the highest importance and novelty from the MDM perspective. Thus, we organize Multimedia Data Mining State of the Art review as shown in Fig. 1. The proposed scheme achieves the following goal—Discussion of the existing preprocessing techniques for multimedia data in MDM literature.

8 Exam Questions

8.1 Yes/No Decision Questions

Please check the following sentences and decide whether the sentence is true = YES; or false = NO; for each correct answer you will be awarded 2 credit points.

01	In data mining tasks we need the knowledge of the domain expert, hence data mining approaches must put the human intelligence into the loop	☐ Yes ☐ No	2 total
02	Artificial Neural networks (ANN) mimic the human neural network and literally "learn" from examples to find patterns in data or to classify data..	☐ Yes ☐ No	2 total
03	Within a neuron the axon collects electrical signals, the cell body integrates the incoming signals and generates outgoing signals to the axon,	☐ Yes ☐ No	2 total
04	Over-fitting occurs when a ANN has too many parameters to be learned from the number of examples available, i.e. when a few points are fitted with a function with too many free parameters.	☐ Yes ☐ No	2 total
05	Hopfield networks are recurrent ANNs serving as content-addressable memory systems with binary threshold nodes; Recurrent mean that the graph may contain cycles.	☐ Yes ☐ No	2 total
06	Important in clinical patient management is to determine a prognosis for a patient, i.e. we define prognosis as the prediction of the future course of a disease process.	☐ Yes ☐ No	2 total
07	A support vector machine (SVM) is a non-deterministic algorithm whereas a typical ANN is deterministic and generalizes well on a given problem.	☐ Yes ☐ No	2 total
08	A support vector machine (SVM) is a machine learning concept for a set of related supervised learning methods that analyze data and recognize patterns.	☐ Yes ☐ No	2 total
09	The PageRank algorithm developed by Brin & Page (1998) is for finding maximum likelihood estimates of parameters in models and one of the most important ensemble methods	☐ Yes ☐ No	2 total
10	Classification And Regression Trees (CART) are predictive models mapping observations about items to conclusions about the goal and are very often used for solving medical problems.	☐ Yes ☐ No	2 total

Sum of Question Block A (max. 20 points)	

8.2 Multiple Choice Questions (MCQ)

The following questions are composed of two parts: the stem, which identifies the question or problem and a set of alternatives which can contain 0, 1, 2, 3, or 4 correct answers, along with a number of distractors that might be plausible—but are incorrect. Please **select the correct answers** by ticking ☒—and do not forget that it can be none. Each question will be awarded 4 points *only if everything is correct*.

01	A well-known ranking algorithm used for information retrieval is called ... ☐ a) ... C 4.5 ☐ b) ... PageRank. ☐ c) ... k-Nearest Neighbor. ☐ d) ... Adaptive Boost.	4 total
02	A semantic method often used for medical text mining is ... ☐ a) ... Latent Semantic Analysis (LSA) ☐ b) ... Latent Dirichlet allocation (LDA) ☐ c) ... Hierarchical Latent Dirichlet Allocation (hLDA). ☐ d) ... Hierarchical Latent Dirichlet Allocation (hLDA).	4 total
03	Data driven challenges in computing include ... ☐ a) ... heterogeneous data. ☐ b) ... massive data sets. ☐ c) ... streaming data. ☐ d) ... data from large sensor networks.	4 total
04	Data preprocessing includes ... ☐ a) ... searching for patterns of interest. ☐ b) ... learning from the application domain. ☐ c) ... choosing an appropriate data mining method. ☐ d) ... removing of noise or outliers and handling of missing data.	4 total
05	A mega challenge in the knowledge discovery process chain is ☐ a) ... to map higher dimensional data into low dimensions (dimension 1 or 2) ☐ b) ... generating data from the data source. ☐ c) ... storing data in-memory. ☐ d) ... to get insight into the data.	4 total
06	Part of the definition of Biomedical Informatics is the ... ☐ ... effective use of biomedical data. ☐ ... motivation to improve computational capacities. ☐ ... effort to expand the technological capabilities. ☐ ... motivation to improve human health.	4 total
07	Unsupervised learning is ... ☐ a) ... also called clustering. ☐ b) ... when new data is classified based on a given training set. ☐ c) ... the process of labeling the training data. ☐ d) ... when the class labels of training data is unknown.	4 total

08	In the clinical decision making strategy the current patient state ... ☐ a) ... includes risk factors. ☐ b) ... is derived from the pathogenesis of the patient. ☐ c) ... is depending on disorders. ☐ d) ... is only inferred from the pathophysiology.	4 total
09	Support Vector Machines (SVM) ...	4 total
	☐ a) ... are non-deterministic. ☐ b) ... are using so-called kernels, which can learn complex functions. ☐ c) ... concern mechanisms other than changes in the underlying DNA ☐ d) ... can easily be learned in incremental fashion.	
10	Data Mining is applied in biomedicine to ... ☐ a) ... Molecular Biology. ☐ b) ... Biosurveillance. ☐ c) ... Epidemiology. ☐ d) ... Clinical purposes.	4 total

Sum of Question Block B (max. 40 points)	

8.3 Free Recall Block

Please follow the instructions below. At each question you will be assigned the credit points indicated if your option is correct (partial points may be given).

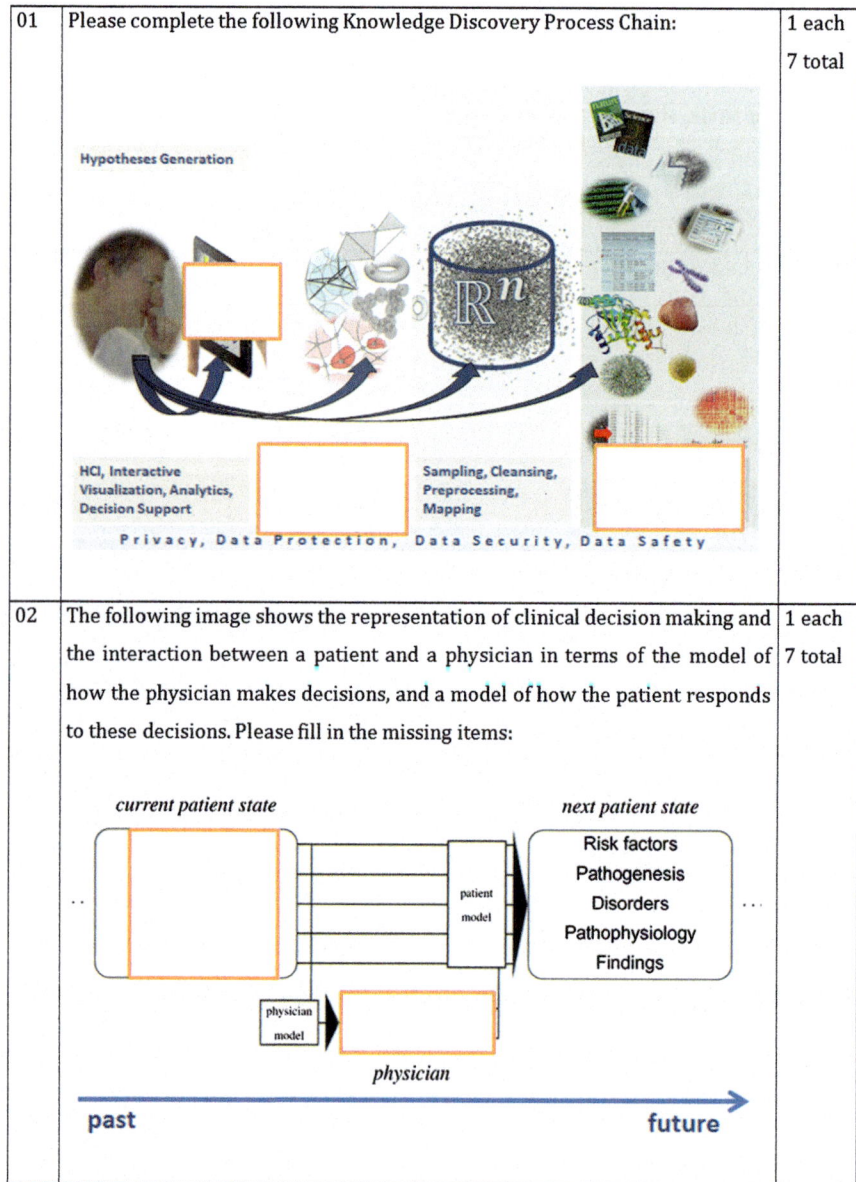

03	In the 1960s, it was shown that networks of such model neurons have properties similar to the brain: they can perform sophisticated pattern recognition, and they can function even if some of the neurons are destroyed. Simple networks of such model neurons are called a perceptron. Please complete the following image: 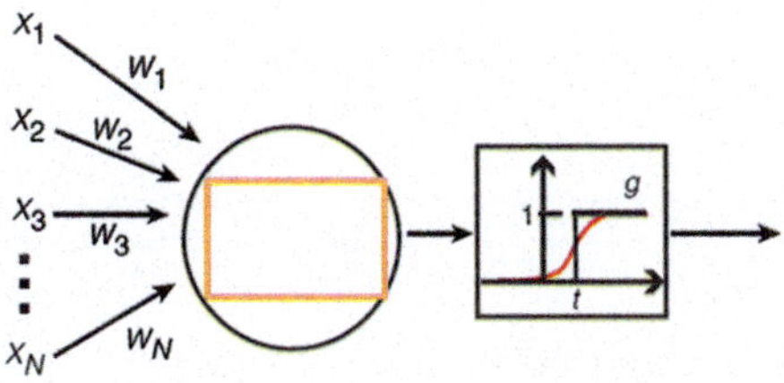	3 each 3 total

04	Please complete the missing terms in the supervised learning process 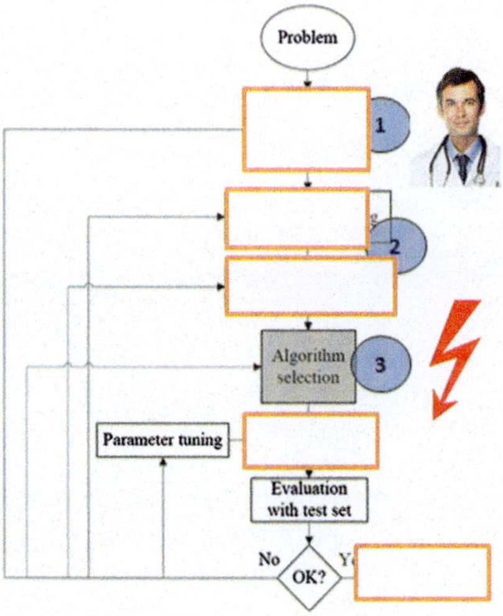	1 each 5 total

05	There is a large variety of different data mining methods following different data mining paradigms. Please assign the correct labels to the methods below:	1 each 3 total
	Bayesian Networks	Verification
	PCA	Classification
	ANOVA	Unsupervised

9 Answers

9.1 Answers to the Yes/No Questions

Please check the following sentences and decide whether the sentence is true = YES; or false = NO; for each correct answer you will be awarded 2 credit points.

01	In data mining tasks we need the knowledge of the domain expert, hence data mining approaches must put the human intelligence into the loop	☒ Yes ☐ No	2 total
02	Artificial Neural networks (ANN) mimic the human neural network and literally "learn" from examples to find patterns in data or to classify data.	☒ Yes ☐ No	2 total
03	Within a neuron the axon collect electrical signals, the cell body integrates the incoming signals and generates outgoing signals to the axon,	☐ Yes ☒ No	2 total
04	Over-fitting occurs when a ANN has too many parameters to be learned from the number of examples available, i.e. when a few points are fitted with a function with too many free parameters.	☒ Yes ☐ No	2 total
05	Hopfield networks are recurrent ANNs serving as content-addressable memory systems with binary threshold nodes; Recurrent mean that the graph may contain cycles.	☒ Yes ☐ No	2 total
06	Important in clinical patient management is to determine a prognosis for a patient, i.e. we define prognosis as the prediction of the future course of a disease process.	☒ Yes ☐ No	2 total
07	A support vector machine (SVM) is a non-deterministic algorithm whereas a typical ANN is deterministic and generalizes well on a given problem.	☐ Yes ☒ No	2 total
08	A support vector machine (SVM) is a machine learning concept for a set of related supervised learning methods that analyze data and recognize patterns.	☒ Yes ☐ No	2 total
09	The PageRank algorithm developed by Brin & Page (1998) is for finding maximum likelihood estimates of parameters in models and one of the most important ensemble methods	☐ Yes ☒ No	2 total
10	Classification And Regression Trees (CART) are predictive models mapping observations about items to conclusions about the goal and are very often used for solving medical problems.	☒ Yes ☐ No	2 total
Sum of Question Block A (max. 20 points)			

9.2 Answers to the Multiple Choice Questions (MCQ)

01	A well-known ranking algorithm used for information retrieval is called ... ☐ a) ... C 4.5 ☒ b) ... PageRank. ☐ c) ... k-Nearest Neighbor. ☐ d) ... Adaptive Boost.	4 total
02	A semantic method often used for medical text mining is ... ☒ a) ... Latent Semantic Analysis (LSA) ☒ b) ... Latent Dirichlet allocation (LDA) ☒ c) ... Hierarchical Latent Dirichlet Allocation (hLDA). ☒ d) ... Hierarchical Latent Dirichlet Allocation (hLDA).	4 total
03	Data driven challenges in computing include ... ☒ a) ... heterogeneous data. ☒ b) ... massive data sets. ☒ c) ... streaming data. ☒ d) ... data from large sensor networks.	4 total
04	Data preprocessing includes ... ☐ a) ... searching for patterns of interest. ☐ b) ... learning from the application domain. ☐ c) ... choosing an appropriate data mining method. ☒ d) ... removing of noise or outliers and handling of missing data.	4 total
05	A mega challenge in the knowledge discovery process chain is ☒ a) ... map higher dimensional data into low dimensions (dimension 1 or 2) ☐ b) ... generating data from the data source. ☐ c) ... storing data in-memory. ☒ d) ... to get insight into the data.	4 total
06	Part of the definition of Biomedical Informatics is the ... ☒ ... effective use of biomedical data. ☐ ... motivation to improve computational capacities. ☐ ... effort to expand the technological capabilities. ☒ ... motivation to improve human health.	4 total
07	Unsupervised learning is ... ☒ a) ... also called clustering. ☐ b) ... when new data is classified based on a given training set. ☐ c) ... the process of labeling the training data. ☒ d) ... when the class labels of training data is unknown.	4 total
08	In the clinical decision making strategy the current patient state ... ☒ a) ... includes risk factors. ☒ b) ... is derived from the pathogenesis of the patient. ☒ c) ... is depending on disorders. ☐ d) ... is only inferred from the pathophysiology.	4 total
09	Support Vector Machines (SVM) ... ☐ a) ... are non-deterministic. ☒ b) ... are using so-called kernels, which can learn complex functions. ☒ c) ... concerns mechanisms other than changes in the underlying DNA ☐ d) ... can easily be learned in incremental fashion.	4 total
10	Data Mining is applied in biomedicine to ... ☒ a) ... Molecular Biology.	4 total
	☒ b) ... Biosurveillance. ☒ c) ... Epidemiology. ☒ d) ... Clinical purposes.	

Sum of Question Block A (max. 20 points)	

9 Answers

9.3 Answers to the Free Recall Questions

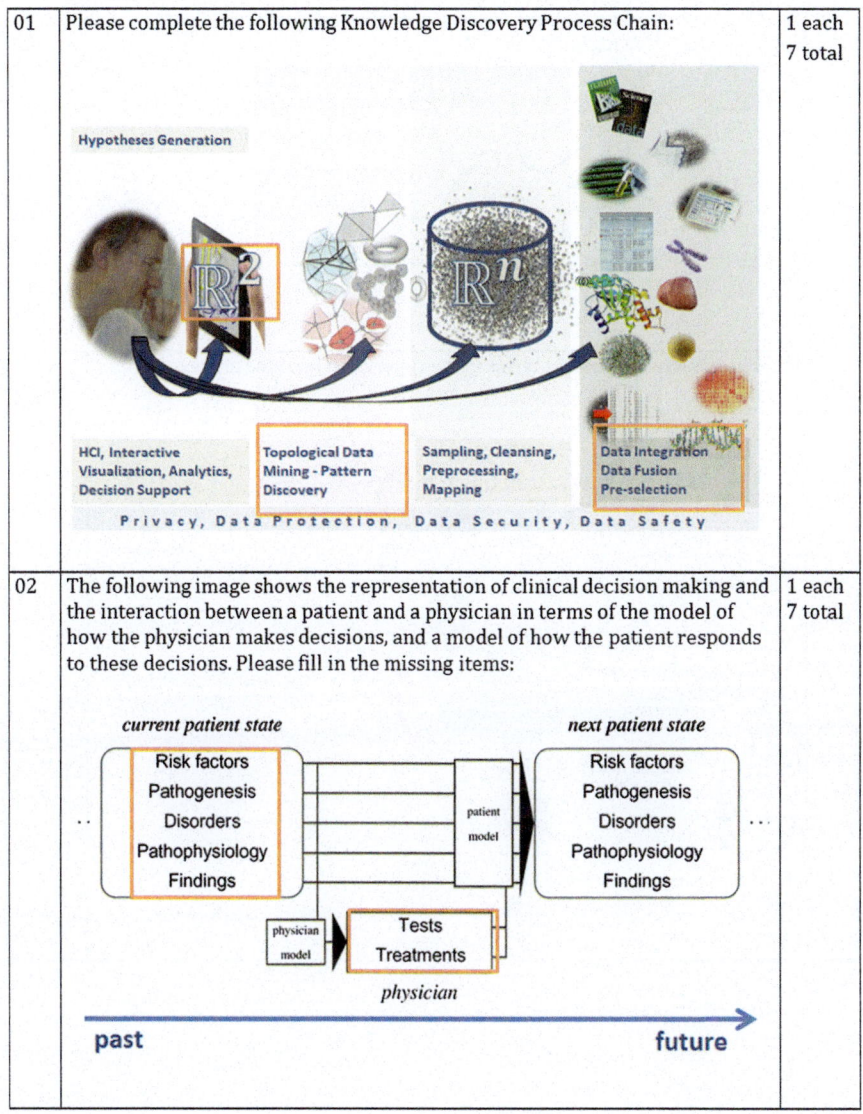

03	In the 1960s, it was shown that networks of such model neurons have properties similar to the brain: they can perform sophisticated pattern recognition, and they can function even if some of the neurons are destroyed. Simple networks of such model neurons are called a perceptron. Please complete the following image:	3 each 3 total

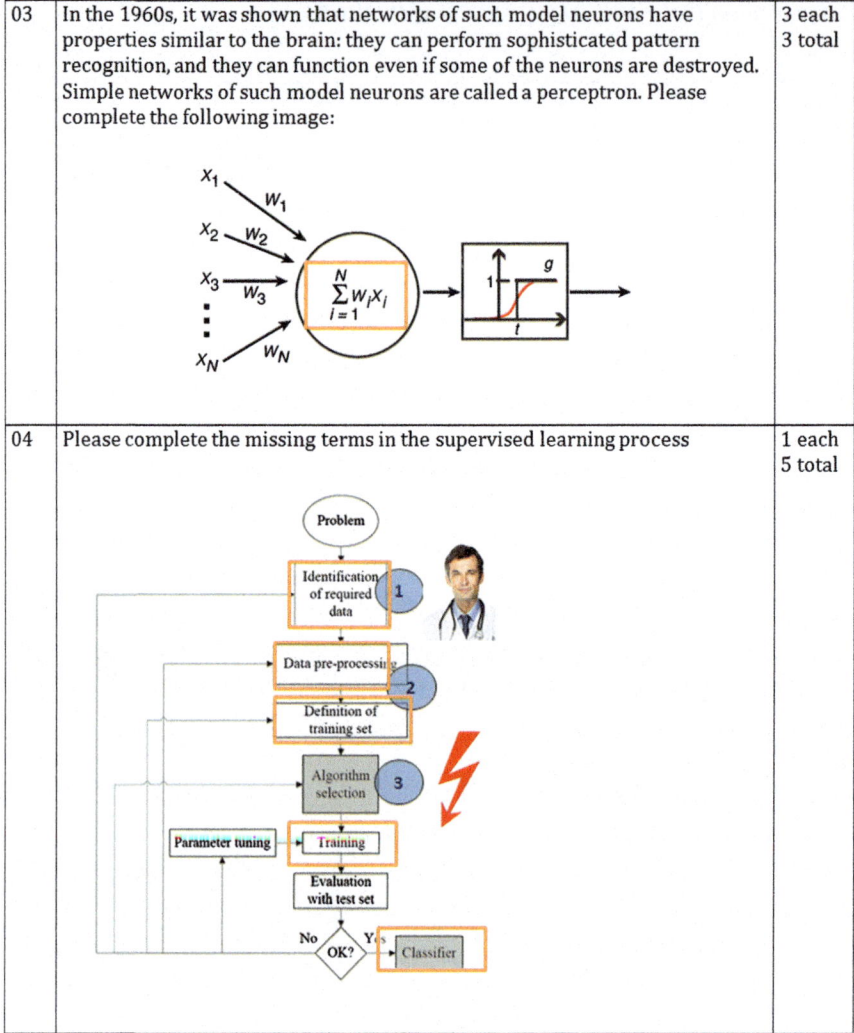

04	Please complete the missing terms in the supervised learning process	1 each 5 total

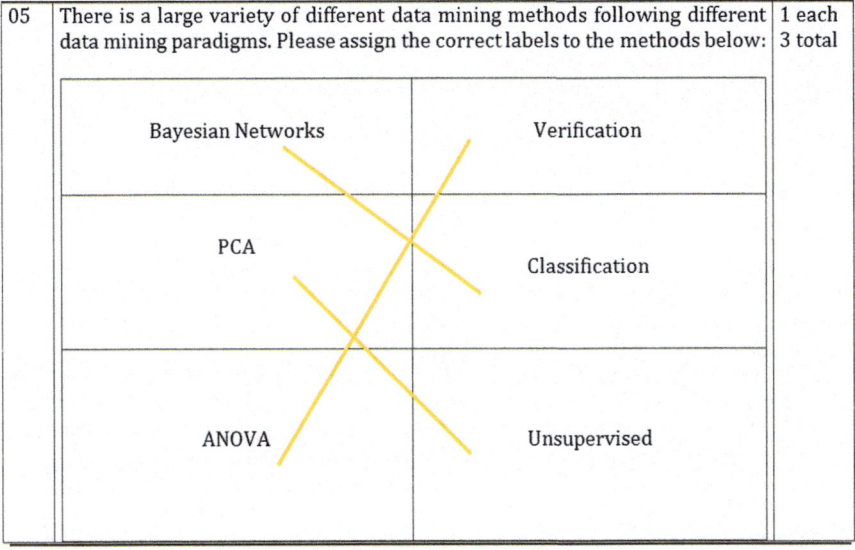

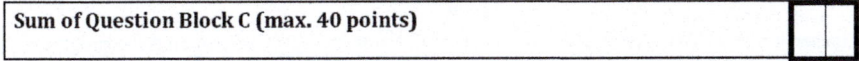

References

Agrawal R, Srikant R (1994) Fast algorithms for mining association rules. In: Proceedings of the 20th International conference on very large data bases, VLDB, pp 487–499

Bandyopadhyay S, Saha S (2013) Unsupervised classification: similarity measures, classical and metaheuristic approaches, and applications. Springer, Heidelberg

Barra A, Bernacchia A, Santucci E, Contucci P (2012) On the equivalence of Hopfield networks and Boltzmann Machines. Neural Netw 34:1–9

Berka P, Rauch J, Tomecková M (2007) Lessons learned from the ECML/PKDD discovery challenge on the atherosclerosis risk factors data. Comput Informatic 26(3):329–344

Blandford A, Attfield S (2010) Interacting with information. Synth Lect Hum Center Informat 3 (1):1–99

Blei DM, Ng AY, Jordan MI (2003) Latent Dirichlet allocation. J Mach Learn Res 3(4–5):993–1022

Boisot M, Canals A (2004) Data, information and knowledge: have we got it right? J Evol Econ 14 (1):43–67

Boser BE, Guyon IM, Vapnik VN (1992) A training algorithm for optimal margin classifiers. Proceedings of the fifth annual workshop on computational learning theory, ACM, pp 144–152

Breiman L, Friedman J, Stone CJ, Olshen RA (1984) Classification and regression trees. Chapman & Hall/CRC, Boca Raton, FL

Brin S, Page L (1998) The anatomy of a large-scale hypertextual Web search engine. Comput Network ISDN Syst 30(1):107–117

Cucchetti A, Vivarelli M, Heaton ND, Phillips S, Piscaglia F, Bolondi L, La Barba G, Foxton MR, Rela M, O'grady J (2007) Artificial neural network is superior to MELD in predicting mortality of patients with end-stage liver disease. Int J Gastroenterol Hepatol GUT 56(2):253

Cortes C, Vapnik V (1995) Support vector networks. Mach Learn 20(3):273–297

Dempster AP, Laird NM, Rubin DB (1977) Maximum likelihood from incomplete data via the EM algorithm. J Roy Stat Soc 39(1):1–38

Domingos P, Pazzani M (1997) On the optimality of the simple Bayesian classifier under zero–one loss. Mach Learn 29(2–3):103–130

Fayyad U, Piatetsky-Shapiro G, Smyth P (1996) The KDD process for extracting useful knowledge from volumes of data. Commun ACM 39(11):27–34

Ferrucci D, Brown E, Chu-Carroll J, Fan J, Gondek D, Kalyanpur AA, Lally A, Murdock JW, Nyberg E, Prager J (2010) Building Watson: an overview of the DeepQA project. AI Maga 31(3):59–79

Ferrucci D, Lally A (2004) Building an example application with the unstructured information management architecture. IBM Syst J 43(3):455–475

Ferrucci D, Levas A, Bagchi S, Gondek D, Mueller ET (2013) Watson: beyond Jeopardy! Artif Intell 199–200:93–105

Fix E, Hodges Jr. JL (1951) Discriminatory analysis-nonparametric discrimination: consistency properties. DTIC document (online open available)

Foltz PW, Kintsch W, Landauer TK (1998) The measurement of textual coherence with latent semantic analysis. Discourse Process 25(2–3):285–307

Freeman S (2008) Biological science. Pearson Education, Upper Saddle River, New Jersey

Freund Y, Schapire RE (1995) A decision-theoretic generalization of on-line learning and an application to boosting. Computational learning theory. Springer, Berlin, pp 23–27

Funk P, Xiong N (2006) Case-based reasoning and knowledge discovery in medical applications with time series. Comput Intell 22(3–4):238–253

Gevaert O, Smet FD, Timmerman D, Moreau Y, Moor BD (2006) Predicting the prognosis of breast cancer by integrating clinical and microarray data with Bayesian networks. Bioinformatics 22(14):184–190

Girolami M, Kaban A (2005) Sequential activity profiling: Latent Dirichlet allocation of Markov chains. Data Min Knowl Disc 10(3):175–196

Han J, Kamber M, Pei J (2011) Data mining: concepts and techniques, 3rd edn, The Morgan Kaufmann series in data management systems. Morgan Kaufmann Publishers, San Francisco, CA

Hirsh H (2008) Data mining research: current status and future opportunities. Stat Anal Data Mining 1(2):104–107

Holzinger A (2013) Human–computer interaction & knowledge discovery (HCI-KDD): what is the benefit of bringing those two fields to work together? In: Cuzzocrea A, Kittl C, Simos DE, Weippl E, Xu L (eds) Multidisciplinary research and practice for information systems. Springer lecture notes in computer science LNCS 8127. Springer, Berlin, pp 319–328

Holzinger A, Scherer R, Seeber M, Wagner J, Müller-Putz G (2012) Computational sensemaking on examples of knowledge discovery from neuroscience data: towards enhancing stroke rehabilitation. In: Böhm C, Khuri S, Lhotská L, Renda M (eds) Information technology in bio- and medical informatics, vol LNCS 7451, Lecture notes in computer science. Springer, Heidelberg, pp 166–168

Hopfield JJ (1982) Neural networks and physical systems with emergent collective computational abilities. Proc Natl Acad Sci U S A 79(8):2554–2558

Hopfield JJ (1987) Learning algorithms and probability distributions in feed-forward and feedback networks. Proc Natl Acad Sci U S A 84(23):8429–8433

Kakkonen T, Myller N, Sutinen E (2006) Applying latent Dirichlet allocation to automatic essay grading, Advances in natural language processing, Proceedings. Springer, Berlin, pp 110–120

Kim SY, Moon SK, Jung DC, Hwang SI, Sung CK, Cho JY, Kim SH, Lee J, Lee HJ (2011) Pre-operative prediction of advanced prostatic cancer using clinical decision support systems: accuracy comparison between support vector machine and artificial neural network. Korean J Radiol 12(5):588–594

References

Knaus WA, Wagner DP, Lynn J (1991) Short-term mortality predictions for critically ill hospitalized adults: science and ethics. Science 254(5030):389

Kohonen T (1982) Self-organized formation of topologically correct feature maps. Biol Cybern 43 (1):59–69

Kotsiantis SB (2007) Supervised machine learning: a review of classification techniques. Informatica 31:249–268

Kreuzthaler M, Bloice MD, Faulstich L, Simonic KM, Holzinger A (2011) A comparison of different retrieval strategies working on medical free texts. J Univer Comput Sci 17(7):1109–1133

Kreuzthaler M, Bloice MD, Simonic K-M, Holzinger A (2010) On the need for open-source ground truths for medical information retrieval systems. In: Tochtermann K, Maurer H (Eds). I-KNOW 2010, 10th international conference on knowledge management and knowledge technologies, Graz, Austria, pp 371-381

Krogh A (2008) What are artificial neural networks? Nat Biotechnol 26(2):195–197

Landauer TK, Dumais ST (1997) A solution to Plato's problem: the latent semantic analysis theory of acquisition, induction, and representation of knowledge. Psychol Rev 104(2):211–240

Landauer TK, Foltz PW, Laham D (1998) An introduction to latent semantic analysis. Discourse Process 25(2–3):259–284

Leitner D, Wassertheurer S, Hessinger M, Holzinger A (2006) A lattice Boltzmann model for pulsatile blood flow in elastic vessels. New Computing in Medical Informatics & Health Care. Special Edition of Springer e&i 123(4):64–68

Liu H, Motoda H, Yu L (2004) A selective sampling approach to active feature selection. Artif Intell 159(1):49–74

Lloyd S (1982) Least squares quantization in PCM. IEEE Trans Inform Theor 28(2):129–137

Maimon O, Rokach L (eds) (2010) Data mining and knowledge discovery handbook, 2nd edn. Springer, New York

Mcculloch WS, Pitts W (1943) A logical calculus of the ideas immanent in nervous activity. Bull Math Biol 5(4):115–133

Mikula S, Trotts I, Stone JM, Jones EG (2007) Internet-enabled high-resolution brain mapping and virtual microscopy. Neuroimage 35(1):9–15

Muth JF (1961) Rational expectations and the theory of price movements. Econometrica 29:315–335

Newell A, Simon H (1976) Computer science as empirical enquiry: symbols and search. Commun Assoc Comput Machine 19:113–126

Overmoyer BA, Lee JM, Lerwill MF (2011) Case 17-2011. A 49-year-old woman with a mass in the breast and overlying skin changes. N Engl J Med 364(23):2246–2254

Pazzani MJ (2000) Knowledge discovery from data? IEEE Trans Intell Syst Their Appl 15 (2):10–12

Pearl J (1988) Probabilistic reasoning in intelligent systems: networks of plausible inference. Morgan Kaufmann, San Francisco, CA

Quinlan JR (1993) C4.5: programs for machine learning. Morgan Kaufmann, San Francisco, CA

Rosenblatt F (1958) The perceptron: a probabilistic model for information storage and organization in the brain. Psychol Rev 65(6):386

Sabah G (2011) Natural language understanding, where are we going? Where could we go? Comput J 54(9):1505–1513

Suh SC, Gurupur VP, Tanik MM (2011) Biomedical engineering: health care systems, technology and techniques. Springer, New York, NY

Tufféry S (2011) Overview of data mining. Data mining and statistics for decision making, Wiley series in computational statistics. Wiley, New York, NY, pp 1–24

Van Gerven MAJ, Taal BG, Lucas PJF (2008) Dynamic Bayesian networks as prognostic models for clinical patient management. J Biomed Inform 41(4):515–529

Vikram K, Upadhayaya N (2011) Data mining tools and techniques: a review. Comput Eng Intell Syst 2(8):31–39

Wall WJ (2007) Liver transplantation for polycystic liver disease. N Engl J Med 356(15):1560
Wang XH, Zheng B, Good WF, King JL, Chang YH (1999) Computer-assisted diagnosis of breast cancer using a data-driven Bayesian belief network. Int J Med Inform 54(2):115–126
Wickens CD (1984) Engineering psychology and human performance. Charles Merrill, Colombus, OH
Xing DS, Girolami M (2007) Employing latent Dirichlet allocation for fraud detection in telecommunications. Pattern Recogn Lett 28(13):1727–1734
Xu L, Tan A, Winslow R, Geman D (2008) Merging microarray data from separate breast cancer studies provides a robust prognostic test. BMC Bioinforma 9(1):125–139

Lecture 7

Knowledge and Decision: Cognitive Science and Human–Computer Interaction

1 Learning Goals

At the end of the seventh lecture, you:

- would be familiar with some principles and elements of human information processing.
- would be able to discriminate between perception, cognition, thinking, reasoning, and problem solving.
- would get an insight into some basics of human decision-making processes.
- would get an overview of the Hypothetico-deductive Method (HDM), versus the PCDA Deming approach.
- would have learned the basics on modelling patient health, differential diagnosis, case-based reasoning, and medical errors.

2 Advance Organizer

Brute force	A trivial very general problem-solving technique that consists of systematically enumerating all possible candidates for the solution and checking whether each candidate satisfies the problem's statement
Cognition	Mental processes of gaining knowledge, comprehension, including thinking, attention, remembering, language understanding, decision making, and problem solving
Cognitive load	According to Sweller (1988) a measure of complexity and difficulty of a task, related to the executive control of the short-term memory, correlating with factors including (human) performance; based on the chunk theory of Miller (1956)

Cognitive science	Interdisciplinary study of human information processing, including perception, language, memory, reasoning, and emotion
Confounding variable	An unforeseen, unwanted variable that jeopardizes reliability and validity of a study outcome
Correlation coefficient	Measures the relationship between pairs of interval variables in a sample, from $r = -1.00$ to 0 (no correlation) to $r = +1.00$
Decision making (DM)	A central cognitive process in every medical activity, resulting in the selection of a final choice of action out of alternatives; according to Shortliffe (2011) DM is still the key topic in medical informatics
Diagnosis	Classification of a patient's condition into separate and distinct categories that allow medical decisions about treatment and prognostic
Differential diagnosis (DDx)	A systematic method to identify the presence of an entity where multiple alternatives are possible, and the process of elimination, or interpretation of the probabilities of conditions to negligible levels
Evidence-based medicine (EBM)	Aiming at the best available evidence gained from the scientific method to clinical decision making. It seeks to assess the strength of evidence of the risks and benefits of treatments (including lack of treatment) and diagnostic tests. Evidence quality can range from meta-analyses and systematic reviews of double-blind, placebo-controlled clinical trials at the top end, down to conventional wisdom at the bottom
External validity	The extent to which the results of a study are generalizable or transferable
Hypothetico-deductive Model (HDM)	Formulating a hypothesis in a form that could conceivably be falsified by a test on observable data, e.g., a test which shows results contrary to the prediction of the hypothesis is the falsification, a test that could but is not contrary to the hypothesis corroborates the theory—then you need to compare the explanatory value of competing hypotheses by testing how strong they are supported by their predictions
Internal validity	The rigor with which a study was conducted (e.g., the design, the care taken to conduct measurements, and decisions concerning what was and was not measured)
PDCA	Plan-Do-Check-Act, the so-called PDCA-cycle or Deming-wheel, can be used to coordinate a

	systematic and continuous improvement. Every improvement starts with a goal and with a plan on how to achieve that goal, followed by action, measurement, and comparison of the gained output
Perception	Sensory experience of the world, involving the recognition of environmental stimuli and actions in response to these stimuli
Qualitative research	Empirical research exploring relationships using textual, rather than quantitative data, e.g., case study, observation, ethnography. Results are not considered generalizable, but sometimes at least transferable
Quantitative research	Empirical research exploring relationships using numeric data, e.g., surveys, quasi-experiments, experiments. Results should be generalized, although it is not always possible
Reasoning	Cognitive (thought) processes involved in making medical decisions (clinical reasoning, medical problem solving, diagnostic reasoning, behind every action)
Receiver-operating characteristic (ROC)	In signal detection theory this is a graphical plot of the sensitivity, or true positive rate, vs. false positive rate (1 − specificity or 1 − true negative rate), for a binary classifier system as its discrimination threshold is varied
Symbolic reasoning	Logical deduction
Triage	Process of judging the priority of patients' treatments based on the severity of their condition

3 Acronyms

CES	Central executive system
DDx	Differential diagnosis
DM	Decision making
DSS	Decision support system
EBM	Evidence-based medicine
fMRI	Functional magnetic resonance image
HDM	Hypothetico-deductive model
HOAC	Hypothesis Oriented Algorithm for Clinicians
IOM	Institute of medicine
LTS	Long-term storage
ME	Medical error

PDCA	Plan-Do-Check-Act
QM	Quality management
ROC	Receiver operating characteristic
RST	Rough set theory
STS	Short-term storage
USTS	Ultrashort-term storage

4 Key Problems

Slide 7-1: Key Challenges

- The short time to make a decision "5 Minutes"
- Limited perceptual, attentive, and cognitive human resources
- Human error

According to Gerd Gigerenzer from the Max Planck Institute in Berlin the average time spend for a patient, and hence to make a decision, in a public hospital is 5 min. This is the famous "5 Minutes" approach (Gigerenzer 2008); however, in essence, time is fundamentally critical in medicine and health care.

5 Human Information Processing

5.1 Decision Making and Reasoning

Slide 7-2: Decision Making Is Central in Biomedical Informatics

According to Shortliffe (2011), decision making is still the major topic in biomedical informatics. Decision making is a cognitive process resulting in the selection of one of several alternatives. Every decision-making process produces a final choice and the output can be an action or an opinion.

The slide shows a typical clinical decision-making process supported by a "second opinion" during a project of the Cognitive Engineering Research Group and the Scottish Centre for Telehealth at Aberdeen Royal Infirmary. The project aims to investigate human errors in carrying out diagnosis of

(continued)

(continued)

dermatological conditions using remote imaging. A particular focus is on the role of color images, i.e., the effect of image quality and characteristics as a function of human color perception. The research will develop a model of human information processing that incorporates the user in a "human-in-the-loop-system." Such a model can form the basis for predicting and preventing human errors relating to color in remote imaging of dermatological problems.

Remark: If it were not for human error there would be little justification for the area of Human Factors Engineering as an applied discipline. Research during the past 20 years has made progress in understanding and controlling human errors in human–computer interaction (HCI) but technology develops fast which poses new challenges for researchers. All users of IT systems make errors and this is accepted.

http://www.comp.rgu.ac.uk/staff/ph/html_docs/Human_error_telehealth.htm

---

Slide 7-3: Reasoning Foundations of Medical Diagnosis

If a physician is asked, "How do you make a medical diagnosis?" his explanation of the process might be as follows. "First, I obtain the case facts from the patient's history, physical examination, and laboratory tests. Second, I evaluate the relative importance of the different signs and symptoms. Some of the data may be of first-order importance and other data of less importance. Third, to make a differential diagnosis I list all the diseases which the specific case can reasonably resemble. Then I exclude one disease after another from the list until it becomes apparent that the case can be fitted into a definite disease category, or that it may be one of several possible diseases, or else that its exact nature cannot be determined." This, obviously, is a greatly simplified explanation of the process of diagnosis, for the physician might also comment that after seeing a patient he often has a "feeling about the case." This "feeling," although hard to explain, may be a summation of his impressions concerning the way the data seem to fit together, the patient's reliability, general appearance, facial expression, and so forth; and the physician might add that such thoughts do influence the considered diagnoses. No one can doubt that complex reasoning processes are involved in making a medical diagnosis. The diagnosis is important because it helps the physician to choose an optimum therapy, a decision which in itself demands another complex reasoning process (Ledley and Lusted 1959).

Slide 7-4: Decision Making Is Central in Medicine

In practice the clinicians use Decision-Making Frameworks to guide patient management, communicate with other health care providers and educate patients and their families. A number of frameworks have been applied to guide clinical practice. Schenkman et al. (2006) have proposed a unifying framework for application to decision making in the management of individuals on the example of neurologic dysfunction. The framework integrates both enablement and disablement perspectives (Schenkman et al. 2006). A good primer of the theory of medical decision making is McNeil et al. (1975). Clinical decisions are dependent on evidence (e.g., patient data, epidemiological research, randomized controlled trials), guidelines, ethics, knowledge, patient/clinicians subjective preferences (cultural beliefs, see for example on the influence of organizational culture and communication Xie et al. 2013), personal values, education and experience; and last but not least: constraints (e.g., formal policies, restrictions, laws, community standards, time, and financial issues) (Hersh 2010). Before we go into more detail let us review how humans process information.

5.2 The Three-Storage Memory Model

Slide 7-5: Human Information Processing Model (Atkinson and Shiffrin 1971)

Information processing is the core of research in decision support and cognitive performance and we must mention that there are dozens of different models available in this scientific community. Here we see the information flow in the famous three-storage memory system developed by Atkinson and Shiffrin (1971): The environmental information is processed by sensory registers in the various physical modalities (visual, auditory, haptic, gustatoric, olfactoric) and entered into the short-term store (STS). The information remains temporarily in STS, the length of stay depending on control processes. Whilst the information remains in STS it may be copied into the long-term store (LTS). While information remains in STS, information in LTS associated with it may also be activated and entered in STS.

Background Information: Lev Vygotsky (1896–1934) and Jean Piaget (1896–1980) were the first who indicated the importance of the end user's knowledge base for interpreting information. New knowledge should not be redundant but should have sufficient reference to what the user already knows. Moreover, sufficient overlap fosters the understandability of

(continued)

(continued)

information, in particular deep understanding (Kintsch 1994), is an important determinant of interest (Schraw and Lehman 2001), and, through both understandability and interest, the knowledge base has an indirect effect on motivation. Concluding, we can say that it is of pivotal importance to have a **model of the user's knowledge** allowing to measure the distance between a topic and the user's Knowledge Model (KM) (van der Sluis and van den Broek 2010).

Slide 7-6: General Model of Human Information Processing

The three-storage model of Atkinson and Shiffrin is up to the present the standard information-processing model, in which each sensory system has its own store.

A current view on the three-stage human memory model (Wickens et al. 2004) makes this obvious: We can determine the four different areas: physics (of information), perception, cognition, and action (motorics)—the system having a feedback loop (e.g., the outcome of a decision is seen and accordingly will be perceived and cognitively processed).

The sensory processing unit is equivalent to the **ultrashort-term storage (USTS)** and holds the incoming sensory information long enough so that unconscious processes can determine whether the input should be let into the working memory, or should immediately be discarded (controlled by attentional resources). Selected stimuli which passes through this USTS will be perceived and hold within the **short-term storage (STM, working memory)**, which is regarded as the center of consciousness, analog to the central processing unit (CPU) of a von Neumann machine (see Lectures 1 and 2).

The capacity of the STM is small, limiting the input to 7 ± 2 chunks (Miller 1956).

Note: A chunk can be any meaningful information—for a little child this might be one single letter of the alphabet—for a medical professional this might be half the page of the medical report—it depends on many factors. Miller (1956) found that the capacity of WM is limited to this certain span of so-called chunks: "With binary items this span is about nine and although it decreases to five with monosyllabic English words, the difference is far less than the hypothesis of constant information would require."

Always remember that a chunk is a flexible amount of information, dependent on the previous knowledge (Simon 1974).

Finally we have the **long-term storage (LTM, permanent memory)** is often compared as the external storage of a von Neumann machine.

5.2.1 Example: Visual and Audial Information Processing

Slide 7-7: Example: Visual Information Processing

We recognize physical objects (color, shapes, etc.), we touch a surface of this object, and we listen to a speech. For the human brain, this is only an information process. It is amazing that literally our physical world is consisting of information. The perceived neural information can be measured (Yu et al. 2010) and although all brain functions require coordinated activity of many neurons, it has been difficult to estimate the amount of information carried by a population of spiking neurons. Let us look for the example of visual information processing.

The visual system is the most complex neural circuitry of all the sensory systems. The auditory nerve contains about 30,000 fibers, but the optic nerve contains over one million. Most of what we know about the functional organization of the visual system is derived from experiments similar to those used to investigate the somatic sensory system. The similarities of these systems allow us to identify general principles governing the transformation of sensory information in the brain as well as the organization and functioning of the cerebral cortex—processing of visual information is still poorly understood (state of 2013). In the slide you can see a contemporary principal sketch of the visual system in the brain. A good primer to understand these processes is still the special issue of Scientific American from September 1979 to still valid!

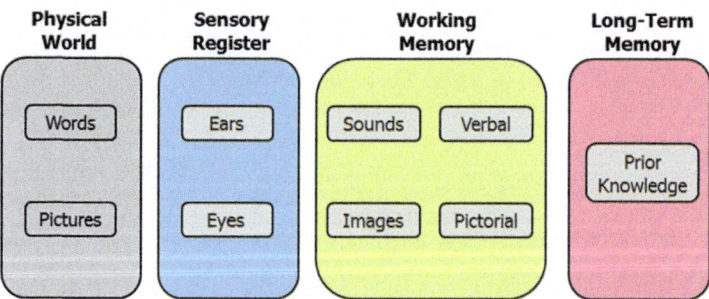

Fig. 1 See Slide 7-8

Slide 7-8: Schematic Information Processing Chain

A simplified schematic process chain consists of the physical world (grey area), the sensory registers (ears, eyes, blue area), the working memory

(continued)

(continued)

(verbal and pictorial, yellow area), and the long-term memory (rose area). This will help us to quickly understand the basic processes of visual and verbal information processing (see next slides) (Holzinger 2002).

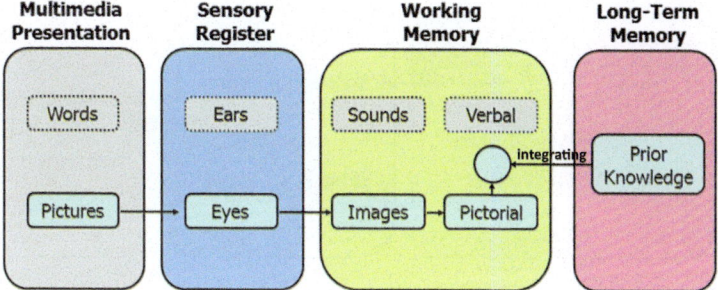

Fig. 2 See Slide 7-9

Slide 7-9: Information Processing of Words/Pictures

In this slide we see the visual information processing chain. The perception of images is similar to the perception of printed words, although within the working memory we have to determine between pictorial/visual information processing and verbal/visual information processing (see next slide).

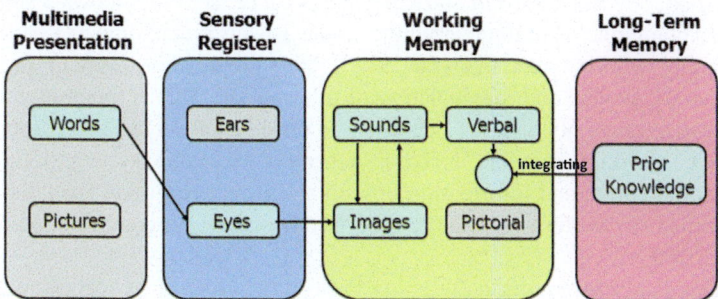

Fig. 3 See Slide 7-10

Slide 7-10: Information Processing of Words/Pictures

Here the processing of printed words, perceived as images but interpreted, processed as verbal information.

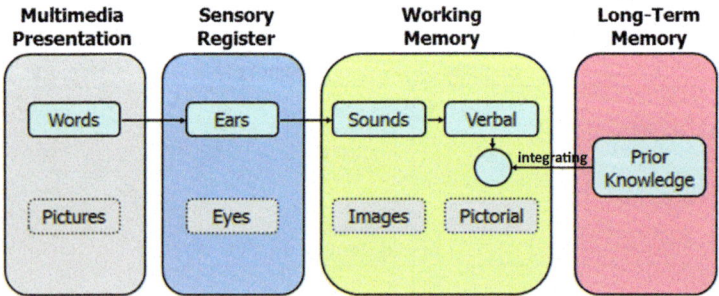

Fig. 4 See Slide 7-11

Slide 7-11: Information Processing of Words/Sounds

Finally, the processing of spoken words, perceived as sounds via the ears and the auditory sensor register and processed as verbal information.

5.2.2 Central Executive System

Slide 7-12: Alternative Model: Baddeley: Central Executive System

The model by Atkinson and Shiffrin which we have seen in Slides 7-5 and 7-6 can explain much, but not everything, some aspects remain unexplained. For example, working memory (the STS) refers to a whole system for the temporary storage and manipulation of information. This is function critical for a wide range of cognitive operations. Consequently, it has been proposed that the working memory includes a central executive system (CES) to control attention and information flow to and from verbal and spatial short-term memory buffers—as can be seen in this slide—a image from Quinette (et al. 2003):

This is a schema originally going back to the concept of "central executive" by Baddeley (1981): A central executive system (CES) has direct influence on all information processes.

5 Human Information Processing

Slide 7-13: Neural Basis for the "Central Executive System"

D'Esposito et al. (1995) have used functional magnetic resonance imaging (fMRI) to examine such brain activations during the concurrent performance of two tasks, which is expected to engage the CES. Activation of the prefrontal cortex was observed when both tasks are performed together, but not when they are performed separately. These results support the view that the prefrontal cortex is involved in human working memory and such a central executive is involved.

In this slide regions significantly activated in the single- and dual-task conditions in individual subjects are shown in a–d. Single-task activations are from the same subject; dual-task activations are from two different subjects. Comparison of the single-task versus baseline conditions show activation in the bilateral parieto-occipital regions for the dot-location task (a) and the left temporal region for the semantic-judgement task (c). Regions significantly activated during the dual-task versus single-task conditions are shown in band d. In the first comparison (dual task minus semantic taSk), activation is seen in prefrontal cortex, anterior cingulate, and premotor cortex (b). In addition, activation is seen in bilateral parieto-occipital regions, which represents activation during the dot-location task. In the second comparison (dual task minus spatial task), activation again is seen in prefrontal cortex and anterior cingulate (d) (D'Esposito et al. 1995).

5.2.3 Selective Attention

Slide 7-14: Central Executive: Selective Attention

Selective Attention—going back to the Central Executive System—is a term originally coined by Broadbent already in the 1950s where he proposed a revised model of the human information-processing system: In the slide we see a classic sketch by Cowan (1988): on the x-axis we have the time since a stimulus reception is represented. The components are arranged in real time and the stimulus information can be present in more than one component at the same time. The STS is represented as an activated subset of LTS and the focus of attention is represented as a subset of the STS. The timing of involvement of the central executive in processing is flexible. The arrows represent the transfer of information from one form to another; these are discrete approximations to continuous processes that can occur in parallel or cascade. Pathways leading to awareness can come from three sources:

1. Changed stimuli for which there is dishabituation

(continued)

(continued)

2. Items selected through effortful processing (whether of sensory origin or not)
3. The spontaneous activation of long-term memory information based on associations (not shown)

Slide 7-15: Human Attention Is Central for Problem Solving

As we can see in this slide, the information received by the receptor cells gets stored in a system of sensory registers (vision, hearing, olfaction, haptic, gustatory). Although the storage capacity would be theoretically enormous, the information is only available a very short time for further processing. Through a process of selective attention (controlled by the central executive system), only **selected subsets** of the vast collection of incoming information become designated for further processing (we call this: perception). Here information can also become meaningful trough comparison with information from the LTS (previous knowledge, hypotheses, models of the world, etc.). A large portion of our conscious effort (especially in making decisions) is dedicated to these processes in the STS—you now understand why the STS is called **working memory** (WM): Rehearsal within the WM enables it finally to become the information encoded into LTS—where it remains a lifetime; otherwise it decays rapidly. Moreover, the WM has severe capacity constraints, governing the amount of information kept active (Wickens et al. 2004 p. 713.)

5.3 Clinical Decision-Making Process

Slide 7-16: Human Decision Making

Human decision making is perception and response execution (outcome) and is an activity tight to fallibility, especially in complex and dynamic situations. If the perceived information is incomplete or fuzzy, intensive interpretation and integration of this information is required. Any hypothesis which the decision maker generates regarding the incoming information is highly dependent on information available in the LTS (=previous knowledge). Constraints often lead to shortcuts in decision making, whereby people opt for choices that are good enough for their purposes and adopt strategies for

(continued)

(continued)

sampling information that they perceive to be most relevant to solve their current problem. Clinical reasoning (problem solving, decision making, judgement) is a central component of the physician's competence (Norman 2005).

To help us to understand the decision-making process we can refer to Wickens (1984): The physical stimuli (cues) are selected by the attentional resources and the perceived information builds working hypotheses H1, H2, ... Those are compared and judged against available hypotheses, already present in the long-term memory. On this basis the best alternative will be chosen and actions A1, A2, ... performed according to likelihoods and consequences of outcomes—which can be perceived again via the feedback loop.

Note: The "nebula of uncertainty" emphasizes what we have already learned in our course: We always deal with probable information! Each information chunk, item, or whatever you call it has always a certain probability aspect (thanks to Bayes ;-)).

The model of Wickens is still the best model to explain clinical decision making.

Slide 7-17: Example: Triage Tags: International Triage Tags

The simplest decision support process is triage, which is the process of determining the priority of patients' treatments based on the severity of their conditions. The idea comes from times where resources are insufficient for everybody to be treated immediately—for example in mass incidents. Triage is also used in hospitals for patients arriving at the emergency department.

Slide 7-18: Clinical DM: Hypothesis-Oriented Algorithm

In practice the clinicians use decision-making frameworks to guide patient management, communicate with other health care providers and educate patients and their families. A number of frameworks have been applied to guide clinical practice. Schenkman et al. (2006) have proposed a unifying framework for application to decision making in the management of

(continued)

(continued)

individuals on the example of neurologic dysfunction. The framework integrates both enablement and disablement perspectives (Schenkman et al. 2006):

The process in Slide 7-18 is patient-centered, which is in contrast to a pathology-driven approach in which the process is disease-centered and culminates in disablement. The patient-centered approach emphasizes roles and functions of which the individual is capable and at the same time identifies limitations in the individual's abilities, with the goal of minimizing barriers to full participation within society or the environment. The purpose of the history and interview is to gain an understanding of the patient as an individual, determine why the individual seeks physical therapy, identify what he or she hopes to achieve through physical therapy, and begin to formulate the examination strategy. The purpose of the systems review is to rule out those body systems with which the physical therapist need not be concerned, identify systems that are resources for the patient, guide choices regarding which aspects of the remaining systems to examine in detail, and determine whether the physical therapist should proceed or should refer the patient to other health care providers. The specific purposes of the examination may vary and depend on the reason for which it is carried out. The level of examination is adjusted to reflect patient-identified problems and goals and can be drawn from elements of the continuum of the ICF model. The evaluation consists of an interpretation of findings in order to develop a realistic plan of care. The plan of care is based on a synthesis of the information from all of the previous steps, including the patient's goals and expectations, task performance, the patient's resources and impairments, and the medical diagnosis and prognosis for the condition. The plan of care is organized around the patient's goals. Goal-directed therapy has been identified as improving motor function and promoting cortical reorganization.

5.4 Reasoning and Problem-Solving Procedures

5.4.1 Hypothetico-deductive Model (HDM) vs. PCDA Deming Wheel

Slide 7-19: Hypothetico-deductive Method vs. PDCA Deming Wheel

The HDM (Newton—Scientific method, in the slide on the right) assumes that properly formed theories arise as generalizations from observable data that they are intended to explain. These hypotheses, however, cannot be

(continued)

(continued)

conclusively established until the consequences, which logically follow from them, are verified through additional observations and experiments. In conformity with the rationalism of Descartes, the HDM treats theory as a deductive system, in which particular empirical phenomena are explained by relating them back to general principles and definitions. This method abandons the Cartesian claim that those principles and definitions are self-evident and valid; it assumes that their validity is determined only by their consequences on previously unexplained phenomena or on actual scientific problems. Scientific methods include techniques for investigating phenomena with the aim of acquiring new knowledge, or correcting and integrating previous knowledge, i.e., going from state of the art to beyond state of the art (Holzinger 2010).

Hypothesis: A working hypothesis (Arbeitshypothese) develops from the formulation of a question, e.g.: *The learning success is greater with the use of the simulation X than with the traditional method Y*. Again, using exact definitions of the terms used. According to the circumstances, the working hypothesis must be continually modified, until it represents, exactly what one wants to examine, e.g.: *by the employment of the simulation X, learning time for the topic Y can be significantly shortened*.

Here, a very careful clarification is necessary: Which variables are to be examined and which expectations exist (again with reference to the theory and/or to preceding work).

Formal scientific hypotheses (Popper 2005) can be brought formally into the form: *When X then Y; X implies Y; $X \rightarrow Y$*.

It is important to understand that *hypotheses are only conditional statements*: The effect occurs, not under all possible circumstances, but only under completely predetermined circumstances. The hypothesis must be evaluated experimentally. Whereby there are exactly three possibilities:

1. **TRUE**: The hypothesis is significantly proved (verified = is true).
2. **FALSE**: The hypothesis is significantly disproved (falsified = is wrong).
3. **UNDECIDED**: The hypothesis can (on the basis of the existing data) neither be verified nor falsified (no statement can be made).

Note: Whether a hypothesis is true (this ALWAYS applies) cannot be confirmed due to the universality of the statement. Scientific hypotheses can therefore only be falsified. Nevertheless, hypotheses can be differentiated with respect to the degree of their confirmation, whereby certain criteria is taken into account, such as how frequently and in which critical places the hypothesis has already been confirmed (Popper 2005).

(continued)

(continued)

PDCA: This concept was developed by Shewhart (1958) as PDSA cycle. The roots can be tracked back to Aristotle (384–322 BC) and Francis Bacon (1561–1626). The PDSA cycle consists of four steps:

1. PLAN: Study the process
2. DO: Make changes on a small scale
3. STUDY: Observe the effects
4. ACT: Identify what you can learn from your observation

William E. Deming (1900–1993) promoted this model effectively and called it PDCA cycle (Deming 1994) and is also known as Deming wheel:

1. PLAN: Clearly define the objectives and processes necessary to gain deliverables in accordance with the *expected output*.
2. DO: Implement the new processes on a *small scale* (e.g., within a trial or pilot project).
3. CHECK: Now *measure the outcome and compare* your results against the expected results and look for differences.
4. ACT: Finally, *analyze the differences* to determine their cause. Each finding can be used as input for a new PDCA cycle.

The PDCA wheel can be used to coordinate your continuous improvement. Every improvement starts with a goal and with a plan on how to achieve that goal, followed by action, measurement, and comparison of the gained output.

The approach in our research group (hci4all.at) fuses both approaches together and our leitmotiv arises from it: "Science is to test ideas (weird ideas ;-))—Engineering is to put these ideas into Business."

Slide 7-20: Modeling Patient health (1/2)

We have discussed modeling already in Lecture 3. Now we see how we can apply this knowledge for decision support: For example, in oncology, one way to represent the patient's health status is in terms of the performance status, which is distinguished into normal, mild complaints, ambulatory, nursing care, intensive care, and death. In this slide we see how we can model the various influences on carcinoid patient health and they can be categorized into the described factors: The variables age, gender, and (past) health determine the patient health independent of the disease. The variable tumor mass represents the direct effect of the tumor on patient health. The

(continued)

(continued)

dashed objects denote past states and square objects denote treatments (van Gerven et al. 2008).

chd = carcinoid heart disease; bmd = bone-marrow depression; plr = partial liver resection; rfa = radiofrequency ablation.

Slide 7-21: Modeling Patient Health (2/2)

The large number of conditioning variables that (partially) determine patient health makes estimation of conditional probabilities for this variable very difficult. However, a large subset of these variables is risk factors that influence health only due to the fact that they may cause immediate patient death, thereby simplifying the specification as follows.

Let $U \subseteq X$ denote this risk factors and
Let $V = X \setminus U$ denote the complement.
The risk of immediate death $p(health(t) = death|X)$ can be expressed as:

$$1 - p(health(t) \neq death | V) \prod_{U \in U} p(health(t) \neq death | U, health(t-1))$$

Further, we obtain:

$$p(health(t) = h|X) = p(h|V) \prod_{U \in U} p(health(t) \neq death | U, health(t-1))$$

for $h \neq death$

These simplifications greatly reduce the number of parameters that need to be estimated (van van Gerven et al. 2008).

To understand these processes we must have a look on some basics of human decision making, and we start with one of the oldest approaches: signal detection theory.

5.4.2 Signal Detection Theory

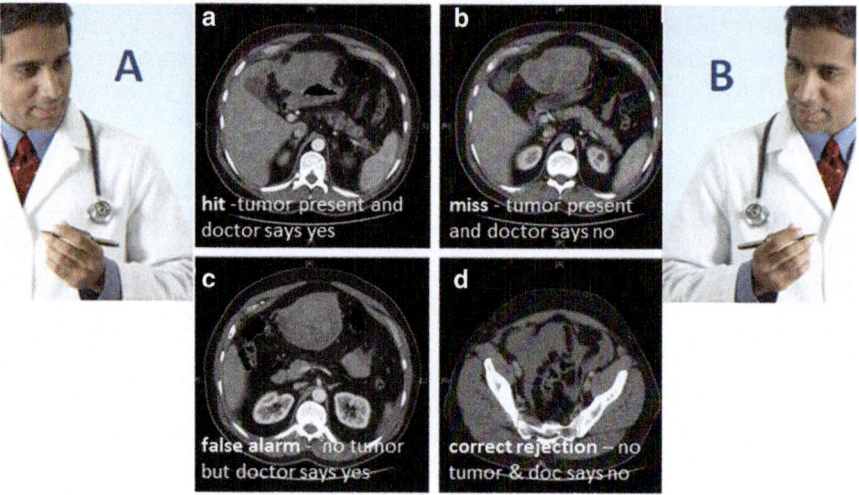

Fig. 5 See Slide 7-22

Slide 7-22: Signal Detection Theory on the MDM Process

Look at Slide 7-22: You see two doctors (doctor A and doctor B), with equally good training, looking at the same CT scan data, so both will have the same information ... but they may gain different knowledge out of the data due to many intervening factors, bias, criteria.

Interpreting CT data is difficult and it takes a lot of training and there is always some uncertainty as to what is correct or not. In principal we have four possibilities: Either there is a tumor (signal present) or there is not (signal absent). Either the doctor sees a tumor (they respond "yes"') or does not (they respond "no").

There are four possible outcomes:

(a) Hit—tumor is present and doctor says "yes"
(b) Miss—tumor is present and doctor says "no"
(c) False alarm—tumor is absent and doctor says "yes"
  and
(d) Correct rejection—tumor is absent and doctor says "no"

Let us look more closely on the information process.

5 Human Information Processing 317

Slide 7-23: Information Acquisition and (Subjective) Criteria—Bias

Information acquisition: First, there is information in the CT data, e.g., healthy lungs have a characteristic shape; the presence of a tumor might distort that shape. Tumors may have different image characteristics: brighter or darker, different texture, etc. With proper training a doctor learns what kinds of things to look for, so with more practice/training they will be able to acquire more (and more reliable) information. Running another test (e.g., MRI) can also be used to acquire more information. Regardless, acquiring more information is good. The effect of information is to increase the likelihood of getting either a hit or a correct rejection, while reducing the likelihood of an outcome in the two error boxes.

Criterion: The second component of the decision process is different: Additionally to technical information acquisition, medical doctors use their own judgment. Different doctors may feel that the different types of errors are not equal. One doctor may feel that missing an opportunity for early diagnosis may mean the difference between life and death; on the other hand a false alarm may result only in a routine biopsy operation. They may choose to err toward yes (tumor present) decisions. Other doctors may feel that unnecessary surgeries are very bad (expensive, stress, etc.). They may choose to be conservative and say no (no tumor) more often. They will miss more tumors, but they will be doing their part to reduce unnecessary surgeries. And they may feel that a tumor, if there really is one, will be picked up at the next check-up. Two doctors, with equal training, looking at the same CT, have the same information; but having a different bias/criteria.

For differences DSS vs. DM see the next slide.

Slide 7-24: Decision-Making Process vs. Data Mining process

The decision-making process described in Slide 7-23 and visualized in this slide can be compared to the data mining process (right in this slide) (Ayed et al. 2010).

Slide 7-25: Decision-Making Process: Signal Detection Theory

Remember our example from Slide 7-22: Two doctors, with equally good training, looking at the same CT scan data, will have the same information... but they may gain different knowledge due to bias/criteria.

(continued)

(continued)

Noise: Detecting a tumor is very difficult and there will always be some amount of uncertainty (as generally in the medical area, always remember that we speak of probable information). There are two kinds of noise factors that contribute to this uncertainty: internal noise and external noise.

There are many possible sources of external noise. There can be noise factors that are part of the imaging process, a smudge, or bad spots. While the doctor makes every effort possible to reduce the external noise, there is little or nothing that they can do to reduce internal noise. Internal noise refers to the fact that neural responses are "noisy" per se. Look at the sketches in this slide—based on the classical signal detection theory (SDT):

Note: The starting point for signal detection theory was that all reasoning and decision making takes place in the presence of uncertainty. Signal detection theory provides a precise language and graphic notation for analyzing decision making under uncertainty.

SDT was developed by Peterson et al. (1954) and applied to psychophysics by Tanner and Swets (1954) and is a very generic and widely used method for drawing inferences from data. The data needed for an SDT analysis are just the counts of hits, false alarms, misses, and correct rejections.

Slide 7-26: Receiver Operating Characteristics (ROC Curve)

Within our course we have seen already examples for ROC curves (for example in Slide 6-40, where we compared the performance of SVM with ANN).

In signal detection theory (SDT), a receiver operating characteristic (ROC) is the curve which illustrates the performance of a binary classifier system as its discrimination threshold is varied. It is created by plotting the fraction of true positives out of the total actual positives (TPR = true positive rate) vs. the fraction of false positives out of the total actual negatives (FPR = false positive rate), at various threshold settings. TPR is also known as sensitivity (also called recall in some fields), and FPR is one minus the specificity or true negative rate.

In classification tasks the more the curve is in the upper left corner, the better.

Slide 7-27: Let Us Consult Reverend Bayes and Price (1763)

A further example on the importance of the work of Bayes (Bretthorst 1990). Let us take a classic clinical example:

D ... acute heart attack
U_+ ... instable chest pain
$p(D)$ ... 37 of 1,000 = 0.037 (heart attack)
$p(\overline{D})$ ... 963 of 1,000 = 0.963 (no heart attack)
40 % of patients report on instable chest pain
$p(U_+|D) = 0.4$
Unfortunately these symptoms also occur in 5 % of the healthy population

$$p(U_+|\overline{D}) = 0.05$$

We find the probability for a heart attack during these symptoms therefore by using Bayes' Rule:

$$p(D|U_+) = \frac{p(U_+|D)*p(D)}{p(U_+|D)*p(D)+p(U_+|\overline{D})*p(\overline{D})} = 0.235$$

5.4.3 Differential Diagnosis

Slide 7-28: Example Clinical Case: Serotonin Toxicity

Let us practice differential diagnosis on a clinical case: the **serotonin syndrome** is a potentially life-threatening **adverse drug reaction** that results from therapeutic drug use, intentional self-poisoning, or **inadvertent interactions between drugs**.

Three features of the serotonin syndrome are critical to an understanding of the disorder. First, the serotonin syndrome is not an idiopathic drug reaction; rather it is a predictable consequence of excess serotonergic agonism of the central nervous system's receptors and peripheral serotonergic receptors. Second, excess serotonin produces a spectrum of clinical findings. Third, clinical manifestations of the serotonin syndrome range from barely perceptible to lethal. The death of an 18-year-old patient in New York City more than 20 years ago, which resulted from coadministration of meperidine and phenelzine, remains the most widely recognized and dramatic example of this preventable condition (Boyer and Shannon 2005).

In this slide we see that the serotonin syndrome encompasses a range of clinical findings. Patients with mild cases may be afebrile but have tachycardia, with a physical examination that is notable for autonomic findings such as shivering, diaphoresis (sweating), or mydriasis (blown pupil).

Slide 7-29: Differential Diagnosis on Serotonin Syndrome

Differential Diagnosis (DD) is principally a **systematic** diagnostic method used to identify the presence of an entity, where multiple alternatives are possible and may also refer to any of the candidate alternatives. It is essentially a process of elimination, or at least, rendering of the probabilities of candidate conditions to negligible levels. The correct medical definition: Differential diagnosis is the process of weighing the p(x) of one disease versus the p(x) of other diseases possibly accounting for a patient's illness. The differential diagnosis of rhinitis (a runny nose) includes allergic rhinitis (hayfever), the abuse of nasal decongestants and the common cold.

In this slide we see Hunter's decision rules for diagnosis of serotonin toxicity (left) from the example in Slide 7-28 and the signs (gamuts, patterns) and symptoms related (Ables and Nagubilli 2010).

Slide 7-30: Differential Diagnosis on Serotonin Syndrome 2/2

The primary differential diagnosis of serotonin syndrome includes anticholinergic syndrome, malignant hyperthermia, and neuroleptic malignant syndrome as shown in this slide, along with the history, vital signs and clinical features (Ables and Nagubilli 2010).

5.4.4 Rough Set Theory (RST)

Slide 7-31: Rough Set Theory for dealing with Incomplete Data

RST is an extension of the Classical Set Theory, for use when representing incomplete knowledge. Rough sets are sets with fuzzy boundaries—sets that cannot be precisely characterized using the available set of attributes, exactly like it is in medical decision making; they are based on two ideas:

1. A given concept can be approximated by partition-based knowledge as upper and lower approximation—which corresponds exactly to the focusing mechanism of differential medical diagnosis: upper approximation as selection of candidates and lower approximation as concluding a final diagnosis.
2. A concept or observations can be represented as partitions in a given data set, where rough sets provides a rule induction method from given data. Consequently, this model can be used to extract rule-based knowledge from medical databases.

(continued)

(continued)

Note: **Medical reasoning** is a focusing mechanism, which is used to select the final diagnosis from many candidates. For example, in the differential diagnosis of headache, more than 60 diseases will be checked (present history, physical examinations and laboratory examinations). In diagnostic procedures, a candidate is excluded if a symptom necessary to diagnose is *not observed*.

This style of reasoning consists of the following two kinds of reasoning processes: exclusive reasoning and inclusive reasoning. The diagnostic procedure is shown in the next slide on the example of symptom "headache."

Slide 7-32: Diagnostic Procedure (Differential Diagnostic)

Exclusive reasoning *excludes a disease from candidates* when a patient does not have a symptom which is necessary to diagnose that disease.

Inclusive reasoning *suspects a disease in the output of the exclusive process* when a patient has symptoms specific to a disease. These two steps are modelled as usage of two kinds of rules, negative rules (or exclusive rules) and positive rules, the former of which corresponds to exclusive reasoning and the latter of which corresponds to inclusive reasoning. In the next two subsections, these two rules are represented as special kinds of probabilistic rules.

In the following slide, we use the notations introduced by Skowron and Grzymala-Busse (1994) (refer also to Pawlak et al. 1995), which are based on rough set theory (Pawlak 1982). These notations are illustrated by a small dataset shown in an example, which includes symptoms exhibited by six patients who complained of headache.

Slide 7-33: Rough Set Theory Example Symptom: Headache 1/3

Let U denote a non-empty, finite set called the universe and A denote a non-empty, finite set of attributes:
$a: U \to Va$ for $a \in A$
where Va is called the domain of a.
Then, the decision table is defined as an information system:

(continued)

(continued)
$$A = (U, A \cup \{d\})$$

Let us have an information system (see table right in the slide) with

$$U = \{1, 2, 3, 4, 5, 6\}$$
$$A = \{age, \ location, \ nature, \ prodrome, \ nausea, \ M1\}$$

and
d = class.
For *location* ∈ A, $V_{location}$ is defined as {*occular, lateral, whole*}.
The table right shows an example data set from Tsumoto (2006).
Note: prodrome = in medicine, a prodrome is an early symptom (or set of symptoms) that might indicate the start of a disease before specific symptoms occur.

Slide 7-34: Rough Set Theory Example Symptom: Headache 2/3

The atomic formula (Tsumoto 2006) over

$$B \subseteq A \cup \{d\}$$

and V are expressions of the form [a = v]
called descriptors over B, where a ∈ B and v ∈ Va.

The set F(B, V) of formulas over B is the least set containing all atomic formulas over B and closed with respect to disjunction, conjunction and negation. For example, [location = occular] is a descriptor of B.

For each f ∈ F(B, V), fA denote the meaning of f in A, i.e., the set of all objects in U with property f, defined inductively as follows.

1. If f is of the form [a = v] then,
   fA = {s ∈ U|a(s) = v}
2. (f ∧ g)A = fA ∩ gA; (f ∨ g)A = fA ∨ gA; (¬f)A = U − fa

For example, f = [location = whole] and fA = {2, 4, 5, 6}. As an example of a conjunctive formula, g = [location = whole] ∧ [nausea = no] is a descriptor of U and fA is equal to glocation, nausea = {2,5}.

Slide 7-35: Classification Accuracy and Coverage

Definition 1. Let R and D denote a formula in F(B, V) and a set of objects which belong to a decision d. Classification accuracy and coverage (true positive rate) for R $\rightarrow$ d is defined as:

$\alpha R(D) = |RA \cap |/|RA|(=P(D|R))$, and
$\kappa R(D) = |RA \cap D|/|D|(=P(R|D))$

where $|S|$, $\alpha R(D)$, $\kappa R(D)$, and $P(S)$ denote the cardinality of a set S, a classification accuracy of R as to classification of D and coverage (a true positive rate of R to D), and probability of S, respectively, which can be seen in this slide (Tsumoto 2006).

Slide 7-36: Probabilistic Rules: Modus Ponens

In the above example, when R and D are set to [nau = yes] and [class = migraine], $\alpha R(D) = 2/3 = 0.67$ and $\kappa R(D) = 2/2 = 1.0$.

It is notable that $\alpha R(D)$ measures the degree of the sufficiency of a proposition, R $\rightarrow$ D, and that $\kappa R(D)$ measures the degree of its necessity. For example, if $\alpha R(D)$ is equal to 1.0, then R $\rightarrow$ D is true. On the other hand, if $\kappa R(D)$ is equal to 1.0, then D $\rightarrow$ R is true. Thus, if both measures are 1.0, then R $\leftrightarrow$ D.

By the use of accuracy and coverage, a probabilistic rule is defined as:
R
$\alpha \rightarrow, \kappa d$ s.t. $R = \wedge j[aj = vk]$, $\alpha R(D) \geq \delta \alpha$
and $\kappa R(D) \geq \delta \kappa$

If the thresholds for accuracy and coverage are set to high values, the meaning of the conditional part of probabilistic rules corresponds the highly overlapped region.

This slide shows the Venn diagram of probabilistic rules with highly overlapped regions. This rule is a kind of probabilistic proposition with two statistical measures, which is an extension of the Ziarko's variable precision model (VPRS) (Tsumoto 2006).

Slide 7-37: Positive Rules

Positive Rules: A positive rule is defined as a rule supported by only positive examples, the classification accuracy of which is equal to 1.0. It is notable that the set supporting this rule corresponds to a subset of the lower approximation of a target concept, which is introduced in rough sets.

(continued)

(continued)

Thus, a positive rule is represented as:
$R \to d$ s.t. $R = \wedge j[a_j = v_k]$, $\alpha_R(D) = 1.0$

This slide shows the Venn diagram of such a positive rule. As shown in this figure, the meaning of R is a subset of that of D. This diagram is exactly equivalent to the classic proposition $R \to d$. In the above example, one positive rule of "m.c.h." (muscle contraction headache) is:
[nausea = no] $\to$ m.c.h. $\alpha = 3/3 = 1.0$.

This positive rule is often called a deterministic rule (Tsumoto 2006).

Slide 7-38: Exclusive Rules

Exclusive Rules: It is also called contrapositive of a negative rule and is defined as a rule supported by all the positive examples, the coverage of which is equal to 1.0. That is, an exclusive rule represents the necessity condition of a decision. It is notable that the set supporting an exclusive rule corresponds to the upper approximation of a target concept, which is introduced in rough sets. Thus, an exclusive rule is represented as:
$R \to d$ s.t. $R = \vee j[a_j = v_k]$, $\kappa_R(D) = 1.0$.

As shown this slide, the meaning of R is a superset of that of D. This diagram is exactly equivalent to the classic proposition $d \to R$. In the above example, the exclusive rule of
"m.c.h." is:
[M1 = yes] $\vee$ [nau = no] $\to$ m.c.h. $\kappa = 1.0$

From the viewpoint of propositional logic, an exclusive rule should be represented as:
$d \to \vee j[a_j = v_k]$

because the condition of an exclusive rule corresponds to the necessity condition of conclusion d. Thus, it is easy to see that a negative rule is defined as the contrapositive of an exclusive rule:
$\wedge j \neg [a_j = v_k] \to \neg d$.

Slide 7-39: Negative Rule

Negative Rule: It is now easy to see that a negative rule is defined as the contrapositive of an exclusive rule:
$\wedge j \neg [a_j = v_k] \to \neg d$,

(continued)

(continued)

which means that if a case does not satisfy any attribute value pairs in the condition of a negative rules, then we can exclude a decision d from candidates. For example, the negative rule of m.c.h. is:
$\neg[M1 = \text{yes}] \land \neg[\text{nausea} = \text{no}] \rightarrow \neg\text{m.c.h.}$
In summary, a negative rule is defined as:
$\land_j \neg[a_j = v_k] \rightarrow \neg d$ s.t. $\forall[a_j = v_k] \; \kappa[a_j = v_k](D) = 1.0$,
where D denotes a set of samples which belong to a class d.

This slide shows the Venn diagram of a negative rule. As shown in this slide, it is notable that this negative region is the "positive region" of "negative concept" (Tsumoto 2006).

Slide 7-40: Example: Algorithms for Rule Induction

The contrapositive of a negative rule, an exclusive rule, is induced as an exclusive rule by the modification of the algorithm introduced in PRIMEROSE-REX (seen in this slide). This algorithm works as follows. (1) First, it selects a descriptor [ai = vj] from the list of attribute-value pairs, denoted by L. (2) Then, it checks whether this descriptor overlaps with a set of positive examples, denoted by D. (3) If so, this descriptor is included into a list of candidates for positive rules and the algorithm checks whether its coverage is equal to 1.0 or not. If the coverage is equal to 1.0, then this descriptor is added to Rer, the formula for the conditional part of the exclusive rule of D. (4) Then, [ai = vj] is deleted from the list L. This procedure, from (1) to (4) will continue unless L is empty. (5) Finally, when L is empty, this algorithm generates negative rules by taking the contrapositive of induced exclusive rules. On the other hand, positive rules are induced as inclusive rules by the algorithm introduced in PRIMEROSE-REX. For induction of positive rules, the threshold of accuracy and coverage is set to 1.0 and 0.0, respectively.

5.4.5 Heuristic Decision Making

Slide 7-41: Science Vol. 185, pp. 1124–1131, Sept. 1974

In 1974 Amos Tversky and Daniel Kahneman wrote an article (meanwhile it received 20k citations) about Judgement under Uncertainty, and they describe the approach of Heuristic Decision Making (Tversky and Kahneman 1974).
There are three major decision-making strategies:

1. Logic
2. Statistics
3. Heuristics

Each strategy is suited to a particular kind of problem. Rules of logic and statistics have been linked to rational reasoning. Heuristics have been linked to error-prone intuitions or irrationality. As Simon (1979) emphasized, the classical model of rationality requires knowledge of all relevant alternatives, their consequences and probabilities, and a predictable world without surprises. These conditions, however, are rarely met for the problems that individuals and organizations face. Savage known as the founder of modern Bayesian decision theory, called such perfect knowledge **small worlds**, to be distinguished from large worlds. Heuristics are strategies that ignore certain information—to make decisions faster.

Slide 7-42: Heuristic Decision Making

1. Heuristics can be more accurate than more complex strategies even though they process less information (less-is-more effects).
2. A heuristic is not good or bad, rational or irrational; its accuracy depends on the structure of the environment (ecological rationality).
3. Heuristics are embodied and situated in the sense that they exploit core capacities of the brain and their success depends on the structure of the environment. They provide an alternative to stable traits, attitudes, preferences, and other internal explanations of behavior.
4. With sufficient experience, people learn to select proper heuristics from their adaptive toolbox.
5. Usually, the same heuristic can be used both consciously and unconsciously, for inferences and preferences, and underlies social as well as nonsocial intelligence.

(continued)

(continued)
6. Decision making in organizations typically involves heuristics because the conditions for rational models rarely hold in an uncertain world (Gigerenzer and Gaissmaier 2011).

In this slide we see a tree which prescribes how emergency physicians can detect acute ischemic heart disease. It only asks up to three yes/no questions, namely, whether the patient's electrocardiogram shows a certain anomaly ("ST segment changes"), whether chest pain is the patient's primary complaint, and whether there is any other factor (Gigerenzer and Gaissmaier 2011); MI = myocardial infarction; N.A. = not applicable; NTG = nitroglycerin; T = T-waves with peaking or inversion.

Slide 7-43: Case Based Reasoning (CBR)

CBR is an analogical reasoning method providing both a methodology for problem solving and a cognitive model. CBR means reasoning from experience or "old cases" in an effort to solve problems, critique solutions, and explain anomalous situations. Some historical interesting medical CBR-Systems include: CASEY that gives a diagnosis for heart disorders, NIMON is a renal function monitoring system, COSYL that gives a consultation for a liver transplanted patient, or ICONS that presents suitable calculated antibiotics therapy advised for intensive care patients (Salem 2007). In this slide we see a typical CBR Methodology. Boxes represent processes and ovals represent KS (Salem 2007).

6 Human Error

Slide 7-44: Medical Errors: The IOM Study

Let us finalize this lecture with a short look on human errors. Preventable medical mistakes and infections are responsible for about 200,000 deaths in the USA each year, according to an investigation by the Hearst media corporation. The report comes 10 years after the Institute of Medicine's "To Err Is Human" analysis, which found that 44,000–98,000 people were dying annually due to these errors and called for the medical community and government to cut that number in half by 2004 (Kohn et al. 1999).

Slide 7-45: Definitions of Medical Errors

Rothschild et al. (2005) provides some definitions of medical errors:

1. Medical error (ME) = any failure of a planned action.
2. Serious ME = causes harm; includes preventable adverse events, intercepted serious errors, and non-intercepted serious errors. Does not include trivial errors with little or no potential for harm or non-preventable adverse events.
3. Intercepted serious error = is caught before reaching patients.
4. Non-intercepted serious error = reaches the patient but of good fortune or sufficient reserves to buffer the error, it did not cause harm.
5. Adverse event = any injury (e.g., a rash caused by an antibiotic, deep vein thrombosis following omission to continue prophylactic subcutaneous heparin orders on transfer to the critical care unit, ventricular tachycardia due to placement of a central venous catheter tip in the right ventricle).
6. Non-preventable adverse event = unavoidable injury due to appropriate medical care.
7. Preventable adverse event = injury due to a non-intercepted serious error in medical care.

Finally in the next slide we see the "Swiss cheese" model—we will discuss more details in Lecture 11.

Slide 7-46: Framework for Understanding Human Error

This is a standard framework for understanding human error, it describes how human error arises and can result in adverse outcomes. There are three major components: (1) Human fallibility, (2) Context, and (3) Barriers (Sharit 2006).

7 Future Outlook

Slide 7-47: Future Outlook

William Stead presented at the IOM Meeting, 8 October 2007 a chart subtitled: "Growth in facts affecting provider decisions versus human cognitive capacity." It shows impressively that the facts per decision increases dramatically, and a modern medical doctor has no chance to consider all facts, this is simply not longer possible. Hence, every attempt to relieve the

(continued)

> (continued)
>
> medical doctors from this avalanche of factual data is acknowledged, so that the doctors can fully concentrate on their relevant core business. Machine intelligence and machine learning approaches are definitely important in that respect and may help in making better, more informed, decisions.

8 Exam Questions

8.1 Yes/No Decision Questions

Please check the following sentences and decide whether the sentence is true = YES; or false = NO; for each correct answer you will be awarded 2 credit points.

01	The facts per decision rate is increasing enormously and there is the problem that a modern doctor is no longer able to consider all facts around the patient.	☐ Yes ☐ No	2 total
02	According to Shortliffe (2011), decision making is still the major topic in biomedical informatics.	☐ Yes ☐ No	2 total
03	Decision making is a cognitive process resulting in the selection of one of several alternatives.	☐ Yes ☐ No	2 total
04	The sensory processing unit is similar to a long term storage and holds the incoming sensory information as long as possible.	☐ Yes ☐ No	2 total
05	The simplest decision support process is triage, which is the process of determining the priority of patients' treatments based on the severity of their conditions.	☐ Yes ☐ No	2 total
06	Differential diagnosis is a systematic diagnostic method used to identify the presence of an entity, where multiple alternatives are possible and may also refer to any of the candidate alternatives.	☐ Yes ☐ No	2 total
07	Case-based reasoning is an analogical reasoning method providing both a methodology for problem solving and a cognitive model.	☐ Yes ☐ No	2 total
08	Heuristics can be more accurate than more complex strategies even though they process less information (less-is-more effects).	☐ Yes ☐ No	2 total
09	In the standard framework for understanding human error, we determined three major components: 1. Machine fallibility, 2. Context and 3. Barriers	☐ Yes ☐ No	2 total
10	A non-intercepted serious error reaches the patient but of good fortune or sufficient reserves to buffer the error, it did not cause harm.	☐ Yes ☐ No	2 total

Sum of Question Block A (max. 20 points)	

8.2 Multiple Choice Questions (MCQ)

The following questions are composed of two parts: the stem, which identifies the question or problem and a set of alternatives which can contain 0, 1, 2, 3, or 4 correct answers, along with a number of distractors that might be plausible—but are incorrect. Please **select the correct answers** by ticking ☒—and do not forget that it can be none. Each question will be awarded 4 points *only if everything is correct*.

01	A non-preventable adverse event is defined as ... ❏ a) ... injury due to a non- intercepted serious error in medical care. ❏ b) ... unavoidable injury due to appropriate medical care. ❏ c) ... caught before reaching any patients. ❏ d) ... injury due to a non- intercepted serious error in medical care.	4 total
02	Rough Set Theory ... ❏ a) ... is suitable for dealing with incomplete data. ❏ b) ... is an extension of the Classical Set Theory. ❏ c) ... is used when representing incomplete knowledge. ❏ d) ... uses sets with fuzzy boundaries.	4 total
03	In tumor detection we have four possible outcomes .... ❏ a) ... a hit – the tumor is present and the doctor says yes. ❏ b) ... a miss – the tumor is absent and the doctor says yes. ❏ c) ... a false alarm – the tumor is present and the doctor says no. ❏ d) ... a correct rejection – the tumor present and the doctor says yes.	4 total
04	Human information processing of printed words follow this path: ❏ a) ... pictures → eyes → pictorial signal → integration in LTM. ❏ b) ... words → eyes → images → pictorial signal → integration in LTM. ❏ c) ... words → ears → images → verbal signal → integration in LTM. ❏ d) ... words → eyes → images → verbal signal → integration in LTM.	4 total
05	The problems in biomedical informatics are rooted mostly in the ... ❏ a) ... high dimensionality and complexity of the data. ❏ b) ... non-standardization of most of the data available. ❏ c) ... lacking memory capacity and computational power. ❏ d) ... heterogeneity and weak structurization of the available data.	4 total
06	The information chunks are stored in average ... ❏ ... a fraction of seconds in the sensory register. ❏ ... half an hour in the working memory. ❏ ... several hours in the long term memory. ❏ ... forever in the long term memory.	4 total
07	Key challenges for decision making research include ... ❏ a) ... limited perceptual human resources. ❏ b) ... in-memory databases to provide information more quickly. ❏ c) ... the presentation of as much of possible information in real time. ❏ d) ... scarce attentional resources and the error prone human.	4 total
08	In signal detection theory the ROC-curve overlapping is "good" if ... ❏ a) ... the two curves do not much overlap. ❏ b) ... the two curves have lots of overlap. ❏ c) ... the two curves have equal overlap. ❏ d) ... the two curves are identical.	4 total

09	Heuristics ... ☐ a) ... are using first order logic. ☐ b) ... are strategies that ignore certain information. ☐ c) ... are neither good or bad, rational or irrational; its accuracy depends on the structure of the environment (ecological rationality). ☐ d) ... instrumentalize the "less-is-more" effect.	4 total
10	The IOM report 1999 reported the average deaths in the US each year with ... ☐ a) ... 200,000. ☐ b) ... 400,000. ☐ c) ... 100,000. ☐ d) ... 300,000.	4 total

Sum of Question Block B (max. 40 points)

8.3 Free Recall Block

Please follow the instructions below. At each question you will be assigned the credit points indicated if your option is correct (partial points may be given).

01	The model of Wickens (1984) is still the most used to explain the cognitive processes of decision making – please complete the missing items:	1 each 4 total
	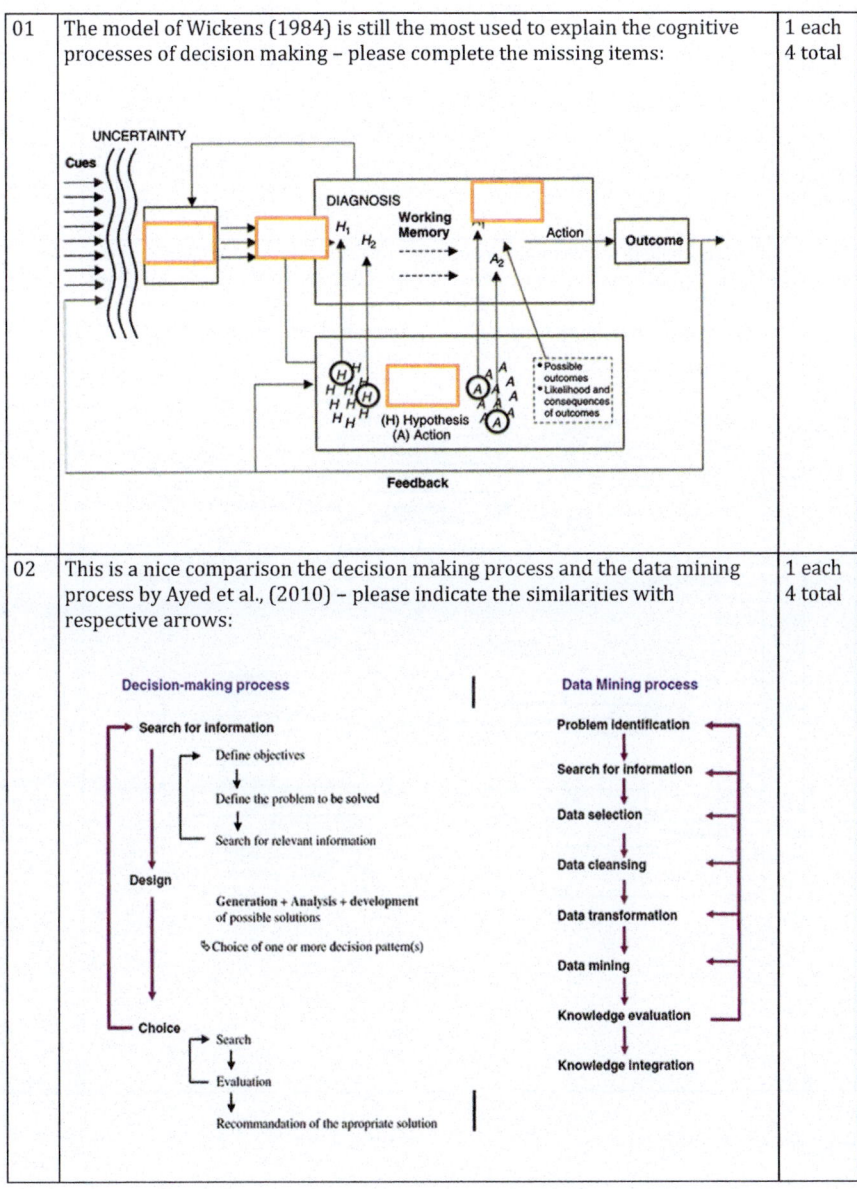	
02	This is a nice comparison the decision making process and the data mining process by Ayed et al., (2010) – please indicate the similarities with respective arrows:	1 each 4 total

03	Signal detection theory provides a precise language and graphic notation for analyzing decision making under uncertainty – please complete the following missing items:	1 each 6 total

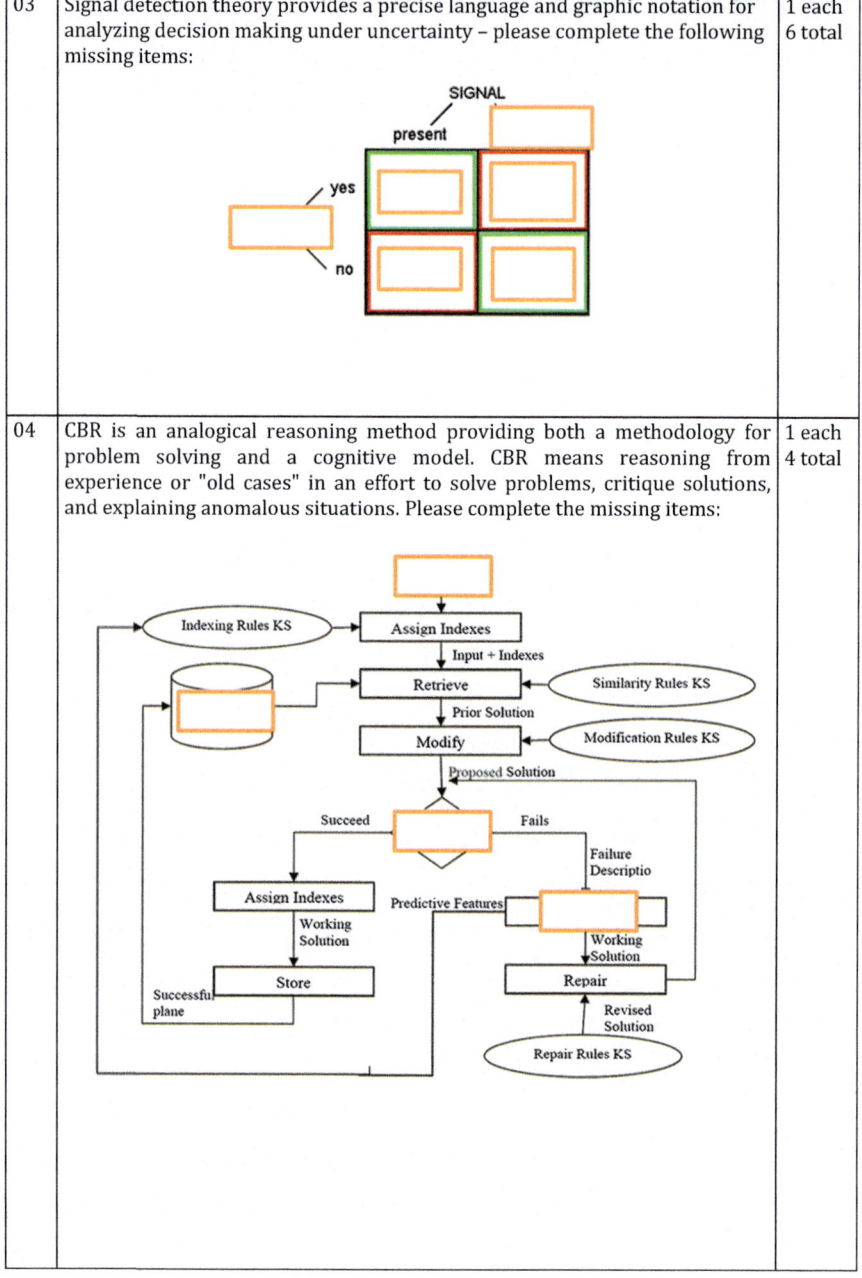

04	CBR is an analogical reasoning method providing both a methodology for problem solving and a cognitive model. CBR means reasoning from experience or "old cases" in an effort to solve problems, critique solutions, and explaining anomalous situations. Please complete the missing items:	1 each 4 total

<table>
<tr><td>05</td><td>Please assign the correct labels to the probabilistic rules below:</td><td>1 each
3 total</td></tr>
<tr><td></td><td>
<table>
<tr><td>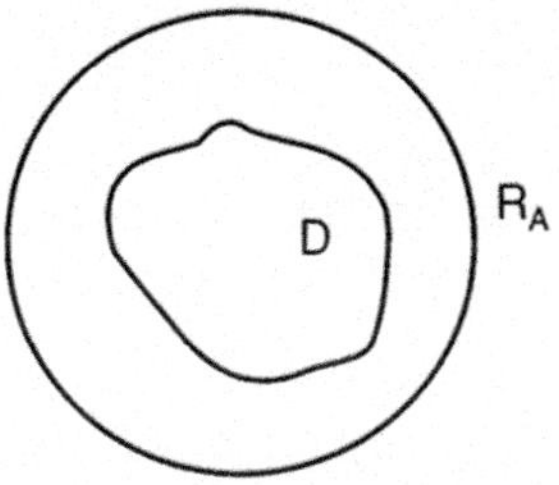
</td><td>Positive rule</td></tr>
<tr><td>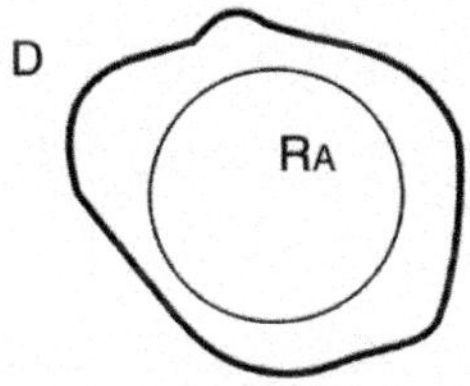
</td><td>Exclusive rule</td></tr>
<tr><td>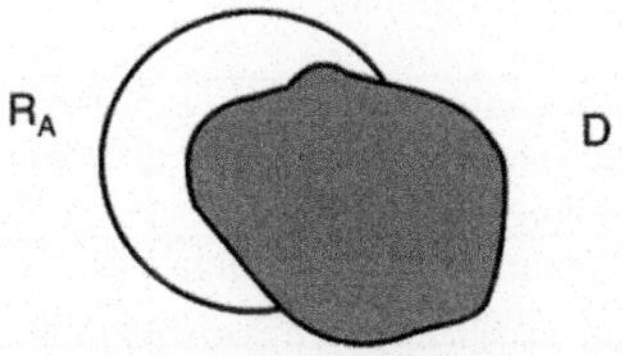
</td><td>Modus ponens</td></tr>
</table>
</td><td></td></tr>
</table>

Sum of Question Block C (max. 40 points)	

9 Answers

9.1 Answers to the Yes/No Questions

Please check the following sentences and decide whether the sentence is true = YES; or false = NO; for each correct answer you will be awarded 2 credit points.

01	The facts per decision rate is increasing enormously and there is the problem that a modern doctor is no longer able to consider all facts around the patient.	☒ Yes ☐ No	2 total
02	According to Shortliffe (2011), decision making is still the major topic in biomedical informatics.	☒ Yes ☐ No	2 total
03	Decision making is a cognitive process resulting in the selection of one of several alternatives.	☒ Yes ☐ No	2 total
04	The sensory processing unit is similar to a long term storage and holds the incoming sensory information as long as possible.	☐ Yes ☒ No	2 total
05	The simplest decision support process is triage, which is the process of determining the priority of patients' treatments based on the severity of their conditions.	☒ Yes ☐ No	2 total
06	Differential diagnosis is a systematic diagnostic method used to identify the presence of an entity, where multiple alternatives are possible and may also refer to any of the candidate alternatives.	☒ Yes ☐ No	2 total
07	Case-based reasoning is an analogical reasoning method providing both a methodology for problem solving and a cognitive model.	☒ Yes ☐ No	2 total
08	Heuristics can be more accurate than more complex strategies even though they process less information (less-is-more effects).	☒ Yes ☐ No	2 total
09	In the standard framework for understanding human error, we determined three major components: 1. Machine fallibility, 2. Context and 3. Barriers	☐ Yes ☒ No	2 total
10	A non-intercepted serious error reaches the patient but of good fortune or sufficient reserves to buffer the error, it did not cause harm.	☒ Yes ☐ No	2 total

Sum of Question Block A (max. 20 points)	

9.2 Answers to the Multiple Choice Questions (MCQ)

01	A non-preventable adverse event is defined as ... ☐ a) ... injury due to a non- intercepted serious error in medical care. ☒ b) ... unavoidable injury due to appropriate medical care. ☐ c) ... caught before reaching any patients. ☐ d) ... injury due to a non- intercepted serious error in medical care.	4 total
02	Rough Set Theory ... ☒ a) ... is suitable for dealing with incomplete data. ☒ b) ... is an extension of the Classical Set Theory. ☒ c) ... is used when representing incomplete knowledge. ☒ d) ... uses sets with fuzzy boundaries.	4 total
03	In tumor detection we have four possible outcomes .... ☒ a) ... a hit – the tumor is present and the doctor says yes. ☐ b) ... a miss – the tumor is absent and the doctor says yes. ☐ c) ... a false alarm – the tumor is present and the doctor says no. ☐ d) ... a correct rejection – the tumor present and the doctor says yes.	4 total
04	Human information processing of printed words follow this path: ☐ a) ... pictures → eyes → pictorial signal → integration in LTM. ☐ b) ... words → eyes → images → pictorial signal → integration in LTM. ☐ c) ... words → ears → images → verbal signal → integration in LTM. ☒ d) ... words → eyes → images → verbal signal → integration in LTM.	4 total
05	The problems in biomedical informatics are rooted mostly in the ... ☒ a) ... high dimensionality and complexity of the data. ☒ b) ... non-standardization of most of the data available. ☐ c) ... lacking memory capacity and computational power. ☒ d) ... heterogeneity and weak structurization of the available data.	4 total
06	The information chunks are stored in average ... ☒ ... a fraction of seconds in the sensory register. ☐ ... half an hour in the working memory. ☐ ... several hours in the long term memory. ☒ ... forever in the long term memory.	4 total
07	Key challenges for decision making research include ... ☒ a) ... limited perceptual human resources. ☐ b) ... in-memory databases to provide information more quickly. ☐ c) ... the presentation of as much of possible information in real time. ☒ d) ... scarce attentional resources and the error prone human.	4 total
08	In signal detection theory the ROC-curve overlapping is "good" if ... ☒ a) ... the two curves do not much overlap. ☐ b) ... the two curves have lots of overlap. ☐ c) ... the two curves have equal overlap. ☐ d) ... the two curves are identical.	4 total

09	Heuristics ... ☐ a) ... are using first order logic. ☒ b) ... are strategies that ignore certain information. ☒ c) ... are neither good or bad, rational or irrational; its accuracy depends on the structure of the environment (ecological rationality). ☒ d) ... instrumentalize the "less-is-more" effect.	4 total
10	The IOM report 1999 reported the average deaths in the US each year with ... ☒ a) ... 200,000. ☐ b) ... 400,000. ☐ c) ... 100,000. ☒ d) ... 300,000.	4 total

Sum of Question Block B (max. 40 points)	

9.3 Answers to the Free Recall Questions

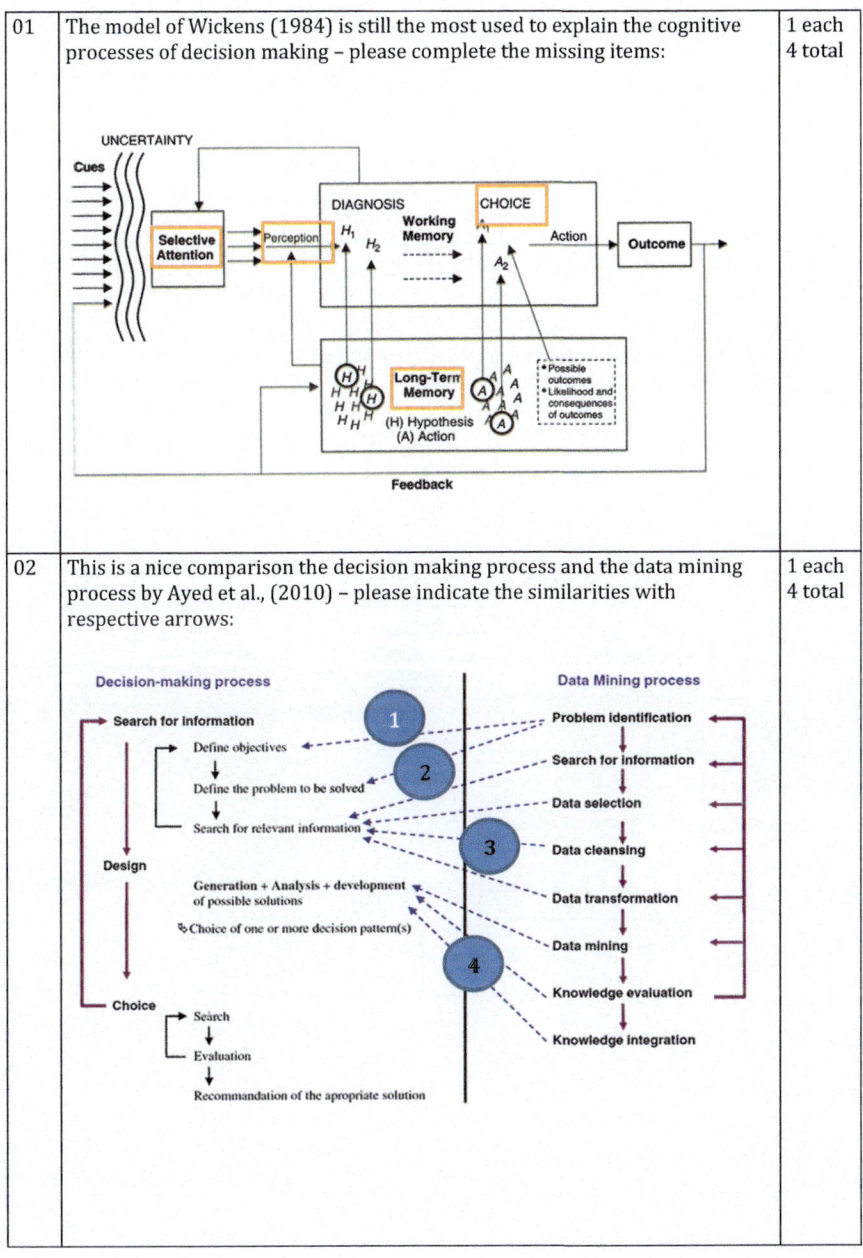

03	Signal detection theory provides a precise language and graphic notation for analyzing decision making under uncertainty – please complete the following missing items:	1 each 6 total

	SIGNAL	
	present	absent
RESPONSE yes	hit	false alarm
RESPONSE no	miss	correct rejection

04	CBR is an analogical reasoning method providing both a methodology for problem solving and a cognitive model. CBR means reasoning from experience or "old cases" in an effort to solve problems, critique solutions, and explaining anomalous situations. Please complete the missing items:	1 each 4 total

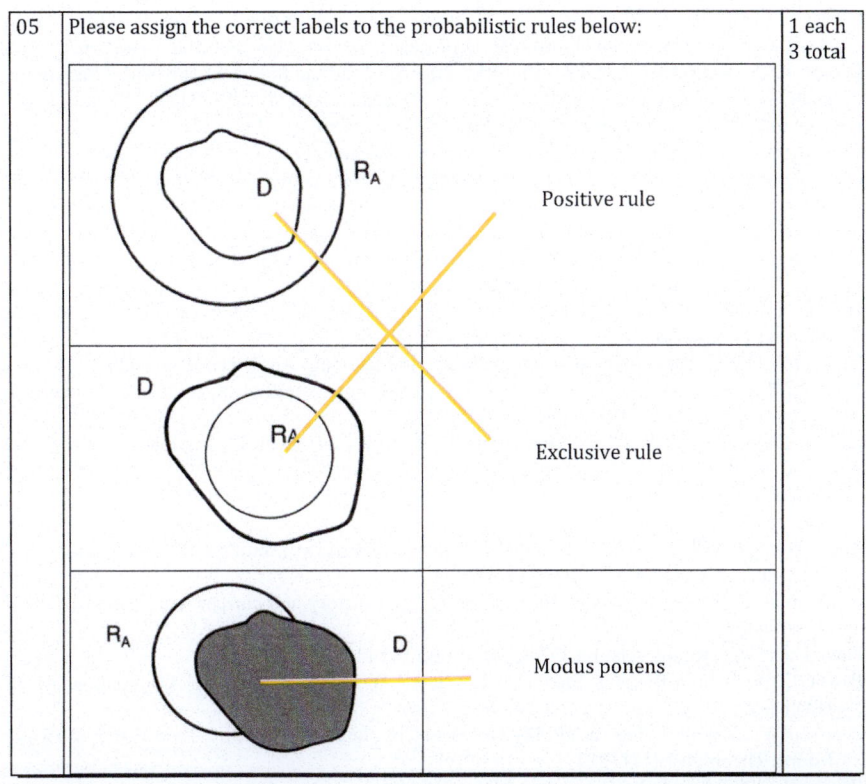

References

Ables AZ, Nagubilli R (2010) Prevention, recognition, and management of serotonin syndrome. Am Fam Phys 81(9):1139

Atkinson RC, Shiffrin RM (1971) The control processes of short-term memory (Technical report 173, April 19, 1971). Institute for Mathematical Studies in the Social Sciences, Stanford University, Stanford

Ayed BM, Ltifi H, Kolski C, Alimi AM (2010) A user-centered approach for the design and implementation of KDD-based DSS: a case study in the healthcare domain. Decis Support Syst 50:64–78

Baddeley A (1981) The concept of working memory: a view of its current state and probable future development. Cognition 10(1–3):17–23

Bayes M, Price M (1763) An essay towards solving a problem in the doctrine of chances. Phil Trans 53:370–418

Boyer EW, Shannon M (2005) The serotonin syndrome. N Engl J Med 352(11):1112–1120

Bretthorst GL (1990) An introduction to parameter estimation using Bayesian probability theory. In: Fougère P (ed) Maximum entropy and Bayesian methods. Springer, Netherlands, pp 53–79

Cowan N (1988) Evolving conceptions of memory storage, selective attention, and their mutual constraints within the human information-processing system. Psychol Bull 104(2):163

D'esposito M, Detre JA, Alsop DC, Shin RK, Atlas S, Grossman M (1995) The neural basis of the central executive system of working memory. Nature 378(6554):279–281

Deming WE (1994) The new economics. MIT Press, Cambridge, MA

Gigerenzer G (2008) Gut feelings: short cuts to better decision making. Penguin, London

Gigerenzer G, Gaissmaier W (2011) Heuristic decision making. In: Fiske ST, Schacter DL, Taylor SE (eds) Annual review of psychology, vol 62. Annual Reviews, Palo Alto, pp 451–482

Hersh W (2010) Information retrieval: a health and biomedical perspective. Springer, New York

Holzinger A (2002) Multimedia basics, volume 2: Learning. Cognitive basics of multimedia information systems. Laxmi-Publications, New Delhi

Holzinger A (2010) Process guide for students for interdisciplinary work in computer science/informatics, 2nd edn. BoD, Norderstedt

Kintsch W (1994) Text comprehension, memory, and learning. Am Psychol 49(4):294

Kohn LT, Corrigan JM, Donaldson MS (1999) To err is human: building a safer health system. The Institute of Medicine, Washington, DC

Ledley RS, Lusted LB (1959) Reasoning foundations of medical diagnosis. Science 130 (3366):9–21

Mcneil BJ, Keeler E, Adelstein SJ (1975) Primer on certain elements of medical decision making. N Engl J Med 293(5):211–215

Miller GA (1956) The magical number seven, plus or minus two: some limits of our capacity for processing information. Psychol Rev 63:81–97

Norman G (2005) Research in clinical reasoning: past history and current trends. Med Educ 39 (4):418–427

Pawlak Z (1982) Rough sets. Int J Comput Inform Sci 11(5):341–356

Pawlak Z, Grzymala-Busse J, Slowinski R, Ziarko W (1995) Rough sets. Commun ACM 38 (11):88–95

Peterson W, Birdsall T, Fox W (1954) The theory of signal detectability. Proc IRE Prof Group Inform Theory 4(4):171–212

Popper K (2005) Logik der Forschung. Mohr, Tübingen

Quinette P, Guillery B, Desgranges B, De La Sayette V, Viader F, Eustache F (2003) Working memory and executive functions in transient global amnesia. Brain 126(9):1917–1934

Rothschild JM, Landrigan CP, Cronin JW, Kaushal R, Lockley SW, Burdick E, Stone PH, Lilly CM, Katz JT, Czeisler CA (2005) The Critical Care Safety Study: the incidence and nature of adverse events and serious medical errors in intensive care. Crit Care Med 33(8):1694

Salem ABM (2007) Case based reasoning technology for medical diagnosis. Proc World Acad Sci Eng Technol 25:9–13

Schenkman M, Deutsch JE, Gill-Body KM (2006) An integrated framework for decision making in neurologic physical therapist practice. Phys Ther 86(12):1681–1702

Schraw G, Lehman S (2001) Situational interest: a review of the literature and directions for future research. Educ Psychol Rev 13(1):23–52

Sharit J (2006) Human Error. In: Salvendy G (ed) Handbook of human factors and ergonomics, 3rd edn. Wiley, Hoboken, NJ, pp 708–760

Shewhart WA (1958) Nature and origin of standards of quality. Bell Syst Tech J 37(1):1–22

Shortliffe EH (2011) Biomedical informatics: defining the science and its role in health professional education. In: Holzinger A, Simonic K-M (eds) Information quality in e-Health. Lecture notes in computer science LNCS 7058. Springer, Heidelberg, pp 711–714

Simon HA (1974) How big is a chunk? Science 183:482–488

Simon HA (1979) Rational decision making in business organizations. Am Econ Rev 69:493–513

Skowron A, Grzymala-Busse J (1994) From rough set theory to evidence theory. In: Yager R, Fedrizzi M, Kasprzyk J (eds) Advances in the Dempster-Shafer theory of evidence. Wiley, New York

Sweller J (1988) Cognitive load during problem solving: effects on learning. Cognit Sci 12 (2):257–285
Tanner WP Jr, Swets JA (1954) A decision-making theory of visual detection. Psychol Rev 61 (6):401
Tsumoto S (2006) Pawlak rough set model, medical reasoning and rule mining. In: Greco S, Hata Y, Hirano S, Inuiguchi M, Miyamoto S, Nguyen H, Slowinski R (eds) Rough sets and current trends in computing. Springer, Berlin, pp 53–70
Tversky A, Kahneman D (1974) Judgment under uncertainty: heuristics and biases. Science 185 (4157):1124–1131
Van Der Sluis F, Van Den Broek E (2010) Modeling user knowledge from queries: introducing a metric for knowledge. In: An A, Lingras P, Petty S, Huang R (eds) Active media technology. Springer, Berlin, pp 395–402
Van Gerven MAJ, Taal BG, Lucas PJF (2008) Dynamic Bayesian networks as prognostic models for clinical patient management. J Biomed Inf 41(4):515–529
Wickens C, Lee J, Liu Y, Gordon-Becker S (2004) Introduction to human factors engineering, 2nd edn. Prentice-Hall, Upper Saddle River, NJ
Wickens CD (1984) Engineering psychology and human performance. Charles Merrill, Columbus, OH
Xie S, Helfert M, Lugmayr A, Heimgärtner R, Holzinger A (2013) Influence of organizational culture and communication on the successful implementation of information technology in hospitals. In: Rau PLP (ed) Cross-cultural design. Cultural differences in everyday life, lecture notes in computer science, LNCS 8024. Springer, Berlin, pp 165–174
Yu Y, Crumiller M, Knight B, Kaplan E (2010) Estimating the amount of information carried by a neuronal population. Front Comput Neurosci 4:1–10

Lecture 8

Biomedical Decision Making: Reasoning and Decision Support

1 Learning Goals

At the end of the eighth lecture, you:

- would be able to apply your knowledge gained in Lecture 7 to some example systems of decision support.
- would have an overview about the core principles and architecture of decision support systems.
- would be familiar with the certainty factors as for example used in MYCIN.
- would be aware of some design principles of DSS.
- would have seen the similarities between DSS and KDD on the example of computational methods in cancer detection.
- would have seen CBR systems.

2 Advance Organizer

Case-based reasoning (CBR)	Process of solving new problems based on the solutions of similar past problems
Certainty factor model (CF)	A method for managing uncertainty in rule-based systems
CLARION	Connectionist Learning with Adaptive Rule Induction ON-line (CLARION) is a cognitive architecture that incorporates the distinction between implicit and explicit processes and focuses on capturing the interaction between these two types of processes. By focusing on this distinction, CLARION has been used to simulate several tasks in cognitive psychology and social psychology. CLARION has

	also been used to implement intelligent systems in artificial intelligence applications
Clinical decision support (CDS)	Process for enhancing health-related decisions and actions with pertinent, organized clinical knowledge and patient information to improve health delivery
Clinical decision support system (CDSS)	Expert system that provides support to certain reasoning tasks, in the context of a clinical decision
Collective intelligence	Shared group (symbolic) intelligence, emerging from cooperation/competition of many individuals, e.g., for consensus decision making
Crowd sourcing	A combination of "crowd" and "outsourcing" coined by Jeff Howe (2006), and describes a distributed problem-solving model; example for crowd sourcing is a public software beta-test
Decision making	Central cognitive process in every medical activity, resulting in the selection of a final choice of action out of several alternatives
Decision support system (DSS)	Is an IS including knowledge-based systems to interactively support decision-making activities, i.e., making data useful
DXplain	A DSS from the Harvard Medical School, to assist making a diagnosis (clinical consultation), and also as an instructional instrument (education); provides a description of diseases, etiology, pathology, prognosis, and up to ten references for each disease
Expert system	Emulates the decision-making processes of a human expert to solve complex problems
GAMUTS in radiology	Computer-supported list of common/uncommon differential diagnoses
ILIAD	Medical expert system, developed by the University of Utah, used as a teaching and testing tool for medical students in problem solving. Fields include pediatrics, internal medicine, oncology, infectious diseases, gynecology, pulmonology, etc.
MYCIN	One of the early medical expert systems (Shortliffe 1970, Stanford) to identify bacteria causing severe infections, such as bacteremia and meningitis, and to recommend antibiotics, with the dosage adjusted for patient's body weight
Reasoning	Cognitive (thought) processes involved in making medical decisions (clinical reasoning, medical problem solving, diagnostic reasoning)

3 Acronyms

DSS   Decision support system
CBR   Case-based reasoning
CDS   Clinical decision support
CF   Certainty factor
KDD   Knowledge discovery from data
LISP LISt processing (programming language)
KE   Knowledge engineering
mRNA Messenger ribonucleic acid

4 Key Problems

> **Slide 8-1: Key Challenges**
>
> Health care Information Networks (HINs) help professionals and patients access the right information at the right time and invite a new design and integration of decision support systems within these collaborative workflow processes. The need to share information and knowledge is increasing (e.g., shared records, professional guidelines, prescriptions, care protocols, public health information, health care networks, etc.). The well-established "Evidence-Based Medicine" (EBM) and "Patient-centered medicine" paradigms representing different visions of medicine are suggesting behaviors, so different that they are also raising dilemmas. Attempts made to standardize care are potentially ignoring the heterogeneity of the patients (Fieschi et al. 2003).
>
> *Challenges in the development of DSS*
>
> The development of medical expert systems is very difficult—as medicine is an extremely complex application domain—dealing most of the time with weakly structured data and probable information (Holzinger 2012).
>
> Some challenges (Majumder and Bhattacharya 2000) are as follows:
>
> (a) Defining general system architectures in terms of generic tasks such as diagnosis, therapy planning and monitoring to be executed for (b) medical reasoning in (a); (c) patient management with (d) minimum uncertainty. Other challenges include (e) knowledge acquisition and encoding; (f) human–computer interface and interaction (HCI); and (g) system integration into existing clinical environments, e.g., the enterprise hospital information system; to mention only a few.
>
> (continued)

(continued)

In the previous lecture we have got an overview about some fundamentals of decision making from the human factors perspective; now we will have a closer look on technological solutions. We follow the definition of Shortliffe (2011) and define a medical DSS as *any computer program designed to support health professionals in their daily decision making processes* (Shortliffe 2011).

In a funny cartoon a physician offers a second opinion from his computer. The patient looks horrified: How absurd to think that a computer could have better judgment than a human doctor! But computer tools can already provide valuable information to help human doctors make better decisions. And there is good reason to wish such tools were broadly available.

5 Decision Support Systems

Slide 8-2: Two Types of Decision: Diagnosis Versus Therapy

In the previous lecture we have got an overview about some fundamentals of decision making from the human factors perspective; now we will have a closer look on technological solutions. We follow the definition of Shortliffe (2011) and define a medical DSS as *any computer program designed to support health professionals in their daily decision making processes*. Dealing with data in the health care process is often accompanied by making decisions. According to Bemmel and Musen (1997) we may determine two types of decision:

Type 1: **Decisions related to the diagnosis**, i.e., computers are used to assist in diagnosing a disease on the basis of the individual patient data. They include the following questions:

(a) What is the probability that this patient has a myocardial infarction on the basis of given data (patient history, ECG)?
(b) What is the probability that this patient has acute appendicitis, given the signs and symptoms concerning abdominal pain?

Type 2: **Decisions related to therapy**, i.e., computers are used to select the best therapy on the basis of clinical evidence, e.g.,

(c) What is the best therapy for patients of age x and risks y, if an obstruction of more than z % is seen in the left coronary artery?
(d) What amount of insulin should be prescribed for a patient during the next 5 days, given the blood sugar levels and the amount of insulin taken during the recent weeks?

(continued)

(continued)

For both types we need medical knowledge. On the basis of the available knowledge we can develop decision models on the basis of the available patient data.

5.1 Decision Models

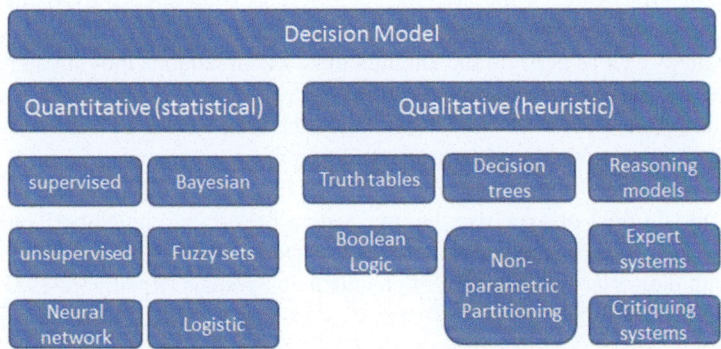

Fig. 1 See Slide 8-3

Slide 8-3: Taxonomy of Decision Support Models

In this slide we see that decision models can be grouped into two main categories:

1. Quantitative: based on **formal statistical methods** to test the probability of the occurrence of an event, e.g., to test that the probability for "healthy" is higher than that for a certain disease as we have seen in differential diagnostics.
2. Qualitative: relying on **symbolic methods**, rather than following a strictly formal mathematical basis. Such models are inspired by insights on human reasoning, thus often called heuristics, and perform deductions on symbolic models using logical operations to conclude a diagnosis based on a case model. According to Van Bemmel we should avoid the word heuristics and use the term symbolic, because such methods may be composed of elementary two-class, single-feature decision units from the first category, e.g., $E = \text{``}x > L\text{''}$.

5.2   Evolution of DSS

Slide 8-4:   History of DSS is a History of Artificial Intelligence

In the early 1950s decision trees and truth tables were used, followed by systems based on statistical methods, finally followed by expert systems. The history of DSS is very closely related to artificial intelligence (AI), the roots can be traced back to attempts to automate chess play. A famous sample was a fake: the Mechanical Turk (see slide, below, left). Built in 1770 by Wolfgang von Kempelen (1734–1804), the device appeared to be able to play against a human, as well as perform the knight's tour, which requires moving a knight to visit every square of a chessboard only once. The "real" start of AI research was in 1955, when John McCarthy coined the term AI and defined it as the science and engineering of making intelligent machines. Edward Feigenbaum was one of first to construct an artificial expert and while looking for an appropriate field of expertise, he met Joshua Lederberg, the Nobel laureate biochemist, who suggested that organic chemists need assistance in determining the molecular structure of chemical compounds (Rheingold 1985).

Slide 8-5:   Evolution of Decision Support Systems

In 1965 Feigenbaum, Lederberg and Buchanan began work on DENDRAL (see top root in the slide), a procedure for non-redundantly enumerating topologically distinct arrangements of any given set of atoms, consistent with the rules of chemical valence (Lindsay et al. 1993). Conventional systems had failed to support organic chemists in forecasting molecular structures. Human chemists know that the possible structure of any chemical compound depends on a number of rules about how different atoms can be bound to one another; and many facts about different atoms in known compounds. By discovering a previously unknown compound, they can gather evidence about the compound by analyzing it with a mass spectroscope, which provides a lot of data, but no clues to what it all means. Look at the slide (Shortliffe and Buchanan 1984): DENDRAL was followed by MYCIN; and actually MYCIN was the inspiration for many other systems.

Slide 8-6:   Early Knowledge-Based System Architecture

DENDRAL was well known to computational chemists who have incorporated many parts of it in their own software. Although it does no longer exist today, it had a major impact on a newly developed field:

(continued)

(continued)

Knowledge engineering (KE), which is both science and engineering of Knowledge-based systems (KBS) and applies methods from artificial intelligence, data mining, expert systems, decision support systems, and mathematical logic, as well as cognitive science. A great amount of work is spent in observing human experts and the design of models of their expertise.

One of the first spinoffs from DENDRAL was Meta-DENDRAL, an expert system for people whose expertise lies in building expert systems. By separating the inference engine from the body of factual knowledge, Buchanan was able to produce a tool for expert-system builders. In this slide we see the first architecture of the basic principle of any expert systems, consisting of a knowledge base, an inference engine, and a dedicated user interface to support the HCI process (Shortliffe and Davis 1975).

Slide 8-7: Static Knowledge Versus Dynamic Knowledge

MYCIN was programmed in Lisp and used judgmental rules with associated elements of **uncertainty**. It was designed to identify bacteria causing severe infections (bacteremia, meningitis), and to recommend antibiotics, with the dosage adjusted for the patient's body weight. Edward Shortliffe, both a physician and computer scientist was confronted with problems associated with diagnosing a certain class of brain infections that was an appropriate area for expert system research and an area of particularly importance, because the first 24 h are most critical for the patients. In the slide we see the idea of the separation of static knowledge (the rules and facts) and dynamic knowledge (the entries made by the human user and deductions made by the system). This is the principle of rule-based systems (Shortliffe and Buchanan 1984).

Slide 8-8: Dealing with Uncertainty in the Real World

We are already well aware about the notion of probable information. The problem is that classical logic permits only exact reasoning: IF A is true THEN A is non-false and IF B is false THEN B is non-true—however, most of our real-world problems do not provide this exact information, mostly is inexact, incomplete, uncertain, noisy, and/or unmeasurable. This is a big problem in the biomedical area.

Slide 8-9: MYCIN: Rule-Based System: Certainty Factors

Shortliffe was aware of the problems involved with classic logic and introduced the **certainty factor (CF)** which is a number between -1 and $+1$ that reflects the **degree of belief in a hypothesis**.

Positive CF's indicate evidence that the hypothesis is valid. If $CF = 1$, the hypothesis is known to be correct (and contrary for $CF = -1$). If $CF = 0$, there is either no evidence regarding the hypothesis or the supporting evidence is equally balanced, suggesting that the hypothesis is not true. MYCIN's hypotheses are statements regarding values of clinical parameters for the various nodes in the context tree.

Let us look on an original example in the next slide.

Slide 8-10: Original Example from MYCIN

This MYCIN example makes the Certainty Factor CF clear (Shortliffe and Buchanan 1984).

Slide 8-11: MYCIN Was Not a Success in the Clinical Practice

MYCIN was not a success in the clinical practice; however, it was a pioneering work for practically each following system; for example ONCOCIN evolved from this work and assisted physicians in managing complex drug regimens for treating cancer patients. It has been built on the results of the MYCIN experiments while gaining experience with regular clinical use of an advice system for use by physicians. The work has also been influenced by data regarding features that may be mandatory if decision support tools are to be accepted by clinicians. Clinical oncology was selected due to the fact that this medical domain meets many of the criteria that have been identified for building an effective consultation tool using AI techniques (Shortliffe 1986). Up to date, the main architecture of a DSS is the same as that developed in the 1970s.

5.3 Design Principles of DSS

Slide 8-12: Basic Design Principles of a DSS

As we have already heard at the very beginning of this lecture, the development of medical expert systems is very difficult—as medicine is a complex application domain—dealing most of the time with weakly structured data (Holzinger 2012). Problems include (Majumder and Bhattacharya 2000): (a) defining general system architectures in terms of generic tasks such as diagnosis, therapy planning and monitoring to be executed for (b) medical reasoning in (a); (c) patient management with (d) minimum uncertainty. Other challenges include (e) knowledge acquisition and encoding, (f) human–computer interface and HCI; and (g) system integration into existing clinical environments, e.g., the enterprise hospital information system.

Slide 8-13: Cybernetic Approach to Medical Diagnostics

This slide shows the typical workflow of a medical reasoning system: Abduction, deduction, and induction represent the basic elements of the inference model of medical reasoning. Clinical patient data is used to generate plausible hypotheses, and these are used as start conditions to forecast expected consequences for matching with the state of the patient in order to confirm or reject these hypotheses (Majumder and Bhattacharya 2000).

Slide 8-14: State-of-the-Art Architecture of DSS

Present-day DSS consist of three main components:

1. Knowledge base, the heart of the system, contains the expert facts, heuristics, judgments, predictions, algorithms, etc., and the relationships—derived from human experts.
2. Inference engine, examines the status of the knowledge base, and determines the order the inferences are made; it also includes the capability of reasoning in the presence of uncertainty (compare with MYCIN).
3. User interface enables effective HCI—additionally there are external interfaces providing access to other databases and data sources (Metaxiotis and Psarras 2003).

Slide 8-15: On the Design and Development of DSS

DSS deal with problems based on *available* knowledge. Some of this knowledge can be extracted using a decision support tool (data mining) which is in fact part of a KDD process (Lecture 6). Data mining tools are usually difficult to exploit because most of the end users are neither experts in computing nor in statistics. It is difficult to develop a KDD system that fits exactly to the end users' needs. Those difficulties can only be tackled by including end users into DSS development. It is necessary to combine methods from Software Engineering (SE) and HCI. Abed et al. (1991) proposed an approach to combine (1) the Unified Process (UP) from SE and (2) the U model from HCI.

The U model (see next slide) considers those steps which do not exist in traditional SE models.

Slide 8-16: Example: Development Following the U Model 1/2

For the effective use in it is necessary to combine methods from Software Engineering (SE) and HCI.

In this U-model we determine two phases:

1. A descending phase for specification and human–computer systems design and development.
2. An ascending phase for the evaluation of the system.

The **validation** consists of comparing the model of the theoretical tasks specified in the descending phase with the model of the real tasks highlighted in the ascending phase, according to the original principles suggested by Abed et al. (1991). The result of the comparison either validates the system or highlights its deficiencies.

Slide 8-17: Improved U Model 2/2

The final model resulting from the assessment allows a generalization of the end users specific behavior *under particular work conditions and context*—what traditional often ignore. Ayed et al. (2010) proposed a modified version of the U-model, specifically adapted to DSS and knowledge discovery (KDD):

1. The analysis of the domain, including the definition of the system objectives, which allows the first functional and structural description of the system to be developed.

(continued)

(continued)

2. The development of the first interface prototypes (models) for the DSS in question, which, by giving future users an idea of the possible solutions, allows them to be implicated as early as possible in the project life cycle.

Slide 8-18: Remember the Similarities Between DSS and KDD

If you look at this slide and compare the DSS process with the KDD (data mining) process, then you will recognize the similarity between decision-making processes and data mining processes.

Slide 8-19: The Design Phases

In this slide we can see the various phases (A to E) of the U-Model approach, which is based on the principle of iterative and incremental development, which allows each task accomplished to be evaluated as soon as the first iterations of the development process have been completed:
 A = Requirements analysis (needs capture)
 B = Analysis and specification
 C = Design and prototyping
 D = Implementation
 E = Test and evaluation
Be aware of the user-centered design process, which will be discussed in Lecture 12!

5.4 Clinical Guidelines

Slide 8-20: Clinical Guidelines as Decision Support and Quality Measure

Guidelines have to be formalized (transformed from natural language to a logical algorithm) and implemented (using the algorithm to program decision support software which is used in practice). Work on formalization has focused on narrative guidelines, which describe a process of care with branching decisions unfolding over time (Medlock et al. 2011). Systematic guidelines have potential to improve the quality of patient care.

(continued)

(continued)

Quality. The demand for increased quality assurance has led to increased interest in performance indicators and other quality metrics. In order for the quality of care to improve as a result of these measures, they must be linked to a process of care. For example, a rule such as "80 % of diabetic patients should have an HbA1c below 7.0" could be linked to processes such as: "All diabetic patients should have an annual HbA1c test" and "Patients with values over 7.0 should be rechecked within 2 months." These measure quality and performance at the population level, but in order to improve the quality of care, action is required at the patient level.

Condition-action rules specify one or a few conditions which are linked to a specific action, in contrast to narrative guidelines which describe a series of branching or iterative decisions unfolding over time. Narrative guidelines and clinical rules are two ends of a continuum of clinical care standards.

Clinical rules represent elementary, isolated care recommendations, while narrative guidelines describe a coherent, unified care process.

Slide 8-21: Clinical Guidelines

Most work in developing computer-interpretable guidelines has focused on the difficult problem of formalizing the *time-oriented structure of guidelines*.

Medlock et al. (2011) propose the **Logical Elements Rule Method (LERM)**, although presented linearly in the text, in practice some steps may be done in parallel, as shown in this slide. Some steps, such as extracting data elements or checking for conflicts between rules, may need to be repeated with the results of later steps as input.

Slide 8-22: Example Exon Arrays

Progress in genomics has increased the data available for conducting **expression analysis**, used in *transcriptomics*. This can be very helpful for decision support. It deals with the study of *mRNA* and the extraction of information contained in the genes. This is reflected in the exon arrays requiring techniques to extract information. This slide shows the correlation of two probe intensities—among 11 tissues (breast, cerebellum, heart, kidney, liver, muscle, pancreas, prostate, spleen, testes, and thyroid): The black boxes represent exons; grey boxes represent introns; (b) Probe design of Exon arrays.

(continued)

(continued)

Four probes target each putative exon; below: The top color bar indicates the probe annotation type, core probes (red), extended probes (blue), full probes (yellow). The signal intensities of core probes tend to have high correlation (top right corner of the heatmap) (Kapur et al. 2007).

Corchado et al. (2009) provided a tool based on a **mixture of experts model** which allows the analysis of the information contained in the exon arrays, from which automatic classifications for decision support in diagnoses of leukemia patients can be made. The proposed model integrates several cooperative algorithms characterized for their efficiency for data processing, filtering, classification and knowledge extraction. This is a mixture of expert tools that integrates different cognitive and statistical approaches to deal with the analysis of exon arrays.

Slide 8-23: Computational Leukemia Cancer Detection 1/6

Exon arrays as seen in Slide 8-22 are chips which allow for a large number of data to be analyzed and classified for each patient (six million features per array). The high dimensionality of data makes it impossible to use standard techniques for expression array analysis (which contain approximately 50,000 probes).

High dimensionality of data from each exon array implies problems in handling and processing, thus making it necessary to improve each of the steps of expression array analysis in order to obtain an efficient method of classification. An expression analysis basically consists of three steps:

1. Normalization and filtering
2. Clustering and classification
3. Extraction of knowledge

These steps can be automated and included within an expert system. Since the problem at hand deals with high dimensional arrays, it is important to have a very good preprocessing technique that can facilitate automatic decision making with regard to selecting the most vitally important variables for the classification process. In light of these decisions, it will be possible to reduce the set of original data. After the organization of groups, patients can be classified and assigned into the group with which they share the most *similarities*. Finally, an extraction of knowledge system facilitates the interpretation of the results obtained after the preprocessing and classification steps, thus making it possible to learn from the information acquired from the

(continued)

(continued)

results. The process of extracting knowledge shapes the knowledge obtained into a set of rules that can be used for improving new classifications.

In this slide we see such an exon array structure: (1) Exon–intron structure of a gene. Grey boxes represent introns, rest represent exons. Introns are not drawn to scale. (2) Probe design of exon arrays. Four probes target each putative exon. (3) Probe design of $3'$ expression arrays. Probe target the $3'$ end of mRNA sequence (Corchado et al. 2009).

Slide 8-24: Computational Leukemia Cancer Detection 2/6

The proposed model by Corchado et al. (2009) incorporates the mixture of three experts in sequential form, having the advantage of integrating different techniques, considered to be optimal for using in the stages of the expression analysis for the problem of classifying leukemia patients. Techniques that offer good results in each phase are combined and the model considers the characteristics of each expert in order to achieve an appropriate integration. The structure of the modules can be seen in Fig. 8-15, the steps include:

1. Preprocessing and filtering
2. Clustering
3. Extraction of knowledge
4. Information representation

The different modules work independently, to facilitate the modification of any of the proposed experts, or to incorporate new techniques (including new experts). This affects the expert of a single module, while the others remain unchanged. This allows a generalization and making it possible to select the expert best suited to apply in each particular problem.

The initial problem description is composed of all the individuals $D = \{d_1, \ldots d_t\}$ together with the n probes. The first expert preprocesses and filters the probes, reducing the set of probes to s elements but maintaining the t individuals. The second expert executes the clustering, creates r groups and assigns the new individual (t + 1) to one of these groups. The third expert explains how the individual elements have been classified into groups by means of a knowledge extraction technique, and by obtaining a graphical representation (a tree). The final module represents the probability of assigning individuals to each of the groups depending on the probes selected, taking into account the knowledge extracted (Corchado et al. 2009).

Slide 8-25: Computational Leukemia Cancer Detection 3/6

This slide shows the classification performed for patients from groups CLL and ALL. The X axis represents the probes used in the classification and the Y axis represents the individuals. Above we can see, represented in black, most of the people of the CLL group are together, coinciding with the previous classification given by the experts. Only a small portion of the individuals departed from the initial classification. Below we see the classification obtained for the ALL patients. It can be seen that, although the ranking is not bad, the proportion of individuals misclassified is higher. Groups that have fewer individuals have a high classification error.

Classification obtained for (a) ALL patients and (b) CLL patients. Each of the values obtained correspond to the fluorescence intensity for an individual. At the bottom of the image is shown the fluorescence scale of values; the lowest level is 2 (blue), while the highest is 12 (red) (for interpretation of this images in color please refer to the original article (Corchdo et al. Bajo 2009)).

Slide 8-26: Computational Leukemia Cancer Detection 4/6

Following the decision tree shown in this slide, the patients were assigned to the expected groups. Only one of the patients was assigned to a different group by both methods. The healthy patients were eliminated in order to proceed with the classification.

The values of the leaf nodes represent the predicted group and the number of elements assigned to each of the groups following the order (ALL, AML, CLL, CML, NOL, MDS). The rest of the nodes represent the probe and the fuzzy value to compare the individual to classify. If the condition is true, then the branch on the left is selected, otherwise, the branch on the right is selected. The tree helps to obtain an explanation of the reason why an individual has been assigned to a group.

Slide 8-27: Computational Leukemia Cancer Detection 5/6

The work of Corchado et al. (2009) demonstrates a model of experts that uses exon arrays to perform an automatic diagnosis of cancer patients. The system incorporates experts at each phase of the microarray analysis, a process that is capable of extracting knowledge from diagnoses that have already been

(continued)

(continued)

performed, and that has been used to increase the efficiency of new diagnoses. The model combines:

1. Methods to reduce the dimensionality of the original set of data.
2. Preprocessing and data filtering techniques.
3. A clustering method to classify patients.
4. Modern extraction of knowledge techniques.

Slide 8-28: Computational Leukemia Cancer Detection 6/6

The system of Corchado et al. (2009) works in a way that is similar to how human specialist teams work in a lab, is also capable of working with big data and making decisions automatically and reduces the time needed for making predictions. The main advantage of this model is the ability to work with exon array data 0; very few tools are capable of working with this type of data because of the high dimensionality. The proposed model resolves this problem by using a technique that detects the importance of the genes for the classification of the diseases by analyzing the available data. For the time being, three experts have been designed, one for each phase of the model.

6 Case-Based Reasoning

Slide 8-29: Thinking–Reasoning–Deciding–Acting

Note: Always remember that Thinking–Reasoning–Decision–Action is intrinsically tied together. A good primer for clinical thinking is Alfaro-LeFevre (2013).

Slide 8-30: Case-Based Reasoning (CBR) Basic principle

CBR is a problem-solving paradigm, different from other AI approaches. Instead of relying solely on general knowledge of a problem domain, or making associations along generalized relationships between problem descriptors and conclusions, CBR is able to utilize the specific knowledge of previously experienced, concrete problem situations (cases). A new

(continued)

(continued)

problem is solved by finding a similar past case, and reusing it in the new problem situation. A second important difference is that CBR also is an approach to incremental, sustained learning, since a new experience is retained each time a problem has been solved, making it immediately available for future problems. The description of a problem defines a new case. This new case is used to RETRIEVE a case from the collection of previous cases. The retrieved case is combined with the new case—through REUSE—into a solved case, i.e., a proposed solution to the initial problem. Through the REVISE process this solution is tested for success, e.g., by being applied to the real world environment or evaluated by a teacher, and repaired if failed. During RETAIN, useful experience is retained for future reuse, and the case base is updated by a new learned case, or by modification of some existing cases (Aamodt and Plaza 1994).

Slide 8-31: The Task-Method Decomposition of CBR

In this slide we see the task-method structure: Tasks have node names in bold letters, while methods are written in italics. The links between task nodes (plain lines) are task decompositions, i.e., part-of relations, where the direction of the relationship is downwards. The top-level task is problem solving and learning from experience and the method to accomplish the task is CBR (indicated in a special way by a stippled arrow). This splits the top-level task into the four major CBR tasks corresponding to the four processes: retrieve, reuse, revise, and retain. All four tasks are necessary in order to perform the top-level task. The relation between tasks and methods (stippled lines) identify alternative methods applicable for solving a task. A method specifies the algorithm that identifies and controls the execution of subtasks, and accesses and utilizes the knowledge and information needed to do this (Aamodt and Plaza 1994).

Slide 8-32: CBR Example: Radiotherapy Planning 1/6

Example: Radiotherapy planning for cancer treatment is a computationally complex problem. An example from Petrovic et al. (2011) shall demonstrate it: Prostate cancer is generally treated in two phases. In phase I, both the prostate and the surrounding area, where the cancer has spread to, will be

(continued)

(continued)

irradiated, while in phase II only the prostate will be irradiated. The total dose prescribed by the oncologist is usually in the range of 70–76 Gy, while the dose ranges in phases I and II of the treatment are 46–64 Gy and 16–24 Gy, respectively. The dose is delivered in fractions, each fraction being usually 2 Gy.

Slide 8-33: CBR Example: Radiotherapy Planning 2/6

In this slide we see the workflow of radiotherapy: (1). CT scanning, (2) Tumor localization, (3) Skin reference marks, (4) Treatment planning, (5) Virtual simulation, (6) Radiotherapy treatment.

Slide 8-34: CBR Example: Radiotherapy Planning 3/6

The patient is first examined and then CT scans or MRI is carried out. Thereafter, the generated scans are passed onto the planning department. In the planning department, first, the tumor volume and the organs at risk are outlined by the medical physicist so that the region that contains the tumor can be distinguished from other parts that are likely to contain microscopic (tiny) tumor cells. Afterwards, the medical physicist in consultation with the oncologist defines the planning parameters including the number of beams to be used in the radiation, the angle between beams, the number of wedges,[1] the wedge angles and generates a Distribution Volume Histogram (DVH) diagram for both phases I and II of the treatment. DVH presents the simulated radiation distribution within a volume of interest which would result from a proposed radiation treatment plan. The next task is to decide the dose in phases I and II of the treatment so that the tumor cells can be killed without impairing the remaining body, particularly the organs lying close to the tumor cells, i.e., rectum and bladder. The organs lying close by should preferably not be impaired at all by the treatment. However, the oncologist usually looks for a compromise of distributing the inevitable dose among the organs. Rectum is a more sensitive organ compared to the bladder and is the primary concern of oncologists while deciding the dose plan. There is a maximum

(continued)

---

[1] Wedge = treatment accessory used to enable a better control of radiation distribution along a beam.

(continued)

dose limit for different volume percentages of the rectum, and it has to be respected by oncologists when prescribing a dose plan. In certain cases, this condition may be sacrificed to some extent so that an adequate dose can be imparted to the cancer cells. Oncologists generally use three groups of parameters to generate a good plan for each patient. The first group of parameters is related to the stage of cancer. It includes Clinical Stage (a labelling system), Gleason Score evaluates the grade of prostate cancer and is a integer between 1 and 10), and Prostate Specific Antigen (PSA) value between 1 and 40. The second group of parameters is related to the potential risk to the rectum (degree of radiation received by different volume percentages of the rectum. It includes the DVH of the rectum for Phases I and II at 66, 50, 25, and 10 % of the rectum volume. Example: the DVH states that 66 % of the rectum will receive 50 % of radiation. It means that if the dose prescribed by the oncologist in the phase I of the treatment is 60 Gy, then the amount of radiation received by 66 % of the rectum is 30 Gy. The final PSA value is a parameter related to the success rate of the patient after the treatment.

Slide 8-35: CBR System Architecture 4/6

In the system developed by Petrovic et al. (2011), the cases which are similar to the new case are retrieved using a fuzzy similarity measure. A modified Dempster–Shafer theory is applied to fuse the information from the retrieved cases and generate a solution as shown in this slide.

1. The clinical stage of the cancer is of ordinal type and can be divided in seven different categories T1a, T1b, T1c, T2a, T2b, T3a, T3b.
2. The value of the Gleason Score is an integer number from [1, 10] interval.
3. PSA is a real numbers from [1, 40].
4. DVH is a real number between [0, 1].

In order to use features of different data type, measurement units and scale together in the similarity measure we need to normalize them. However, it would not be easy to define a preferably linear mapping in the [0, 1] interval. Instead, we define fuzzy sets low, medium, and high for each feature. They are normalized fuzzy sets whose membership functions take value from [0, 1] interval. In addition, fuzzy sets enable expression of preference of the oncologist. An example of membership functions of fuzzy sets low, medium and high Gleason score is given in Figure Slide 8-36.

Slide 8-36: Membership Function of Fuzzy Sets in Gleason Score 5/6

The parameters of these membership functions are set in collaboration with the oncologist. Each attribute l (Gleason score (l = 1), PSA (l = 2)) of case cp is represented by a triplet (vpl1, vpl2, vpl3), where vplm, m = 1, 2, 3 are membership degrees of attribute l in the corresponding fuzzy sets low (m = 1), medium (m = 2), and high (m = 3).

Slide 8-37: Case-Based Reasoning 6/6

This final slide demonstrates the adaptation mechanism. In this example, the final outcome of the Dempster–Shafer theory is a dose plan having 62 Gy and 10 Gy of radiation in phases I and II of treatment, respectively. This is not a feasible dose plan because the dose received by 10 % of the rectum is 56.2 Gy which is larger than the prescribed maximum dose limit (55 Gy). Hence, in order to generate a feasible dose plan, the repair mechanism is performed. The dose corresponding to the phase II of the treatment is decreased by 2 Gy, which leads to the new dose plan 62 and 8 Gy, which is a feasible dose plan.

Note: The Dempster–Shafer theory (DST) is a mathematical theory of evidence and allows the combination of evidence from different sources resulting in a degree of belief (represented by a belief function) that takes into account all the available evidence (Zadeh 1986).

7 Future Outlook

Slide 8-38: Future Outlook

Modern approaches, as for example the IBM Watson Technologies (Holzinger et al. 2013) may considerably assist future medical doctors. Following the hypothesis that medicine is a data problem; the answer to your problem is "out there" in the masses of data, but literally, a human has little chance to find it. Hence, we need machine intelligence to help to harness these large amounts of data and to provide decision support for the clinician. A long debate is ongoing whether such technological approaches will replace medical doctors. To answer such a question we can remember that the same question was asked in the 1960s, when computer was first used for teaching and learning purposes—and up to now, still human teachers are in the classroom ...

8 Exam Questions

8.1 Yes/No Decision Questions

Please check the following sentences and decide whether the sentence is true = YES; or false = NO; for each correct answer you will be awarded 2 credit points.

01	Decisions related to the diagnosis means that computers are used to select the best therapy on the basis of clinical evidence.	☐ Yes ☐ No	2 total
02	MYCIN is a rule-based Expert System, which is used for therapy planning for patients with bacterial infections.	☐ Yes ☐ No	2 total
03	In MYCIN, if CF = 0, there is either no evidence regarding the hypothesis or the supporting evidence is equally balanced.	☐ Yes ☐ No	2 total
04	Clinical rules represent elementary, isolated care recommendations, while narrative guidelines describe a coherent, unified care process.	☐ Yes ☐ No	2 total
05	Case-based reasoning is a problem solving paradigm, relying solely on general knowledge of a problem domain as in typical AI applications.	☐ Yes ☐ No	2 total
06	"What is the probability that this patient has a myocardial infarction on the basis of given data?" is an example for Type 1: Decisions related to the diagnosis	☐ Yes ☐ No	2 total
07	MYCIN was programmed in Fortran and used judgmental rules with associated elements of uncertainty.	☐ Yes ☐ No	2 total
08	Qualitative DSS models are relying on symbolic methods, rather than following a strictly formal mathematical basis.	☐ Yes ☐ No	2 total
09	One of the first spinoffs from DENDRAL was ONCOCIN, an expert system for people whose expertise lies in building expert systems.	☐ Yes ☐ No	2 total
10	In the early 1950ies decision trees and truth tables were used, followed by systems based on statistical methods.	☐ Yes ☐ No	2 total
Sum of Question Block A (max. 20 points)			

8.2 Multiple Choice Questions (MCQ)

The following questions are composed of two parts: the stem, which identifies the question or problem and a set of alternatives which can contain 0, 1, 2, 3, or 4 correct answers, along with a number of distractors that might be plausible—but are incorrect. Please **select the correct answers** by ticking ☒ - and do not forget that it can be none. Each question will be awarded 4 points *only if everything is correct*.

01	The Dempster-Shafer theory ... ☐ a) ... is a theory for checking evidence the genes for the classification of the diseases by analyzing the available data. ☐ b) ... is a evidence theory and allows the combination of evidence from different sources. ☐ c) ... is based on statistical methods to test the probability of the occurrence of an event. ☐ d) ... results in a degree of belief (represented by a belief function) that takes into account all the available evidence.	4 total
02	Qualitative Decision Models include ... ☐ a) ... truth tables. ☐ b) ... neural networks ☐ c) ... decision trees. ☐ d) ... boolean logic.	4 total
03	MYCIN .... ☐ a) ... was designed to recommend antibiotics, with the dosage adjusted for the patient's body weight. ☐ b) ... has been integrated in hospital information systems as part of the standard decision support. ☐ c) ... applies the principles of rule-based systems. ☐ d) ... had achieved high acceptance amongst clinicians in their daily routine patient work.	4 total
04	The architecture of typical Decision Support Systems include ☐ a) ... a knowledge base for storing facts and heuristics. ☐ b) ... a data integration and data fusion engine. ☐ c) ... a interactive user interface for consultation. ☐ d) ... an inference engine with reasoning mechanism.	4 total
05	The Model of Corchado et al. (2009) combines ... ☐ a) ... dimensionality reduction of the data. ☐ b) ... pre-processing and data filtering techniques. ☐ c) ... a system integration into clinical information systems. ☐ d) ... a clustering method to classify patients.	4 total
06	Boolean Logic can be used as ... ☐ ... qualitative decision model. ☐ ... quantitative decision model. ☐ ... statistical decision model. ☐ ... heuristic decision model.	4 total
07	A certainty factor CF[h,E] of 0,8 means ... ☐ a) ... strongly suggestive evidence that the result is true. ☐ b) ... it is definite that the result is true. ☐ c) ... it is weakly suggestive evidence that the result is true. ☐ d) ... it is weakly suggestive evidence that the result is false.	4 total

8 Exam Questions

08	For design and development of a decision support system ... ☐ a) ... to keep the user acceptance into the focus is a necessity. ☐ b) ... it is necessary to understand the context and the end-user. ☐ c) ... it helps to combine and integrate methods from SE and HCI. ☐ d) ... the knowledge of sophisticated data mining algorithms is sufficient.	4 total
09	Case-based reasoning ... ☐ a) ... is an approach for incremental, sustained learning. ☐ b) ... is able to utilize knowledge gained from previous problem situations. ☐ c) ... solves a new problem by finding a similar past case. ☐ d) ... is a theoretical approach and has not yet been used in clinical practice.	4 total
10	The development of medical expert systems is difficult, because ... ☐ a) ... medicine is per se an extremely difficult application domain. ☐ b) ... there are no implementable evidence-based guidelines available. ☐ c) ... the computational power available is insufficient. ☐ d) ... typically we deal with uncertain and probable information.	4 total

Sum of Question Block B (max. 40 points)

8.3 Free Recall Block

Please follow the instructions below. At each question you will be assigned the credit points indicated if your option is correct (partial points may be given).

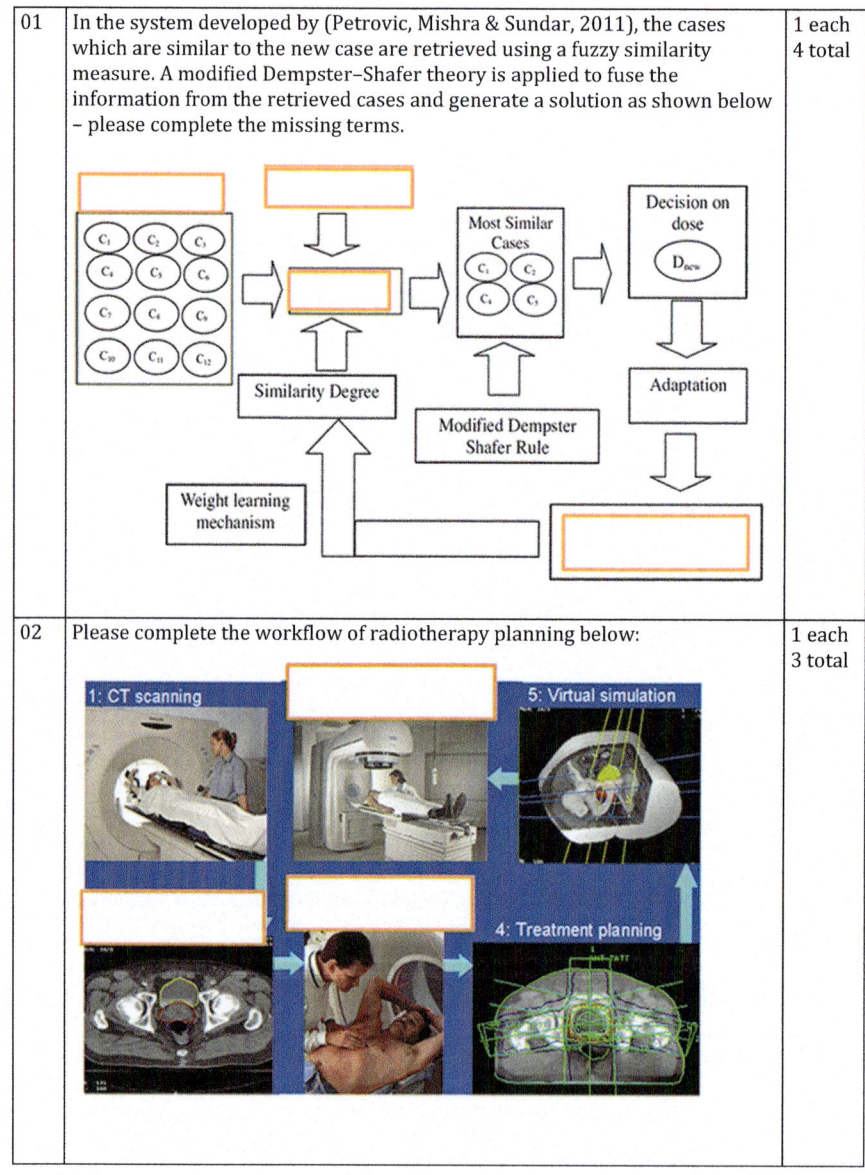

01	In the system developed by (Petrovic, Mishra & Sundar, 2011), the cases which are similar to the new case are retrieved using a fuzzy similarity measure. A modified Dempster–Shafer theory is applied to fuse the information from the retrieved cases and generate a solution as shown below – please complete the missing terms.	1 each 4 total
02	Please complete the workflow of radiotherapy planning below:	1 each 3 total

03	The proposed model by (Corchado, De Paz, Rodriguez & Bajo, 2009) incorporates the mixture of three experts in sequential form, having the advantage of integrating different techniques, considered to be optimal for using in the stages of the expression analysis for the problem of classifying leukaemia patients – please complete the missing terms	1 each 4 total

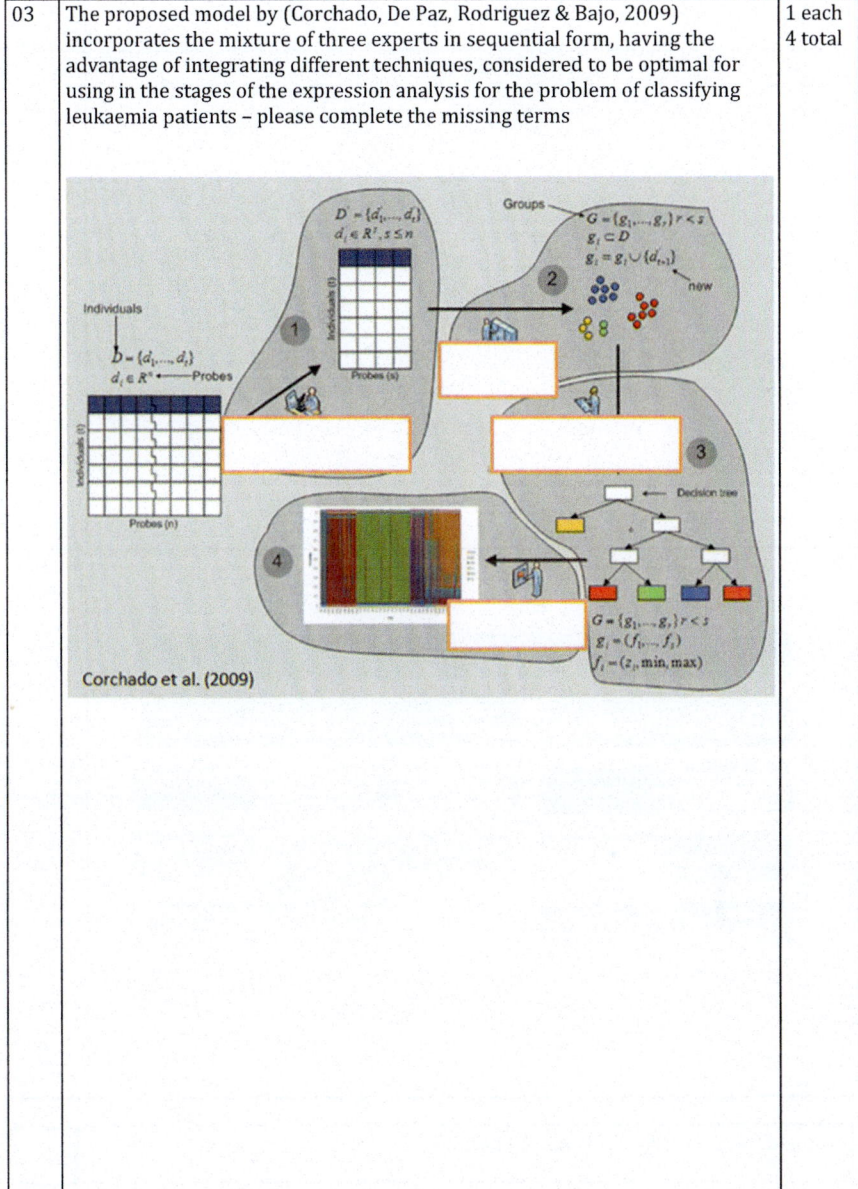

Corchado et al. (2009)

04	Please sketch the basic architecture of a Decision Support System:	1 each 5 total

05	Please complete the table below	1 each 4 total
	Decision Model — Quantitative (statistical) / Qualitative (heuristic); boxes: [ ], Bayesian, [ ], [ ], Reasoning models, unsupervised, [ ], Boolean Logic, Non-parametric Partitioning, Expert systems, Neural network, Logistic, Critiquing systems	

Sum of Question Block C (max. 40 points)

9 Answers

9.1 Answers to the Yes/No Questions

Please check the following sentences and decide whether the sentence is true = YES; or false = NO; for each correct answer you will be awarded 2 credit points.

01	Decisions related to the diagnosis means that computers are used to select the best therapy on the basis of clinical evidence.	☐ Yes ☒ No	2 total
02	MYCIN is a rule-based Expert System, which is used for therapy planning for patients with bacterial infections.	☒ Yes ☐ No	2 total
03	In MYCIN, if CF = 0, there is either no evidence regarding the hypothesis or the supporting evidence is equally balanced.	☒ Yes ☐ No	2 total
04	Clinical rules represent elementary, isolated care recommendations, while narrative guidelines describe a coherent, unified care process.	☒ Yes No	2 total
05	Case-based reasoning is a problem solving paradigm, relying solely on general knowledge of a problem domain as in typical AI applications.	☐ Yes ☒ No	2 total
06	"What is the probability that this patient has a myocardial infarction on the basis of given data?" is an example for Type 1: Decisions related to the diagnosis	☒ Yes ☐ No	2 total
07	MYCIN was programmed in Fortran and used judgmental rules with associated elements of uncertainty.	☐ Yes ☒ No	2 total
08	Qualitative DSS models are relying on symbolic methods, rather than following a strictly formal mathematical basis.	☒ Yes ☐ No	2 total
09	One of the first spinoffs from DENDRAL was ONCOCIN, an expert system for people whose expertise lies in building expert systems.	☐ Yes ☒ No	2 total
10	In the early 1950ies decision trees and truth tables were used, followed by systems based on statistical methods.	☒ Yes ☐ No	2 total
Sum of Question Block A (max. 20 points)			

9.2 Answers to the Multiple Choice Questions (MCQ)

01	The Dempster-Shafer theory ... ☐ a) ... is a theory for checking evidence the genes for the classification of the diseases by analyzing the available data. ☒ b) ... is a evidence theory and allows the combination of evidence from different sources. ☐ c) ... is based on statistical methods to test the probability of the occurrence of an event. ☒ d) ... results in a degree of belief (represented by a belief function) that takes into account all the available evidence.	4 total
02	Qualitative Decision Models include ... ☒ a) ... truth tables. ☐ b) ... neural networks ☒ c) ... decision trees. ☒ d) ... boolean logic.	4 total
03	MYCIN .... ☒ a) ... was designed to recommend antibiotics, with the dosage adjusted for the patient's body weight. ☐ b) ... has been integrated in hospital information systems as part of the standard decision support. ☒ c) ... applies the principles of rule-based systems. ☐ d) ... had achieved high acceptance amongst clinicians in their daily routine patient work.	4 total
04	The architecture of typical Decision Support Systems include ☒ a) ... a knowledge base for storing facts and heuristics. ☐ b) ... a data integration and data fusion engine. ☒ c) ... a interactive user interface for consultation. ☒ d) ... an inference engine with reasoning mechanism.	4 total
05	The Model of Corchado et al. (2009) combines ... ☒ a) ... dimensionality reduction of the data. ☒ b) ... pre-processing and data filtering techniques. ☐ c) ... a system integration into clinical information systems. ☒ d) ... a clustering method to classify patients.	4 total
06	Boolean Logic can be used as ... ☒ ... qualitative decision model. ☐ ... quantitative decision model. ☐ ... statistical decision model. ☒ ... heuristic decision model.	4 total
07	A certainty factor CF[h,E] of 0,8 means ... ☒ a) ... strongly suggestive evidence that the result is true. ☐ b) ... it is definite that the result is true. ☐ c) ... it is weakly suggestive evidence that the result is true. ☒ d) ... it is weakly suggestive evidence that the result is false.	4 total
08	For design and development of a decision support system ... ☒ a) ... to keep the user acceptance into the focus is a necessity. ☒ b) ... it is necessary to understand the context and the end-user. ☒ c) ... it helps to combine and integrate methods from SE and HCI. ☐ d) ... the knowledge of sophisticated data mining algorithms is sufficient.	4 total

09	Case-based reasoning ... ☒ a) ... is an approach for incremental, sustained learning. ☒ b) ... is able to utilize knowledge gained from previous problem situations. ☒ c) ... solves a new problem by finding a similar past case. ☐ d) ... is a theoretical approach and has not yet been used in clinical practice.	4 total
10	The development of medical expert systems is difficult, because ... ☒ a) ... medicine is per se an extremely difficult application domain. ☐ b) ... there are no implementable evidence-based guidelines available. ☐ c) ... the computational power available is insufficient. ☒ d) ... typically we deal with uncertain and probable information.	4 total

Sum of Question Block B (max. 40 points)

9.3 Answers to the Free Recall Questions

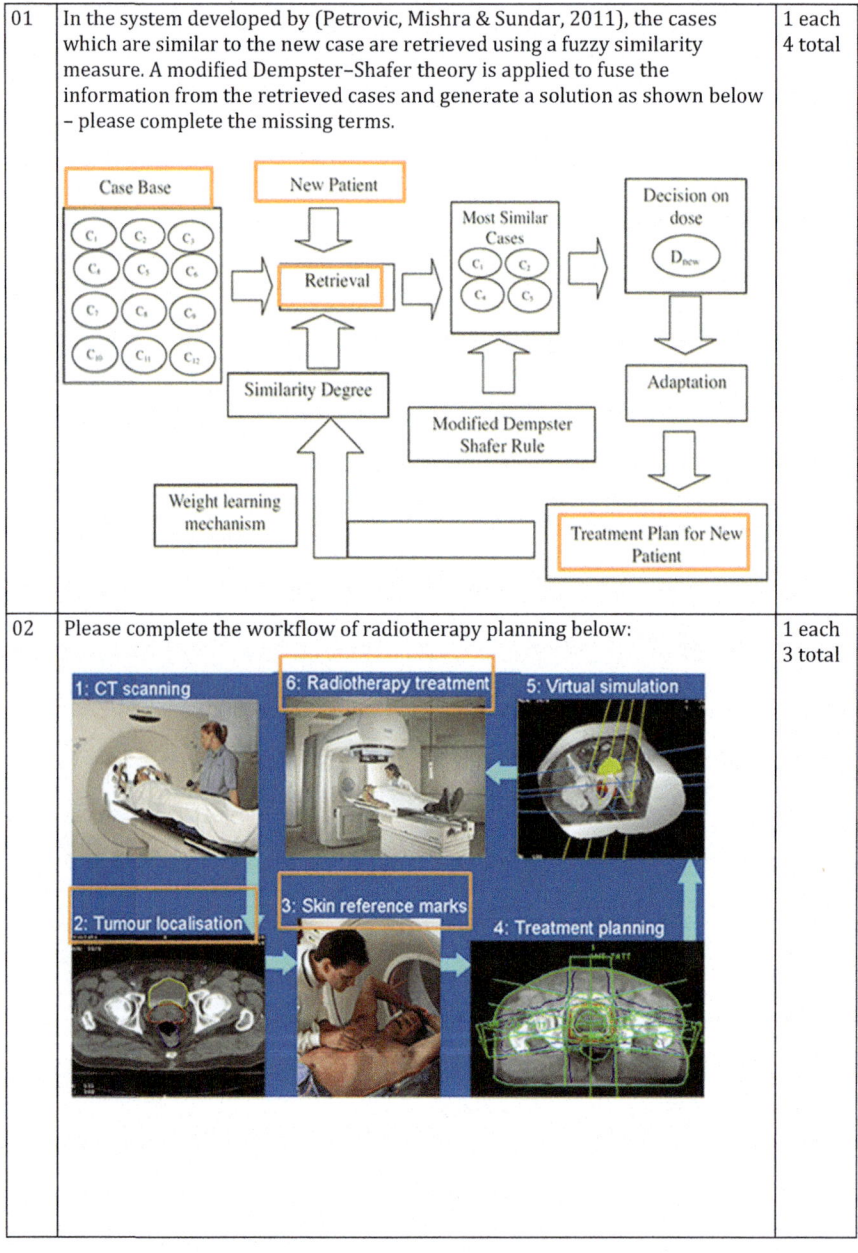

9 Answers

03	The proposed model by (Corchado, De Paz, Rodriguez & Bajo, 2009) incorporates the mixture of three experts in sequential form, having the advantage of integrating different techniques, considered to be optimal for using in the stages of the expression analysis for the problem of classifying leukaemia patients – please complete the missing terms	1 each 4 total

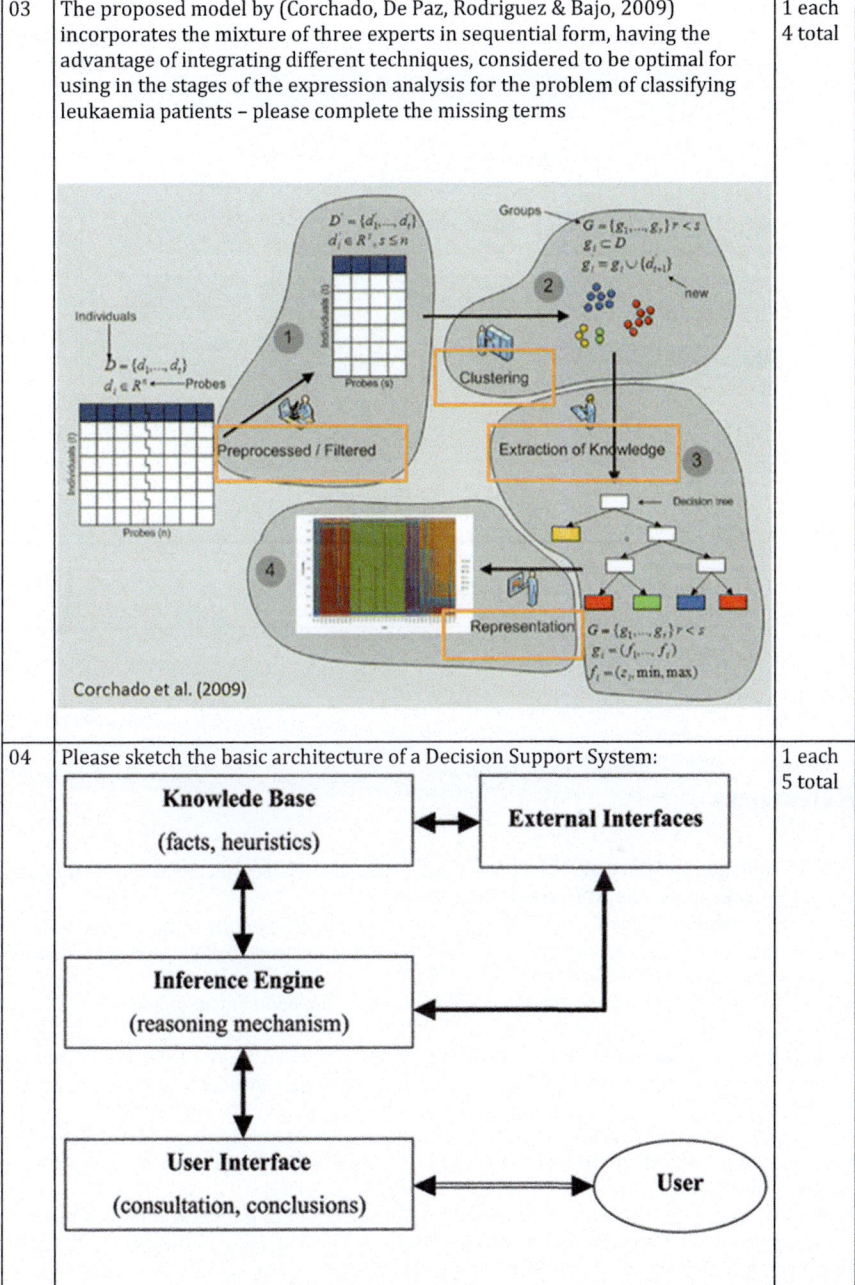

Corchado et al. (2009)

04	Please sketch the basic architecture of a Decision Support System:	1 each 5 total

05	Please complete the table below	1 each 4 total
	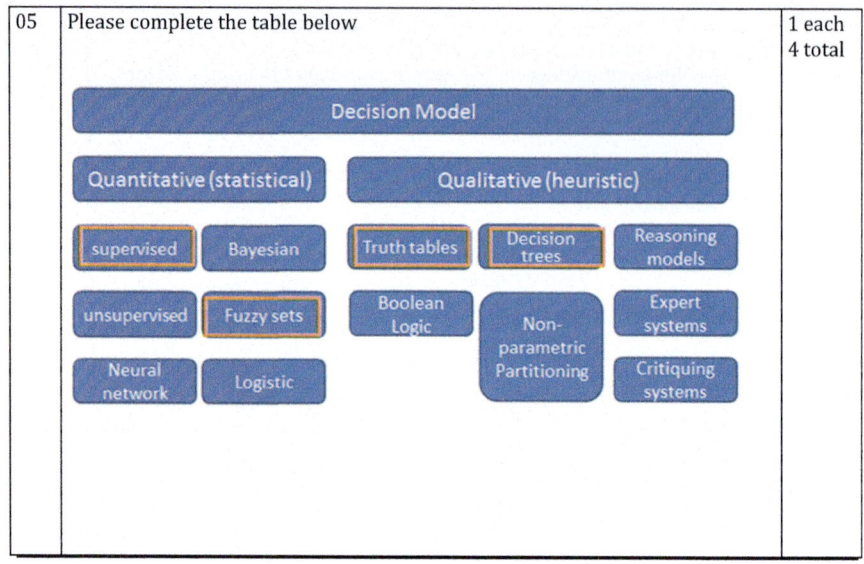	

Sum of Question Block C (max. 40 points)	

References

Aamodt A, Plaza E (1994) Case-based reasoning: foundational issues, methodological variations, and system approaches. AI Commun 7(1):39–59

Abed M, Bernard J, Angué J (1991) Task analysis and modelization by using SADT and Petri Networks. Tenth European annual conference on human decision making and manual control, Liege. pp. 11–13

Alfaro-Lefevre R (2013) Critical thinking, clinical reasoning, and clinical judgment: a practical approach. Elsevier Saunders, St Louis

Ayed BM, Ltifi H, Kolski C, Alimi AM (2010) A user-centered approach for the design and implementation of KDD-based DSS: a case study in the healthcare domain. Decis Support Syst 50:64–78

Bemmel JHV, Musen MA (1997) Handbook of medical informatics. Springer, Heidelberg

Corchado JM, De Paz JF, Rodriguez S, Bajo J (2009) Model of experts for decision support in the diagnosis of leukemia patients. Artif Intell Med 46(3):179–200

Fieschi M, Dufour J, Staccini P, Gouvernet J, Bouhaddou O (2003) Medical decision support systems: old dilemmas and new paradigms. Methods Inf Med 42(3):190–198

Holzinger A (2012) On knowledge discovery and interactive intelligent visualization of biomedical data: challenges in human–computer interaction & biomedical informatics. In: Helfert M, Fancalanci C, Filipe J (eds) DATA—international conference on data technologies and applications. INSTICC, Rome, pp 5–16

Holzinger A, Stocker C, Ofner B, Prohaska G, Brabenetz A, Hofmann-Wellenhof R (2013) Combining HCI, natural language processing, and knowledge discovery—potential of ibm content analytics as an assistive technology in the biomedical domain. Springer lecture notes in computer science LNCS 7947. Springer, Heidelberg, pp 13–24

Howe J (2006) The rise of crowd sourcing. Wired Mag 14(6):1–4

Kapur K, Xing Y, Ouyang Z, Wong W (2007) Exon arrays provide accurate assessments of gene expression. Genome Biol 8(5):R82

Lindsay RK, Buchanan BG, Feigenbaum EA, Lederberg J (1993) DENDRAL: a case study of the first expert system for scientific hypothesis formation. Artif Intell 61(2):209–261

Majumder DD, Bhattacharya M (2000) Cybernetic approach to medical technology: application to cancer screening and other diagnostics. Kybernetes 29(7/8):871–895

Medlock S, Opondo D, Eslami S, Askari M, Wierenga P, De Rooij SE, Abu-Hanna A (2011) LERM (Logical Elements Rule Method): a method for assessing and formalizing clinical rules for decision support. Int J Med Inform 80(4):286–295

Metaxiotis K, Psarras J (2003) Expert systems in business: applications and future directions for the operations researcher. Ind Manag Data Syst 103(5):361–368

Petrovic S, Mishra N, Sundar S (2011) A novel case based reasoning approach to radiotherapy planning. Exp Syst Appl 38(9):10759–10769

Rheingold H (1985) Tools for thought: the history and future of mind-expanding technology. Simon & Schuster, New York

Shortliffe EH (1986) Update on ONCOCIN: a chemotherapy advisor for clinical oncology. Inform Health Soc Care 11(1):19–21

Shortliffe EH (2011) Biomedical informatics: defining the science and its role in health professional education. In: Holzinger A, Simonic K-M (eds) Information quality in e-Health. Lecture notes in computer science LNCS 7058. Springer, Heidelberg, pp 711–714

Shortliffe EH, Buchanan BG (1984) Rule-based expert systems: the MYCIN experiments of the Stanford Heuristic Programming Project. Addison-Wesley, Reading, MA

Shortliffe T, Davis R (1975) Some considerations for the implementation of knowledge-based expert systems. ACM SIGART Bull 55:9–12

Zadeh LA (1986) A simple view of the Dempster-Shafer theory of evidence and its implication for the rule of combination. AI Mag 7(2):85

Lecture 9

Interactive Information Visualization and Visual Analytics

1 Learning Goals

At the end of this ninth lecture, you:

- would have some theoretical background on visualization and visual analytics.
- would get an overview about various possible visualization methods for various data.
- would get an introduction of the work of and possibilities with parallel coordinates.
- would have seen the principles of RadViz mappings and algorithms.
- would be aware of the possibilities of Star Plots.
- would have seen that visual analytics is intelligent human–computer interaction (HCI).

2 Advance Organizer

Biological data visualization	In bioinformatics the visualization of sequences, genomes, alignments, phylogenies, macromolecular structures, systems biology, etc.
Business Intelligence (BI)	All issues in a company which provides historical, current, and predictive views of business operations; methods include data mining, process mining, content analytics, and particularly visual analytics; BI is directly connected with decision support; any BI system is also a decision support system
Classification	The problem of identifying to which set of categories a new observation belongs, on the basis of a training set of data containing observations (or instances) whose category membership is known

Clustering	Mapping objects into disjoint subsets to let appear similar objects in the same subset; it is a main task in exploratory data mining in bioinformatics
Content analytics	A general term addressing so-called "unstructured" data—mainly text—by using methods from visual analytics in business intelligence
Data visualization	Visual representation of complex data, to communicate information clearly and effectively, making data useful and usable
Information visualization	The interdisciplinary study of the visual representation of large-scale collections of non-numerical data, such as files and software, databases, networks, etc., to allow users to see, explore, and understand information at once
Multidimensional	Containing more than three dimensions and data are multivariate
Multidimensional scaling	Mapping objects into a low-dimensional space (plane, cube, etc.) in order to let appear similar objects close to each other
Multi-variate	Encompassing the simultaneous observation and analysis of more than one statistical variable (antonym: univariate = one-dimensional)
Parallel coordinates	For visualizing high-dimensional and multivariate data in the form of N parallel lines, where a data point in the n-dimensional space is transferred to a polyline with vertices on the parallel axes
RadViz	Radial visualization method, which maps a set of m-dimensional points in the 2D space, similar to Hooke's law in mechanics
Semiotics	Deals with the relationship between symbology and language, pragmatics, and linguistics. Information and Communication Technology deals not only in words and pictures but also in ideas and symbology
Semiotic engineering	A process of creating a semiotic system, i.e., a model of human intelligence and knowledge and the logic for communication and cognition
Star plot	Aka radar chart, spider web diagram, star chart, polygon plot, polar chart, or Kiviat diagram, for displaying multivariate data in the form of a 2D chart of three or more quantitative variables represented on axes starting from the same point
Visual analytics	Focuses on analytical reasoning of complex data facilitated by interactive visual interfaces
Visualization	A method of computer science to transform the symbolic into the geometric, to form a mental model and foster unexpected insights

| Visualization mantra | "Overview first, zoom and filter on demand" (Shneiderman 1996) |

3 Acronyms

InfoVis	Information visualization
‖-Coords	Parallel coordinates
PCP	Parallel coordinate plot
RadViz	Radial coordinate visualization
UIMI	User interface model for InfoVis
VA	Visual analytics

4 Key Problems

Slide 9-1: Key Challenges

- How to understand high-dimensional spaces?
- The transformation of results from high-dimensional space $\mathbb{R}^N$ into $\mathbb{R}^2$
- From the complex to the simple
- Low integration of visual analytics techniques into the clinical workplace
- Sampling, modelling, rendering, perception, cognition, decision making
- Trade-off between time and accuracy
- How to model uncertainty

Information visualization is the study of visual representations of abstract data to reinforce human cognition; hence, it is very important for decision making. A lot of challenges are involved: The human perceptual system can handle large quantities of data of few dimensions but has great difficulty as the data dimensionality increases (again: the curse of dimensionality (Donoho 2000)). The grand challenge is to focus not simply on computational methods of displaying large quantities of data but on **both perception and cognition of such large amounts of data**. One aspect is to focus on how the process of computer visualization can be improved to mirror the process of natural visualization. Our perceptual systems were designed specifically for survival

(continued)

(continued)

in and understanding of the surrounding external environment, *not* abstract objects and images (Grinstein et al. 1998).

Visual analysis is becoming an essential component of medical visualization due to the rapidly growing role and availability of complex multidimensional, time-varying, mixed-modality, simulation, and multi-subject datasets. The magnitude, complexity, and heterogeneity of the data necessitate the use of visual analysis techniques for diagnosis and medical research and, even more importantly, treatment planning and evaluation, e.g., radiotherapy planning and post-chemotherapy evaluation (Childs et al. 2013).

5 Fundamentals of Visualization Science

5.1 Verbal Information Versus Visual Information

Slide 9-2: Verbal Information Versus Visual information

Letting aside olfactory information (smell), gustatory information (taste), and haptic information (touch), and following the **dual-coding theory** (Paivio and Csapo 1973) we separate visual information (images) and verbal information (spoken and written natural language). In the slide we see the title page of the Journal Cell (www.cell.com): the Latin letters C-e-l-l describe the basic structural, functional, and biological unit of all known living organisms. Cells are the smallest unit of life and therefore we may call it the "basic building block of life." However, a huge problem with verbal information is that we are confronted with **semantic ambiguity**, which means that a word has often more than one meaning, see Slide 9-3.

5 Fundamentals of Visualization Science

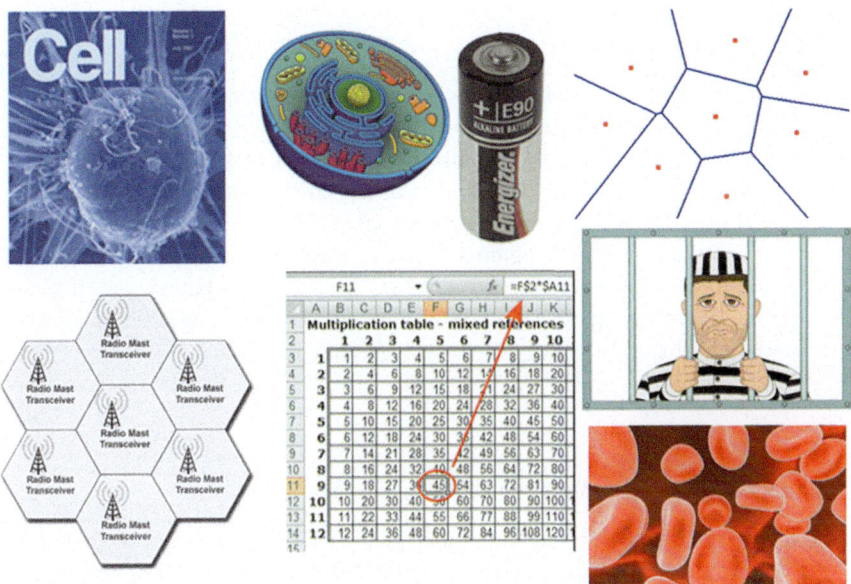

Fig. 1 See Slide 9-3

Slide 9-3: Semantic Ambiguity of Verbal Information

The word "cell" has a lot of different meanings: the famous Journal, but also the basic building block of life, a battery cell, a Voronoi cell in mathematical topology, a prisoner's cell, a cell of a radio network, a blood cell, a cell of a spreadsheet, an cell in aircrafts or car manufacturing, a foam cell, cellulose (in German: "Zell-Stoff"), etc. This is the most difficult problem: the semantic ambiguity of our natural language (noses can run and feet can smell—without **context** this is unsolvable for any computer). To better understand these processes let us review more detailed the already learned human information processing.

Slide 9-4: Visual Information Processing (Pictures)

According to the theory of Multimedia Learning by Mayer (2001) perceived physical visual stimuli (e.g., images) are pictorially processed and are thus cognitive "similar" to the original physical real-world data, as can be seen in this slide: Pictures are physically perceived intentionally by the eyes and then briefly hold in the so-called visual sensory register. Only if there is a certain amount of attention (for the importance of selective attention, please refer to

(continued)

(continued)

Lecture 7, Sect. 5.2.3), the pictures will become represented within the working memory. Once the working memory is full of image pieces (cognitive overload), the next active cognitive processing involves organizing those pieces into a coherent structure. The resulting knowledge representation is a pictorial model, that is, the person builds an organized visual representation of the main parts of the picture. Finally, active cognitive processing is required to connect this new representation with previous knowledge. If this happens then the picture will be memorized in the long-time memory (Holzinger 2002b).

Slide 9-5: Verbal Information Processing: Written Text

Written text, i.e., written natural language (words) are perceived as images, but processed symbolically as text. Consequently our "natural" language is a kind of **artificial concept** to represent real-world data: The presentation of printed text creates an information-processing challenge for the **dual-channel system**: Although the words are presented visually so they are initially perceived through the eyes and thus brought into the working memory as image, they must be mentally processed by the auditory part of the working memory, thus processed like spoken words. Consequently, when verbal material must enter through the visual channel, the words must take a complex route through the system, and must also compete for attention with images which might be perceived in parallel through the visual channel. The consequences of this problem are addressed in the modality principle (Moreno and Mayer 1999).

Slide 9-6: Verbal Information Processing: Spoken Text

Natural language (text) can also be perceived directly as spoken words, consequently directly auditory processed: In this case the piece of text (word) is picked up as sound by the ears and held temporarily in the auditory sensory memory. If the person pays attention to the sounds coming into the ears, some of the incoming sounds will be selected for inclusion in the so-called word sound base. The words in the word base are disorganized fragments, so the next step is to build them into a coherent mental structure. In this process, the words change from being represented based on sound to

(continued)

(continued)

being represented based on word meaning. Again, the person may use prior knowledge to integrate the new words into the word base—this is ensuring the context—and exactly this is missing in our computational approaches so far, but future advances in machine learning, as for example in the IBM Watson technologies, may bring a significant step forwards, because these approaches "learn" from the environment. A good example in this respect is the Never Ending Learning Machine (NELL), running at the group of Tom Mitchell (Carlson et al. 2010).

5.2 Is a Picture Really Worth a Thousand Words?

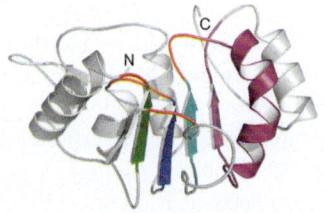

Fig. 2 See Slide 9-7

Slide 9-7: Example: Ribbon Diagram of a Protein Structure

The famous proverb "*a picture is worth a thousand words*" refers to the concept that a complex idea can be conveyed with just one single image and infers a central goal of visualization: to make it possible to perceive and cognitively process large amounts of data quickly. Look at this image. It is a good example on how a picture can explain a complex idea: A ribbon diagram aka Richardson diagram, (Richardson 2000), is a standard method of schematic protein representation. The ribbon shows the overall path and organization of the protein backbone and is generated by interpolating a smooth curve through the polypeptide backbone. So-called α-helices are shown as curly ribbons, β-strands as arrows, and thin lines for non-repetitive coils or loops. The direction of the polypeptide chain is shown locally by the arrows, and may be indicated overall by a color ramp along the length of the ribbon. Such diagrams are useful for expressing the molecular structure (twist, fold, and unfold). Remember: A protein is a single chain of amino acids, which

(continued)

(continued)

folds into a globular structure and the Thermodynamics Hypothesis states that a protein always folds into a state of minimum energy. Computationally, the protein folding problem becomes an optimization problem: We are looking for a path to the global minimum in a very high-dimensional energy landscape. First ribbon diagrams were hand drawn (Richardson 2000; Magnani et al. 2010).

Slide 9-8: "Is a Picture Really Worth a Thousand Words?"

Whether and to what extent the proverb above is true is a long debate and there is no clear evidence to date. The best answer to the question "Is a picture worth a thousand words?" is: "It depends!". Some researchers are arguing that sometimes a picture might be worth a billion words (Michel et al. 2011), whereas others are arguing that sometimes text is better than an image.

A more profane example: what is the difference between tortoise and turtle? If you are not good in animal biology, you maybe have problems with the words, but if you look at a picture showing a turtle and at one showing a tortoise, you immediately understand that one is a land animal and the other a sea animal.

Be aware, that most information in hospitals is only available in text format, and that text is *the* communication media in patient findings, and the amount of this unstructured data is immensely increasing (Holzinger et al. 2008, 2013). Consequently, text mining is a huge area of biomedical informatics (refer to Lecture 6).

5.3 *Informatics as Semiotics Engineering*

Slide 9-9: Three Examples for Visual Languages (Ware 2004)

Computer science lacks a reliable concept of the human mind, whereas the psychological science lacks solid concepts for algorithms and data structures; consequently, there is a need for a theory in which both domains find a place (Andersen 2001): A so-called **sign** (in German: "Zeichen") integrates these two sides: the physical is called signifier and the psychological is called signified, see this slide.

(continued)

(continued)

The first example is a cave painting: Paleolithic cave art is an exceptional archive of early human symbolic behavior (Pike et al. 2012). We can readily interpret human figures and infer that the people are using bows and arrows to hunt deer;

The second example is a schematic diagram showing modern interaction between a human and a computer in a virtual environment system: the brain in this diagram is a simplified picture, but can easily be recognized as part of the human anatomy. The arrows show data flows and are arbitrary conventions, as are the printed words.

The third example is the expression of a mathematical equation that may be obscure to the non-mathematical trained, but is immediately obvious to an expert.

These examples clearly show that some visual languages are easier to perceive and to understand than others. But why? Perhaps it is simply that we have more experience with the kind of pictorial image shown in the cave painting and less with the abstract mathematical notation.

Slide 9-10: Informatics as Semiotics Engineering

Semiotics is the study of signs and therefore can describe representations (algorithms and data structures as signifiers) and the interpretation by the end user (domain concepts as the signified). However, only those parts of the computational process that influence the interpretation, and only those parts of the interpretations that are influenced by the computation, can be analyzed by semiotic methods (Holzinger et al. 2011; Holzinger 2002a). Semiotics can be divided into three branches:

Syntactics: Relations among signs in formal structures.
Semantics: Relations between signs and their meaning.
Pragmatics: Relations between signs, and the effects these may have on the end users who use them.

A relatively new field is biosemiotics, which is a synthesis of biology and semiotics, and studying the origins, action, and interpretation of signs and biological codes (Barbieri 2008): Life is essentially about creating new organic codes and conserving those which have been created (macroevolution). For example, biosemiotics claims that language has biological roots and must be studied as a natural phenomenon, not following the divide between nature and culture. Or another example: The study of protein synthesis has revealed that genes and proteins are not formed

(continued)

(continued)

spontaneously in the cell but are manufactured by a system of molecular machines based on RNAs. In 1981, the components of this manufacturing system were called ribosoids, and the system itself was given the collective name of ribotype. The cell was described in this way as a structure made of genes, proteins, and ribosoids, i.e., as a trinity of genotype, phenotype, and ribotype (Barbieri 2010).

A different example: Burton-Jones et al. (2005) proposed metrics to assess the quality of an ontology by drawing upon semiotic theory; their metrics assess the syntactic, semantic, pragmatic, and social aspects of the quality of an ontology.

5.4 Visualization Process

Before we go on to the visualization process we summarize some definitions.

Slide 9-11: Definitions of the Term "Visualization"

Visualization is a method of computer science **to transform the symbolic into the geometric**, to support the formation of a mental model and foster insights; as such it is an essential component of the knowledge discovery process (refer to Lecture 6, Slide 6-3) (2007).

Information visualization is the interdisciplinary study of the visual representation of large-scale collections of non-numerical data, such as files and software, databases, networks, etc., to allow users to see, explore, and understand information at once (Ware 2004, 2012).

Data visualization is the visual representation of complex data, to communicate information clearly and effectively, making data useful and usable (Ward et al. 2010).

Visual Analytics focuses on analytical reasoning of complex data facilitated by interactive visual interfaces (Aigner et al. 2007).

Content Analytics is a general term addressing so-called "unstructured" data—mainly text—by using mixed methods from visual analytics and business intelligence (Holzinger et al. 2013).

5 Fundamentals of Visualization Science

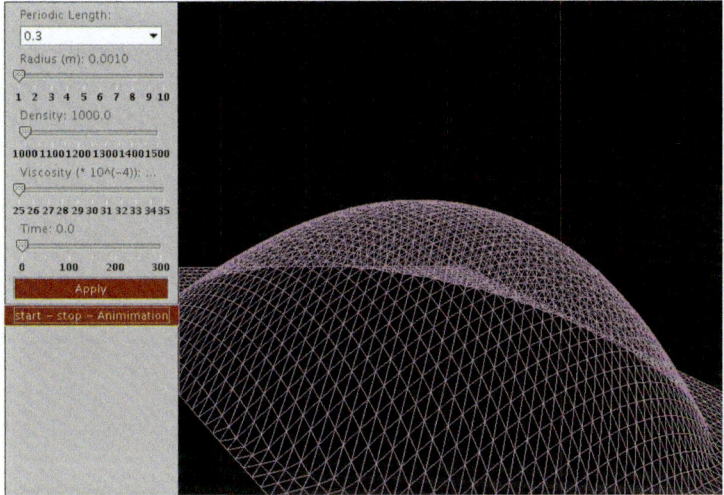

Fig. 3 See Slide 9-12

Slide 9-12: Process of Interactive (Data) Visualization

Interactive visualizations provide the ability to comprehend data and to interactively analyze information properties. The process of data visualization includes four steps:

1. The data itself and interactive data exploration.
2. The preprocessing and transformation of the data.
3. The graphical algorithms to produce the corresponding image on a screen.
4. The human perceptual and cognitive system;

McCormick (1987) defined the science of visualization by a taxonomy diagram, wherein he stated that images and signals (captured from cameras, sensors, etc.) are transformed by image processing and presented pictorially. Abstractions of these visual representations can then be transformed by computer vision to create symbolic representations in the form of symbols and structures. Finally, by using computer graphics the symbols or structures can be synthesized into visual representations. McCormick concluded that the common denominator of the computational sciences is visualization and indeed the research opportunities and engineering applications of visualization are sheer endless.

In this example we see, how an interactive visualization can enhance student understanding of complex data (Holzinger et al. 2009): Generally, learning in the area of physiology is difficult for medical students for several reasons. Medical students often do not have sufficient mathematical background for

(continued)

(continued)

understanding physiological models and the dynamics of complex mathematical rules related to these models. Moreover, learning is often without recourse to patients due to ethical restrictions (Simon 1972). Simulations are assumed to offer various benefits, especially to novice medical students learning theoretical concepts, processes, relationships, as well as invasive procedural skills, which is extremely important within decreasing clinical exposure. Consequently, students can acquire knowledge in a safe environment (Kneebone 2005) and apply the new knowledge in practice (Weller 2004).

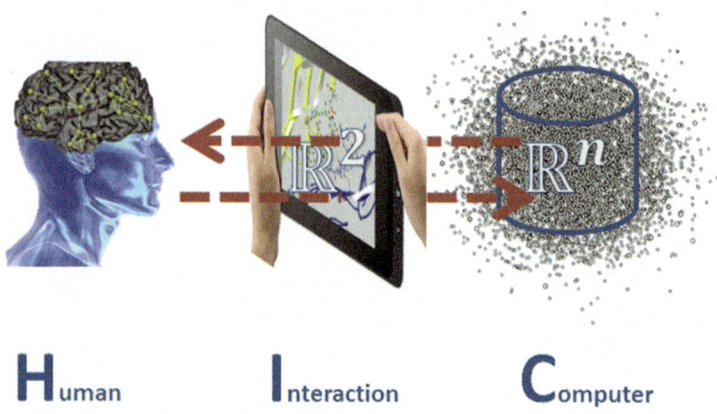

Fig. 4 See Slide 9-13

Slide 9-13: Visualization is a Typical HCI Topic

Ward et al. (2010) follow the notion that there is no distinction between data and information visualization—both provide representations of data; however, the datasets might be different. To interactively making the data understandable for the end user is a typical task from HCI (Holzinger 2013).

Slide 9-14: We Can Conclude that Visualization Is:

1. the common denominator of computational sciences
2. the transformation of the symbolic into the geometric

(continued)

(continued)

3. the support of human perception
4. facilitating knowledge discovery in data

(McCormick 1987) defined the science of visualization by a taxonomy diagram (see in the slide right), wherein he stated that images and signals (captured from cameras, sensors, etc.) are transformed by **image processing** and presented pictorially. Abstractions of these visual representations can then be transformed by **computer vision** to create symbolic representations in the form of symbols and structures. Finally, by using **computer graphics** the symbols or structures can be synthesized into visual representations. McCormick concluded that the common denominator of the computational sciences is visualization and indeed the research opportunities and engineering applications of visualization are sheer endless.

Slide 9-15: Visualization as a Knowledge-Eliciting Process

Mental models can be seen as abstractions of visualizations. The first step between mental models and external representations is internalization. The formation of a mental model happens ontogenetically after the appearance of the original external phenomenon. This insight has been formulated by Vygotsky, who argues that in each individual's development, every higher order cognitive function appears twice: first between people, as an inter-psychological process, then inside an individual, as an intra-psychological process. In information visualization it makes sense to understand the role of visualization in human cognitive activities from a developmental perspective. Mental models can serve as the cognitive basis of creativity and innovation. The construction and simulation of mental models can give rise to new concepts and designs including novel visual representations. Some visual representations such as the scatter plot and line graph have existed for centuries. In the slide we see that visualization is an interactive process (internalize, process, augment, create) across representational media (Liu and Stasko 2010).

Slide 9-16: Model of Perceptual Visual Processing

Ware (2004) summarized the human perceptual visual process in a three-stage model.

(continued)

(continued)

Stage 1: Parallel processing to extract low-level properties of the visual scene, i.e., billions of neurons in the eye and visual cortex work in parallel, extracting features from every part of the visual field simultaneously. The information at this stage is of transitory nature, briefly held in an iconic store (refer to memory models).

Stage 2: Pattern perception, the visual field is divided into regions and patterns (simple contours, regions of same color, patterns of motion, etc.). The information at this stage is slowly serially processed in a state of flux.

Stage 3: Sequential goal-directed processing, here objects are held in the visual working memory by demands of active attention.

5.5 The Case of John Snow

Slide 9-17: A Look Back into History

Visualization starts with paleolithic cave art, an early human symbolic information processing (Pike et al. 2012). A good example for the start of scientific visualization is the case of John Snow, a medical doctor in London, who identified a water pump as the source for the many deaths around Broad Street in the 1854 Cholera epidemic (see next slide).

Slide 9-18: Medical Visualization by John Snow (1854)

There are many famous visualization examples from the past, and one important for medicine is the work done by John Snow in 1854 during the Cholera epidemic in London. He identified the Broad Street pump as the source of cholera by plotting the location of cholera deaths on a map. After he removed the pump handle the epidemic ended (McLeod 2000).

Slide 9-19: Systematic Visual Analytics > Content Analytics

In this slide we see a nice remaking of John Snow maps, showing the significant higher death rates around the Broad Street pump (Koch and Denike 2009).

6 Visualization Methods

Slide 9-20: A Periodic Table of Visualization Methods

The periodic table of the chemical elements is a tabular form of displaying the chemical elements in categories, first devised in 1869 by the Russian chemist Dmitri Mendeleev; a similar approach to categorize visualization methods was done by Lengler and Eppler (2007) (accessible online: www.visual-literacy.org/periodic_table/periodic_table.html).

They have subdivided the application area dimension ("groups") into the following categories and distinguished them by background color—see next slide.

Slide 9-21: A Taxonomy of Visualization Methods

1. **Data Visualization** includes standard quantitative formats such as pie charts, area charts, or line graphs. They are visual representations of quantitative data in schematic form (either with or without axes), they are all-purpose, mainly used for getting an overview of data. They have mapped them to the Alkali Metals which most easily form bonds with non-metals, a correspondence might be the combination between data visualization (answering "how much" questions) and visual metaphors (answering how and why questions).
2. **Information Visualization**, such as semantic networks or tree-maps, is defined as the use of interactive visual representations of data to amplify cognition. This means that the data is transformed into an image; it is mapped to screen space. The image can be changed by users as they proceed working with it.
3. **Concept Visualization**, such as a concept map or a Gantt chart; these are methods to elaborate (mostly) qualitative concepts, ideas, plans, and analyses through the help of rule-guided mapping procedures. In concept visualization knowledge is usually presented in a 2D graphical display where concepts (usually represented within boxes or circles), connected by directed arcs encoding brief relationships (linking phrases) between pairs of concepts. These relationships usually consist of verbs, forming propositions or phrases for each pair of concepts.
4. **Metaphor Visualization** such as metro maps or story template can be used as effective and simple templates to convey complex insights. Visual Metaphors fulfill a dual function, first they position information

(continued)

(continued)

   graphically to organize and structure it. Second they convey an insight about the represented information through the key characteristics of the metaphor that is employed.
5. **Strategy Visualization**, such as a Strategy Canvas or technology roadmap, is defined "as the systematic use of complementary visual representations to improve the analysis, development, formulation, communication, and implementation of strategies in organizations." This is the most specific of all groups, as it has achieved great relevance in management.
6. **Compound Visualization** consists of several of the aforementioned formats. They can be complex knowledge maps that contain diagrammatic and metaphoric elements, conceptual cartoons with quantitative charts, or wall sized infomurals. This label thus typically designates the complementary use of different graphic representation formats in one single schema or frame. According to Tufte they result from two (or more) spatially distinct different data representations, each of which can operate independently, but can be used together to correlate information in one representation with that in another.

6.1 Overview

Slide 9-22: Visualizations for Multivariate Data Overview 1/2

For a first overview let us summarize some important visualization methods: **Scatterplots** (SP) are the oldest, point-based techniques, and projects (maps) data from an n-dimensional space into an arbitrary k-dimensional display space (usually it will be the 2D space ;-)). To verify cluster separation in high-dimensional data, analysts often reduce the data with a dimension reduction technique, and then visualize it with 2D Scatterplots, interactive 3D Scatterplots, or Scatterplot Matrices (SPLOMs) (Sedlmair et al. 2013).

Parallel Coordinates (PCP) is best suited for the study of high-dimensional geometry, where each data point is plotted as a polyline.

Radial Coordinate Visualization (RadViz) is a "force-driven" point layout technique, based on Hooke's law for equilibrium.

A visual survey of visualization techniques for time-oriented data can be found here: http://survey.timeviz.net.

6 Visualization Methods

Slide 9-23: Visualizations for Multivariate Data Overview 2/2

Radar Chart aka star plot, spider web, polar graph, polygon plot is a radial axis technique.

Heatmap is a tabular display technique using color instead of figures for the entities.

Glyph is a visual representation of the entity, where its attributes are controlled by data attributes.

Chernoff face is a face glyph which displays multivariate data in the shape of a human face.

6.2 Parallel Coordinates

Slide 9-24: Parallel Coordinates: Multidimensional Visualization

The so-called ∥-coords have been developed in the context of modern visualization by Alfred Inselberg in the 1950s and are excellent for visualizing high-dimensional and multivariate data in the form of N parallel lines, where a data point in the N-dimensional space is represented as a polyline with vertices on the parallel axes. We follow the paper of Inselberg (2005) and the book of Inselberg (2009), which contains an excellent compact disk providing a lot of interactive material.

On the plane with xy-Cartesian coordinates a vertical line, labeled $\overline{X}_i$ is placed at each $x = i - 1$ for $i = 1, 2, \ldots N$. These are the axes of the parallel coordinate system for $\mathbb{R}^N$. A point $C = (c_1, c_2, \ldots, c_N) \in \mathbb{R}^N$ is now mapped into the polygonal line $\overline{C}$ whose N-vertices with xy-coords $(i - 1, c_i)$ are on the parallel axes. In $\overline{C}$ the full lines and not only the segments between the axes are included.

∥-coords are constructed by placing axes in parallel with respect to the embedding 2D Cartesian coordinate system in the plane (the parallel-coordinates domain). While the orientation of axes can be chosen freely, the most common use is either horizontal (parallel to the x-axis) or vertical (parallel to the y-axis), see next slide.

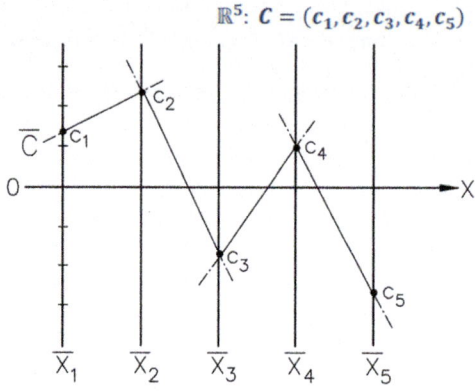

Fig. 5 See Slide 9-25

Slide 9-25: Polygonal Line $\overline{C}$ Is Representing the Single Point in $\mathbb{R}^5$:
$C = (c_1, c_2, c_3, c_4, c_5)$

Note that each point in the n-dimensional space is represented as a polyline with vertices on the parallel axes; the position of the vertex on the i-th axis corresponds to the i-th coordinate of the point (Inselberg 2005).

Slide 9-26: Heavier Polygonal Lines Represent End Points

A polygonal line $\overline{P}$ on the $N - 1$ points represents a point $P = (p_1, \ldots p_{i-1}, p_i \ldots p_N) \in \ell$ since the pair of values $\ldots p_{i-1}, p_i$ marked on the $\overline{X}_{i-1}$ and $\overline{X}_i$ axes. We can see several polygonal lines, intersecting at $\ell_{(i-1),i}$ representing data points on a line $\ell \subset \mathbb{R}^{10}$. If we plot this result we get the line interval in $\mathbb{R}^{10}$ as can be seen in Slide 9-27.

Note: When thinking of visualizing data with ‖-coords, especially in the biomedical domain, one is immediately confronted with some challenges (Heinrich and Weiskopf 2012):

- Overplotting occurs in parallel coordinates if lines potentially occlude patterns in the data.
- The order of axes implicitly defines which patterns emerge between adjacent axes.
- The line-tracing problem occurs if two or more lines intersect an axis at the same position.
- Nominal and ordinal data such as sets and clusters have to be mapped to a metric scale before it can be visualized in parallel coordinates.

(continued)

(continued)

- Time series are special in that time points, if interpreted as dimensions, have a fixed order.
- Uncertain data is another challenge for visualization, and there are approaches for the visualization of uncertainty in parallel coordinates.

The indexing is essential and is important for the visualization of proximity properties such as the minimum distance between a pair of lines, see next slide.

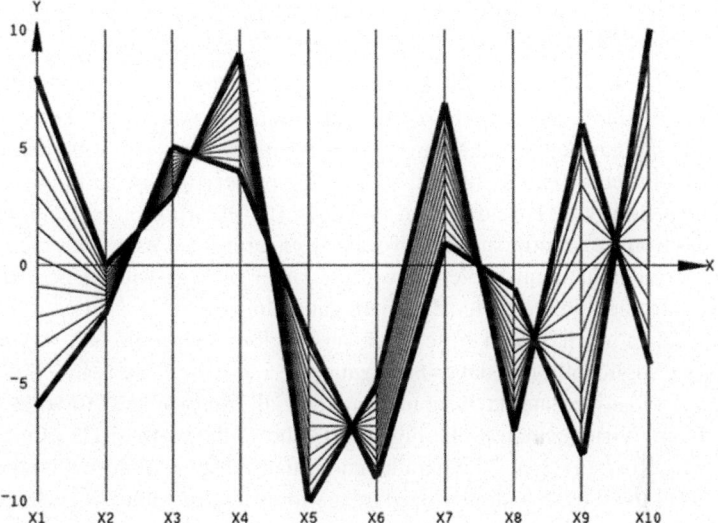

Fig. 6 See Slide 9-27

Slide 9-27: Line Interval in $\mathbb{R}^{10}$

Again: The indexing is essential and is important for the visualization of proximity properties such as the minimum distance between a pair of lines.

Slide 9-28: Example: Par Coords in a Vis Software in R

This is an example of an implementation in R by Graham Williams, see: http://datamining.togaware.com/survivor/Scatterplot.html.

Slide 9-29: Par Coords → Knowledge Discovery in Big Data

This shows an interesting example of (Mane and Börner 2007): The multiple coordinated views help a medical practitioner to gain a good insight about the medical data variations among selected patients. The matrix view (A) helps to quickly identify similar patterns and worst case conditions. The parallel coordinates view (B) helps to quickly identify and compare trends shared by groups of patients. So the matrix view and parallel coordinate view complement each other to help the medical practitioners gain an understanding of the data.

Slide 9-30: Ensuring Data Protection with k-Anonymization

An important issue of all real-world datasets is that many of their attributes can identify individuals, or the data are proprietary and valuable. The field of data mining has developed a variety of ways for dealing with such data, and has established an entire subfield for privacy-preserving data mining.

The **k-anonymity model** is an approach (NP-hard) to protect individual records from re-identification. It works by ensuring that each data record in the table is indistinguishable from $k - 1$ other records with respect to the quasi-identifiers in the table (El Emam and Dankar 2008).

Visualization has seen little work on handling sensitive data. With the growing applicability of data visualization in real-world scenarios, the handling of sensitive data has become a nontrivial issue we need to address in developing visualization tools. Figure 203 shows the work of Dasgupta and Kosara (2011): (a) shows the parallel coordinates display of the original data, and (b) to (d) show the anonymized image for different values of k. Since we see aggregated values instead of single ones, represented by polygons instead of clusters, it is not possible to point to particular values on the axes. At the same time, much of the overall structure is still visible in the visualization, even though individual records cannot be identified anymore.

Slide 9-31: Decision Support with Par Coords in Diagnostics

Decision Support with Parallel Coordinates in Diagnostics. Visual inspection of the tongue is an important diagnostic method of Traditional Chinese Medicine (TCM). Observing the abnormal changes in the tongue and the tongue coating can support in diagnosing diseases. Pham and Cai (2004) demonstrate nicely how proper visualization can contribute to medical decision making.

6.3 Radial Coordinate Visualization

Slide 9-32: RadViz: Idea Based on Hooke's Law

RadViz is a method for mapping a set of n-dimensional points into a plane and to identify *relations* among data. Its main advantage is that it needs no projections and provides a global view on the multidimensional data. This method is following Hooke's law from classical mechanics (spring laws).

A nice tool can be found at: http://orange.biolab.si/.

This is an open source data visualization and analysis tool for novice and experts for data mining through visual programming and Python scripting including components for machine learning and add-ons for bioinformatics and text mining (Demšar et al. 2013).

Slide 9-33: RadViz Principle

Each RadViz mapping of points from n-dimensional space into a plane is uniquely defined by position of the corresponding n anchors (points S_j), which are placed in a single plane. Let us consider a point $y_i = (y_1, y_2, \ldots y_n)$ from the n-dimensional space. This point is now mapped into a single point u in the plane of anchors: for each anchor j the stiffness of its spring is set to y_j and the Hooke's law is used to find the point u, where all the spring forces reach equilibrium (means they sum to 0). The position of $u = [u_1, u_2]$ is now derived by:

$$\sum_{j=1}^{n}(\vec{S}_j - \vec{u})y_i = 0 \qquad \sum_{j=1}^{n}\vec{S}_j y_j = \vec{u}\sum_{j=1}^{n} y_j$$

$$\vec{u} = \frac{\sum_{j=1}^{n}\vec{S}_j y_j}{\sum_{j=1}^{n} y_j} \qquad u_1 = \frac{\sum_{j=1}^{n} y_j \cos(\alpha_j)}{\sum_{j=1}^{n} y_j} \qquad u_2 = \frac{\sum_{j=1}^{n} y_j \sin(\alpha_j)}{\sum_{j=1}^{n} y_j}$$

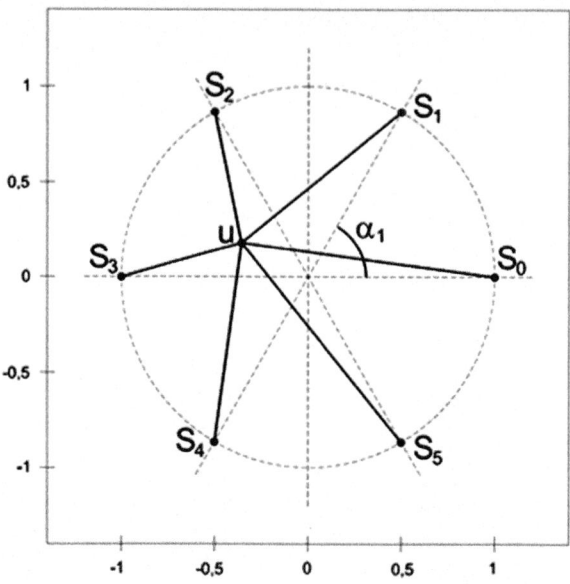

Fig. 7 See Slide 9-34

Slide 9-34: RadViz Mapping Principle and Algorithm

The algorithm using the RadViz process (Novakova and Stepankova 2009) is:

1. Normalize the data to the interval $\langle 0,1 \rangle$

$$\overline{x}_{ij} = \frac{x_{ij} - \min_j}{\max_j - \min_j}$$

2. Now place the dimensional anchors
3. Now calculate the point to place each record and to draw it:

$$y_i = \sum_{j=1}^{n} \overline{x}_{ij} \qquad \vec{u}_i = \frac{\sum_{j=1}^{n} \vec{S}_j \overline{x}_{ij}}{y_i}$$

Slide 9-35: RadViz for Showing the Existence of Clusters

Here we see some RadViz Examples from Novakova and Stepankova (2009):
A = ata projection in 3D space; B = the example from Slide 9-34; C = RadVizS clustering; D = same from another angle; E = Mirror projection; F = after normalization.

6.4 Star Plots

Slide 9-36: Star Plots/Radar Chart/Spider-Web/Polygon Plot

StarPlots aka radar chart, spider web diagram, polygon plot, polar chart, or Kiviat diagram, are graphical methods for displaying multivariate data in the form of a 2D chart of three or more quantitative variables represented on axes starting from the same point.

Despite of their usefulness, such diagrams have not been widely used in the biomedical domain. One example of their use include distinguishing metabolic profiles of different cancer classes with star plots by the work of (Vion-Dury et al. 1993).

In this slide we see an example of gender differences in death rate by treatment overlaid (Saary 2008).

Slide 9-37: Star Plot Production

Each multivariate observation can be seen as a data point in an n-dimensional vector space:

- Arrange N axes on a circle in $\mathbb{R}^N$
- $3 \leq N \leq N_{max}$
- Map coordinate vectors $P \in \mathbb{R}^N$ from $\mathbb{R}^N \to \mathbb{R}^2$
- $P = \{p_1, p_2, \ldots, p_N\} \in \mathbb{R}^N$ where each p_i represents a different attribute with a different physical unit
- Each axis represents one attribute of data
- Each data record, or data point P is visualized by a line along the data points

(continued)

(continued)

- A line is perceived better than just points on the axes

There are commercial software tools which can help, e.g., the SAS statistical software package (http://www.sas.com/) and Microsoft Excel (http://office.microsoft.com/en-us/excel-help/present-your-data-in-a-radar-chart-HA010218672.aspx).

In SAS you can use the GRADAR procedure; the data can consist of one or more group variables and one or more outcome variables for which there is a count or frequency for each level of the group variable. The vertices of the radar plot are determined by the levels of a single variable that is given in the CHART statement. The spokes in the chart are positioned much like a clock starting at the 12-o'clock position and moving in a clockwise direction.

In Excel, the radar plot is generated by using the Insert function that allows the option to insert one of a variety of charts. Radar plots are among a number of other less commonly used graphing styles available on this menu including surface, doughnut, bubble, and stock plots. In this example, Excel does not prepare the chart by performing calculations on the raw data; hence, the sums (or means) of the raw-data groups must be represented in a separate new table (Saary 2008).

```
angle_sector = 2 * π / N
for each a_i from axes[]
{
angle_i = i * angle_sector
x_i = mid.x + r * cos(angle_i)
y_i = mid.y + r * sin(angle_i)
DrawLine(midpoint.x, midpoint.y, x_i, y_i)




max_i = a_i.upperBound()

scaled_val_i = a_i.value() * r / max_i
x_val_i = mid.x + scaled_val_i * cos(angle_i)

y_val_i = mid.y + scaled_val_i * sin(angle_i)
DrawLine(x_val_i, y_val_i, x_val_i-1, y_val_i-1)
}
```

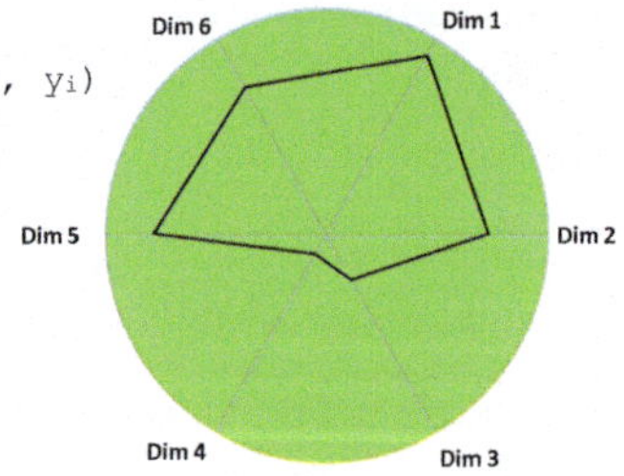

Fig. 8 See Slide 9-38

Slide 9-38: Algorithm for Drawing the Axes and the Lines

Here we see an simple algorithm for drawing the axes and lines of a star plot diagram.

7 Visual Analytics

Slide 9-39: Visual Analytics is Intelligent HCI

Visual analytics seeks to support an "intelligent" interaction discourse with the end user through images, to stimulate curiosity and a penchant to decipher the unknown. In this slide we see a representative visual analytics process (Mueller et al. 2011)—which is also a nice example for HCI: The computer supports the user in analytical reasoning, constructing a formal model of the given data and enabling insight. Validation and refinement of this computational model of insight can occur only in the human domain expert's mind. The human user must guide the computer in the formalization (learning process) of sophisticated models that capture what the human desires, cf. with (Holzinger 2012).

Slide 9-40: Design of Interactive Information Visualization

A recent example is from Ren et al. (2010), who developed DaisyViz, which is based on the idea of model-based interface development, which uses a declarative model of user interfaces to enhance the development process. Basically, a user interface model abstracts the features of a user interface and represents all the relevant aspects of the user interface in a formal language. The user interface model, the core of development process, is then parsed according to knowledge bases to generate applications. From the perspective of end users, their only design concern is to construct an interface model. The slide shows the user interface model for Infovis (UIMI), developed by Ren et al. (2010). In this framework, one can construct a model by simply answering several questions:

1. What facets of the target information should be visualized?
2. What data source should each facet be linked to and what relationships these facets have?
    The answers to these two questions result in the data model.
3. What layout algorithm should be used to visualize each facet?

(continued)

(continued)

The answer to this question results in the visualization model.
4. What interactive techniques should be used for each facet and for which infovis tasks?

The answer to this question results in the control model.

Slide 9-41: Overview First: Then Zoom and Filter on Demand

There are many visual design guidelines but the basic principle might be summarized as the Visual Information Seeking Mantra by Shneiderman (1996): Overview first, zoom and filter, then details-on-demand, which has been further elaborated by Keim et al. (2008):

1. Overview: Gain an overview about the entire dataset ("Know your data!").
2. Zoom: Zoom in on items of interest; (Remember: The question "What is interesting?" is a very hard one.
3. Filter: Filter out uninteresting items—get rid of distractors—eliminate irrelevant information.
4. Details-on-demand: Select an item or group and provide details when needed.
5. Relate: View relationships among items.
6. History: Keep a history of actions to support undo, replay, and progressive refinement.
7. Extract: Allow extraction of sub-collections and of the query parameters.

Slide 9-42: Letting the User Interactively Manipulate the Data

Ward et al. (2010) summarize how a developer can let the user interactively manipulate the data.

1. Focus Selection = via direct manipulation and selection tools, e.g., multi-touch (in data space a n-dim location might be indicated).
2. Extent Selection = specifying extents for an interaction, e.g., via a vector of values (a range for each data dimension or a set of constraints.
3. Interaction type selection = for example a pair of menus: one to select the space, and the other to specify the general class of the interaction.
4. Interaction level selection = for example the magnitude of scaling that will occur at the focal point (via a slider, along with a reset button).

Slide 9-43: Rapid Graphical Summary of Patient Status

Powsner and Tufte (1994) presented a graphical summary of the patient status, which maps findings and treatments over time. The example shows that no tests of serum glucose were done during the 12 months before admission, although many were made more than 1 year earlier. The nonlinear time-scale compresses years of data into a context for assessing recent trends.

Slide 9-44: Example Project LifeLines

LifeLines was a very early project on a general visualization environment for personal patient histories. A Java user interface presents a one-screen overview of a computerized patient record using timelines. Problems, diagnoses, test results or medications can be represented as dots or horizontal lines. Zooming provides more details; line color and thickness illustrate relationships or significance. The visual display acts as a giant menu, giving direct access to the data (Plaisant et al. 1996).

Slide 9-45: Temporal Analysis Tasks

The analysis of time-oriented data is very important in the biomedical domain. In recent years, a variety of techniques for visualizing such data are available, e.g.,

1. Classification = given a set of classes: the aim is to determine which class the dataset belongs to; a classification is often necessary as preprocessing.
2. Clustering = grouping data into clusters based on similarity; the similarity measure is the key aspect of the clustering process.
3. Search/Retrieval = look for a priori specified queries in large datasets (query-by-example), can be exact matched or approximate matched (similarity measures are needed that define the degree of exactness).
4. Pattern discovery = automatically discovering relevant patterns in the data, e.g., local structures in the data or combinations thereof.
5. Prediction = foresee likely future behavior of data—to infer from the data collected in the past and present how the data will evolve in the future (autoregressive models, rule-based models, etc.)

For an excellent overview consult the book by Aigner et al. (2011).

8 Future Outlook

Slide 9-46: Future Outlook

Visualization is most important, because this is what the end user "experiences". Very important in the future are visualizations of time (e.g., entropy) and space (e.g., topology).

As we know from the famous Forrester Reports, more than 80 % of all data contain "unstructured" elements, hence content analytics techniques along with advanced interactive visualizations are essential.

Amazingly, there are only very few of the sophisticated visualization methods integrated in "real-world" (e.g., hospital information system) and we are lacking of visualization methods for mobile computers. Finally, two major questions will be important in the future:

1. How can we measure the benefits of visual analysis as compared to traditional methods?
2. Can (and how can) we develop powerful visual analytics tools for the nonexpert end user?

Future research questions include finding proper visual encodings, understanding multidimensional spaces, facilitating sensitivity, understanding of uncertainty, and facilitating trade-offs (Saad et al. 2010; Tory and Möller 2004).

9 Exam Questions

9.1 Yes/No Decision Questions

Please check the following sentences and decide whether the sentence is true = YES; or false = NO; for each correct answer you will be awarded 2 credit points.

01	By answering the question "What interactive techniques should be used for each facet and for which infovis tasks?" we get as a result the corresponding data model.	☐ Yes ☐ No	2 total
02	Scatterplot is a point-based technique, where each data point is plotted as a polyline.	☐ Yes ☐ No	2 total
03	Metaphor visualization include semantic networks, tree-maps and radar-charts for example.	☐ Yes ☐ No	2 total
04	Syntactic aspects of data includes meaning, proposition and truth to answer the question: "can it be understood?".	☐ Yes ☐ No	2 total
05	Computer science lacks a reliable concept of the human mind, whereas the psychological sciences lack solid concepts for algorithms and data structures.	☐ Yes ☐ No	2 total
06	The Church–Turing thesis states that Turing machines capture the informal notion of effective methods in logic and mathematics, and provide a precise definition of an algorithm	☐ Yes ☐ No	2 total
07	Past research in information visualization proofed that "A picture is always worth a 1000-words" is true.	☐ Yes ☐ No	2 total
08	Information visualization is the study of visual representations of abstract data to reinforce human cognition; hence it is very important for decision making.	☐ Yes ☐ No	2 total
09	Mental models can be seen as abstractions of visualizations. The first step between mental models and external representations is internalization.	☐ Yes ☐ No	2 total
10	RadViz is for mapping a set of n-dimensional points into a plane and to identify relations among data. Its main advantage is that it needs no projections and provides a global view on the data.	☐ Yes ☐ No	2 total

Sum of Question Block A (max. 20 points)

9.2 Multiple Choice Questions (MCQ)

The following questions are composed of two parts: the stem, which identifies the question or problem and a set of alternatives which can contain 0, 1, 2, 3, or 4 correct answers, along with a number of distractors that might be plausible—but are incorrect. Please **select the correct answers** by ticking ☒—and do not forget that it can be none. Each question will be awarded 4 points *only if everything is correct*.

01	The original visual information seeking mantra by Shneiderman (1996) is … ☐ a) … Provide details first, show the big picture last. ☐ b) … Overview first, zoom and filter, then details-on-demand. ☐ c) … Overview first, details-on demand, keep history of actions. ☐ d) … Overview first, zoom and filter, extract sub-collections.
02	Future research topics of information visualization include … ☐ a) … time. ☐ b) … space. ☐ c) … uncertainty. ☐ d) … trade-offs.
03	Semiotics is the study of …. ☐ a) … signs and their interpretation by the end users. ☐ b) … relations among signs in formal structures. ☐ c) … relations between signs and their meaning. ☐ d) … signs and their physical represenations.
04	The "Quality Era" of biomedical informatics is mainly characterized by ☐ a) … focus on data processing and storage. ☐ b) … health care networks, telemedicine and CPOE-Systems. ☐ c) … pervasive and ubiquitous computing technologies. ☐ d) … patient empowerment and individual molecular medicine.
05	Visualization can be defined by … ☒ a) … transforming the symbolic into the geometric. ☐ b) … forming mental models and foster unexpected insights. ☐ c) … the study of large collections of complex data sets. ☐ d) … analytical reasoning facilitated by visual interfaces.
06	The following method belongs to classical data visualization … ☐ … pie chart. ☐ … PERT diagram. ☐ … metro map. ☐ … line graph.
07	Glyphs are … ☐ a) … representations, where its attributes are controlled by data attributes. ☐ b) … another name for heatmaps. ☐ c) … a typical radial axis technique. ☐ d) … used for displaying multivariate data.
08	A Kiviat diagram is another name for … ☐ a) … star plot. ☐ b) … spider web. ☐ c) … polygon plot. ☐ d) … RadViz.

09	Extent Selection can be provided to the end user by ... ❐ a) ... direct manipulation and selection tools, e.g. multi-touch. ❐ b) ... specifying extents for an interaction, e.g. via a vector of values. ❐ c) ... a pair of menus: one to select the space, and the other the interaction. ❐ d) ... the magnitude of scaling that will occur at the focal point.
10	Overplotting occurs in parallel coordinates ... ❐ a) ... if two or more lines intersect an axis at the same position. ❐ b) ... if the data is noisy. ❐ c) ... if all data have a fixed order. ❐ d) ... if lines potentially occlude patterns in the data..

Sum of Question Block B (max. 40 points)	

9.3 Free Recall Block

Please follow the instructions below. At each question you will be assigned the credit points indicated if your option is correct (partial points may be given).

01	Visual analytics seeks to support an "intelligent" interaction discourse with the end user through images, to stimulate curiosity and a penchant to decipher the unknown. In this slide we see a representative visual analytics process (Mueller et al., 2011) – which is also a nice example for human–computer interaction – please complete the missing items:	1 each 5 total
02	Please draw the principle of a star plot diagram for 5 positive values	1-28 1 each 6 total

03	Please draw the principle of a Parallel Coordinate plot for $\mathbb{R}^4$: $C = (c_1, c_2, c_3, c_4)$, where the scalars c1 = 1,4; c2=2,2; c3=-1,8, c4 = 1	1/42 1 each 7 total
04	Please sketch examples for visualization methods for multivariate data. 1) scatterplot, 2) RadViz, 3) Glyph, 4) Chernoff-face 1)　　　　　　　　　　　2) 3)　　　　　　　　　　　4)	1 each 4 total

05	Temporal analysis and temporal data mining are very important for medical informatics and especially concerned with extracting useful information from time-oriented medical data. Please describe the following temporal analysis tasks:	1 each 5 total

Sum of Question Block C (max. 40 points)

10 Answers

10.1 Answers to the Yes/No Questions

Please check the following sentences and decide whether the sentence is true = YES; or false = NO; for each correct answer you will be awarded 2 credit points.

01	By answering the question "What interactive techniques should be used for each facet and for which infovis tasks?" we get as a result the corresponding data model.	☐ Yes ☒ No	2 total
02	Scatterplot is a point-based technique, where each data point is plotted as a polyline.	☐ Yes ☒ No	2 total
03	Metaphor Visualizations include semantic networks, tree-maps and radar-charts for example.	☐ Yes ☒ No	2 total
04	Syntactic aspects of data includes meaning, proposition and truth to answer the question: "can it be understood?".	☐ Yes ☒ No	2 total
05	Computer science lacks a reliable concept of the human mind, whereas the psychological sciences lack solid concepts for algorithms and data structures.	☒ Yes ☐ No	2 total
06	The Church–Turing thesis states that Turing machines capture the informal notion of effective methods in logic and mathematics, and provide a precise definition of an algorithm	☒ Yes ☐ No	2 total
07	Past research in information visualization proofed that "A picture is always worth a 1000-words" is true.	☐ Yes ☒ No	2 total
08	Information visualization is the study of visual representations of abstract data to reinforce human cognition; hence it is very important for decision making.	☒ Yes ☐ No	2 total
09	Mental Models can be seen as abstractions of visualizations. The first step between mental models and external representations is internalization.	☒ Yes ☐ No	2 total
10	RadViz is for mapping a set of n-dimensional points into a plane and to identify relations among data. Its main advantage is that it needs no projections and provides a global view on the data.	☒ Yes ☐ No	2 total

Sum of Question Block A (max. 20 points)

10.2 Answers to the Multiple Choice Questions (MCQ)

01	The original visual information seeking mantra by Shneiderman (1996) is ... ☐ a) ... Provide details first, show the big picture last. ☒ b) ... Overview first, zoom and filter, then details-on-demand. ☐ c) ... Overview first, details-on demand, keep history of actions. ☐ d) ... Overview first, zoom and filter, extract sub-collections.	4 total
02	Future research topics of information visualization include ... ☒ a) ... time. ☒ b) ... space. ☒ c) ... uncertainty. ☒ d) ... trade-offs.	4 total
03	Semiotics is the study of .... ☒ a) ... signs and their interpretation by the end users. ☐ b) ... relations among signs in formal structures. ☐ c) ... relations between signs and their meaning. ☐ d) ... signs and their physical represenations.	4 total
04	The "Quality Era" of biomedical informatics is mainly characterized by ☐ a) ... focus on data processing and storage. ☐ b) ... health care networks, telemedicine and CPOE-Systems. ☐ c) ... pervasive and ubiquitous computing technologies. ☒ d) ... patient empowerment and individual molecular medicine.	4 total
05	Visualization can be defined by ... ☒ a) ... transforming the symbolic into the geometric. ☒ b) ... forming mental models and foster unexpected insights. ☐ c) ... the study of large collections of complex data sets. ☒ d) ... analytical reasoning facilitated by visual interfaces.	4 total
06	The following method belongs to classical data visualization ... ☒ ... pie chart. ☐ ... PERT diagram. ☐ ... metro map. ☒ ... line graph.	4 total
07	Glyphs are ... ☒ a) ... representations, where its attributes are controlled by data attributes. ☐ b) ... another name for heatmaps. ☐ c) ... a typical radial axis technique. ☒ d) ... used for displaying multivariate data.	4 total
08	A Kiviat diagram is another name for ... ☒ a) ... star plot. ☒ b) ... spider web. ☒ c) ... polygon plot. ☐ d) ... RadViz.	4 total
09	Extent Selection can be provided to the end user by ... ☐ a) ... direct manipulation and selection tools, e.g. multi-touch. ☒ b) ... specifying extents for an interaction, e.g. via a vector of values. ☐ c) ... a pair of menus: one to select the space, and the other the interaction. ☐ d) ... the magnitude of scaling that will occur at the focal point.	4 total

10	Overplotting occurs in parallel coordinates ... ☐ a) ... if two or more lines intersect an axis at the same position. ☐ b) ... if the data is noisy. ☐ c) ... if all data have a fixed order. ☒ d) ... if lines potentially occlude patterns in the data..	4 total

Sum of Question Block B (max. 40 points)

10.3 Answers to the Free Recall Questions

01	Visual analytics seeks to support an "intelligent" interaction discourse with the end user through images, to stimulate curiosity and a penchant to decipher the unknown. In this slide we see a representative visual analytics process (Mueller et al., 2011) – which is also a nice example for human–computer interaction – please complete the missing items: 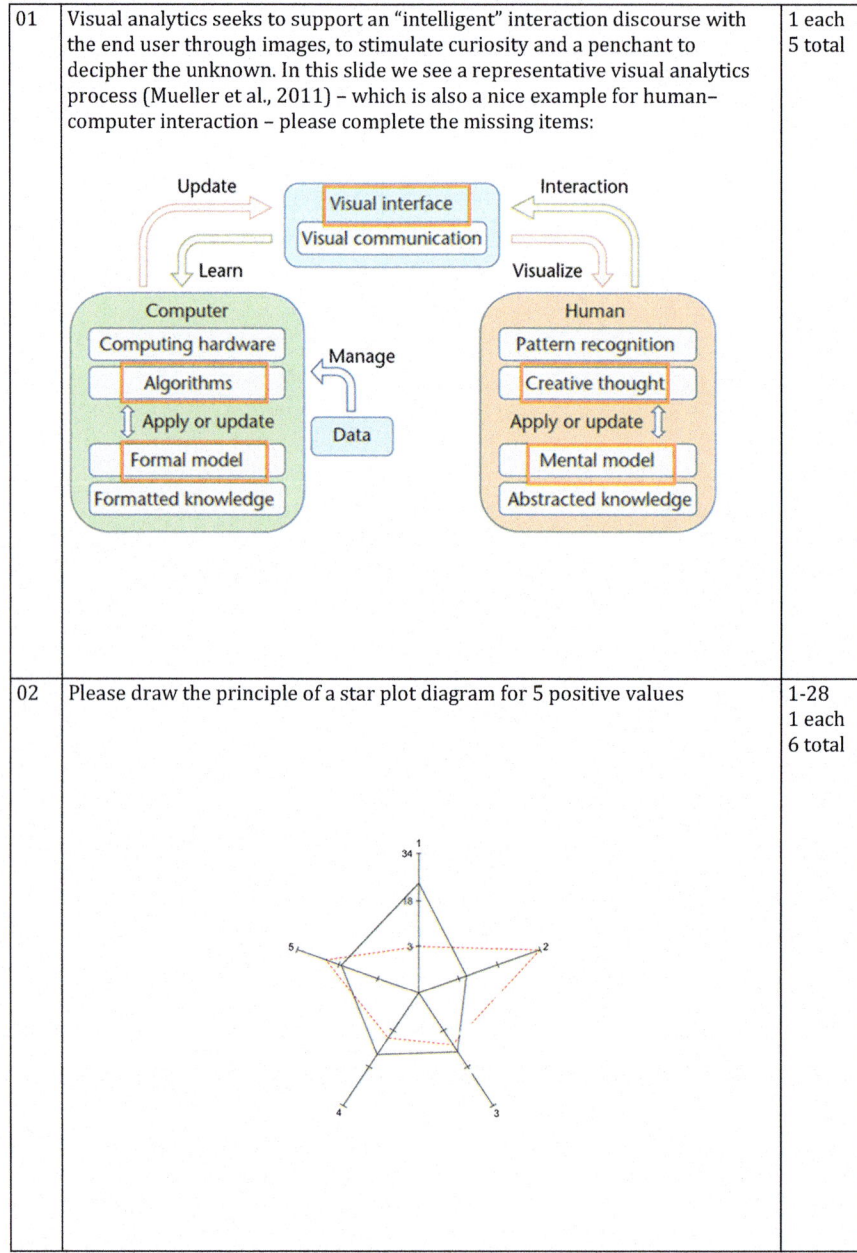	1 each 5 total
02	Please draw the principle of a star plot diagram for 5 positive values	1-28 1 each 6 total

10 Answers

| 03 | Please draw the principle of a Parallel Coordinate plot for $\mathbb{R}^4$: $C = (c_1, c_2, c_3, c_4)$, where the scalars c1 = 1,4; c2=2,2; c3=-1,8, c4 = 1 | 1/42 1 each 7 total |

| 04 | Please sketch examples for visualization methods for multivariate data. 1) scatterplot, 2) RadViz, 3) Glyph, 4) Chernoff-face | 1 each 4 total |

05	Temporal analysis and temporal data mining are very important for medical informatics and especially concerned with extracting useful information from time-oriented medical data. Please describe the following temporal analysis tasks:		1 each 5 total
		Classification	
		Clustering	
		Search/Retrieval	
		Pattern discovery	
		Prediction	

Sum of Question Block C (max. 40 points)	

References

(2007) Visualization basics. Software visualization. Springer, Berlin. pp. 15–33

Aigner W, Bertone A, Miksch S (2007) Tutorial: introduction to visual analytics. In: Holzinger A (ed) HCI and usability for medicine and health care: third symposium of the workgroup human-computer interaction and usability engineering of the Austrian Computer Society, USAB 2007. Lecture notes in computer science (LNCS 4799). Springer, Graz, pp 453–456

Aigner W, Miksch S, Schumann H, Tominski C (2011) Visualization of time-oriented data. Human–computer interaction series. Springer, London

Andersen PB (2001) What Semiotics can and cannot do for HCI. Knowledge-Based Syst 14(8):419–424

Barbieri M (2008) What is biosemiotics? Biosemiotics 1(1):1–3

Barbieri M (2010) On the origin of language. Biosemiotics 3(2):201–223

Burton-Jones A, Storey VC, Sugumaran V, Ahluwalia P (2005) A semiotic metrics suite for assessing the quality of ontologies. Data Knowledge Eng 55(1):84–102

Carlson A, Betteridge J, Kisiel B, Settles B, Hruschka ER Jr, Mitchell TM (2010) Toward an architecture for never-ending language learning. AAAI, Bellevue

Childs H, Geveci B, Schroeder W, Meredith J, Moreland K, Sewell C, Kuhlen T, Bethel EW (2013) Research challenges for visualization software. Computer 46(5):34–42

Dasgupta A, Kosara R (2011) Privacy-preserving data visualization using parallel coordinates. Visualization and data analysis 2011. SPIE, San Francisco

Demšar J, Curk T, Erjavec A, Gorup Č, Hočevar T, Milutinović M, Možina M, Polajnar M, Toplak M, Starič A (2013) Orange: data mining toolbox in Python. J Mach Learn Res 14:2349–2353

Donoho DL (2000) High-dimensional data analysis: the curses and blessings of dimensionality. AMS Math Challenges Lecture. pp. 1–32

El Emam K, Dankar FK (2008) Protecting privacy using k-anonymity. J Am Med Inform Assoc 15 (5):627–637
Grinstein G, Inselberg A, Laskowski S (1998) Key problems and thorny issues in multidimensional visualization. Ninth IEEE visualization 1998 (VIS'98). IEEE Computer Society. pp. 505–506
Heinrich J, Weiskopf D (2012) State of the art of parallel coordinates. Eurographics 2013-state of the art reports. The Eurographics Association. pp. 95–116
Holzinger A (2002a) Multimedia basics, volume 1: technology. Technological basics of multimedia information systems. Laxmi-Publications, New Delhi, Translation of the original German Second Edition (with assistance from Gig Searle)
Holzinger A (2002b) Multimedia basics, volume 2: learning. Cognitive basics of multimedia information systems. Laxmi-Publications, New Delhi
Holzinger A (2012) On knowledge discovery and interactive intelligent visualization of biomedical data: challenges in human–computer interaction & biomedical informatics. In: Helfert M, Fancalanci C, Filipe J (eds) DATA—international conference on data technologies and applications. INSTICC, Rome, pp 5–16
Holzinger A (2013) Human–computer interaction & knowledge discovery (HCI-KDD): what is the benefit of bringing those two fields to work together? In: Alfredo Cuzzocrea CK, Dimitris E Simos, Edgar Weippl, Lida Xu (Eds.) Multidisciplinary research and practice for information systems, springer lecture notes in computer science LNCS 8127. Springer, Heidelberg. pp. 319–328
Holzinger A, Geierhofer R, Modritscher F, Tatzl R (2008) Semantic information in medical information systems: utilization of text mining techniques to analyze medical diagnoses. J Univ Comp Sci 14(22):3781–3795
Holzinger A, Kickmeier-Rust MD, Wassertheurer S, Hessinger M (2009) Learning performance with interactive simulations in medical education: lessons learned from results of learning complex physiological models with the HAEMOdynamics SIMulator. Comput Educ 52 (2):292–301
Holzinger A, Searle G, Auinger A, Ziefle M (2011) Informatics as semiotics engineering: lessons learned from design, development and evaluation of ambient assisted living applications for elderly people. Universal access in human–computer interaction. Context diversity, Lecture notes in computer science (LNCS 6767). Springer, Berlin, pp 183–192
Holzinger A, Stocker C, Ofner B, Prohaska G, Brabenetz A, Hofmann-Wellenhof R (2013) Combining HCI, Natural language processing, and knowledge discovery—potential of IBM content analytics as an assistive technology in the biomedical domain Springer Lecture Notes in Computer Science LNCS 7947. Springer, Heidelberg, pp 13–24
Inselberg A (2005) Visualization of concept formation and learning. Kybernetes Int J Syst Cybern 34(1/2):151–166
Inselberg A (2009) Parallel coordinates: visual multidimensional geometry and its applications (foreword by Ben Shneiderman). Springer, Heidelberg
Keim D, Mansmann F, Schneidewind J, Thomas J, Ziegler H (2008) Visual analytics: scope and challenges. In: Simoff S, Böhlen M, Mazeika A (eds) Visual data mining, lecture notes in computer science 4404. Springer, Berlin, pp 76–90
Kneebone R (2005) Evaluating clinical simulations for learning procedural skills: a theory-based approach. Acad Med J Assoc Am Med Coll 80(6):549–553
Koch T, Denike K (2009) Crediting his critics' concerns: remaking John Snow's map of Broad Street cholera, 1854. Soc Sci Med 69(8):1246–1251
Lengler R, Eppler MJ (2007) Towards a periodic table of visualization methods for management. Proceedings of the conference on graphics and visualization in engineering (GVE 2007), Clearwater, FL
Liu Z, Stasko JT (2010) Mental models, visual reasoning and interaction in information visualization: a top-down perspective. IEEE Trans Vis Comput Graph 16(6):999–1008

Magnani R, Dirk LMA, Trievel RC, Houtz RL (2010) Calmodulin methyltransferase is an evolutionarily conserved enzyme that trimethylates Lys-115 in calmodulin. Nat Commun 1:43

Mane KK, Börner K (2007) Computational diagnostic: a novel approach to view medical data. Los Alamos National Laboratory, Los Alamos, NM

Mayer RE (2001) Multimedia learning. Cambridge University Press, Cambridge, UK

Mccormick B (1987) Scientific and engineering research opportunities. Comput Graph 21(6)

Mcleod KS (2000) Our sense of Snow: the myth of John Snow in medical geography. Soc Sci Med 50(7–8):923–935

Michel JB, Shen YK, Aiden AP, Veres A, Gray MK, Pickett JP, Hoiberg D, Clancy D, Norvig P, Orwant J (2011) Quantitative analysis of culture using millions of digitized books. Science 331 (6014):176

Moreno R, Mayer RE (1999) Cognitive principles of multimedia learning: the role of modality and contiguity. J Educ Psychol 91(2):358

Mueller K, Garg S, Nam JEJ, Berg T, Mcdonnell KT (2011) Can computers master the art of communication? A focus on visual analytics. IEEE Comput Graph Appl 31(3):14–21

Novakova L, Stepankova O (2009) RadViz and identification of clusters in multidimensional data. 13th international conference on information visualisation, 15–17 July 2009. pp. 104–109

Paivio A, Csapo K (1973) Picture superiority in free recall: imagery or dual coding? Cogn Psychol 5(2):176–206

Pham BL, Cai Y (2004) Visualization techniques for tongue analysis in traditional Chinese medicine. IEEE Trans Biomed Eng 51(10):1803–1810

Pike AWG, Hoffmann DL, García-Diez M, Pettitt PB, Alcolea J, De Balbín R, González-Sainz C, De Las Heras C, Lasheras JA, Montes R, Zilhão J (2012) U-series dating of paleolithic art in 11 caves in Spain. Science 336(6087):1409–1413

Plaisant C, Milash B, Rose A, Widoff S, Shneiderman B (1996) Life lines: visualizing personal histories. ACM CHI '96, September 1995. Vancouver, BC, Canada. 13–18 April 1996

Powsner SM, Tufte ER (1994) Graphical summary of patient status. Lancet 344(8919):386–389

Ren L, Tian F, Zhang X, Zhang L (2010) DaisyViz: s model-based user interface toolkit for interactive information visualization systems. J Vis Lang Comput 21(4):209–229

Richardson JS (2000) Early ribbon drawings of proteins. Nat Struct Mol Biol 7(8):624–625

Saad A, Hamarneh G, Möller T (2010) Exploration and visualization of segmentation uncertainty using shape and appearance prior information. IEEE Trans Vis Comput Graph 16(6):1365–1374

Saary MJ (2008) Radar plots: a useful way for presenting multivariate health care data. J Clin Epidemiol 61(4):311–317

Sedlmair M, Munzner T, Tory M (2013) Empirical guidance on scatterplot and dimension reduction technique choices. IEEE Trans Vis Comput Graph 19(12):2634–2643

Shneiderman B (1996) The eyes have it: a task by data type taxonomy for information visualizations. Proceedings of the 1996 IEEE symposium on visual languages, pp 336–343

Simon W (1972) Mathematical techniques for physiology and medicine. Academic, New York

Tory M, Möller T (2004) Human factors in visualization research. IEEE Trans Vis Comput Graph 10(1):72–84

Vion-Dury J, Favre R, Sciaky M, Kriat M, Confort-Gouny S, Harle J, Grazziani N, Viout P, Grisoli F, Cozzone P (1993) Graphic-aided study of metabolic modifications of plasma in cancer using proton magnetic resonance spectroscopy. NMR Biomed 6(1):58–65

Ward M, Grinstein G, Keim D (2010) Interactive data visualization: foundations, techniques and applications. Peters, Natick, MA

Ware C (2004) Information visualization: perception for design (interactive technologies), 2nd edn. Morgan Kaufmann, San Francisco

Ware C (2012) Information visualization: perception for design, 3rd edn. Elsevier, Amsterdam

Weller JM (2004) Simulation in undergraduate medical education: bridging the gap between theory and practice. Med Educ 38(1):32–38

Lecture 10

Biomedical Information Systems and Medical Knowledge Management

1 Learning Goals

After the end of this tenth lecture, you:

- would have an overview about workflows and workflow modeling in health care.
- would get an overview of typical architectures of hospital information systems for patient records as already discussed in Lecture 4.
- would have understood the principles of Picture Archiving and Communication Systems (PACS).
- would be aware of how important multimedia in medicine is.
- would have a basic understanding of DICOM and HL7.
- would be aware of the constraints of open source software in the medical domain.
- would have got an idea of possible future system solutions.

2 Advance Organizer

Bioinformatics workflow management system	designed specifically to compose and execute a series of computational and/or data manipulation steps and/or workflows in the domain of bioinformatics
Business process reengineering (BPR)	analysis and design of workflows and processes within an organization (=hospital). According to Davenport and Short (1990) a BP is a set of logically related tasks performed to achieve a defined result
Clinical Pathway	aka care map, a tool used to manage the quality in health care concerning the

	standardization of care processes and promote organized and efficient patient care based on EBM
Digital Imaging and Communications in Medicine (DICOM)	standard for handling, storing, printing, and transmitting data in medical imaging (also file format and a network communications protocol using TCP/IP)
Evidence-based medicine (EBM)	aiming at developing mathematical estimates of benefit and harm from population-based research and apply these in the clinical routine, claiming that best research evidence on medical interventions comes from experiments (e.g., randomized controlled trials)
Health Level Seven (HL 7)	a Standardization Organization accredited by the American National Standards Institute (ANSI) to push consensus-based standards representing health-care stakeholders
Hospital Information System (HIS)	integrated information system for (administrative, financial, clinical, etc.) information management in a hospital
Integrating the Healthcare Enterprise (IHE)	initiative by health-care professionals and industry to improve interoperability of computer systems in health care (i.e., promotes coordinated use of standards, e.g., DICOM and HL7)
National Electrical Manufacturers Association (NEMA)	holds copyright of DICOM
Paradigm	according to Kuhn (1962) a shared view of a group of researchers, comprising four elements: concepts, theories, methods, and instruments
Picture Archiving and Communication System (PACS)	system for handling images from imaging instruments, including ultrasound (US), magnetic resonance (MR), positron emission tomography (PET), computed tomography (CT), endoscopy (ENDO), mammographs (MG), Digital radiography (DR), computed radiography (CR) ophthalmology, etc.
Workflow	consists of a sequence of connected steps, succeeding the flow paradigm, where each step follows the precedent

3 Acronyms

BPR	Business Process Reengineering
CT	Computer Tomography
DICOM	Digital Imaging and Communications in Medicine
EBM	Evidence-based medicine
EHR	Electronic Health Record
ENDO	Endoscopy
FMA	Foundational Model of Anatomy
FOL	First-order logic
GO	Gene Ontology
ICD	International Classification of Diseases
IHE	Integrating the Healthcare Enterprise
IOM	Institute of Medicine
KIF	Knowledge Interchange Format
LOINC	Logical Observation Identifiers Names and Codes
MeSH	Medical Subject Headings
MRI	Magnetic Resonance Imaging
HL7	Health Level 7
LOINC	Logical Observation Identifier Names and Codes
NEMA	National Electrical Manufacturers Association
PET	Positron Emission Tomography
PHR	Personal Health Record
PACS	Picture Archiving and Communication System
SaaS	Software as a Service
UML	Unified Modeling Language
US	Ultrasound

4 Key Problems

Slide 10-1: Key Challenges

- Lack of integrated systems
- Clinical workplace efficiency
- Cloud computing (Privacy, Security, Safety issues)
- Software as a Service (as electricity is already)

(continued)

(continued)

Although we write the year 2014—and we are now in the computing business since 1970—for more than four decades—there is still the lack of integrated systems, and moreover the lack of data integration. A major issue is in data fusion. The next problem is in the clinical workplace efficiency. It is not enough just to deliver more and more data to the clinician—they are already overwhelmed by the masses of data. They are interested in **relevant data, and relevant information to support knowledge and decision making**. There is a lot to do in the coming years.

5 Workflow Modeling

Slide 10-2: Workflow Modeling

We have already heard about workflows and how important they are. Please remember that a von Neumann machine is a mathematical processor, and hence is good in processing numbers—mathematically. Modeling is an approach to mathematize our world. What cannot be modeled in a mathematical way cannot be processed by a computer. In this slide we see a typical medical workflow on the example of radiology—from admission of the patient to the discharge.

Use case diagrams describe sequences of actions, providing measurable values to actors. A good overview is given by (Juan et al. 2005).

Slide 10-3: Workflow > Interaction > Decision > Action

Remember what we have learned about the relationships of data, information, and knowledge and that the knowledge gained is the essential part in decision making, which results in an appropriate action. Do never forget that clinical medicine is not a theoretical, but an action science (in German: "Handlungswissenschaft"). In this slide on the left, we see the "knowledge elements" (gained clinical information, policies, standard operating procedures, clinical guidelines, etc.) as a basis for decision making.

5.1 Workflow and Decision Support

Slide 10-4: Various Levels of Decision Support

The core element of Slide 10-3 is applied in every stage of workflows from the large scale (world health management, country policies etc.) to the small scale (event handling).

Slide 10-5: Workflow Modeling in a Nutshell

A workflow consists of a sequence of connected steps, with the emphasis on the flow paradigm. A workflow can be seen as an abstraction of the real world. Workflow modeling is basically the process of simplifying reality (with all the problems and dangers of oversimplification involved). The modeling is based on facts gathered during observations and we need to accept that this representation can never be perfect. Expectations from a model should be limited to the intentions with which it is designed for, be it problem solving or understanding of system intricacies. A workflow management system is designed specifically to compose and execute a series of computational and/or data manipulation steps. In the research area there is the vision of e-Science, i.e., distributed scientists being able to collaborate on conducting large scale experiments and knowledge discovery applications using distributed systems of computing resources, datasets, and devices. Workflow modeling is the process of *simplifying the real-world*; Modeling is based on facts gathered during observations, and we need to accept that this representation can never be perfect; Expectations from a model should be **limited** to the intentions with which it is designed for, be it problem solving or understanding of system intricacies, i.e., elaborately complex details (Malhotra et al. 2007).

5.2 Formal Modeling

Slide 10-6: Example: Formal Workflow Modeling 1/2

A workflow W is defined as a process that contains tasks T, and the respective rules on how those tasks are executed:
  $W := (T, P, C, A, S_0)$ where
  $T = \{T_1, T_2, \ldots T_m\}$ A set of tasks, $m \geq 1$

(continued)

(continued)

P = $(p_{ij})_{m \times m}$ **Precedence matrix of the task set**
C = $(c_{ij})_{m \times m}$ **Conflict matrix of the task set**
A = AT1, AT2, …, ATM Precondition set for each task
$S_0 \in \{0, 1, 2, 3_m\}$ is the initial state
For more information please refer to: (Wang et al. 2008).

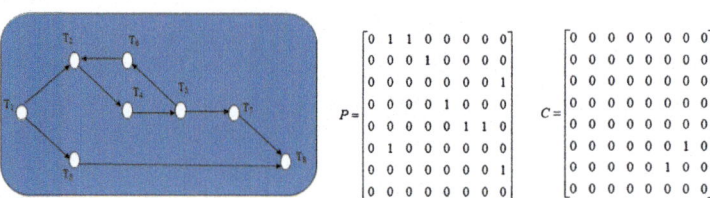

Fig. 1 See Slide 10-7

Slide 10-7: Example: Formal Workflow Modeling 2/2

Here you can see the model of the defined workflow from Slide 10-6 and the mathematical representation (Wang et al. 2008).

P = $(p_{ij})_{m \times m}$ Precedence matrix of the task set
C = $(c_{ij})_{m \times m}$ Conflict matrix of the task set

Slide 10-8: Example: Modeling in UML—Use Case Diagram

In this slide you see the description of a real-world example by the use of the UML and a corresponding use case diagram (Holzinger et al. 2010).

The Unified Modeling Language (UML) is a standardized (ISO/IEC 19501:2005), modeling language developed from the software engineering domain (Oestereich 1999). UML includes a set of graphic notation techniques to create visual models of object-oriented systems. Originally, it was developed by (Booch 1994) and (Booch et al. 1999). It was adopted as a standard in 1997 by the OMG (Object Management Group) and has been managed by this organization ever since. The current version of the UML is 2.4.1 published by the OMG in August 2011.

5 Workflow Modeling

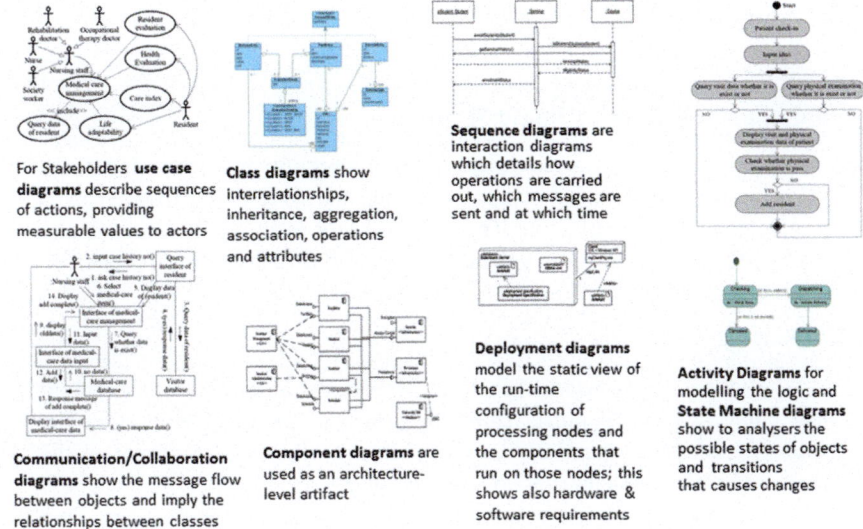

Fig. 2 See Slide 10-9

Slide 10-9: Modeling: Different Views for Different People

Although this slide is overcrowded and hard to read—the advantage is having a good overview on the usage of different diagrams—enabling different views for different people.

For Stakeholders use case diagrams describe sequences of actions, providing measurable values to actors.

Class diagrams show interrelationships, inheritance, aggregation, association, operations and attributes.

Sequence diagrams are interaction diagrams which details how operations are carried out, which messages are sent and at which time.

Communication/Collaboration diagrams show the message flow between objects and imply the relationships between classes.

Component diagrams are used as an architecture-level artifact

Deployment diagrams model the static view of the run-time configuration of processing nodes and the components that run on those nodes; this shows also hardware and software requirements.

Activity Diagrams for modeling the logic and State Machine diagrams show to analysts the possible states of objects and transitions that cause changes.

5.3 Example Clinical Workflow

Slide 10-10: Example Clinical Workflow

This is an example of how a typical hospital workflow looks like and in the following we will see what computer science can do to help to make such a workflow much more effective. This example is from an outpatient university clinic where questionnaires are very important for supporting research in fighting skin cancer (Holzinger et al. 2006, 2011).

Look at the slide:

When the patients arrive at the central administration desk of the outpatient clinic, a medical nurse hands them over a paper questionnaire. The patient is asked to complete the questionnaire alone and return it to the nurse, where they are collected together for the medical doctor. Theoretically, the doctor can peruse the paper questionnaire during the treatment, but concentration is usually limited to the patient and the Electronic Patient Record (EPR) displayed on the clinical workplace monitor. Most often, the questionnaires are just gathered and manually typed into a separate electronic database after the clinic is closed—in the spare time of the doctor. Moreover, the separate database has no direct connection to the electronic patient record of the Enterprise Hospital Information System (EHIS). Consequently, a lot of effort is spent just on typing. Let us look to the next slide, what we can do from a computer science perspective to save time and effort.

Slide 10-11: Example: WF-Optimization with Mobile Computing

In the new workflow, the patient reports to the central administration desk of the outpatient clinic. There, they are registered anyway via the MEDOCS hospital information system administration program. This will now be used as a patient identifier. At the clinical workplace, a list of the waiting patients, who have been registered already in the system but not yet released by a medical doctor, is displayed, flagged to show whether or not they have already filled out a questionnaire. This is indicated by means of text and/or a symbol and differentiates between a questionnaire which has been made available to the patient, a questionnaire filled out on the current day; and a questionnaire, which was completed by the patient during a previous visit (non-current date) and is still available. When no questionnaire is available the flag column remains empty. The medical doctor or the nursing staff of the clinic can decide whether the patient should complete a questionnaire and whether this should be the long or the short version. Clicking on the relevant

(continued)

(continued)

icon generates the empty questionnaire and registers it in MEDOCS using a unique identification code which identifies the patient unequivocally as the user. Using an XML communication, the patient identification number (PID), the unique number of the document (document number at the top of the questionnaire), and any further data (e.g., name, date of birth) considered necessary by MoCoMED-Graz are transmitted. At the terminal, the patient is equipped with a touch-based Tablet PC and a code, with which he/she can login to MoCoMED-Graz and complete the questionnaire following a touch-based application. The authentication at MoCoMED-Graz is necessary for data security reasons, so that no patient can access other data and patients avoid mistakes or errors. The code is a unique identification number, generated by and linked with the enterprise hospital information systems patient record MEDOCS, ensuring that the proper patient enters the data. An incorrect code entry prompts an error message and the data is not accepted by the system.

After the questions have been answered and the questionnaire is completed, MoCoMED-Graz transfers the questionnaire directly into the MEDOCS system. The corresponding column in MEDOCS now shows the status "Questionnaire was filled out on the current day". As soon as the patient has completed the questionnaire, the XML file containing the answers is stored on the server and subsequently transferred to MEDOCS by using a remote function call (RFC). The XML document containing all answers of a patient includes, of course, the unique identification of each. This project serves as an example of how computer application can benefit all three groups of people: patients, medical professionals and hospital managers. Patients were very satisfied with the front end of the application. Medical professionals could save up to 90 % (!) of their formerly wasted time; which ultimately saves money for the hospital manager. Most important, the quality of the medical service is increased, since the newly created workflow brings together patients and doctors in front of the clinical workplace, to check whether all entries are correct.

Since robust, reliable, light and unobtrusive, uncomplicated, appealing hardware is an essential part of the success of such an application in a real-life hospital, the selection of appropriate hardware is an essential success factor. For more details refer to (Holzinger et al. 2011).

Slide 10-12: Mobile Computing Project MoCoMed-Graz

Here you see the architecture of this application. PAT = Patient, MED = medical doctor. For details see (Holzinger et al. 2011).

Slide 10-13: Important: Macrolevel—The View of the Manager

From the hospital managers' point of view, this method may serve as a valuable tool to achieve extensive surveillance data with limited resources, to reduce human errors to a minimum, and to create standardized datasets for further analyses, thus improving skin cancer prevention and quality of care. As patients were actively integrated into their treatment process and provided with a useful way of spending their waiting time, this may also lead to a substantial improvement in patient care services, and therefore increased patient satisfaction., Apart from the time saving benefit, an important aspect for the health system in general, which is also relevant to the hospital managers in terms of quality of patient treatment is the increased patient empowerment. For clearly depicting the benefit for the hospital management, a scenario analysis has been made by calculating the facts in this particular outpatient clinic on the basis of $N = 30$ patients a day, 250 days a year.

Scenario 1 (without mobile computers): A doctor costs the hospital approx. 44 EUR per hour which is 0.73 EUR per minute. With an average of 10 min per patient, the 300 min of extra effort required by the doctors to transcribe the questionnaires are equivalent to 219 EUR per day = 55 kEUR per year and there is the additional risk of lost and missing data.

We also tested the alternative of letting copy typists transcribe the data: They are faster median = 7 min (min: 5 min; max: 8 min), however, since they are not medical professionals they make more errors, which cannot be corrected without extra costs. Calculating on the basis of their average hourly wages results in a cost of 19 EUR per hour = 0.32 EUR per minute, we have costs of approx. 17 kEUR per annum.

1. Scenario 2 (using two mobile computers):
2. Development Costs: 7 Person Months = 28 kEUR
3. Hardware Costs: 3.2 kEUR
4. Original cost of acquisition = 31.2 kEUR
5. Annual service costs: 0.5 Person mpy = 2 kEUR
6. The reduction to only 30 min of the medical doctors' time per day is equivalent to 22 EUR per day or approx. 5.5 kEUR per year.
7. Total annual costs = 7.5 kEUR
8. Assuming an average life span of 4 years, this amounts to a total of $31.2 + (4*7.5) = 61.2$ kEUR, or 15.3 kEUR per annum.
9. Maximum total possible cost saving per year: 55 kEUR—15.3 kEUR = 39.7 kEUR
10. Minimum total possible cost saving per year: 23.5 kEUR—15.3 kEUR = 8.2 kEUR
11. You clearly see the benefit of such an application—reached through careful workflow optimization.

Slide 10-14

Finally, have a view on the front end of this application—from the patient side.

Slide 10-15

What can we learn out of this project: Most of all you must acknowledge that different people have different views, different expectations, and different goals. To provide a clear benefit, an application should satisfy different requirements and most of all should bring a clear—measurable (!)—benefit. Time to perform a task is still a major issue, which is easy to measure and pays off. Time is money. If you save time of a medical professional, you save money, but at the same time you enhance quality!

For evaluation purposes the ecological systems theory by (Bronfenbrenner 1977) is useful, which originates from child development theory and describes the interaction between various environments using a layer structure—each having an effect on a child's development. Originally, Bronfenbrenner described five such layers:

(a) The microsystem, which encompasses relationships and interactions within the person's direct environment—from the individual persons' point of view.
(b) The mesosystem, which describes the interaction between the microsystem and the macrosystem.
(c) The exosystem, which defines the larger social system in which the person does not function directly.
(d) The macrosystem, which includes the cultural values, customs, and laws.
(e) The chronosystem, which encompasses the dimension of time as it relates to a persons' environment.

The essence is that interactions at outer levels have always impact on the inner structures.

For our purpose, we instrumentalized only three of Bronfenbrenner's layers: microsystem, mesosystem, and macrosystem and we incorporate them into our context:

1. The microsystem relates to the end user and their immediate environment (end-user-centered human–computer interaction (HCI): user tasks, patient empowerment, . . .).
2. The mesosystem relates to the medical professionals and their environment (professional process-centered HCI: work tasks, medical processes, social context, i.e., discussing the medical data together with the patient— strong influence on patient empowerment, . . .).

5.4 Workflows in Bioinformatics

Slide 10-16: Summary: Workflows in Bioinformatics

We can summarize: A data-driven procedure consisting of one or more transformation processes can be seen as nodes and thus can be represented as a directed graph, where the direction is time—i.e., the order of transformations and a set of transformation rules. The data flow origins from a source to a destination (or result) via a series of data manipulations and the specification is designed in a Workflow Design System (modeling component) and then run by a Workflow Management System (execution component) (Hasan et al. 2006).

Slide 10-17: Bioinformatics Workflow Management System

Romano (2008) described the components of a WMS in the bioinformatics domain.

Note: WMS = a system that defines, creates, and manages the execution of workflows.

Its main components include:

1. A graphical interface for composing workflows, entering data, watching execution, displaying results.
2. An data archive to store workflow descriptions, results of executions, and related traces.
3. A registry of available services, either local or remote.
4. A scheduler able to invoke services included in the workflow at the appropriate time.
5. A set of programming interfaces able to dialogue with remote services.
6. A monitor tool for controlling the execution of the workflow.
7. A set of visualization capabilities for displaying different types of results.

6 Hospital Information Systems

Slide 10-18: Remember: Classic Conceptual Model of a HIS

Remember the classic architecture of a hospital information systems (HIS) which we have already discussed in Lecture 4. Managing all clinical

(continued)

(continued)

information is a challenge implying huge constraints, because health-care providers use technology in a ubiquitous manner. Medical information systems (e.g., Hospital Information Systems (HIS), Picture Archiving and Communication Systems (PACS), Radiology Information Systems (RIS), and Laboratory Information Systems (LIS)) are used in parallel together and are often heterogeneous. A **process-oriented** HIS has to be modified whenever the business process changes, i.e., it must be adjustable to process changes and to changing organizational structures very quickly and at reasonable costs (Dadam et al. 2000).

6.1 Architectures

Slide 10-19: Architectures of Hospital Information Systems

There is a need to integrate the various theoretical frameworks and formalisms for modeling clinical guidelines, workflows, and pathways, in order to move beyond providing support for individual clinical decisions and toward the provision of process-oriented, patient-centered HIS that formally model guidelines, workflows, and care pathways (Gooch and Roudsari 2011).

Historically, we determine between three different system architectures:

1. **1970+ "Vertical Approach"**—monolithic mainframes: central computer systems mainly for accounting, typical "data processing" (EDV, Elektronische Datenverarbeitung).
2. **1985+ "Horizontal Approach"**—evolutionary systems: departmental clinical information systems, local area networks, distributed systems.
3. **2000+ "Integrated Approach"**—open, distributed systems: hospital intranets, networked electronic patient/health record, mobile computing, "information quality focus" (Holzinger 2002).

Slide 10-20: Basic Architecture of a Standard HIS

A standard universal integrated architecture contains a data center—fusing together the data from the main databases (patient records, laboratory information system, PACS storage, etc.) and a integrated view in terms of a clinical workplace (El Azami et al. 2012).

6.2 Process-Oriented Information Systems

Slide 10-21: Model of Process-Oriented Health Info Systems

Coming back to a grand challenge: data integration and data fusion in the life sciences: There is a need to integrate the various theoretical frameworks and formalisms for modeling clinical guidelines, workflows, and pathways, in order to move beyond providing support for individual clinical decisions and toward the provision of process-oriented, patient-centered HIS that formally model guidelines, workflows, and care pathways. One approach can be seen in this slide, where different workflow models and data models and knowledge models are combined and used for the overall clinical process model (Gooch and Roudsari 2011).

7 Multimedia in the Hospital

Slide 10-22: Multimedia Throughout the Hospital

Various medical departments generate images or videos for medical documentation. Although these images are lightweight from a technological perspective, the workflows for image generation and viewing are very diverse (Bellon et al. 2011).

In primary diagnosis images must often be acquired in a separate step before diagnosis is possible, e.g., when images are generated using ionizing radiation, radio waves, laser beams (optical coherence tomography, retinal thickness analysis, etc.), or fluorescence (fluorescein angiography, etc.). In contrast, in pathology using visual light microscopy diagnosis can be performed by direct observation, but even for this application there are advantages obtaining an image first by using the computer as a digital microscope. Such **virtual microscopy** images contain many thousand pixels along each direction. This huge storage (much higher than in radiology) makes routine use difficult. Workflows in many cases are similar to radiology, however, not in the completely different class of applications in which images or short video sequences are not needed for primary diagnosis, but are obtained for later reference, e.g., to facilitate discussion with colleagues, to document the case, or for communication with the patient (left in the slide). For such **reference images,** quality usually is not a concern and technological needs are low. In many cases the acquisition devices are consumer products such as digital photo cameras that do not support DICOM related concepts. For example, many systems in ophthalmology output reports in the form of a

(continued)

(continued)

document containing text, numbers, graphs, and some images (right in the slide). Even if some new equipment might be able to generate a DICOM structured report, most systems just generate output on paper. An easy but efficient way to capture the latter documents into an electronic system is to replace the traditional paper printer by a virtual printer that generates PDF. In a third class of applications there is a **real-time aspect**, for example in surgery (endoscopic cameras for keyhole surgery, fluoroscopic imaging equipment, vital signs monitors, surgical robots, image processing applications for intra-surgical guidance, etc.).

7.1 PACS

Slide 10-23: First Commercial Picture Archiving and Communication System

A PACS (Picture Archiving and Communication System) is still primarily associated with radiology, but can handle images from various medical sources, including ultrasound (US), magnetic resonance (MR), positron emission tomography (PET), computational tomography (CT), endoscopy (ES), mammograms (MG), etc. In the slide above we see the first commercial PACS installation in the USA—a Fuji CR-101 system at the Osner Clinics, New Orleans. Below we see a PACS console from 1990 (Huang 2011).

Note: Although the concept of PACS was developed in Europe during the late 1970s, no commercial system was completed at that time. The first PACS implementations took place in the USA in the early 1980s, e.g., at Pennsylvania University, UCLA, Kansas City University, and the Osner Clinics, New Orleans. Some more or less successful PACS developments also took place in Europe in the 1980s—and Graz was amongst the first PACS installations. Most systems could be characterized by their focus on a single department, such as radiology or nuclear medicine. Hospital-wide PACS evolved in the early 1990s in London (Hammersmith Hospital) and Vienna (SMZO) (Lemke 2003, 2011).

Slide 10-24: Typical Workflow in a European PACS

A typical workflow can be seen in this slide: The dark oval symbols indicate the activity steps of the clinical process from examination to therapy planning on a low granulation level. Bright rectangular symbols indicate the personnel involvement in the activity steps, light beige oval symbols indicate IT system support for the given activities, and dark rectangular symbols indicate the functionalities of medical workstations supporting the given activities (Lemke and Berliner 2011; Lemke 2011).

Slide 10-25: Generic PACS Components and Data Flow

In this slide we see the PACS basic components and data flow; HIS: Hospital Information System; RIS: Radiology Information System. System integration and clinical implementation are two other necessary components during the implementation after the system is physically connected. Application servers and Web servers connected to the PACS controller enrich the PACS infrastructure for other clinical, research and education applications (Huang 2011).

Note: Imaging **modalities** are the different image generating devices, e.g., CT, PET, MR

(has nothing to do with the modality known from multimedia theory)

7.2 Data Standards (DICOM, HL7, LOINC)

Slide 10-26: Data Standards for Communication and Representation

In the context of hospital information systems in general and with medical imaging in particular, three standards have a significant importance: DICOM, HL7, and LOINC (Bui and Taira 2010):
 DICOM 3.0—Digital Imaging and Communication in Medicine (1993) is ...

1. A set of protocols for network communication.
2. A syntax and semantics for commands and info.
3. A set of media storage services (standard compliant).

(continued)

(continued)

HL 7—Health Level 7 ...

1. HL 7 v2.x is a messaging protocol, to provide exchange of textual healthcare data between hospital information systems.
2. Reference Implementation Model (RIM) contains data types, classes, state diagrams, use case models, ant terminology to derive domain-specific information models.
3. Clinical Document Architecture (CDA) is a document markup standard to specify structure and semantics of clinical documents in XML.

LOINC—Logical Observation Identifier Names and Codes is used for ...

1. Laboratory data (e.g., molecular pathology observations used for identification of genetic mutations, tumor genes, gene deletions, etc.).
2. Clinical Observations (e.g., non-laboratory diagnostic studies, critical care, nursing measures, patient history, instrument surveys).
3. Claims attachments (e.g., handles the definition of new LOINC terms and codes to manage claims-related data etc.).

Slide 10-27: DICOM and HL 7

In this compact overview we see the ISO OSI Reference Layer model showing the most common protocols and data standards, from the physical level up to the application level (= level 7).

HL7 (Health Level 7) is both an organization and a standard, which was founded in 1987 aiming at a general standard for hospital information systems. The organization was accredited in 1994 by the American National Standards Institute. The name is a reference to the seventh layer of the ISO OSI Reference layer model aka application layer and indicates that it focuses on that layer, independent of the layers. All relevant information can be retrieved from the Web site: http://www.hl7.org

Slide 10-28: DICOM Object Containing Multiple Data Elements

Communication of images between medical imaging systems and among their applications has always been difficult because of the multiple platforms and vendor-specific communication protocols and data formats.

(continued)

(continued)

The DICOM standard, developed in 1992 by a joint committee formed by the American College of Radiology (ACR) and the National Electrical Manufacturers Association (NEMA), is intended to provide connectivity and interoperability for multivendor imaging equipment, allowing for communication of images and exchange of information among these individual systems. Medical images are defined in DICOM as information objects or datasets. An information object represents an instance of a real-world information object (i.e., an image) and is composed of multiple data elements that contain the encoded values of attributes of that object. Each data element is made of three fields: the data element tag, the value length, and the value field. The data element tag is a unique identifier consisting of a group number and an element number in hexadecimal notation and is used to identify the specific attribute of the element (Wong and Lou 2009).

Look at the slide (left side): Each data element is uniquely identified by its corresponding tag composed of a group number and an element number. Pixel data of the image is stored in element 0010 within group 7FE0. Look at the right side: A DICOM file consists of a file meta-information header and an information object (image dataset). The file meta-information header is made of a file preamble, a DICOM prefix, and multiple file meta-elements (Wong and Lou 2009).

Slide 10-29: DICOM Instances Allowing the Storage of Meshes

This is just an example: A special DICOM supplement defines a way to store and communicate meshes as DICOM instances, allowing the storage of the meshes together with the images on a PACS server—meshes are important for computer assisted surgery (Lemke and Berliner 2011).

Slide 10-30: Image Distribution Workflow

In the last years many changes in computer and communication technology have pushed the limits of PACS beyond the traditional system architecture providing new perspectives and innovative approaches to a traditionally conservative medical community. Technologies such as the web, wireless networking, open source software and recent emergence of cyber communities and social networks have imposed an accelerated pace in the progress of

(continued)

(continued)

computer and technology infrastructure applicable to medical imaging applications.

In this slide we see some modern approaches: Industry has started to adopt OSIRIX as a base for new business models where they provide the support and integration of services as well as training and customization of generic platforms. Several certified versions for Europe and for FDA in the USA have already appeared on the market recently. And finally, and probably most importantly, the academic community has started to regroup its efforts to support and promote open source initiatives in medical imaging and medical informatics (Ratib et al. 2011).

OSIRIX software program and its source code are available under open source licensing agreement and can be downloaded free of charge at: http://www.osirix-viewer.com

Slide 10-31: HL7 Use Case: Serum Glucose Lab Result

This slide show an example business use case related to a laboratory results message, as well as a V2.4 and a v3 representation. The v3 message is based upon the normative XML ITS 1.0 and schema from the May 2006 Informative Edition of HL7 v3. The use case is the completion of a serum glucose laboratory result of 182 mg/dL authored by Howard H. Hippocrates. The laboratory test was ordered by Patricia Primary for patient Eve E. Everywoman. The use case takes place in the US Realm. See more details via: http://www.ringholm.com/docs/04300_en.htm

Slide 10-32: PACS Consoles and Modern Workplaces

This slide shows a PACS console and a typical PACS-Viewer, which runs meanwhile on standard platforms such as Mac or Windows.

Slide 10-33: Multimodal Imaging Techniques: Hybrid PET-CT

In this slide we see advanced image display and analysis tools adapted to multimodality imaging techniques such as hybrid PET-CT images showing

(continued)

(continued)

multi-planar reformatting of fused images (back window) as well as volume rendering and segmentation of tumor outline from PET images (front window).

Development of commercial systems driven by medical imaging manufacturers lags behind the increasing demand for such advanced processing tools. Industry usually follows research and clinical validation slowly and requires several years before releasing new products on the market. The certification processes within the medical area are long lasting and costly; therefore, each vendor tries to sell their available products as long as possible.

Although open source and free software is becoming more widely adopted in the medical community, it is still rare that a hospital (especially a public one) adopt these to their daily routine. Private hospitals are more open for advanced technologies in general as they have usually a shorter decision chain in the hospital management.

Open source would not only provide a cost effective alternative, but, because they are being developed by a community of developers from the field, the software tools can be customized to match the needs and specific usage in clinical setups outside radiology. Also open source have less restrictions on providing new innovative and challenging viewing and analysis tools that respond to users demands even before industry and commercial vendors identify these new trends as potential source of revenue (Ratib et al. 2011).

Slide 10-34: Example: OsiriX Integrated RIS–PACS Interface

This slide shows an OsiriX integrated RIS–PACS interface: A query list for retrieval from PACS and image list for a given PACS-retrieved examination. Note: The Radiological Information System (RIS) is often part of the general Hospital Information System (HIS).

The increasing availability of such integrated RIS–PACS solutions for image reading and reporting of medical imaging examinations has fuelled the development of integrated tools for digital image processing. This evolution represents a further step towards total integration of all instruments needed for interpretation and reporting of diagnostic imaging examinations in a health service environment, i.e.m at a universal clinical workplace. Technical problems have been faced, especially in early times, due to stringent hardware and software requirements for accomplishing such tasks (Faggioni et al. 2011).

More information about the open-source software OsiriX can be found here: http://www.osirix-viewer.com

Slide 10-35: Example: Application of VR in Medical Imaging

Radiologists were ever since the drivers in imaging technologies and they constantly are adopting advanced technologies into routine. Here, an example of the application of virtual reality in medical imaging (Faggioni et al. 2011).

Slide 10-36: Bone Age Assessment DSS Workflow of Skeletal Maturity

Meanwhile many preprocessing and imaging processing techniques are integrated in the radiological workplace as can be seen here on the example of a workflow of the automated assessment of the skeletal maturity. Regions that may deliver false positives are excluded from the analysis. Due to the individual variability, some landmarks may be pointed in order to precisely define the size of the Region of Interest (RoI). Typically, two types of RoIs are defined: The first one, usually of a rectangular shape, shrinks the area to be searched and permits robust segmentation procedures. These regions are pointed by radiologists as areas of a strong discriminative power that feature significant changes during the development of the process to be described. As an example, distal, middle, and proximal regions in the phalanges have been pointed by radiologist as areas very sensitive to the development of the skeletal system. "The more distal the better" is their opinion. Another region pointed by radiologists includes the carpal bones. They happen to be good indicators at the early stage of the skeletal development. The other type of the RoIs is related to anatomical structures. If an abnormal tissue is searched within a certain structure (lung, liver, brain, etc.), this structure has to be segmented in the first place. The required accuracy of the delineation depends on a possible location of the pathological tissue within this structure. If the tissue is attached to the border and, in addition, its pixel (voxel) intensity is similar to the intensity of the neighborhood, a problem dependent procedure has to be developed in order to delineate the structure (Piętka et al. 2011).

Slide 10-37: Open Source Software: Legal and Regulatory Constraints

In Slide 10-33 we have discussed the advantages of Open Source software.
 Attention: Medical certifications such as FDA in the USA and CE marking in Europe do not apply to open source software!

(continued)

(continued)

These certifications require a legal commercial entity to be identified as the owner of the product and warrant the legal liability of its distribution and commercial support.

Open source software being often developed outside commercial enterprises, such as academic groups or university research labs, do not have the proper legal structure to apply for such certifications.

Also, most open source products being distributed free of charge lack the legal binding between the provider and the user that is required for software distribution under FDA and CE certification.

8 Future Outlook

Slide 10-38: Future Outlook

For sure the future hospital information systems will have three components:

1. The world (even the hospital world) of tomorrow will be mobile.
2. The data will be stored in the cloud as well as software-as-a-service will be used (Key problem: privacy, security, safety, and data protection).
3. Content analytics will be integrated as tools within the clinical workplace.

Let us look on a few examples in the following slides.

Slide 10-39: Future Cloud Computing Solutions

Innovative solutions for rapid and robust transfer of very large data files are emerging and will probably change significantly the current limitations in the next few years. Online free file transfer providers such as yousendit (http://www.yousendit.com) have already become very popular for convenient transfer of large files, notifying the recipient by Email when the file is available for him to download. However, these public services lack the performance and security that would be required for transferring medical data. Examples are open source software solutions such as the Xebra software framework for Web-based distribution and visualization of medical imaging (http://www.hxti.com/technology/xebra.html). Major standardization efforts are being deployed as part of the IHE (Integrating the Healthcare Enterprise) initiative to develop new communication profiles and guidelines for

(continued)

(continued)

exchanging medical data and images. The Cross-Document Sharing (XDS) an IHE profile for exchanging documents has led to an extension specific for medical images (XDS-i) that is more suitable for exchange of large medical imaging files. In compliance with DICOM standards, an extension for image transfer called WADO (Web Access to DICOM Objects) is now being adopted in replacement of traditional C-move (a DICOM service for moving data), allowing for more flexibility and better performance. These emerging standards are already being adopted in large metropolitan or even national projects. A recent example of such deployments is the Canadian national diagnostic imaging repository (DI-r) which consolidates imaging results and provide a shared PACS application as part of the national interoperable electronic health record (Ratib et al. 2011).

Slide 10-40: Future System Architecture: Example Cloud Ultrasound

Cloud-based computing follows the SaaS-Paradigm "Software as a Service" and offers a pay-per-use based virtual infrastructure, made dynamically scalable by the ability to spawn and destroy virtual machine instances on demand. This allows virtual servers to be created as needed to meet current demand and then scaled back when strain on the system is reduced, lowering resource usage and cost. In this example we see such an system architecture: (a) Overall system architecture includes the mobile console component and the remote processing server (Expert System) which performs the computation-extensive work. (b) Mobile Console Architecture: The console has one or more data acquisition devices, a communication module and a display capability. (c) Server Architecture: Contains a communication module, a processing engine, a visualization engine, and an expert assessment mechanism (Mohammed et al. 2011).

The usefulness of mobile computing applications in the medical domain is commonly accepted. Such systems are able to facilitate efficient and effective patient care information input and access at the point of patient care resulting in an improvement of the quality of services (Holzinger et al. 2011). Problems include issues of usability, privacy, and security.

Slide 10-41: Future: Software as a Service (SaaS)

This image illustrates the delivery of health data and information from a cloud computing environment (Botts et al. 2010).

The main problems involved are legal ones: privacy, security, safety and data protection (Weippl et al. 2006).

Slide 10-42: Future Content Analytics

Hospital Information Systems contain—to a large part—unstructured information. Such unstructured information is the subset of information, where the information itself describes parts of what constitutes as significant within it, or in other words—structure and information are not completely separable. The best example for such unstructured information is text (in the medical area often—not quite correctly called "free text"). Although such text can easily be created by medical professionals, the support of automatic analyses for knowledge discovery is extremely difficult. We follow the definition that knowledge consists of a set of hypotheses, and knowledge discovery is the process of finding or generating new hypotheses by medical professionals with the aim of getting insight into the data. In the future the HCI-KDD approach will be most important, i.e., with the human expert in the loop matching the best of two worlds: human intelligence with computational intelligence (Holzinger et al. 2013).

One goal for the future is to combine the three worlds: Watson Technology (= synonym for machine learning) on mobile computers and cloud computing. The vision is to allow to state questions to the data, i.e., a medical doctor might say: "Watson, what was conspicuous within the history of patient X within the last two years?"

IBM—with Watson and with its Jeopardy experience—is developing the technology to be able to handle complex questions across health care with the aim (strategically put towards 2020) to provide a medical assistant. Making Watson a capable assistant will require much research and advances in several areas, because the current Watson needs ten racks of IBM Power750 servers located in Yorktown Heights, NY. That compares to approx. 6,000 desktop computers. You can find more information via: http://www-01.ibm.com/software/ecm/offers/programs/icpa.html

9 Exam Questions

9.1 Yes/No Decision Questions

Please check the following sentences and decide whether the sentence is true = YES; or false = NO; for each correct answer you will be awarded 2 credit points.

01	Workflow modeling is basically the process of simplifying reality with all the problems and dangers of oversimplification involved.	☐ Yes ☐ No	2 total
02	UML includes a set of graphic notation techniques to create visual models of object-oriented systems.	☐ Yes ☐ No	2 total
03	HL7 (Health Level 7) is both an organization and a standard, which was founded in 1987 aiming at a general standard for hospital information systems.	☐ Yes ☐ No	2 total
04	Cloud based computing follows the SaaS-Paradigm "Software as a service" and offers a pay-per-use based virtual infrastructure.	☐ Yes ☐ No	2 total
05	A workflow management system is a system in the hospital that defines, creates and manages the execution of workflows.	☐ Yes ☐ No	2 total
06	Bronfenbrenner's macrosystem includes the cultural values, customs, and laws in a given environment.	☐ Yes ☐ No	2 total
07	Medical certifications such as FDA in the US and CE marking in Europe are also applicable to Open Source software.	☐ Yes ☐ No	2 total
08	Integration of theoretical frameworks and formalisms for modeling clinical guidelines, workflows, and pathways is already be done on a standard business basis.	☐ Yes ☐ No	2 total
09	The vertical approach in hospital information systems builds on open, distributed systems, hospital intranets and networked electronic patient/health records.	☐ Yes ☐ No	2 total
10	Although medical images are lightweight from a technological perspective, the workflows for image generation and viewing are still very diverse.	☐ Yes ☐ No	2 total

Sum of Question Block A (max. 20 points)

9.2 Multiple Choice Questions (MCQ)

The following questions are composed of two parts: the stem, which identifies the question or problem and a set of alternatives which can contain 0, 1, 2, 3, or 4 correct answers, along with a number of distractors that might be plausible—but are incorrect. Please **select the correct answers** by ticking ☒—and do not forget that it can be none. Each question will be awarded 4 points *only if everything is correct*.

01	Future hospital information systems will ... ☐ a) ... be stand-alone vertical systems. ☐ b) ... use software-as-a-service. ☐ c) ... have no access to databases outside the environment. ☐ d) ... offer mobile computing services.	4 total
02	Privacy, security, safety and data protection... ☐ a) ... is a nice to have add-on. ☐ b) ... is an additional and voluntary service. ☐ c) ... is a central and important issue. ☐ d) ... is a matter of future research.	4 total
03	Health Level 7 .... ☐ a) ... includes a CDA to specify structure and semantics of clinical docs. ☐ b) ... is a set of protocols for image network communication. ☐ c) ... provides codes for the identification of genetic mutations. ☐ d) ... contains data types to derive domain-specific information models.	4 total
04	Within the ISO/OSI layer model, DICOM "uses" the ... ☐ a) ... application, presentation and physical layer. ☐ b) ... application, network and session layer. ☐ c) ... application, data link and session layer. ☐ d) ... application, presentation and session layer.	4 total
05	First commercial PACS system were introduced ... ☐ a) ... straight with the beginning of medical computing in the 1970ies. ☐ b) ... already with the upcoming of mainframe computers in the 1960ies. ☐ c) ... with the introduction of the WorldWideWeb in the 1990ies. ☐ d) ... with delay based on academic pioneer work in the 1980ies.	4 total
06	Part of the definition of Biomedical Informatics is the ... ☐ ... effective use of biomedical data. ☐ ... motivation to improve computational capacities. ☐ ... effort to expand the technological capabilities. ☐ ... motivation to improve human health.	4 total
07	A basic architecture of a standard HIS contains always ... ☐ a) ... PACS, HIS and LIS. ☐ b) ... clinical terminals. ☐ c) ... patient ID-services. ☐ d) ... a data center.	4 total
08	The use of monolithic mainframes is called ... ☐ a) ... Vertical approach. ☐ b) ... Horizontal approach. ☐ c) ... Integrated approach. ☐ d) ... Distributed approach.	4 total

09	Use case diagrams... ☐ a) ... are used as an architecture-level artifact. ☐ b) ... show to analysts possible states of objects and transitions. ☐ c) ... describe sequences of actions, providing measurable values to actors. ☐ d) .... model the view of the run-time configuration of processing nodes.	4 total
10	Workflow modelling ... ☐ a) ... can be seen as an abstraction of the real world. ☐ b) ... is needed to demonstrate the complexity of the reality. ☐ c) ... is based on heuristics and feelings gathered during interviews. ☐ d) ... is based on facts gathered during observations and we need to accept that this representation can never be perfect.	4 total

Sum of Question Block B (max. 40 points)	

9.3 Free Recall Block

Please follow the instructions below. At each question you will be assigned the credit points indicated if your option is correct (partial points may be given).

01	Please complete the missing items in the following overview of the ISO OSI Reference Layer:	1 each 4 total
02	Please make a hand-drawn sketch of the classic architecture of a hospital information system according to Reichertz (1984). Note that at this time PACS systems were standalone and NOT integrated in the whole system structure. (Note: Points are awarded if at least "central data base" – "central communication system" – "patient records" – "order entry" ... appears)	1 each 4 total

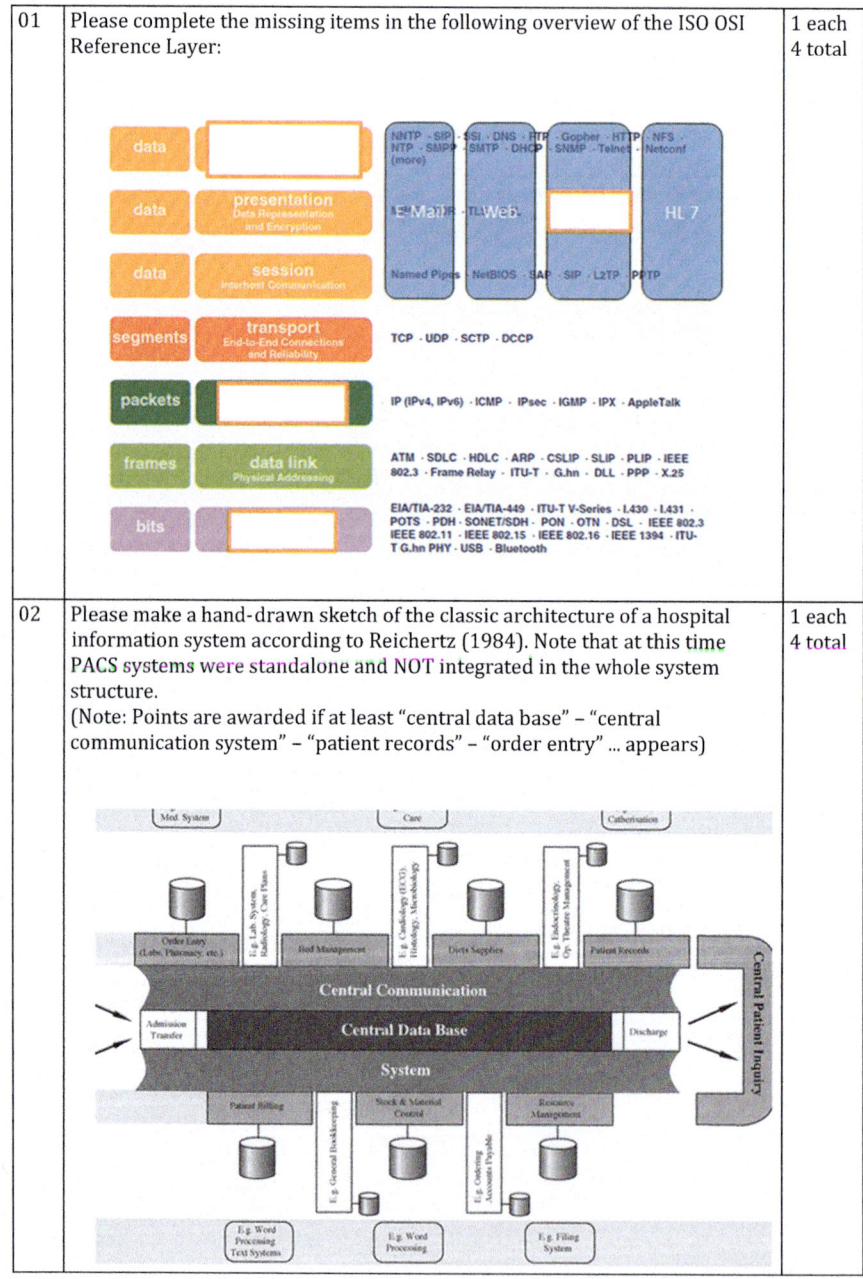

03	A standard universal integrated architecture contains a data center – fusing together the data from the main data bases (patient records, laboratory information system, PACS storage, etc.) and a integrated view in terms of a clinical workplace – please complete the missing items:	1 each 3 total
04	In the image below, showing the PACS basic components and data flow please complete the missing items:	1 1 each 4 total

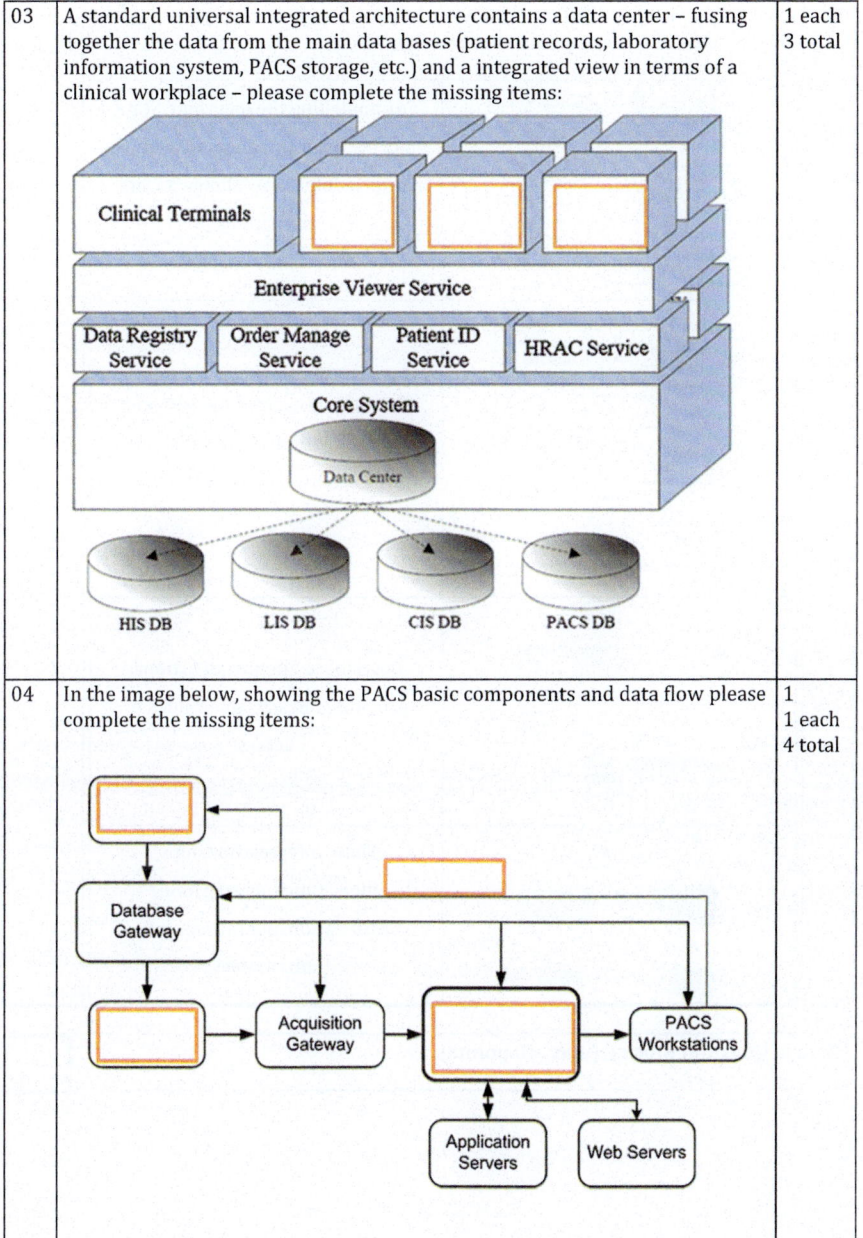

05	Modeling provides a different view for different people: please assign the diagrams to the correct action what you want to do.	1 each 4 total
	[Use case diagram]	For modelling the logic and State Machine diagrams show to analysts the possible states of objects and transitions that causes changes
	[Collaboration/communication diagram]	Show the message flow between objects and imply the relationships between classes
	[Class diagram]	Describe sequences of actions, providing measurable values to actors
	[State machine diagram]	Show interrelationships, inheritance, aggregation, association, operations and attributes

Sum of Question Block C (max. 40 points)

10 Answers

10.1 Answers to the Yes/No Questions

Please check the following sentences and decide whether the sentence is true = YES; or false = NO; for each correct answer you will be awarded 2 credit points.

01	Workflow modeling is basically the process of simplifying reality with all the problems and dangers of oversimplification involved.	☒ Yes ☐ No	2 total
02	UML includes a set of graphic notation techniques to create visual models of object-oriented systems.	☒ Yes ☐ No	2 total
03	HL7 (Health Level 7) is both an organization and a standard, which was founded in 1987 aiming at a general standard for hospital information systems.	☒ Yes ☐ No	2 total
04	Cloud based computing follows the SaaS-Paradigm "Software as a service" and offers a pay-per-use based virtual infrastructure.	☒ Yes ☐ No	2 total
05	A workflow management system is a system in the hospital that defines, creates and manages the execution of workflows.	☒ Yes ☐ No	2 total
06	Bronfenbrenner's macrosystem includes the cultural values, customs, and laws in a given environment.	☒ Yes ☐ No	2 total
07	Medical certifications such as FDA in the US and CE marking in Europe are also applicable to Open Source software.	☐ Yes ☒ No	2 total
08	Integration of theoretical frameworks and formalisms for modeling clinical guidelines, workflows, and pathways is already be done on a standard business basis.	☐ Yes ☒ No	2 total
09	The vertical approach in hospital information systems builds on open, distributed systems, hospital intranets and networked electronic patient/health records.	☐ Yes ☒ No	2 total
10	Although medical images are lightweight from a technological perspective, the workflows for image generation and viewing are still very diverse.	☒ Yes ☐ No	2 total

Sum of Question Block A (max. 20 points)	

10.2 Answers to the Multiple Choice Questions (MCQ)

01	Future Hospital Information Systems will ... ☐ a) ... be stand-alone vertical systems. ☒ b) ... use software-as-a-service. ☐ c) ... have no access to databases outside the environment. ☒ d) ... offer mobile computing services.	4 total
02	Privacy, Security, Safety and Data Protection ... ☐ a) ... is a nice to have add-on. ☐ b) ... is an additional and voluntary service. ☒ c) ... is a central and important issue. ☒ d) ... is a matter of future research.	4 total
03	Health Level 7 .... ☒ a) ... includes a CDA to specify structure and semantics of clinical docs. ☐ b) ... is a set of protocols for image network communication. ☐ c) ... provides codes for the identification of genetic mutations. ☒ d) ... contains data types to derive domain-specific information models.	4 total
04	Within the ISO/OSI layer model, DICOM "uses" the ... ☐ a) ... application, presentation and physical layer. ☐ b) ... application, network and session layer. ☐ c) ... application, data link and session layer. ☒ d) ... application, presentation and session layer.	4 total
05	First commercial PACS system were introduced ... ☐ a) ... straight with the beginning of medical computing in the 1970ies. ☐ b) ... already with the upcoming of mainframe computers in the 1960ies. ☐ c) ... with the introduction of the WorldWideWeb in the 1990ies. ☒ d) ... with delay based on academic pioneer work in the 1980ies.	4 total
06	Part of the definition of Biomedical Informatics is the ... ☒ ... effective use of biomedical data. ☐ ... motivation to improve computational capacities. ☐ ... effort to expand the technological capabilities. ☒ ... motivation to improve human health.	4 total
07	A basic architecture of a standard HIS contains always ... ☒ a) ... PACS, HIS and LIS. ☒ b) ... clinical terminals. ☒ c) ... patient ID-services. ☒ d) ... a data center.	4 total
08	The use of monolithic mainframes is called ... ☒ a) ... Vertical approach. ☐ b) ... Horizontal approach. ☐ c) ... Integrated approach. ☐ d) ... Distributed approach.	4 total
09	Use Case Diagrams ... ☐ a) ... are used as an architecture-level artifact. ☐ b) ... show to analysts possible states of objects and transitions. ☒ c) ... describe sequences of actions, providing measurable values to actors. ☐ d) .... model the view of the run-time configuration of processing nodes.	4 total

10	Workflow modelling ... ☒ a) ... can be seen as an abstraction of the real world. ☐ b) ... is needed to demonstrate the complexity of the reality. ☐ c) ... is based on heuristics and feelings gathered during interviews. ☒ d) ... is based on facts gathered during observations and we need to accept that this representation can never be perfect.	4 total

Sum of Question Block B (max. 40 points)

10.3 Answers to the Free Recall Questions

01	Please complete the missing items in the following overview of the ISO OSI Reference Layer:	1 each 4 total
	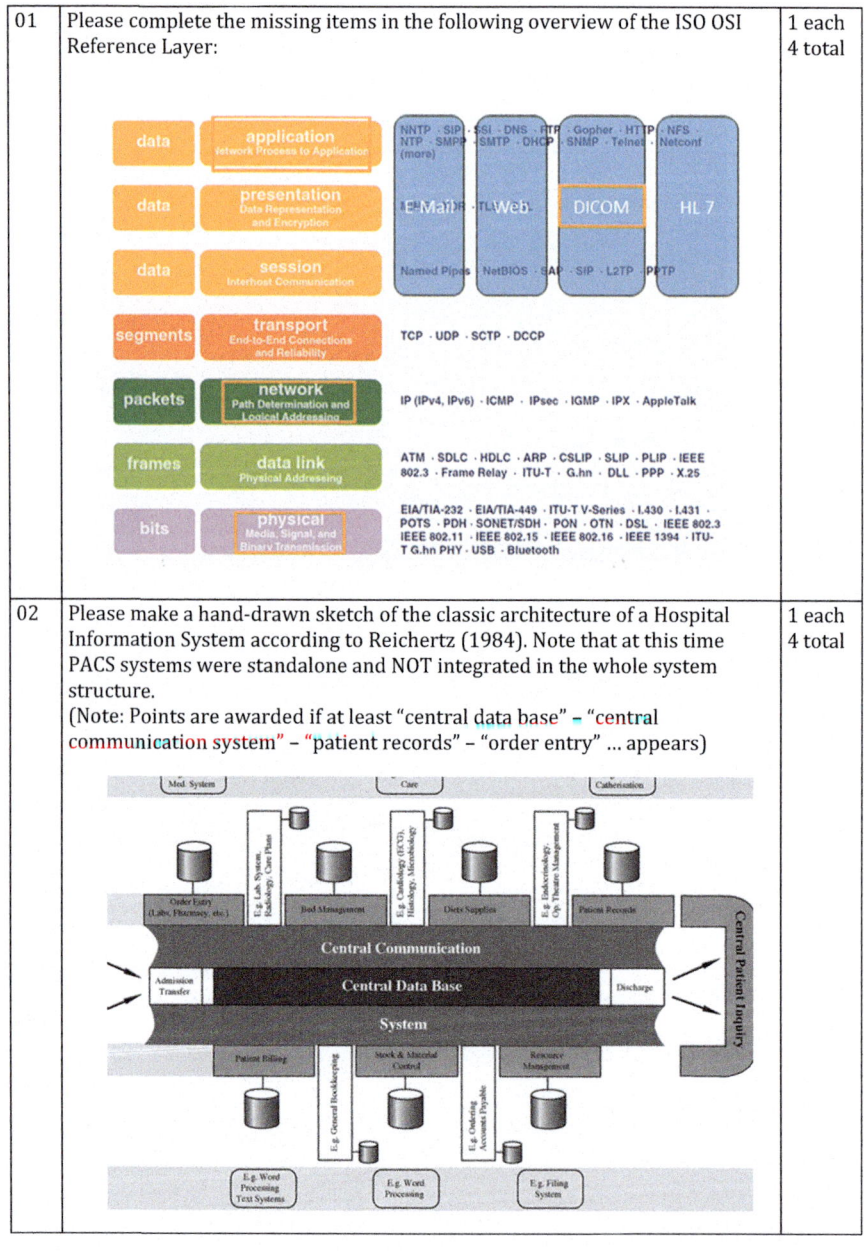	
02	Please make a hand-drawn sketch of the classic architecture of a Hospital Information System according to Reichertz (1984). Note that at this time PACS systems were standalone and NOT integrated in the whole system structure. (Note: Points are awarded if at least "central data base" – "central communication system" – "patient records" – "order entry" ... appears)	1 each 4 total

03	A standard universal integrated architecture contains a data center – fusing together the data from the main data bases (patient records, laboratory information system, PACS storage, etc.) and a integrated view in terms of a clinical workplace – please complete the missing items:	1 each 3 total
04	In the image below, showing the PACS basic components and data flow please complete the missing items:	1 1 each 4 total

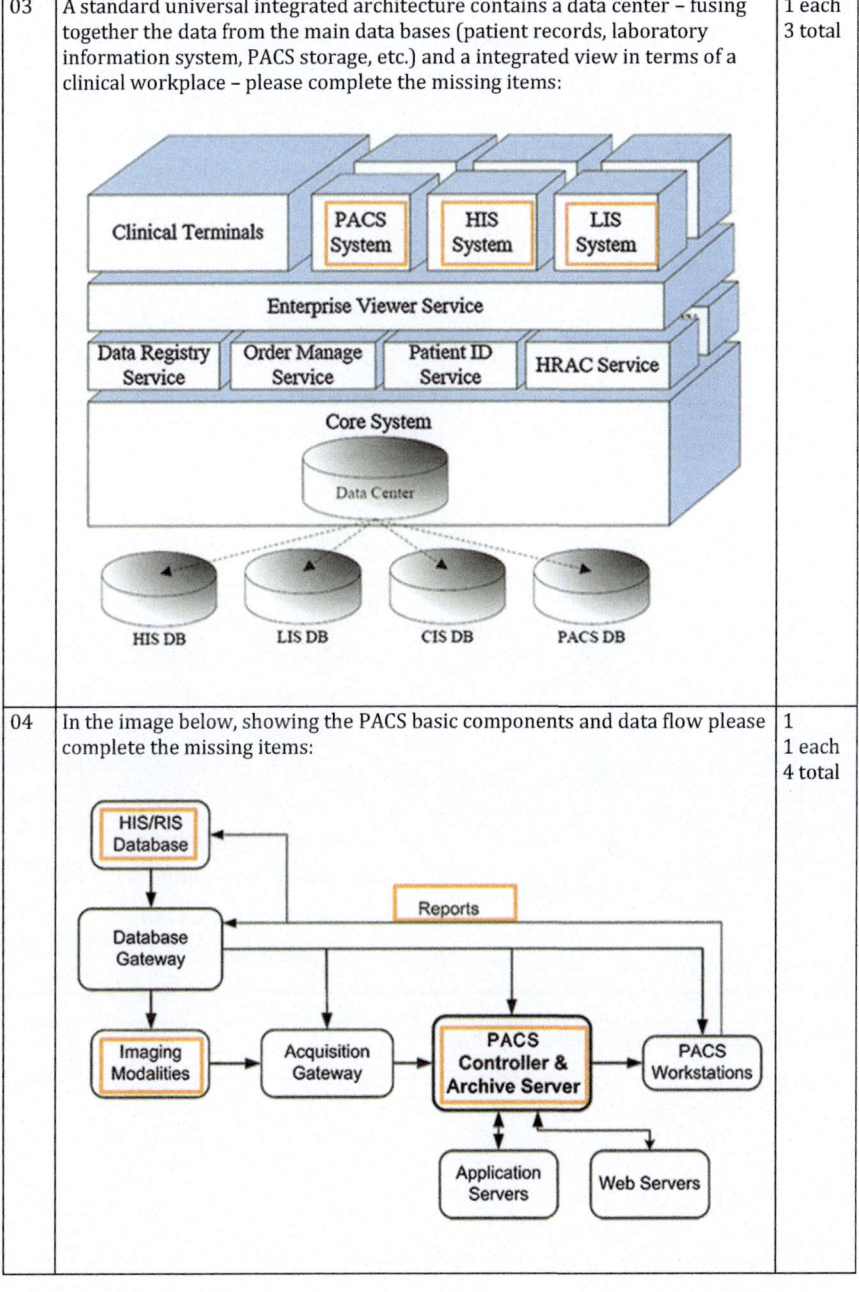

05	Modeling provides a different view for different people: please assign the diagrams to the correct action what you want to do.	1 each 4 total

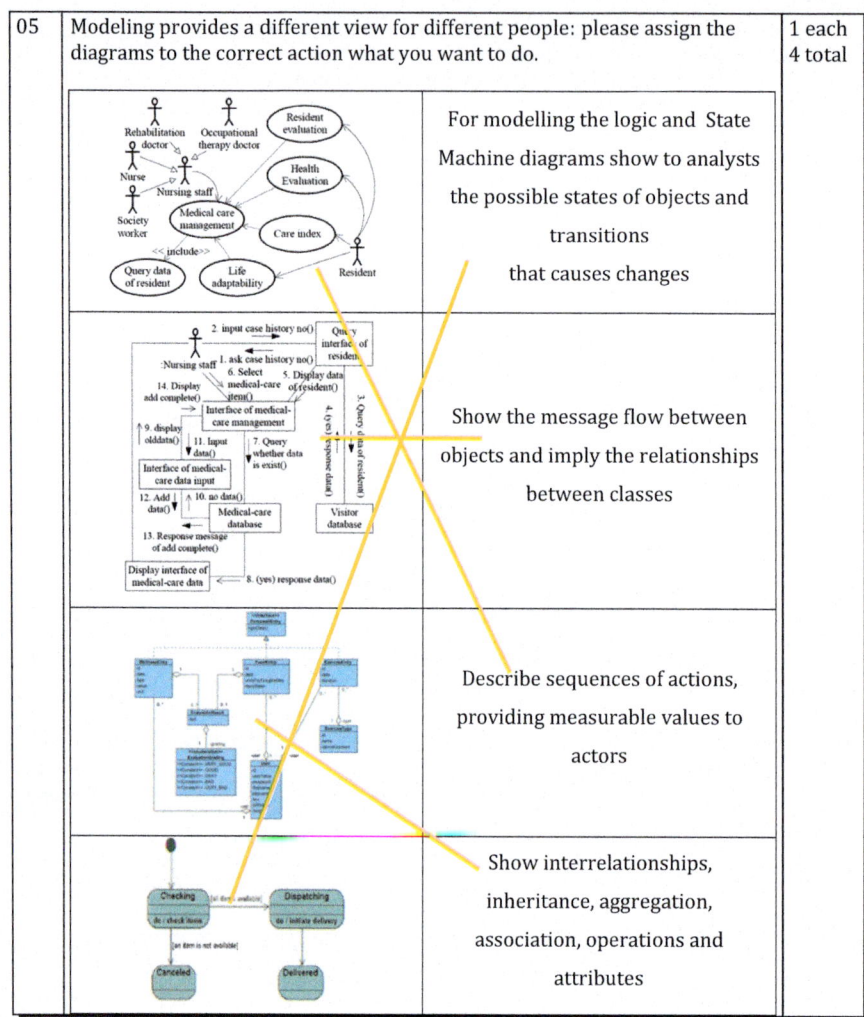

	For modelling the logic and State Machine diagrams show to analysts the possible states of objects and transitions that causes changes
	Show the message flow between objects and imply the relationships between classes
	Describe sequences of actions, providing measurable values to actors
	Show interrelationships, inheritance, aggregation, association, operations and attributes

Sum of Question Block C (max. 40 points)	

References

Bellon E, Feron M, Deprez T, Reynders R, Van Den Bosch B (2011) Trends in PACS architecture. Eur J Radiol 78(2):199–204

Booch G (1994) Object-oriented analysis and design with applications. Benjamin, Redwood City, CA

Booch G, Rumbaugh J, Jacobson I (1999) The unified modeling language user guide. Addison Wesley, Reading, MA

References

Botts N, Thoms B, Noamani A, Horan TA (2010) Cloud computing architectures for the underserved: Public Health Cyberinfrastructures through a Network of HealthATMs. 43rd International conference on system sciences (HICSS), pp 1–10

Bronfenbrenner U (1977) Toward an experimental ecology of human-development. Am Psychol 32(7):513–531

Bui AAT, Taira RK (2010) Medical imaging informatics. Springer, New York

Dadam P, Reichert M, Kuhn K (2000) Clinical workflows-the killer application for process-oriented information systems? 4th International conference on business information systems (BIS '00), 2000 Poznan (Poland)

Davenport TH, Short JE (1990) The new industrial engineering: information technology and business process redesign. Sloan Manag Rev 31(4):11

El Azami I, Cherkaoui Malki M, Tahon C (2012) Integrating hospital information systems in healthcare institutions: a mediation architecture. J Med Syst 36(5):3123–3134

Faggioni L, Neri E, Castellana C, Caramella D, Bartolozzi C (2011) The future of PACS in healthcare enterprises. Eur J Radiol 78(2):253–258

Gooch P, Roudsari A (2011) Computerization of workflows, guidelines, and care pathways: a review of implementation challenges for process-oriented health information systems. J Am Med Inform Assoc. doi:10.1136/amiajnl-2010-000033

Hasan S, Daugelat S, Rao PSS, Schreiber M (2006) Prioritizing genomic drug targets in pathogens: application to *Mycobacterium tuberculosis*. PLoS Comput Biol 2(6):e61

Holzinger A (2002) Basiswissen IT/Informatik Band 1: Informationstechnik. Vogel Buchverlag, Wuerzburg

Holzinger A, Dorner S, Födinger M, Valdez A, Ziefle M (2010) Chances of increasing youth health awareness through mobile wellness applications. In: Leitner G, Hitz M, Holzinger A (eds) HCI in work and learning, life and leisure, vol 6389, Lecture notes in computer science, LNCS. Springer, Berlin, pp 71–81

Holzinger A, Kosec P, Schwantzer G, Debevc M, Hofmann-Wellenhof R, Frühauf J (2011) Design and development of a mobile computer application to reengineer workflows in the hospital and the methodology to evaluate its effectiveness. J Biomed Inform 44(6):968–977

Holzinger A, Sammer P, Hofmann-Wellenhof R (2006) Mobile computing in medicine: designing mobile questionnaires for elderly and partially sighted people. In: Miesenberger K, Klaus J, Zagler WL, Karshmer AI (eds) Computers helping people with special needs, ICCHP 2006, vol 4061, Lecture Notes in Computer Science (LNCS). Springer, Berlin, pp 732–739

Holzinger A, Stocker C, Ofner B, Prohaska G, Brabenetz A, Hofmann-Wellenhof R (2013) Combining HCI, natural language processing, and knowledge discovery—potential of IBM Content Analytics as an assistive technology in the biomedical domain, vol 7947, Lecture notes in computer science (LNCS). Springer, Heidelberg, pp 13–24

Huang HK (2011) Short history of PACS.Part I: USA. Eur J Radiol 78(2):163–176

Juan YC, Ma CM, Chen HM (2005) Applying UML to the development of medical care process management system for nursing home residents. Int J Elect Bus Manage 3(4):322–330

Kuhn TS (1962) The structure of scientific revolutions. University of Chicago Press, Chicago, IL

Lemke HU (2003) PACS developments in Europe. Comput Med Imaging Graph 27(2–3):111–120

Lemke HU (2011) Short history of PACS (Part II: Europe). Eur J Radiol 78(2):177–183

Lemke HU, Berliner L (2011) PACS for surgery and interventional radiology: features of a therapy imaging and model management system (TIMMS). Eur J Radiol 78(2):239–242

Malhotra S, Jordan D, Shortliffe E, Patel VL (2007) Workflow modeling in critical care: piecing together your own puzzle. J Biomed Inform 40(2):81–92

Mohammed S, Servos D, Fiaidhi J (2011) Developing a secure distributed OSGI cloud computing infrastructure for sharing health records. In Autonomous and intelligent systems, pp 241–252

Oestereich B (1999) Developing software with UML: object-oriented analysis and design in practice. Addison Wesley, Harlow

Piętka E, Kawa J, Spinczyk D, Badura P, Więcławek W, Czajkowska J, Rudzki M (2011) Role of radiologists in CAD life-cycle. Eur J Radiol 78(2):225–233

Ratib O, Rosset A, Heuberger J (2011) Open source software and social networks: disruptive alternatives for medical imaging. Eur J Radiol 78(2):259–265

Romano P (2008) Automation of in-silico data analysis processes through workflow management systems. Brief Bioinform 9(1):57–68

Wang J, Rosca D, Tepfenhart W, Milewski A, Stoute M (2008) Dynamic workflow modeling and analysis in incident command systems. IEEE Trans Syst Man Cybernet A Syst Hum 38(5):1041–1055

Weippl E, Holzinger A, Tjoa AM (2006) Security aspects of ubiquitous computing in health care. Elektrotech Informationstech 123(4):156–162

Wong A, Lou SL (2009) Medical image archive, retrieval, and communication. In: Isaac NB (ed) Handbook of medical image processing and analysis, 2nd edn. Academic, Burlington, pp 861–873

Lecture 11

Biomedical Data: Privacy, Safety, and Security

1 Learning Goals

After this lecture, you:

- would be able to determine the differences between privacy, safety, security, and data protection.
- would be able know the famous IOM "Why do accidents happen" report and its influence on safety engineering.
- would have a basic understanding of human error and are able to determine types of adverse events in medicine and health care.
- would have seen some examples on how ubiquitous computing might contribute to enhancing patient safety.
- would get an idea of the principles of context-aware patient safety.
- would have seen a recent approach on pseudonymization for privacy in e-health.
- would be aware of the security characteristics of personal health records.

2 Advance Organizer

Acceptable risk	the residual risk remaining after identification/reporting of hazards and the acceptance of those risks
Adverse event	harmful, undesired effect resulting from a medication or other interventions such as surgery
Anonymization	important method of de-identification to protect the privacy of health information (antonym: reidentification)
Authentication	to verify the identity of a user (or other entities, could also be another device), as a prerequisite to allow access to the system; also to verify the integrity of the stored data to possible unauthorized modification

Confidentiality	the rule dates back to at least the Hippocratic Oath: "Whatever, in connection with my professional service, or not in connection with it, I see or hear, in the life of man, which ought not to be spoken of abroad, I will not divulge, as reckoning that all such should be kept secret"
Data protection	ensuring that personal data is not processed without the knowledge and the consent of the data owner (e.g., patient)
Data security	includes confidentiality, integrity, and availability of data, and helps to ensure privacy
Hazard	the potential for adverse effects, but not the effect (accident) itself; hazards are just contributory events that might lead to a final adverse outcome
Human fallibility	addresses the fundamental sensory, cognitive, and motor limitations of humans that predispose them to error
k-Anonymity	an approach to counter linking attacks using quasi-identifiers, where a table satisfies k-anonymity if every record in the table is indistinguishable from at least $k-1$ other records with respect to every set of quasi-identifier attributes; hence, for every combination of values of the quasi-identifiers in the k-anonymous table, there are at least k records that share those values, which ensures that individuals cannot be uniquely identified by linking attacks
Medical error	any kind of adverse effect of care, whether or not harmful to the patient; including inaccurateness, incompleteness of a diagnosis, treatment, etc.
Nomen nescio (N.N)	used to signify an anonymous nonspecific person
Patient safety	in health care this is the equivalent of system safety in industry
Personally identifying information	can be used to connect a medical record back to an identified person
Prevention	any action directed to preventing illness and promoting health to reduce the need for secondary or tertiary health care; including the assessment of disease risk and raising public health awareness
Privacy	(US pron. "prai ..."; UK pron. "pri ..."; from Latin: privatus "separated from the rest") is the individual rights of people to protect their personal life and matters from the outside world
Privacy policy	organizational access rules and obligations on privacy, use and disclosure of data

Protected health information (PHI)	any info on for example health status, treatments, or even payment details for health care which may be linked back to a particular person
Pseudonymization	procedure where (some) identifying fields within a data record are replaced by artificial identifiers (pseudonyms) in order to render the patient record less identifying
Quasi-Identifiers	sets of attributes (e.g., gender, date of birth, and zip code) that can be linked with external data so that it is possible to identify individuals out of the population
Safety	any protection from any harm, injury, or damage
Safety engineering	is an applied science strongly related to systems engineering/industrial engineering and the subset system safety engineering. Safety engineering assures that a life-critical system behaves as needed even when components fail.
Safety risk management	follows the process defined in the ISO 14971 standard (see Lecture 12)
Safety-critical systems research	interdisciplinary field of systems research, software engineering and cognitive psychology to improve safety in high-risk environments; such technologies cannot be studied in isolation from human factors and the contexts and environments in which they are used
Security	(in terms of computer, data, information security) means protecting from unauthorized access, use, modification, disruption, destruction, etc.
Sensitive data	According to EC definition it encompasses *all* data concerning health of a person
Swiss-Cheese model	used to analyze the causes of systematic failures or accidents in aviation, engineering, and health care; it describes accident causation as a series of events which must occur in a specific order and manner for an accident to occur

3 Acronyms

AERFMI	Adverse Events Reporting Forms in Medical Imaging
AERMMI	Adverse Events Manager Reports in Medical Imaging
AEKMMI	Adverse Events Knowledge Manager in Medical Imaging
AHRQ	Agency for Healthcare Research and Quality
ASA	American Society of Anesthesiologists
CRC	Cyclic Redundancy Check

ECM	Eindhoven Classification Model
FAA	Federal Aviation Administration
HIPAA	Health Insurance Portability and Accountability Act
HRO	High Reliability Organization
HSM	Hardware Security Module
ID	Identification
IOM	Institute of Medicine
MAC	Message authentication code
PHI	Protected Health Information
PHR	Personal Health Record
PIPE	Pseudonymization of Information for Privacy in e-Health
RFID	Radiofrequency Identification
QI	Quality Improvement

4 Key Problems

Slide 11-1: Key Challenges

- Data in the cloud.
- Mobile solutions, the trend towards "software-as-a-service".
- The massive increase in the amount of data ...

... in the medical area require a lot of future effort in privacy, data protection, security, and safety.

The challenges of data integration, data fusion, and the increased use of data for secondary use put these issues from a "nice-to-have" into the key interest.

Example: In January 2013, the US Department of Health and Human Services released the Omnibus Final Rule, which significantly modified the privacy and security standards under the Health Insurance Portability and Accountability Act (HIPAA). These new regulations were driven by a need to ensure the confidentiality, integrity, and security of patients' protected health information (PHI) in electronic health records (EHRs) and addresses these concerns by expanding the scope of regulations and increasing penalties for PHI violations (Wang and Huang 2013).

5 Standardization and Health Care

5.1 What Is Risk?

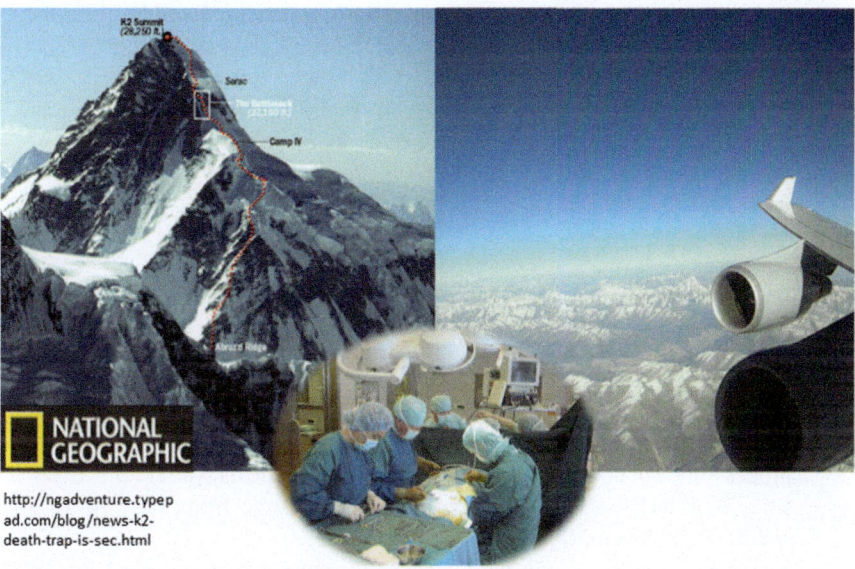

Fig. 1 See Slide 11-2

Slide 11-2: We Start with Thinking About Safety First ...

According to a classic survey by (Amalberti et al. 2005) we can determine between very risky enterprises, typically Himalaya mountaineering and relatively save enterprises with low risk, typically commercial large-jet aviation. The medical area is in between, with a tendency to the Himalaya depending on the health area.

Slide 11-3: Exposure of Catastrophes: Associated Deaths

In this slide we see the average rates per exposure of catastrophes and associated deaths in various industries and human activities (Himalaya mountaineering versus commercial large-jet aviation).

The size of each box represents the range of risk in which a given barrier is active. Reduction of risk beyond the maximum range of a barrier

(continued)

(continued)

presupposes crossing this barrier. Shaded boxes represent the five system barriers (ASA = American Society of Anesthesiologists).

In health care, placed on autonomy, the drive for productivity, and the economics of the system may lead to severe safety constraints and adverse medical events. In many clinical domains, such as trauma surgery, the rate of serious complications is relatively high, but not all complications are related to medical errors. In contrast, some health-care sectors, e.g., gastroenterologic endoscopy, are inherently very safe (Amalberti et al. 2005).

Slide 11-4: Definitions: Privacy, Security, Safety

1. **Privacy** = include the individual rights of people to protect their personal life and matters from the outside world.
2. **Safety** = any protection from harm, injury, or damage; a weighting process reflects how comfortable an organization deals with its risk exposure. Accident rates in health care currently range from 10^{-1} to 10^{-7} events per exposure (Amalberti et al. 2005).
3. **Security** = (in terms of computer, data, information security) means protecting from unauthorized access, use, modification, disruption, destruction, etc.

A good example for these issues is the electronic health record in Slide 11-26: The patient data must be confidential, secure, and safe, whilst at the same time it must be usable, useful, accurate, up to date, and accessible.

5.2 The IOM Report

Slide 11-5: The Famous Report "Why Do Accidents Happen"

As we have already heard in Lecture 7, the Institute of Medicine (IOM) released a report in 1999 entitled "To Err is Human: Building a Safer Health System." The IOM report called for a 50 % reduction in medical errors over 5 years. Its goal was to break the cycle of inaction regarding medical errors by advocating a comprehensive approach to improve patient safety. The health-care industry responded with a wide range of patient safety efforts and safety was a topic for researchers. Hospital information systems vendors adopted safer practices and emphasized that safety was also now a priority for them (Stelfox et al. 2006).

(continued)

(continued)

However, so far no comprehensive nationwide monitoring system exists for patient safety, and a recent effort by the Agency for Healthcare Research and Quality (AHRQ) to get a national estimate by using existing measures showed little improvement (Leape and Berwick 2005).

Slide 11-6: The Impact of the "To Err Is Human" IOM Study

A large shift in the number of patient safety publications followed the release of the IOM report as we can see in this slide. Approximately 60 papers on patient safety were published per 100,000 MEDLINE publications in the 5 years before the IOM report; this increased to approximately 165 articles per 100,000 MEDLINE publications in the 5 years after the report (this is a significant increase). Rates of patient safety publications in the top general medical journals mirrored those in MEDLINE indexed journals, averaging four articles per 100,000 MEDLINE publications before the IOM report and 13 articles per 100,000 MEDLINE publications after the IOM report (Stelfox et al. 2006).

Slide 11-7: Research Activities Stimulated by the IOM Report

This slide impressively shows the large increase in patient safety research followed by the release of the IOM report (Stelfox et al. 2006).

Slide 11-8: Deaths from Medical Error (2009)

But still there is the issue of deaths from medical errors as this blog issue of Scientific American discusses. On Aug 10, 2009, Katherine Harmon reported: Preventable medical mistakes and infections are responsible for about 200,000 deaths in the USA each year, according to an investigation by the Hearst media corporation. The report comes 10 years after the Institute of Medicine's "To Err Is Human" analysis, which found that 44,000–98,000 people were dying annually due to these errors and called for the medical community and government to cut that number in half by 2004.

The precise number of these deaths is still unknown because many states lack a standard or mandatory reporting system for injuries due to medical

(continued)

(continued)

mistakes. The investigative team gathered disparate medical records, legal documents, personnel files and reports and analyzed databases to arrive at its estimate.

Many, including President Barack Obama, have advocated for a broader adoption of electronic medical records as both a life- and cost-saver. But not everyone is convinced that current technology will help doctors and nurses who already have set ways of handling patient information. "The systems as they stand now are still fairly clunky and user unfriendly," Robert Wachter, a professor of hospital medicine at the University of California, San Francisco, told Hearst. "In the last several years, we've seen a literature emerge of medical errors caused by computer systems."

Some think that despite the grim numbers, patient safety has improved overall since the 1999 report. "Now, you have checklists prior to surgery; you mark the spot on which limb you were going to operate on," Mary Stefl, dean of health-care administration at Trinity University in San Antonio, Texas, told Hearst. "And afterwards, they count the surgical sponges and instruments so they presumably don't leave anything inside. But it still happens."

Source: http://www.scientificamerican.com/blog/post.cfm?id=deaths-from-avoidable-medical-error-2009-08-10

5.3 Medical Error

Slide 11-9: Medical Error Example: Wrong-Site Surgery

As you can still read in the newspapers wrong-site surgery is still a big issue, or as (Manjunath et al. 2010) put it forward it is *a clear and constant fear*.

5.3.1 Eindhoven Classification Model

Slide 11-10: Deal with Errors: Eindhoven Classification Model

A large number of different systems have been used to classify events regarding to patient safety and many of the methods used to analyze patient safety were adapted from **risk-management techniques in industries**, especially in high-risk industries such as the chemical, nuclear power and most of all the aviation industry.

(continued)

(continued)

Remark: Often medicine is compared with aviation—this is only partly true: Imagine fog at the airport—the captain can decide not to start; but in an emergency a surgeon has to start an operation, regardless the difficulties involved.

In this slide we see the Eindhoven Classification Model (ECM), which was originally developed to manage human error in the chemical process industry and was then applied to various other industries, finally in health care. The ECM medical version consists of 20 codes, divided into four categories, frequently used in a medical environment to classify the underlying causes of the adverse events (Rodrigues et al. 2010). In the next slide we see an application of it.

5.3.2 Adverse Event Reporting

Slide 11-11: Adverse Event Reporting and Learning System

Here we see the AEMI (Adverse Events in Medical Imaging) system developed by (Rodrigues et al. 2010), which intends to reduce the amount of time and manual labor required for analysis. The AEMI architecture includes tree modules:

1. Adverse Events Reporting Forms in Medical Imaging (AERFMI).
2. Adverse Events Manager Reports in Medical Imaging (AERMMI).
3. Knowledge Manager Adverse Events in Medical Imaging (AEKMMI).

AERFMI provides the Web interface for adverse events registration. The effort on this interface was focused in its usability. AERMMI is also web-based and aims to enable the individual analysis of each adverse event recorded by AERFMI and provides some relevant statistics related to the various events registered. AEKMMI is a Java application. This module uses the data from the system database to create a Knowledge Base (KB) based on the EECM using the logic programming language Prolog (Rodrigues et al. 2010).

5.3.3 Human Error

Slide 11-12: Review: Framework for Understanding Human Error

In Lecture 7 we discussed a framework for demonstrating how human error—resulting in adverse events—arise. Remember, the framework consists of three components:

1. **Human fallibility** addresses the fundamental sensory, cognitive, and motor limitations of humans that predispose them to error.
2. **Context** refers to situational variables that can affect the way in which human fallibility becomes manifest.
3. **Barriers** concerning the various ways in which human errors can be contained.

We will now focus on one particular issue in the third component: The next slide shows the famous "Swiss cheese" model of accident causation.

Slide 11-13: Reason (1997) Swiss Cheese Model

The "Swiss cheese" model of accident causation emphasizes that adverse events occur when active failures align with gaps or weaknesses in the systems permitting an error to go untrapped and uncompensated (Sundt et al. 2005). The model was originally developed by Reason 1997, and a good reading is Reason 2000.

5.3.4 Risk Management

Slide 11-14: Risk Management: FAA System Safety

We will talk about risk management also in the last lecture, but we need the definitions now for a common understanding, and look at the image top right in the slide:

1. **Total risk** = identified + unidentified risks.
2. **Identified risk** = determined through various analysis techniques. The first task of system safety is to identify, within practical limitations, all

(continued)

(continued)

   possible risks. This step precedes determining the significance of the risk (severity) and the likelihood of its occurrence (hazard probability). The time and costs of analysis efforts, the quality of the safety program, and the state of technology impact the number of risks identified.
3. **Unidentified risk** is the risk not yet identified. Some unidentified risks are subsequently identified when a mishap occurs. Some risks are never known.
4. **Unacceptable risk** is that risk which cannot be tolerated by the managing activity. It is a subset of identified risk that must be eliminated or controlled.
5. **Acceptable risk** is the part of identified risk that is allowed to persist without further engineering or management action. Making this decision is a difficult yet necessary responsibility of the managing activity. This decision is made with full knowledge that it is the user who is exposed to this risk.
6. **Residual risk** is the risk left over after system safety efforts have been fully employed. It is not necessarily the same as acceptable risk. Residual risk is the sum of acceptable risk and unidentified risk. This is the total risk passed on to the user.

Slide 11-15: Improving Safety with IT: Example Mobile

Patient safety in health care is the equivalent of systems safety in industry, which is usually built in four steps:

1. Measuring risk and planning the ideal defense model.
2. Assessing the model against the real behavior of professionals, and modifying the model or inducing a change in behavior when there are gaps.
3. Adopting a better micro- and macro-organization.
4. Gradually reintroducing within the rather rigid, prescriptive system built in steps 1–3 some level of resilience, enabling it to adapt to crises and exceptional situations.

In this slide we see an example of a mobile system screening for laboratory abnormalities, for example, hypokalemia and a decreasing hematocrit, would require urgent action but occur relatively infrequently, often when a clinician is not at hand, and such results can be buried amid less critical data.

Such mobile systems can identify and rapidly communicate these problems to clinicians automatically (Bates and Gawande 2003).

5.3.5 Ubiquitous Devices

Slide 11-16: Enhancing Patient Safety with Ubiquitous Devices

This is another example on how for example wrong site surgery can be avoided: Patients check in at the Hospital—in addition to an ordinary wristband an RFID transponder is supplied. Patient data is entered via our application at the check-in point, any previous patient data can be retrieved from the HIS. From this information, uncritical but important data (such as name, blood type, allergies, vital medication etc.) is transferred to the wristband's RFID transponder. The Electronic Patient Record (EPR) is created and stored at the central server. From this time the patient is easily and unmistakably identifiable. All information can be read from the wristband's transponder or can be easily retrieved from the EPR by identifying the patient with a reader. In contrast to manual identification, automatic processes are less error-prone. Unlike barcodes, RFID transponders can be read without line of sight, through the human body and most other materials. This enables physicians and nurses to retrieve, verify and modify information in the hospital accurately and instantly. In addition, this system provides patient identification and patient data—even when the network is crashed (Holzinger et al. 2005).

Slide 11-17: Security Problems of Ubiquitous Computing

Security requires confidentiality (aka secrecy), integrity and availability. All other requirements such as non-repudiation can be traced back to one of these three requirements. Non-repudiation, for instance, can be seen as a special case of integrity, i.e., the integrity of log data recording.

The most well-known security requirement is confidentiality. It means that users may obtain access only to those objects for which they have received authorization, and will not get access to information they must not see.

The integrity of the data and programs is just as important as confidentiality but in daily life it is frequently neglected. Integrity means that only authorized people are permitted to modify data (or programs). Secrecy of data is closely connected to the integrity of programs of operating systems. If the integrity of the operating system is compromised, then the integrity of the data can no longer be guaranteed. The reason is that a part of the operating system (i.e., the reference monitor) checks for each access to a resource whether the subject is authorized to perform the requested operation. Since the operating system is compromised the reference monitor is no

(continued)

(continued)

longer trustworthy. It is then obvious that secrecy of information cannot be guaranteed any longer if this mechanism is not working. For this reason it is important to protect the integrity of operating systems just as properly as the secrecy of information.

It is through the Internet that many users have become aware that availability is one of the major security requirements for computer systems. Availability is defined as the readiness of a system for correct service.

With growing ubiquitous computing in health care security problems are increasing (Weippl et al. 2006):

1. Protection precautions: Vulnerability to eavesdropping, traffic analysis, spoofing, and denial of service. Security objectives such as confidentiality, integrity, availability, authentication, authorization, nonrepudiation, and anonymity are not achieved unless special security mechanisms are integrated into the system.
2. Confidentiality: The communication between reader and tag is unprotected, except of high-end systems (ISO 14443). Consequently, eavesdroppers can listen in if they are in immediate vicinity.
3. Integrity: With the exception of high-end systems which use message authentication codes (MACs), the integrity of transmitted information cannot be assured. Checksums (cyclic redundancy checks, CRCs) are used, but protect only against random failures. The writable tag memory can be manipulated if access control is not implemented.

5.3.6 Context-Aware Patient Safety

Slide 11-18: Clinical Example: Context-Aware Patient Safety 1/2

Bardram and Norskov (2008) developed a context aware patient safety and information system (CAPSIS) designed for use during surgery, designed to monitor what is going on in the operating room (OR). This information is used to display medical data to the clinicians at the appropriate time, and to issue warnings if any safety issues are detected. CAPSIS was implemented using the Java Context-Awareness Framework (JCAF) and monitors such information as the status of the operation; the status and location of the patient; the location of the clinicians in the operating team; and equipment, medication, and blood bags used in the operating room. This information is acquired and handled by the JCAF context awareness infrastructure, and a special safety service, implemented by means of the Java Expert System Shell (Jess), is used for overall reasoning on what actions should be taken or what

(continued)

(continued)

warnings should be issued. CAPSIS differs from other patient safety systems in being designed to monitor everything (or as many things as possible) in the OR, and therefore to be capable of reasoning across the entire gamut of facts pertaining to the situation in the OR. It thus supplements human vigilance on safety by providing a machine counterpart that is capable of drawing inferences (Bardram and Norskov 2008).

Slide 11-19: Clinical Example: Context Aware Patient Safety 2/2

This slide shows the user interface of the CAPSIS system, which consists of four windows:

(A) Is the main patient safety window, which provides an overview of the patient's safety status for the operation in question.
(B) Shows the patient's medical record.
(C) Shows the patient's medical images.
(D) Shows the relevant checklist for the given surgical procedure.

The patient safety window (A) is composed of three panels: the patient panel, the staff panel, and the patient safety panel. The patient panel aggregates important information about the current patient and surgery, including the patient's name, social security number (SSN), allergies (CAVE), picture, scheduled surgery, and current status and location. The main purpose of this frame is to help the surgical staff avoid the three big wrongs: wrong patient, wrong procedure, and wrong surgical site, as well as presenting vital information on the safety of the patient such as the CAVE list and patient status (Bardram and Norskov 2008).

Slide 11-20: Patient Safety

Patient safety in health care is the equivalent of system safety in industry, which is usually built in four steps: (1) measuring risk and planning the ideal defense model, (2) assessing the model against the real behavior of professionals, and modifying the model or inducing a change in behavior when there are gaps, (3) adopting a better micro- and macro-organization, (4) gradually reintroducing within the rather rigid, prescriptive system built in steps 1–3 some level of resilience enabling it to adapt to crises and

(continued)

(continued)

exceptional situations. The development of patient safety has nowhere near reached step 4 except in specific areas such as blood transfusion or laboratory testing. Even step 1 has not been completed (Amalberti et al. 2011).

Slide 11-21: Types of Adverse Events in Medicine and Care

An error may or may not cause an adverse event. Adverse events are injuries that result from a medical intervention and are responsible for harm to the patient (death, life-threatening illness, disability at the time of discharge, prolongation of the hospital stay, etc.). For example, a near miss (Number 6 in this slide) is an adverse event that either resolves spontaneously or is neutralized by voluntary action before the consequences have time to develop. Adverse events may be due to medical errors, in which case they are preventable, or to factors that are not preventable; so, the occurrence is always a combination of human factors and system factors (Garrouste-Orgeas et al. 2012).

5.4 Safety, Security and Technical Dependability

Slide 11-22: Safety, Security → Technical Dependability

Dependability consists of three parts: the threats to, the attributes of, and the means by which dependability is attained, as shown in this slide.

Computing systems are characterized by five fundamental properties: functionality, usability, performance, cost, and dependability. Dependability of a computing system is the ability to deliver service that can justifiably be trusted. The trust-factor is perceived by the users (remember the Previous Exposure to Technology, PET-Factor (Holzinger et al. 2011)), and a user is another system (human) that interacts with the former at the service interface. The function of a system is what the system is intended to do, and is described by the functional specification. Correct service is delivered when the service implements the system function. A system failure is an event that occurs when the delivered service deviates from correct service. A failure is thus a transition from correct service to incorrect service, i.e., to not implementing the system function. The delivery of incorrect service is a system outage. A transition from incorrect service to correct service is service restoration. Based on the definition of failure, an three alternate definition of

(continued)

(continued)

dependability, which complements the initial definition in providing a criterion for adjudicating whether the delivered service can be trusted or not: the ability of a system to avoid failures that are more frequent or more severe, and outage durations that are longer, than is acceptable to the user(s). In the opposite case, the system is no longer dependable: it suffers from a dependability failure, i.e., a meta-failure (Avizienis et al. 2001).

Slide 11-23: Types of faults: Design, Physical, Interaction

Combining the elementary fault classes leads to the tree in this slide: The leaves of the tree lead into three major fault classes for which defenses need to be devised: design faults, physical faults, interaction faults. The boxes in this slide point at generic illustrative fault classes. Non-malicious deliberate faults can arise during either development or operation. During development, they result generally from tradeoffs, either (a) aimed at preserving acceptable performance and facilitating system utilization, or (b) induced by economic considerations; such faults can be sources of security breaches, in the form of covert channels. Non-malicious deliberate interaction faults may result from the action of an operator either aimed at overcoming an unforeseen situation, or deliberately violating an operating procedure without having realized the possibly damaging consequences of his or her action. Non-malicious deliberate faults share the property that often it is recognized that they were faults only after an unacceptable system behavior, thus a failure, has ensued; the specifier(s), designer(s), implementer(s), or operator(s) did not realize that the consequence of some decision of theirs was a fault (Avizienis et al. 2001).

Slide 11-24: A Two-Tiered System of Medicine

This table by Amalberti et al. (2005) show a detailed comparison of these two possible tiers of health care. Physician training would have to accommodate this two-tiered approach, and patients would have to understand that aggressive treatment of high-risk disease may require acceptance of greater risk and number of medical errors during clinical treatment.

Slide 11-25: Toward a Strategic View on Safety in Health Care

An improved vision by leadership of the safety and dangers of health care is needed to optimize the risk–benefit ratio. Stratification could lead to two tiers or "speeds" of medical care, each with its own type and level of safety goals. This two-tier system could distinguish between medical domains that are stable enough to reach criteria for ultrasafety and those that will always deal with unstable conditions and are therefore inevitably less safe. For medicine, high-reliability organizations may offer a sound safety model and high-reliability organizations are those that have consistently reduced the number of expected or "normal" accidents (according to the normal accident theory) through such means as change to culture and technologic advances, despite an inherently high-stress, fast-paced environment (Amalberti et al. 2005).

6 Patient Data Privacy

Slide 11-26: Requirements of an Electronic Patient Record

Remember the requirements to a patient record from the viewpoint of ensuring privacy: The patient data must be confidential, secure, and safe, while at the same time be usable, useful, accurate, up to date, and accessible.

Slide 11-27: Pseudonymization of Information for Privacy 1/8

An excellent paper by (Neubauer and Heurix 2011) shall provide a good teaching example, in the following consisting of eight slides.

Protection of the patients' data privacy can be achieved with two different techniques, anonymization and encryption, which unfortunately both suffer from major drawbacks: While anonymization—the removal of the identifier from the medical data—cannot be reversed and therefore prevents primary use of the records by health care providers who obviously need to know the corresponding patient (as a minor point, patients cannot benefit from the results gained in clinical studies because they cannot be informed about new findings etc.), encryption of the medical records prevents them from being used for clinical research (secondary use of clinical data). At least without the explicit permission of the patient, who has to decrypt the data and, in doing so, reveals her identity. Considering that some medical records can be very large, encryption can also be seen as a time-consuming operation.

(continued)

(continued)

A method that resolves these issues is **pseudonymization**, where identification data is transformed and then replaced by a specifier that cannot be associated with the identification data without knowing a certain secret. Pseudonymization allows the data to be associated with a patient only under specified and controlled circumstances (Neubauer and Heurix 2011).

Aimed to provide a pseudonymization service, PIPE (Pseudonymization of Information for Privacy in e-Health) can be applied to different scenarios: In the local scenario, the PIPE server pseudonymizes only records stored in the local (health) data repository and makes them available to a local (health care provider's)workstation where both patient and health care provider interact with the pseudonymization server as part of a health care provider environment (e.g., with a hospital information system). In an alternative central scenario, the PIPE pseudonymization server is responsible for providing linking information to different health records stored at distributed locations. In the slide two separate health care provider environments exist where the individual workstations have direct access to their local data repositories. Via the pseudonymization service, the health-care providers are able to access records of other domains if they are explicitly authorized to do so. In this scenario, the patient also has the opportunity to retrieve the records at home (Neubauer and Heurix 2011).

Slide 11-28: Pseudonymization of Information for Privacy 2/8

The PIPE protocol uses a combination of symmetric and asymmetric cryptographic keys to realize a logical multi-tier hull model with three different layers, where each layer is responsible for one step in the data access process. The user has to pass all layers in order to retrieve the actual health data records. The outer public and outer private keys form the outer layer, the authentication layer, which is responsible for unambiguously identifying the corresponding user. Together with the user's identifier, the outer private key represents the authentication credentials, which are stored along with the server's public key on the user's smart card. In combination with the correct PIN, the smart card provides two-factor authentication, where the authentication procedure involves both the user's and the PIPE server's outer keypair, the user's identifier, and two randomly selected challenges. The middle layer, the authorization layer, consists of the user's inner asymmetric keypair and the inner symmetric key. While the user's outer private key is created on the smart card when the card is issued to the user and never

(continued)

(continued)

actually leaves the card, the other keys are stored in the pseudonymization database where the secret keys are stored encrypted: the inner symmetric key is encrypted with the inner public key, while the inner private key is encrypted with the outer public key (Neubauer and Heurix 2011).

Slide 11-29: Pseudonymization of Information for Privacy 3/8

Access rights. In contrast to authorized users, an affiliated user, e.g., a close relative, is entrusted with the data owner's inner private key and is therefore able to decrypt the data owner's inner symmetric key, granting the affiliated user full access to all. (1) Pseudonyms are stored in cleartext when mapped to a particular record while the link between them is hidden by storing the pseudonyms encrypted in a single relation. (2) By affiliations via key-sharing, the affiliated user is granted access to the root pseudonyms as well data corresponding to the data owner. Therefore, the affiliated user is able to decrypt the links between all root and shared pseudonyms related to the data owner. The conceptual data model is depicted in this slide: The identification and health pseudonyms always form a 1:1 relationship and are referenced with their corresponding document type where this reference is stored in cleartext (record/pseudonym mapping). The link between the identification and health pseudonyms is stored encrypted with the user's inner symmetric key (pseudonym/pseudonym mapping): while the root pseudonyms are encrypted with the data owner's (patient's) inner symmetric key only, the shared pseudonyms are encrypted with both the data owner's and the authorized user's (health professional's) inner symmetric key so that both users are able to decrypt them using their corresponding ciphertexts. The link between the identification and health record is hidden and represented by the link between identification and health pseudonyms. Each health record is assigned exactly one root health pseudonym while each identification record has multiple root pseudonyms, depending on the number of health records, due to the 1:1 relationship. The health record is assigned a number of shared health pseudonyms according to the number of individual authorizations for that particular health record (Neubauer and Heurix 2011).

Slide 11-30: Pseudonymization of Information for Privacy 4/8

This slide shows the User authentication, which involves the mutual authentication of the user using the smart card and the server, involving their outer keypair and two nonces (randomly selected numbers used once) as user/server challenges. Once both identities are confirmed, the user's inner private key is retrieved from the pseudonymization database and transferred to the user's smart card to be decrypted with the user's outer private Transport Layer Security key. With the decrypted inner private key, the user's inner symmetric key can be decrypted within the HSM at the pseudonymization server and be cached for further operations along with the user's inner private key. In addition, a session key is generated at the HSM and securely (via encryption) transported to the user's smart card so that the key appears in cleartext only on the smart card and HSM (Neubauer and Heurix 2011).

Slide 11-31: Pseudonymization of Information for Privacy 5/8

To retrieve a particular health record, the user first needs to query for the particular encrypted pseudonyms by creating a keyword using the keyword templates, retrieving the corresponding keyword identifier, and querying for the encrypted identifier to find matching encrypted pseudonyms, i.e., the encrypted pseudonym mappings associated with the encrypted keyword identifier. The pseudonym pairs are then decrypted with the user's inner symmetric key and the plaintext pseudonyms then used to retrieve the corresponding identification and health records, which are transferred to the user to be displayed (possibly merged). Optionally, the pseudonyms and keyword identifier are also transferred to the user (root pseudonyms for authorizations). The record retrieval procedure is the same for the patient as data owner, health care provider as authorized user, and relative as affiliated user, with the difference that the patient and relative both query for the patient's root pseudonyms, while the health care provider relies on the shared pseudonyms (Neubauer and Heurix 2011).

Slide 11-32: Pseudonymization of Information for Privacy 6/8

To provide a trusted health care provider with the knowledge of the link between the patient's identification record and a particular health record, a new shared pseudonym pair is created as authorization relation. The patient

(continued)

(continued)

first has to retrieve the root pseudonym pair and keyword identifier corresponding to the health record he or she intends to share with the health-care provider. Furthermore, both the patient as data owner and the health-care provider as authorized user have to be authenticated at the same workstation so that both user identifiers are available at the client side, while both inner symmetric keys are cached at the HSM of the pseudonymization server. The root pseudonym pair is then transferred to the pseudonymization server along with both user identifiers and the keyword identifier, and the corresponding record identifiers retrieved using the cleartext record/pseudonym mappings. The server then randomly selects a newshared pseudonym pair, which is first encrypted with both users' inner symmetric keys (along with both identifiers and the keyword identifier) and then stores them in the database as authorization relation. Finally, the cleartext pseudonyms are then referenced with the retrieved record identifiers to create two new record/pseudonym mappings (Neubauer and Heurix 2011).

Slide 11-33: Pseudonymization of Information for Privacy 7/8

As with authorizations, a user affiliation requires that both the patient as data owner and the trusted relative as affiliated user are authenticated at the same workstation. Then both user identifiers are transferred to the pseudonymization server where they are encrypted with both users' inner symmetric keys. In addition, the patient's inner private key is also encrypted with the relative's inner symmetric key, and all elements are stored in the pseudonymization metadata storage as affiliation relation (Neubauer and Heurix 2011).

Slide 11-34: Pseudonymization of Information for Privacy 8/8

Finally, from the viewpoint of the patient as data owner, health data storage first requires that an "old" root identification pseudonym is retrieved as reference to the identification record. Furthermore, the patient creates a new keyword and enters the new health record into the workstation. Then the pseudonym, new keyword, new health record, and user identifier are transferred to the pseudonymization server, where the keyword is stored (and its identifier determined by the database engine) and the identification record

(continued)

(continued)

identifier retrieved. The new record is stored in the health records database and its record identifier returned to the server. Then, the server creates a new root pseudonym pair and stores it encrypted with the keyword identifier and user identifier as root access, as well as the cleartext record/pseudonym mappings (Neubauer and Heurix 2011).

7 Private Personal Health Record

Slide 11-35: Example: Private Personal Health Record

As the awareness of patients for their medical data increases, there is a trend of private personal health records, sometimes called health vaults. An example can be seen in http://healthbutler.com

In the following four slides we look at the technological concept of such a personal health record system. In this concept we will get to know a very interesting concept: mashups.

Slide 11-36: Example: Concept of a Personal Health Record System 1/4

PHRs that use centralized data stores do not offer stakeholders a choice in services, data storage, or user requirements. However, various stakeholders have varying skills, requirements, and responsibilities, which a single application can not satisfy. Consequently, personalization is required where such a heterogeneous mix of stakeholders exists. The concept of **mashups** (Auinger et al. 2009) let users create applications to suit their individual requirements. End users can use **mashup makers** to integrate various resources. Mashup makers let users create personalized applications with lower costs than traditional integration projects, in which a single application must incorporate many users' needs. As the explosion of Web mashups available on the Programmable Web (www.programmableweb.com) show, many users are finding new and diverse ways to satisfy individual requirements.

This slide shows the conceptual architecture of a system called Sqwelch (Fox et al. 2011). Within the architecture, there are three components:

(continued)

(continued)
1. Composition services provide mechanisms for modeling widgets and engaging with the stakeholder community in developing mashups.
2. Hosting services provide mechanisms for managing the environment, customizing mashup containers, and deploying mashups.
3. Infrastructure services form the basis of the mashup maker, including discovery services, social networking capabilities, security and trust, widget interaction, and management.

Slide 11-37: Example for Component Relationships 2/4

Here we see the Sqwelch component relationships. The components work in cooperation and fulfill specific roles to enable heterogeneous widgets and users to collaborate in a trusted way: When registering widgets, developers create model references that are stored for future use in the discovery and mediation components. During a mashup's execution, the social networking component determines the destinations for data if users are collaborating, which in turn uses trust and importance as a means of controlling data access. Model references are used to transform data, and component interaction is provided as publish–subscribe to loosely couple the remote resources (Web widgets) (Fox et al. 2011).

Slide 11-38: Widget Collaboration Sequence 3/4

Here we see the Widget collaboration sequence. Widgets communicate with the Sqwelch server using HTML 5 standards. Sqwelch alerts users if widgets aren't trusted.

The diagram shows the calls to be made by widgets, the execution host (Sqwelch default.html), and the server (Sqwelch. com) in enabling trusted publish–subscribe between heterogeneous widgets. In our example, the publishing widget could be the sensor viewer widget and the subscribing widget could be the sensor filter widget. We must consider some important points (Fox et al. 2011):

1. The HTML 5 postMessage syntax is used to publish data payloads from widgets and from the Sqwelch main page. HTML 5 event listener functions are required in subscribing widgets to listen for incoming payloads.

(continued)

(continued)
2. The payloads sempublishpost returns are those expected by the subscribing widgets (payload), based on the original published payload.
3. Payload as received by the subscribing widget will be a combination of default values the user specifies and real values, depending on the importance associated with the real data and the trust specified for the subscribing widget.
4. If the widget isn't trusted, Sqwelch alerts the user and provides a view of the data elements the subscribing widget has requested. This will happen only once for each widget in the current session.

Slide 11-39: User Collaboration Sequence 4/4

Finally, here the user collaboration sequence is depicted: Polling is used by subscribing mashups deployed by caregivers to retrieve data published by the patient. Sqwelch alerts the caregiver if the patient doesn't trust him or her. The sequences include (Fox et al. 2011):

1. The polling code is run on the hosting mashup webpage, retrieving data for all social widgets in the current page using getsocialsubscriptions.
2. The hosting mashup webpage returns with the latest heart rate readings for Mary.
3. If Mary doesn't trust either the widget or John, the payload will contain static, user-defined information, and Mary will be alerted.

Slide 11-40: Security and Privacy of Some PHR's

This work by (Carrión et al. 2011) is interesting for two reasons: (1) it provides a good overview of some personal health records and (2) it shows to what extent they addressed security and privacy issues.

The figure shows scores as two overlapping histograms: In general, quite a good level can be observed in the characteristics analyzed. Nevertheless, some improvements could be made to current PHR privacy policies to enhance specific capabilities such as: the management of other users' data, the notification of changes in the privacy policy to users and the audit of accesses to users' PHRs. The characteristics on how they reached these scores can be inferred from the following slides.

Slide 11-41: Nine Security Characteristics to Analyze PHR's 1/2

Carrión et al. (2011) defined nine characteristics to analyze the personal health records: privacy policy, location, data source, data managed, access management, access audit, data accessed without the user's permission, security measures, changes in privacy policy and standards:

Privacy policy location: This characteristic is related to the question *Where is the Privacy Policy on the PHR Web site*? PHRs should provide a privacy policy which describes how users' data are used in order for users to be informed. The privacy policy should be easily accessible by users. The difficulty of privacy policy access is assessed by counting the number of links clicked. The values that this characteristic may take are: 1. The privacy policy is not visible or not accessible. 2. The privacy policy is accessed by clicking one link. 3. The privacy policy is accessed by clicking two or more links.

Data source: This characteristic is related to the question *Where do users' PHR data proceed from*? Generally, the user is his/her data source, but there are PHRs which do not only use this source. Some contact the users' health-care providers, while others allow other users and different programs to enter users' data and some others use self-monitoring devices to obtain users' data. The values that this characteristic may take are: 1. Not indicated. 2. User. 3. User health-care provider. 4. User and his/her health-care providers. 5. User, other authorized users and other services/programs. 6. Self-monitoring devices connected with the user.

Data Managed: This characteristic is related to the question *Who do the data managed by the users belong to*? The users can manage their own data, but they can sometimes manage other users' data, such as that of their family. The values that this characteristic may take are: 1. Not indicated. 2. Data user. 3. Data user and his/her family data.

Access management: This characteristic is related to the question *Who can obtain access granted by the users*? The users decide who can access their PHR data. The PHR systems analyzed allow access to be given to different roles. The values that this characteristic may take are: 1. Not indicated. 2. Other users and services/programs. 3. Health-care professionals. 4. Other users. 5. Other users, health-care professionals and services/programs. To be continued on the next slide.

Slide 11-42: Nine Security Characteristics to Analyze PHR's 2/2

Access audit: This characteristic is related to the question *Can users see an audit of accesses to their PHRs*? The values that this characteristic may take are: 1. No. 2. Yes. **Data accessed without the user's permission**. This characteristic is related to the question *What data are accessed without the user's explicit consent*? The PHR systems typically access certain data related to the users in order to verify that everything is correct. The values that this characteristic may take are: 1. Not indicated. 2. Information related to the accesses. 3. De-identified user information. 4. Information related to the accesses and de-identified user information. 5. Information related to the accesses and identified user information.

Security measures: This characteristic is related to the question *What security measures are used in PHR systems*? There are two types of security measures: physical measures and electronic measures. The physical security measures are related to the protection of the servers in which the data are stored. The electronic security measures are related to how stored and transmitted data are protected, for example, by using a Secure Sockets Layer (SSL) scheme. The values that this characteristic may take are: 1. Not indicated. 2. Physical security measures. 3. Electronic security measures. 4. Physical security measures and electronic security measures.

Changes in privacy policy: This characteristic is related to the question *Are changes in privacy policy notified to users*? Changes in privacy policy should be notified to users in order to make them aware of how their data are managed by the PHR system. The values that this characteristic may take are: 1. Not indicated. 2. Changes are notified to users. 3. Changes are announced on home page. 4. Changes are notified to users and changes are announced on home page. 5. Changes may not be notified.

Standards. This characteristic is related to the question *Are PHR systems based on privacy and security standards*? The PHR systems analyzed use or are based on two standards: the *Health Insurance Portability and Accountability Act* (HIPAA) and the *Health On the Net Code of Conduct* (HONcode). The values that this characteristic may take are: Usable privacy and security in personal health records 41. 1. Not indicated. 2. HIPAA is mentioned. 3. System is covered by HONcode. 4. HIPAA is mentioned and system is covered by HONcode (Carrión et al. 2011).

Slide 11-43: Overview Personal Health Records (PHR)

The last slide shows the summary of the researched personal health records (Carrión et al. 2011). Note: By 2013 the Google Health record is not longer in operation: Google Health has been permanently discontinued. All data remaining in Google Health user accounts as of January 2, 2013 has been systematically destroyed, and Google is no longer able to recover any Google Health data for any user, see: http://www.google.com/intl/en_us/health/about

See also this blog: http://googleblog.blogspot.co.at/2011/06/update-on-google-health-and-google.html

Slide 11-44: Ethical Issues: During Quality Improvement

Here a summary of ethical issues by a work of (Tapp et al. 2009): They identified the experiences of professionals involved in planning and performing QI programs in European family medicine on the ethical implications involved in those processes. For this purpose the used four focus groups with 29 general practitioners (GPs) and administrators of general practice quality work in Europe. Two focus groups comprised EQuiP members and two focus groups comprised attendees to an invitational conference on QI in family medicine held by EQuiP in Barcelona. Four overarching themes were identified, including implications of using patient data, prioritizing QI projects, issues surrounding the ethical approval dilemma and the impact of QI. Each theme was accompanied by an identified solution. Practical implications—prioritizing is necessary and in doing that GPs should ensure that a variety of work is conducted so that some patient groups are not neglected. Transparency and flexibility on various levels are necessary to avoid harmful consequences of QI in terms of bureaucratization, increased workload, and burnout on part of the GP and harmful effects on the doctor–patient relationship. There is a need to address the system of approval for national QI programs and QI projects utilizing more sophisticated methodologies (Tapp et al. 2009).

8 Future Outlook

Slide 11-45: Future Outlook

Privacy, security, safety and data protection are of enormous increasing interest in the future. Due to the trend to mobile and cloud computing approaches and the omnipresence of data it is of vital importance. Electronic Health Records (EHR) are the fastest growing example of an application which concern data privacy and patient consent. Increasing amounts of personal health data are being stored in databases for the purpose of maintaining a life-long health record of an individual.

A further big issue is secondary use of data, providing patient data for clinical or medical research. For most secondary data use, it is possible to use deidentified data, but for the remaining data protection issues are very important (Safran et al. 2007). The secondary use of data involves the linkage of datasets to bring different modalities of data together, which raises more concerns over the privacy of the data. The publication of the human genome gave rise to new ways of finding relationships between clinical disease and human genetics. The increasing use and storage of genetic information also impacts the use of familial records, since the information about the patient also provides information on the patient's relatives. The issues of data privacy and patient confidentiality and the use of the data for medical research are made more difficult in this post-genomic age.

Another issue is the production of anonymized open dataset to support international joint research efforts.

9 Exam Questions

9.1 Yes/No Decision Questions

Please check the following sentences and decide whether the sentence is true = YES; or false = NO; for each correct answer you will be awarded 2 credit points.

01	Privacy is defined as a means for protecting from unauthorized access, use, modification, disruption or destruction.	☐ Yes ☐ No	2 total
02	Privacy, security, safety and data protection are of enormous increasing interest in the future.	☐ Yes ☐ No	2 total
03	With ubiquitous computing there is the potential danger that eavesdroppers can listen in if they are in immediate vicinity.	☐ Yes ☐ No	2 total
04	Unidentified risks are the risks left over after system safety efforts have been fully employed.	☐ Yes ☐ No	2 total
05	Human fallibility includes affect and personality traits as well as fatigue and sleep deprivation.	☐ Yes ☐ No	2 total
06	A patient record must be up to date, usable, useful, accessible and confidential.	☐ Yes ☐ No	2 total
07	The Eindhoven Classification Model (ECM) was originally developed to manage human error in air traffic control and was later adapted to healthcare.	☐ Yes ☐ No	2 total
08	The concept of mashups let users create applications to suit their individual requirements. End users can use mashup makers to integrate various resources.	☐ Yes ☐ No	2 total
09	Due to the huge progress in IT safety, wrong-site surgery is no longer an issue.	☐ Yes ☐ No	2 total
10	Non-repudiation can be seen as a special case of integrity, i.e. the integrity of log data recording.	☐ Yes ☐ No	2 total

Sum of Question Block A (max. 20 points)	

9.2 Multiple Choice Questions (MCQ)

The following questions are composed of two parts: the stem, which identifies the question or problem and a set of alternatives which can contain 0, 1, 2, 3, or 4 correct answers, along with a number of distractors that might be plausible—but are incorrect. Please **select the correct answers** by ticking ☒—and do not forget that it can be none. Each question will be awarded 4 points *only if everything is correct*.

01	"The event causes harm on body of patient, extents hospital stay, loses abilities, or death – but is not coming from original disease ..." is clearly a ... ❏ a) ... sentinel event. ❏ b) ... accident. ❏ c) ... near miss. ❏ d) ... medical adverse event.	4 total
02	Dependability means ... ❏ a) ... fault prevention. ❏ b) ... fault tolerance ❏ c) ... fault removal. ❏ d) ... fault forecasting.	4 total
03	Ultra safe systems are .... ❏ a) ... commercial jet aviation. ❏ b) ... nuclear industry. ❏ c) ... chartered jet aviation. ❏ d) ... chemical industry.	4 total
04	Protection of the patients' data privacy can be achieved with ... ❏ a) ... anonymization. ❏ b) ... removal of identifiers from the medical data. ❏ c) ... encryption. ❏ d) ... pseudonymization.	4 total
05	Pseudonymization is a procedure by which ... ❏ a) ... identifying fields within a data record are replaced by one or more artificial identifiers. ❏ b) ... relevant data is replaced by pseudonyms. ❏ c) ... an electronic trail is destroyed. ❏ d) ... does not take away the original data.	4 total
06	Patient safety in healthcare is ... ❏ ... measuring risk and planning the ideal defense model. ❏ ... part of usability engineering. ❏ ... not related to information systems. ❏ ... the equivalent of systems safety in industry.	4 total
07	The IOM report of 1999 ... ❏ a) ... called for a 50% reduction in medical errors over 5 years. ❏ b) ... was politically relevant but did not change much. ❏ c) ... was a theoretical study with no measurable impact. ❏ d) ... was very effective in triggering health safety research.	4 total
08	The Swiss Cheese Model (Reason, 1997) discriminates between ... ❏ a) ... latent failures. ❏ b) ... latent and active failures. ❏ c) ... active failures. ❏ d) ... non latent failures.	4 total

09	In the framework for understanding human error, so called barriers are ... ☐ a) ... training and workgroup culture. ☐ b) ... sensors and automatic shutdown. ☐ c) ... warning and alarms. ☐ d) ... physical capabilities of the individual.	4 total
10	A risk of 10^{-2} on the risk scale means ... ☐ a) ... very unsafe system. ☐ b) ... blood transfusion. ☐ c) ... typical road safety. ☐ d) ... Himalaya mountaineering.	4 total

Sum of Question Block B (max. 40 points)

9.3 Free Recall Block

Please follow the instructions below. At each question you will be assigned the credit points indicated if your option is correct (partial points may be given).

01	Please complete the missing requirements of an electronic patient record: 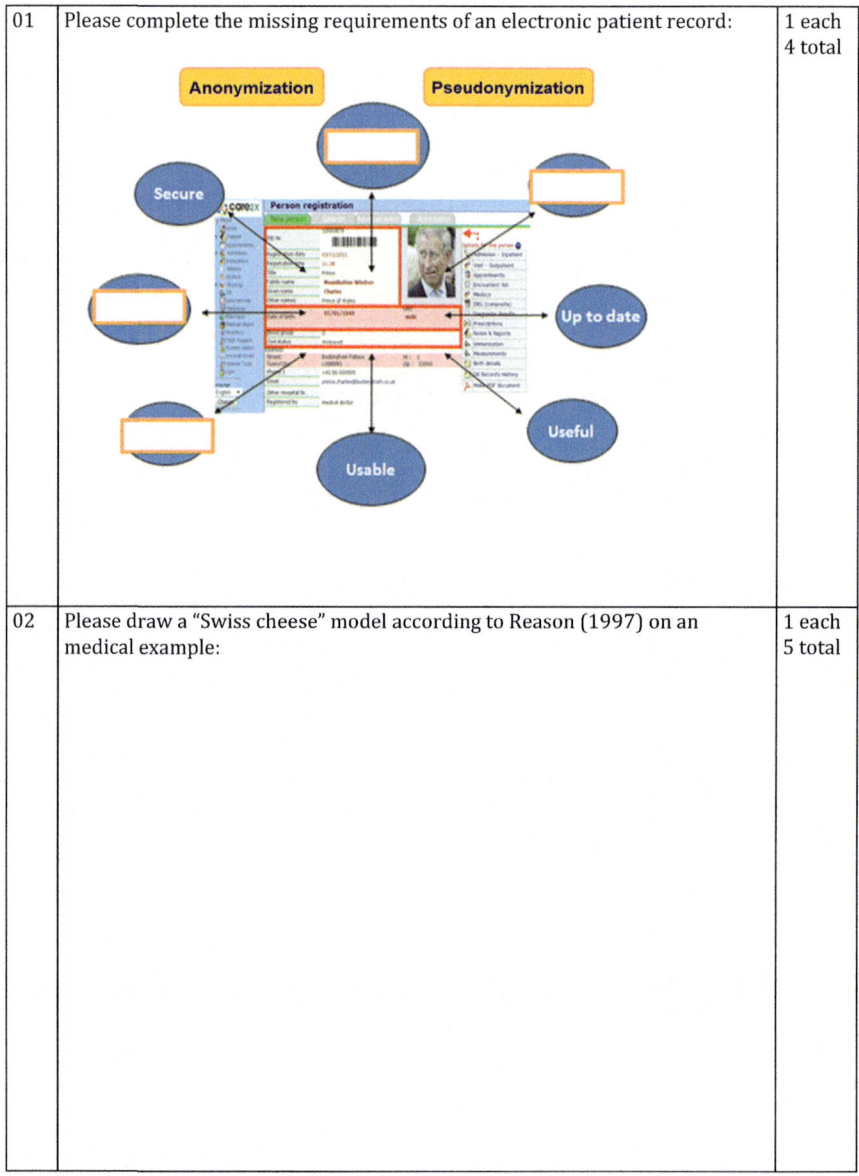	1 each 4 total
02	Please draw a "Swiss cheese" model according to Reason (1997) on an medical example:	1 each 5 total

03	Please make a sketch to explain the term "residual risk":	1 each 2 total
04	The technical dependability consists of three categories. Please complete them in the following image: 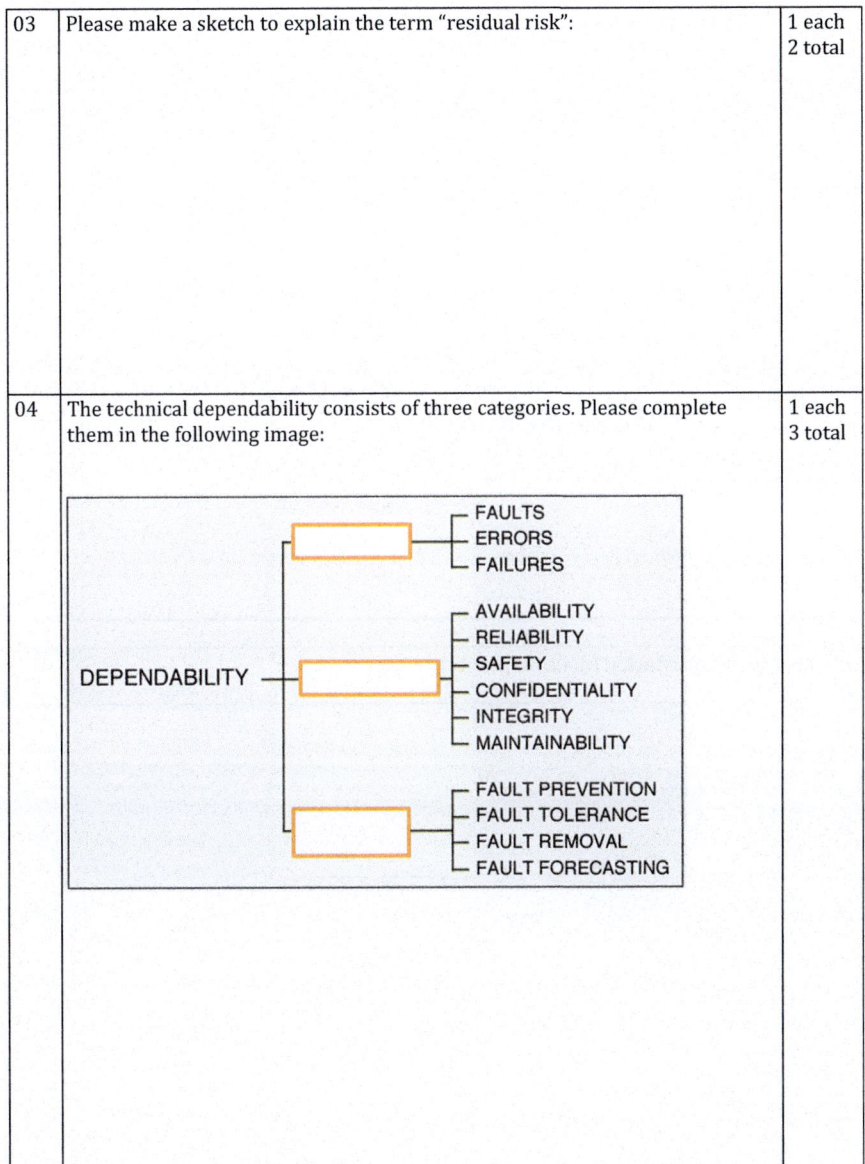	1 each 3 total

| 05 | Please complete the curves to indicate the effect of the IOM report 1999: | 1 each 2 total |

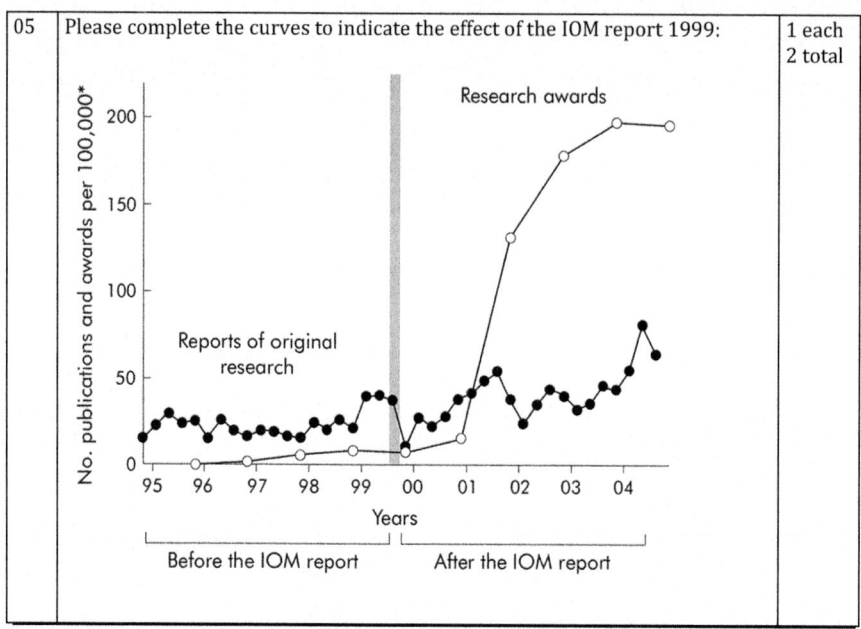

| Sum of Question Block C (max. 40 points) | |

10 Answers

10.1 Answers to the Yes/No Questions

Please check the following sentences and decide whether the sentence is true = YES; or false = NO; for each correct answer you will be awarded 2 credit points.

01	Privacy is defined as a means for protecting from unauthorized access, use, modification, disruption or destruction.	☐ Yes ☒ No	2 total
02	Privacy, security, safety and data protection are of enormous increasing interest in the future.	☒ Yes ☐ No	2 total
03	With ubiquitous computing there is the potential danger that eavesdroppers can listen in if they are in immediate vicinity.	☒ Yes ☐ No	2 total
04	Unidentified risks are the risks left over after system safety efforts have been fully employed.	☐ Yes ☒ No	2 total
05	Human Fallibility includes affect and personality traits as well as fatigue and sleep deprivation.	☒ Yes ☐ No	2 total
06	A patient record must be up to date, usable, useful, accessible and confidential.	☒ Yes ☐ No	2 total
07	The Eindhoven Classification Model (ECM) was originally developed to manage human error in air traffic control and was later adapted to healthcare.	☐ Yes ☒ No	2 total
08	The concept of mashups let users create applications to suit their individual requirements. End users can use mashup makers to integrate various resources.	☒ Yes ☐ No	2 total
09	Due to the huge progress in IT safety, wrong-site surgery is no longer an issue.	☐ Yes ☒ No	2 total
10	Non-repudiation can be seen as a special case of integrity, i.e. the integrity of log data recording.	☒ Yes ☐ No	2 total

Sum of Question Block A (max. 20 points)	

10.2 Answers to the Multiple Choice Questions (MCQ)

01	"The event causes harm on body of patient, extents hospital stay, loses abilities, or death – but is not coming from original disease ..." is clearly a ... ☐ a) ... sentinel event. ☐ b) ... accident. ☐ c) ... near miss. ☒ d) ... medical adverse event.	4 total
02	Dependability means ... ☒ a) ... fault prevention. ☒ b) ... fault tolerance ☒ c) ... fault removal. ☒ d) ... fault forecasting.	4 total
03	Ultra safe systems are .... ☒ a) ... commercial jet aviation. ☒ b) ... nuclear industry. ☐ c) ... chartered jet aviation. ☐ d) ... chemical industry.	4 total
04	Protection of the patients' data privacy can be achieved with ... ☒ a) ... anonymization. ☒ b) ... removal of identifiers from the medical data. ☒ c) ... encryption. ☒ d) ... pseudonymization.	4 total
05	Pseudonymization is a procedure by which ... ☒ a) ... identifying fields within a data record are replaced by one or more artificial identifiers. ☒ b) ... relevant data is replaced by pseudonyms. ☐ c) ... an electronic trail is destroyed. ☐ d) ... does not take away the original data.	4 total
06	Patient safety in healthcare is ... ☒ ... measuring risk and planning the ideal defense model. ☐ ... part of usability engineering. ☐ ... not related to information systems. ☒ ... the equivalent of systems safety in industry.	4 total
07	The IOM report of 1999 ... ☒ a) ... called for a 50% reduction in medical errors over 5 years. ☐ b) ... was politically relevant but did not change much. ☐ c) ... was a theoretical study with no measurable impact. ☒ d) ... was very effective in triggering health safety research.	4 total
08	The "Swiss cheese" model (Reason, 1997) discriminates between ... ☒ a) ... latent failures. ☒ b) ... latent and active failures. ☒ c) ... active failures. ☐ d) ... non latent failures.	4 total
09	In the framework for understanding human error, so called barriers are ... ☐ a) ... training and workgroup culture. ☒ b) ... sensors and automatic shutdown. ☒ c) ... warning and alarms. ☐ d) ... physical capabilities of the individual.	4 total

10	A risk of 10^{-2} on the risk scale means ... ☒ a) ... very unsafe system. ☐ b) ... blood transfusion. ☐ c) ... typical road safety. ☒ d) ... Himalaya mountaineering.	4 total

Sum of Question Block B (max. 40 points)

10.3 Answers to the Free Recall Questions

01	Please complete the missing requirements of an electronic patient record:	1 each 4 total
	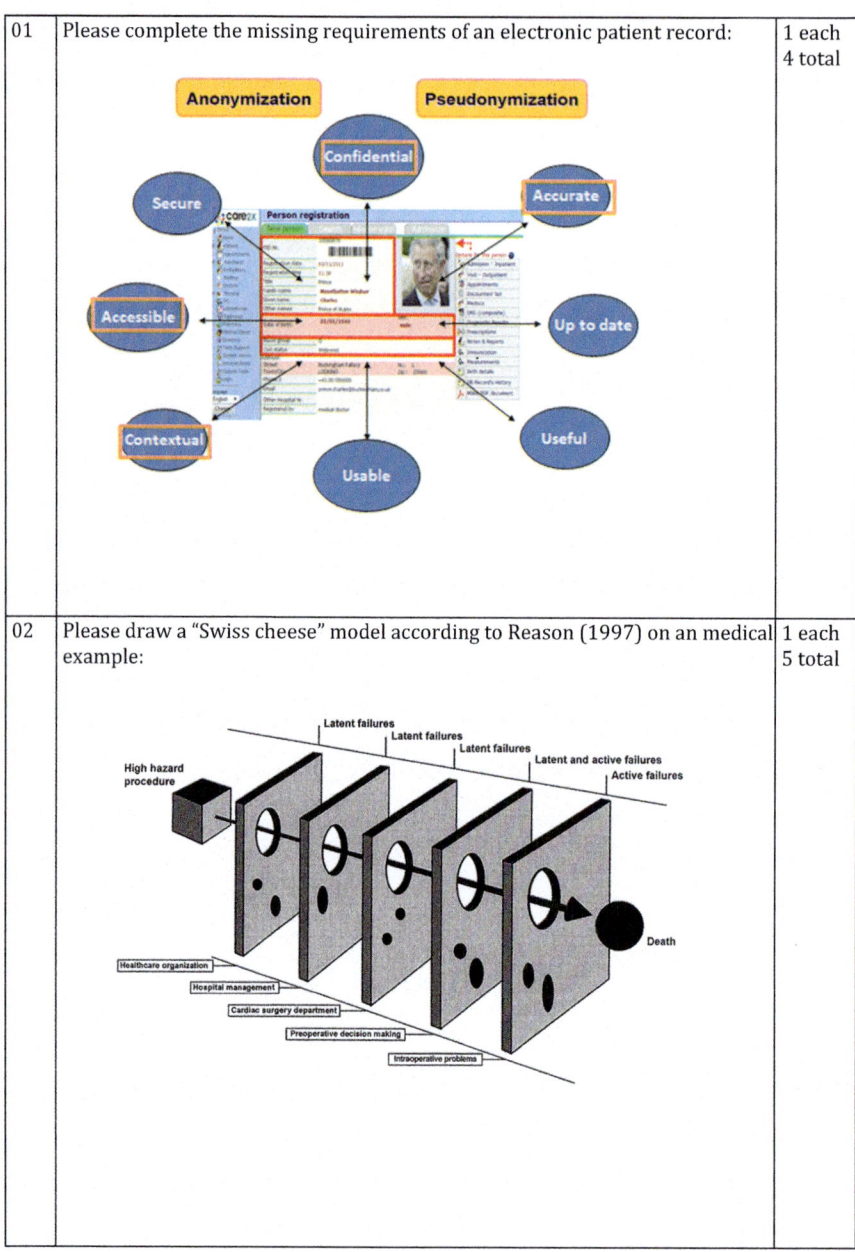	
02	Please draw a "Swiss cheese" model according to Reason (1997) on an medical example:	1 each 5 total

10 Answers

03	Please make a sketch to explain the term "residual risk":	1 each 2 total
	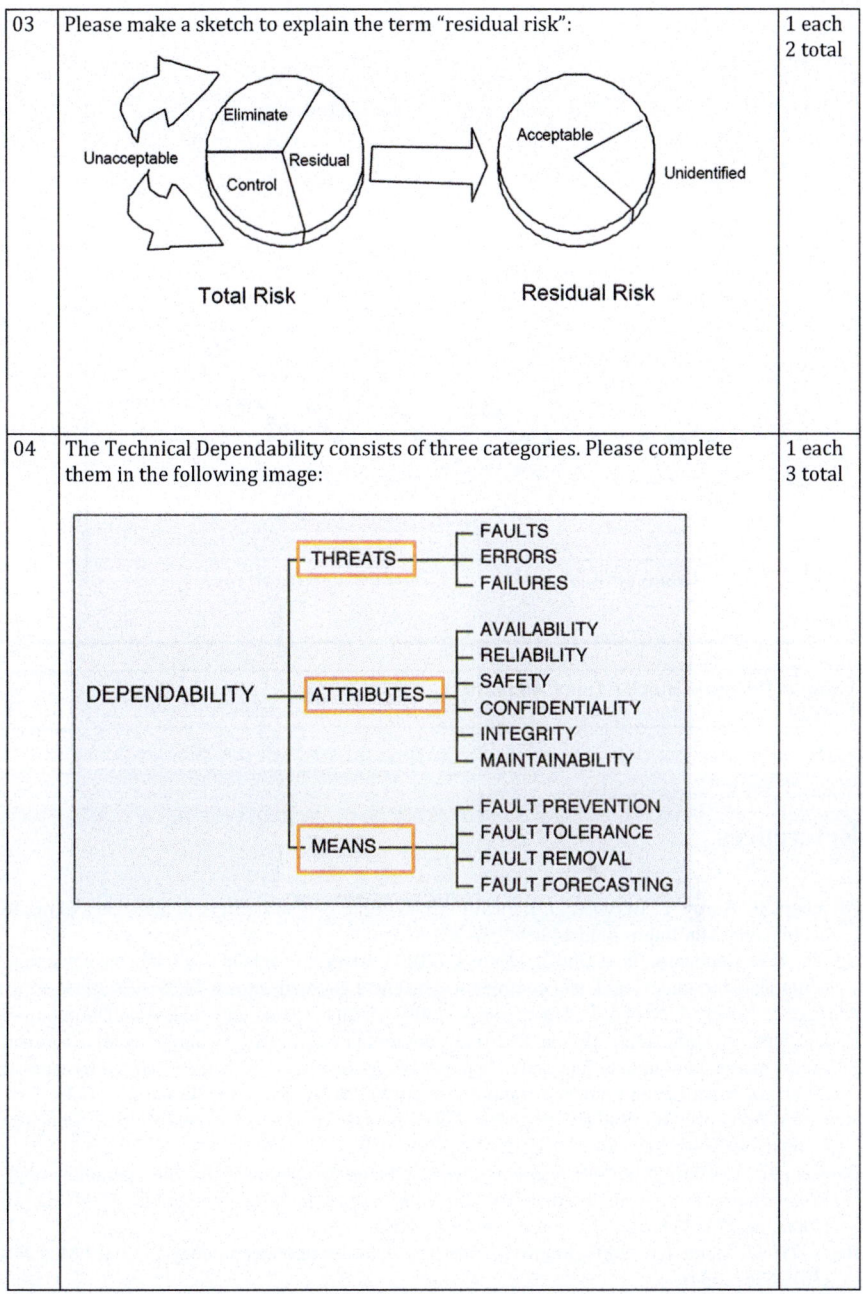	
04	The Technical Dependability consists of three categories. Please complete them in the following image:	1 each 3 total

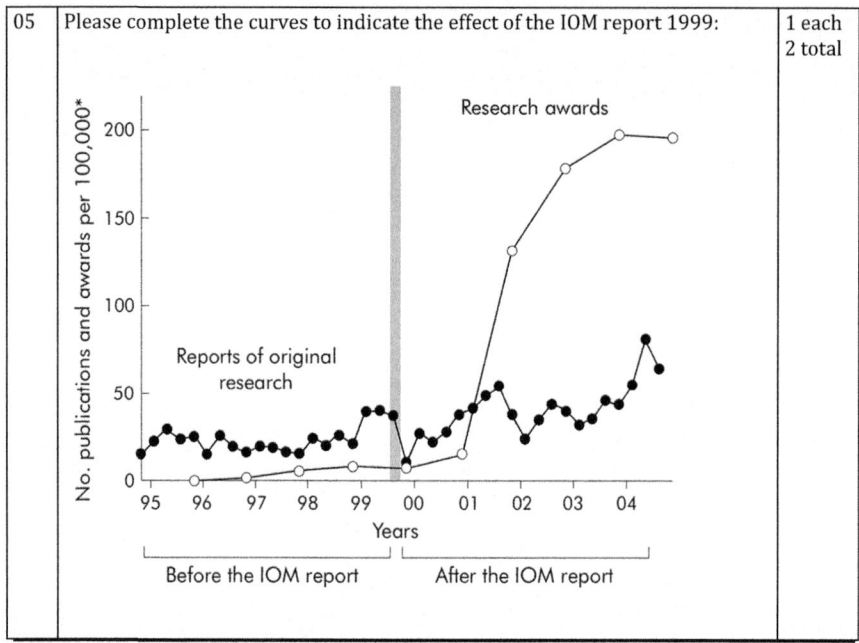

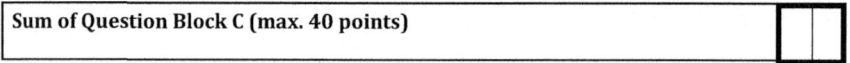

References

Amalberti R, Auroy Y, Berwick D, Barach P (2005) Five system barriers to achieving ultrasafe health care. Ann Intern Med 142(9):756–764

Amalberti R, Benhamou D, Auroy Y, Degos L (2011) Adverse events in medicine: easy to count, complicated to understand, and complex to prevent. J Biomed Inform 44(3):390–394

Auinger A, Ebner M, Nedbal D, Holzinger A (2009) Mixing content and endless collaboration—mashups: towards future personal learning environments. In: Stephanidis C (ed) Universal access in human-computer interaction hci, part III: applications and services, HCI International 2009, vol 5616, Lecture notes in computer science (LNCS). Springer, Berlin, pp 14–23

Avizienis A, Laprie JC, Randell B (2001) Fundamental concepts of dependability. Tech Rep Comput Sci Univ Newcastle 1145(CS-TR-739):7–12

Bardram JE, Norskov N (2008) A context-aware patient safety system for the operating room. Proceedings of the 10th International conference on ubiquitous computing. ACM, Seoul, Korea, pp 272–281

Bates DW, Gawande AA (2003) Improving safety with information technology. N Engl J Med 348 (25):2526–2534

Carrión I, Fernández-Alemán J, Toval A (2011) Usable privacy and security in personal health records. In: Campos P, Graham N, Jorge J, Nunes N, Palanque P, Winckler M (eds) INTERACT 2011, vol 6949, Lecture notes in computer science (LNCS). Springer, Berlin, pp 36–43

Fox R, Cooley J, Hauswirth M (2011) Creating a virtual personal health record using mashups. IEEE Internet Comput 15(4):23–30

References

Garrouste-Orgeas M, Philippart F, Bruel C, Max A, Lau N, Misset B (2012) Overview of medical errors and adverse events. Ann Int Care 2(1):2

Holzinger A, Schwaberger K, Weitlaner M (2005) Ubiquitous computing for hospital applications: RFID-applications to enable research in real-life environments. 29th Annual international conference on computer software & applications (IEEE COMPSAC). IEEE, Edinburgh, pp 19–20

Holzinger A, Searle G, Wernbacher M (2011) The effect of previous exposure to technology (PET) on acceptance and its importance in usability engineering. Univ Access Inf Soc Int J 10 (3):245–260

Leape LL, Berwick DM (2005) Five years after to err is human. JAMA 293(19):2384–2390

Manjunath PS, Palte H, Gayer S (2010) Wrong site surgery—a clear and constant fear. Br Med J 341:c1103

Neubauer T, Heurix J (2011) A methodology for the pseudonymization of medical data. Int J Med Inform 80(3):190–204

Reason J (1997) Management of the risks of organizational accidents. Ashgate, Brookfield, VT

Reason J (2000) Human error: models and management. Br Med J 320(7237):768–770

Rodrigues S, Brandao P, Nelas L, Neves J, Alves VA (2010) Logic Programming Based Adverse Event Reporting and Learning System. IEEE/ACIS 9th International conference on computer and information science (ICIS), 18–20 Aug 2010, pp 189–194

Safran C, Bloomrosen M, Hammond WE, Labkoff S, Markel-Fox S, Tang PC, Detmer DE (2007) Toward a national framework for the secondary use of health data: an American Medical Informatics Association white paper. J Am Med Inform Assoc 14(1):1–9

Stelfox HT, Palmisani S, Scurlock C, Orav E, Bates D (2006) The "To Err is Human" report and the patient safety literature. Qual Saf Health Care 15(3):174–178

Sundt TM, Brown JP, Uhlig PN (2005) Focus on patient safety: good news for the practicing surgeon. Ann Thorac Surg 79(1):11–15

Tapp L, Elwyn G, Edwards A, Holm S, Eriksson T (2009) Quality improvement in primary care: ethical issues explored. Int J Health Care Qual Assur 22(1):8–29

Wang C, Huang DJ (2013) The HIPAA conundrum in the era of mobile health and communications. JAMA 310(11):1121–1122

Weippl E, Holzinger A, Tjoa AM (2006) Security aspects of ubiquitous computing in health care. Elektrotech Informationstech 123(4):156–162

Lecture 12

Methodology for Information Systems: System Design, Usability, and Evaluation

1 Learning Goals

At the end of this last lecture, you:

- would be able to understand the concepts of usability and the importance of usability engineering for medical information systems.
- would be aware that medical software is now included within the Medical Device Act (Medizinproduktegesetz (MPG)).
- would have a feeling for quality and can determine between product quality, process quality and information quality.
- would be familiar with some important ISO standards for quality and usability of medical software and systems.
- would be able to understand the user-centered design process, from concept phase till verification and validation.
- would be able to apply some usability engineering methods and evaluation methods applicable in the medical domain.

2 Advance Organizer

Accessibility	The degree to which a system or service is available to a diverse set of end users
Accreditation	A formal declaration by the Accreditation Authority that a system is approved to operate in the defined standards with accuracy, completeness and traceability
Act	A formal law passed by a legislative body
Audit	Is performed to verify conformance to standards by review of objective evidence (e.g., ISO 9001),

Certification	it is an independent examination of the life cycle processes within the audited organization A (product/software) qualification to verify that performance tests and quality assurance tests or qualification requirements are certified
Cognitive modeling	Aka mental modeling = producing a computational model for how people perform tasks and solve problems, based on psychological principles. These models may be outlines of tasks written on paper or computer programs which enable us to predict the time it takes for people
Cognitive walkthrough	An approach to evaluating a user interface based on stepping through common tasks that a user would need to perform and evaluating the user's ability to perform each step
Consistency	Principle that things that are related should be presented in a similar way and things that are not related should be made distinctive
Consistency inspection	A quality control technique for evaluating and improving a user interface. The interface is methodically reviewed for consistency in design, both within a screen and between screens, in graphics (color, typography, layout, icons), text (tone, style, spelling)
Effectiveness	The degree to which a system facilitates a user in accomplishing a specific task, measured by task completion rate; often confused with efficiency
Efficiency	A measurable concept, determined by the ratio of output to input; it is the ability to accomplish a task in minimum time with a minimum of effort (once the end users have learned to use the system); often confused with effectiveness
Emotion	A mental and physiological state associated with a wide variety of feelings, thoughts, and behaviors, very important for usability
End user	The primary target user of a system, assumed to be the least computer-literate user
End-user programming (EUP)	Making computational power fully accessible to expert end users, e.g., to medical professionals with no specific computer programming knowledge; usually done by a user interface which enables easy programming (e.g., visual programming, natural-language syntax, wizard-based programming, mash-up programming)

2 Advance Organizer

Errors	An important measurement of usability on how many errors do end users make, how severe are these errors, and how easily they can recover from the errors
Evaluation	Is the systematic process of measuring criteria against a set of standards
Formative Evaluation	Usability evaluation that helps to "form" the design process, i.e., evaluation is taking place parallel and iteratively to the development process
Heuristic Evaluation	Method to identify any problems associated with the design of user interfaces
ISO 13407	Human Centred Design Processes for Interactive Systems
ISO 13485 (2003)	Represents the requirements for a comprehensive management system for the design and manufacture of medical devices
ISO 14971 (2007)	Risk management for medical devices
ISO 62304 (2006)	Medical device software
ISO 9001	The ISO 9000 international standards family is for quality management and guidelines as a basis for establishing effective and efficient quality management systems
ISO 9241	Software usability standard
ISO 9241-10	Ergonomic requirements for office work with visual display terminals (VDTs): Dialogue principles (1996)
ISO 9241-11	Ergonomic requirements for office work with visual display terminals (VDTs): Guidance on usability specifications and measures (1998)
ISO/HL7	Joint ISO and HL7 (Health Level Seven) International Standard
ISO/IEEE	Joint ISO and IEEE (Institute of Electrical and Electronics Engineers) International Standard
ISO/OECD	Joint ISO and OECD (Organisation for Economic Cooperation and Development) International Standard
Learnability	Degree of which a user interface can be learned quickly and effectively by measure of learning time
Learning curve	The amount of time an end user needs to fulfill a previously unknown task
Mash-up	The use of existing functionalities to create new functionalities, Mash-up composition tools are

	usually simple enough to be used by end users without programming skills (e.g., by supporting visual wiring of GUI widgets, services, and/or components together); The concepts of mash-up are combination, visualization, and aggregation in order to make data useful
Medical Safety Design	Process including usability engineering and risk management to make the product compliant to EN 60601 and EN 62366 which is no longer a nice to have, but a requirement; the developer must provide a documentation on the usability engineering process
Medizinproduktegesetz (MPG)	Medical device act = Valid law in Austria, based on European law (in Germany: Medizinproduktegesetz MPG in der Fassung der Bekanntmachung vom 7. 8. 2002 (BGBl. I S. 3146), das durch Artikel 13 des Gesetzes vom 8. 11. 2011 (BGBl. I S. 2178) geändert worden ist)
Memorability	The measure of when an end user returns to the system after a period of not using it, and how easily can he reestablish efficiency
Mental model	The internal model of an end user on how something works; can be used by the designer for aligning his design strategy with human behavior
Methodology	Systematic study of methods that are, can be, or have been applied within a discipline
Participatory design	A common approach to design that encourages participation in the design process by a wide variety of stakeholders, such as designers, developers, management, users, customers, salespeople, distributors, etc.
Performance	Measurement of output or behavior in both engineering and computing
Performance measure	A quantitative rating on how someone performed a task, such as the time it took to complete, the number of errors they made in doing it, their success rate or the time spent in a particular phase of a process
Satisfaction	A subjective degree of how much an end user enjoys using a system (joy of use, enjoyability)
Semiotics	The study of signs and symbols and their use in communicating meaning, especially useful in analyzing the use of icons in software, but also

	appropriate to the analysis of how screen design as a whole communicates
Software Usability Measurement Inventory (SUMI)	A rigorously tested and proven method of measuring software quality from the end user's point of view; consistent method for assessing the quality of use of a software product or prototype
Software Usability Scale (SUS)	A ten-item attitude Likert scale providing a single score reflecting the overall view of subjective assessments of usability, developed by Brooke (1996), the power is in its simplicity
Task analysis	A set of methods for decomposing people's tasks in order to understand the procedures better and to help provide computer support for those tasks
Thinking aloud	Direct observation, where end users are asked to speak out loud everything they do, think, feel in each moment during execution of a task; the only method to gain insight into the thinking, helpful at early stages of design for determining expectations and identifying what aspects of a system are confusing
Usability engineering	A methodical approach to user interface design and evaluation involving practical, systematic approaches to developing requirements, analyzing a usability problem, developing proposed solutions, and testing those solutions
User Interface (UI), Graphical User Interface (GUI)	Input/output possibilities of a system—for the end user, the interface actually *is* the system
Validation	Is a (external) quality process to demonstrate (to the stakeholder) that the system complies with the original specifications
Verification	Is a (internal) quality process, used to evaluate whether and to what extent the system complies with the original specifications

3 Acronyms

ATT	Attitude towards use
CE	Conformité Européenne (European Conformity)
GUI	Graphical user interface
HCI	Human–computer interaction

IEC   International Electro-technical Commission
ISO   International Organization for Standardization
SUMI   Software usability measurement inventory
SUS   System usability scale
TAM   Technological acceptance model
PU   Perceived usefulness
PEoU   Perceived ease of use
UCD   User-centered design
UE   Usability engineering
UI   User interface

4 Key Problems

Slide 12-1: Key Challenges

- Usability is still underestimated in health applications
- User-Centered Designs are rarely applied in medical information systems
- Evaluation is still a small part in medical information systems research

Usability is still underestimated in the design and development of applications for medicine and health care although they are proven to be often a matter of life or death. Jakob Nielsen reported this very impressive in his blog, under the title: "How to kill patients through bad design",
http://www.nngroup.com/articles/medical-usability

Medical systems have provided many well-documented killer designs, such as the radiation machines that fried six patients because of complex and misleading operator consoles. What's less known is that usability problems in the medical sector's good old-fashioned office automation systems can harm patients just as seriously as machines used for treatment. A further problem is that traditional approaches of HCI are essential, but they are unable to cope with the complexity of typical modern interactive devices in the safety critical context of medical devices. The broad scale of typical devices means that conventional user-centered approaches, while still necessary, are insufficient to contribute reliably to safety-related interaction issues (Thimbleby 2007).

5 A Framework for Understanding Usability

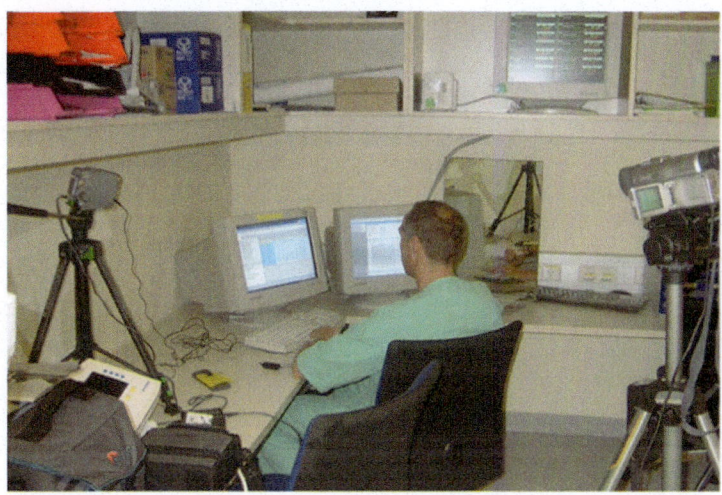

Fig. 1 See Slide 12-2

Slide 12-2: Medical Workplace Usability as Key to Enhance Quality

In this slide we see a typical usability survey setting: Experiments with end users in real-world settings in the hospital often show that the whole workflow, along with human factors of the workplace, including hardware usability, e.g., seating and environmental issues must be considered. To ensure a total acceptance of a system, satisfaction, and most of all the reduction in **time to perform task** is essential. To reach that, all issues of the whole workflow must be considered which is called Total Workplace Usability (TWU) (Holzinger and Leitner 2005). The issues to consider include:

- On-site training to familiarize the end users with the tools having one big goal in mind: reduce time to perform a task. Time is precious for the medical doctors and each time saved is also a cost saved for the hospital.
- Electronic tutoring to assist part-time end users to find their way through the workflow quickly and to help them to solve their problems rapidly. Benefit: Once developed, it would constantly assist many end users at all workplaces, independent of end users' training efforts.

(continued)

(continued)
- Customizing and proper adaptation of third-party software; in practice most tools separately used to the clinical workplace, especially when rarely used, cause a serious effort by the end users.
- Ergonomic aspects of the workplace include: proper distance to the screen, correct table height, proper chair (easily adjustable for height, to enable the user to quickly achieve a comfortable angle from eyes to the screen), proper mouse pad location and working space. Benefit: End users can concentrate and feel more comfortable during the strenuous information processing process; results are immediate in lower task performance time and consequently a better output.

Slide 12-3: A Framework for Understanding Usability

Usability can be defined via a combination of efficiency, effectiveness, and satisfaction: Each of so-called **usage indicators** contributes to the aspects of the higher level, e.g., low error rate increases effectiveness; good performance indicates good efficiency. The indicators are measured using a set of metrics. One level lower is the level of means, which can be used in "heuristics" for improving the usage indicators and are not goals by themselves, e.g., consistency may have a positive effect on learnability, as warnings may reduce errors. On the other hand, high adaptability may have a negative effect on memorability while having a positive effect on performance. In order to find optimal levels for all means, the designer has to apply the three knowledge domains: humans, design, and task. For example, design knowledge such as guidelines should include how changes in use of the means affect the usage indicators (Veer and Welie 2004).

System Characteristic	Corresponding Quality factor(s)
Safety-critical (medical) Systems	Reliability, Correctness, Verifiability
Classified (patient) data	Security
Real-time operation	Efficiency
Heterogeneity of system landscape	Portability
Diverse set of (medical) end users	Usability
Possible further (hospital) development	Expandability

Fig. 2 See Slide 12-4

Slide 12-4: System Characteristic Versus Quality Factor

Quality can be seen as the key success factor and if we look at our classic system characteristics we can determine six characteristics with corresponding quality factors, as seen in this slide. From top to bottom it includes reliability, correctness, and verifiability; security, efficiency, portability, usability, and expandability—and the nature of the system dictates the prioritizing of the features (Cosgriff 1994).

6 Standards

Slide 12-5: ISO Standards for Healthcare

In Lecture 3 we have already heard about the advantages and disadvantages of standardization. Now we will elaborate on standardization efforts. One organization is of eminent importance: ISO, which stands short for International Organization for Standardization and is since 1947 the world's largest developer of (voluntary) international standards, which are intended to provide state of the art specifications for products, services, and good practices, helping to make industry more efficient and effective. Currently there are more than 19,500 standards available covering almost all aspects of technology and business, from food safety to computers, and agriculture to healthcare (see: www.iso.org).

6.1 EU Directive: Medical Device Directive

Slide 12-6: EU Directive 93/42/EEC Medical Device Directive (MDD)

The EU directive 93/42/EEC1 states criteria to define medical devices. For systems and devices that fall under these definitions, the directive states requirements that have to be met.

Medical devices in the sense of this directive are devices that serve the following purposes (Neuhaus et al. 2011):

1. Diagnosis, prevention, monitoring, treatment, or alleviation of disease.
2. Diagnosis, monitoring, treatment, alleviation of or compensation for an injury or handicap.
3. Investigation, replacement, or modification of the anatomy or of a physiological process.
4. Control of conception:
   The important aspect for IT systems is that software of medical devices is explicitly included in this definition. Every device classified a medical device under the above criteria has to bear a CE 2 (conformité européenne) mark that indicates conformity with the requirements on medical devices of this directive. These requirements are defined in Annex I of the directive and include:
5. Device may not compromise the clinical condition or the safety of patients when used in the intended way.
6. Risks have to be minimized (elimination of risks through security by design, alerts have to warn about dangerous conditions, users have to be informed about residual risks).

Further detailed requirements concern sterility, used materials in manufacturing, influence or emittance of radiation etc.

Devices are classified into risk categories I, IIa, IIb, and III depending on the typical duration of use, degree of invasiveness and inherent risk. Category III indicates the highest risk. The requirements for the attainment of a CE mark depend of the risk category the device is classified into. Class III-devices must be approved by the corresponding authority in a EU country prior to market placement and may involve clinical trials (Neuhaus et al. 2011)

6.2 ISO Standards

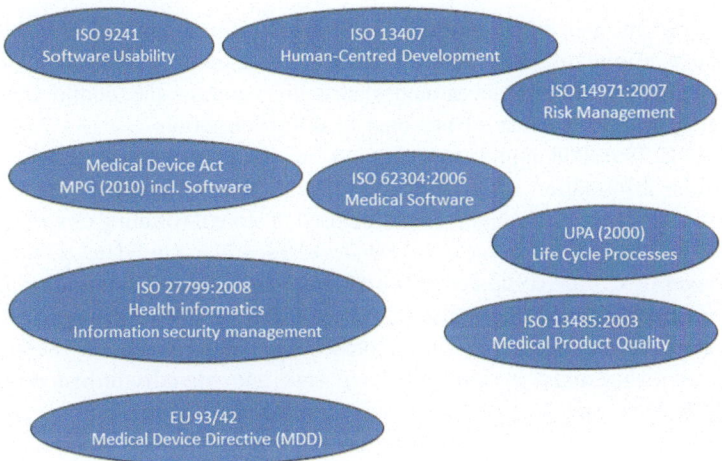

Fig. 3 See Slide 12-7

Slide 12-7: Quality of Medical Software: Some Standards to Know

The International Organization for Standardization (ISO and the International Electro-technical Commission (IEC)) provides best practice recommendations on information security risks, management and controls through its ISO/IEC 27000-series standards. The standards cover the fundamental requirements of information management systems; provide 13 guidelines and principles for the implementation of such systems. Among the standards, ISO 27799:2008 and ISO/TR 27809:2007 meant for health informatics and provides guidelines for designing health sector specific information management systems.

ISO/IEC 27002 provides control guidelines for patient safety within such systems. ISO/IEC Joint Technical Committee 1 (JTC1) deals with all matters of Information Technology including develop, maintain, promote and facilitate IT standards by enterprises and users concerning the security of IT systems and information.

SO 27799:2008 defines guidelines to support the interpretation and implementation in health informatics of ISO/IEC 27002 and is a companion to that standard.

ISO 27799:2008 specifies a set of detailed controls for managing health information security and provides health information security best practice guidelines. By implementing this international standard, healthcare

(continued)

(continued)

organizations and other custodians of health information will be able to ensure a minimum requisite level of security that is appropriate to their organization's circumstances and that will maintain the confidentiality, integrity and availability of personal health information.

ISO 27799:2008 applies to health information in all its aspects; whatever form the information takes (words, numbers, sound recordings, drawings, video, images etc.), whatever means are used to store it (printing or writing on paper or electronic storage) and whatever means are used to transmit it (by hand, fax, over computer networks, mail etc.), as the information must always be appropriately protected.

ISO 13485, published in 2003, represents the requirements for a comprehensive management system for the design and manufacture of medical devices. This standard supersedes earlier documents. ISO 13485 is generally harmonized with ISO 9001. A fundamental difference, however, is that ISO 9001 requires the organization to demonstrate **continual improvement**, whereas ISO 13485 requires only that they demonstrate the quality system is implemented and maintained.

ISO 14971 is an international standard that is quickly being recognized as one of the best processes to ensure that all aspects of risk management are considered throughout the product lifecycle for medical devices. Compliance to this standard is required to sell medical devices in the European Economic Area, as indicated in the Medical Devices Directive (MDD), which covers most implants, sets conformity assessment procedures depending on the medical device class type, and requires risk analysis to be performed. The use of this standard is also required in Canada and Australia. Within the USA, the standard is recognized by the Food and Drug Administration (FDA) as a way to meet the intent of the quality system regulation requirements for the development of safe medical products. ISO 14971 concerns the application of risk management and it is designed to help manufacturers introduce safe medical devices into the healthcare market. The manufacturer is responsible for identifying and controlling not only the risks associated with their medical device, but evaluating interactions with other devices. The standard also allows for other healthcare manufacturing organizations to use the process and obtain certification. This might include human tissue, animal care products, pharmaceutical manufacturers, etc., who may choose to use this standard (Catelani et al. 2011).

Slide 12-8: Medical Devices Act (Medizinproduktegesetz, MPG)

Here we see a medical device act on the example of Austria. Since 30th December 2009 software is included in the law, i.e., before this time it was relatively easy to declare that a specific software is *not* a medical product—it is just software; now it is, and the very strict rules of this law applies for software; this includes of course "apps" for smartphones as well (see next slide).

Slide 12-9: Medical Product Law and Mobile Apps

On September 25, 2013, the FDA (Food and Drug Administration, see: http://www.fda.gov/medicaldevices/productsandmedicalprocedures/connectedhealth/mobilemedicalapplications) released a (non-binding) document on mobile medical applications recommendations.

The widespread adoption of mobile computing in medicine and in particular the success of mobile applications (apps) is opening new and innovative ways to improve medicine, health, and health care delivery (Peischl et al. 2013; Breitwieser et al. 2013; Novak et al. 2012; Holzinger et al. 2011a).

Apps can also help to manage personal health and wellness and promote healthy living (Alagoez et al. 2010; Holzinger et al. 2010). According to industry estimates, 500 million smartphone users worldwide will be using a health care application by 2015, and by 2018, 50 % of the more than 3.4 billion smartphone and tablet users will have downloaded mobile health applications (http://www.research2guidance.com/500m-people-will-be-using-healthcare-mobile-applications-in-2015). These users include health care professionals, consumers, and patients.

The FDA encourages the development of mobile medical apps that improve health care and provide consumers and health care professionals with valuable health information. The FDA also has a public health responsibility to oversee the safety and effectiveness of medical devices—including mobile medical apps, for this purpose the FDA issued the Mobile Medical Applications Guide, which explains the oversight of mobile medical apps as devices and our focus only on the apps that present a greater risk to patients if they don't work as intended and on apps that cause smartphones or other mobile platforms to impact the functionality or performance of traditional medical devices.

6.3 Quality Management Process Cycle

Slide 12-10: ISO 13485:2003 Quality Management Process Cycle

ISO 13485:2003 represents the requirements for a comprehensive management system for the design and manufacture of medical devices. More specific it describes the requirements for a **quality management system** where an organization (regardless of size or type) needs to demonstrate its ability to provide medical devices and related services that consistently meet customer requirements and regulatory requirements. In this slide we see the main idea behind it: The quality management process cycle. The customers (aka end users) specify the requirements as active input for the product realization. Within the cycle we have a consequent iteration, forth back checking if the requirements are met, similar to the PDCA cycle (Holzinger 2011)—see next slide. The ideal output is in satisfactory addressing all end user requirements.

Slide 12-11: Quality Improvement Cycle

The quality improvement cycle is based on the original PDCA-Cycle aka Deming Cycle (Deming 1994)—Plan, Do, Study (the results), Act (incorporate your improvements). In this slide it is extended to a seven-step improvement process, which applies to any organization. In hospitals this approach brought for example enormous reductions of waste of supplies (Cleary 1995); the steps include:

1. Defining the system
2. Assessing the current situation
3. Analyzing causes
4. Applying an improvement process
5. Studying the results
6. Planning continuous improvement: No improvement process is ever finished!
7. Standardize the improvements!

This process is widely adopted in the medical area (Cleary 1995).

6.4 Software Product Quality Model

Slide 12-12: Product Quality Versus Process Quality

Before we concentrate on the software quality model, we emphasize again the difference between product quality (which is defined in ISO 9126) and process quality (which is defined in ISO 25000) and the important insight that both of them are important for the goal: quality in use (see next slide).

Slide 12-13: The Goal: Quality of Use = Measured USABILITY

The important insight which we shall always consider is that the quality in use is the goal and that *the* quality of use measure is "usability" (Bevan 1995, 1997, 2009; Holzinger et al. 2009) and this is always taking place within a **context** wherein the user constantly *interacts* with the product. Software in that sense is also a product.

Slide 12-14: ISO/IEC 9126-1 Software Product Quality

Usability is important but only a small part within the whole software product quality life cycle. ISO 9126-1 defines six large areas, each containing a set of important issues:

1. Functionality: accuracy, suitability, interoperability, security.
2. Reliability: maturity, fault tolerance, recoverability, availability.
3. Efficiency: time behavior (especially critical in the clinical domain!), utilization.
4. Maintainability: analyzability, changeability, stability, testability.
5. Portability: adaptability, installability, co-existence, replaceability.
6. Usability: understandability, learnability, operability, attractiveness.

Let us look closer on issue Nr. 5 Portability: this is particularly important with apps: Making apps useable on different platforms—the shipment of smartphones exceeded that of personal computers in 2011. However, the screen sizes and display resolutions of different devices vary to a large degree, along with different aspect ratios and the complexity of mobile tasks. These obstacles are a major challenge for software developers, especially when they try to reach the largest possible audience and develop for

(continued)

(continued)

multiple mobile platforms or device types. On the other side, the end users' expectations regarding the usability of the applications are increasing. Consequently, for a successful mobile application the user interface needs to be well-designed, thus justifying research to overcome these obstacles. In this paper, we report on experiences during an industrial project on building user interfaces for database access to a business enterprise information system for professionals in the field.

Holzinger et al. (2012) discuss a systematic analysis of standards and conventions for design of user interfaces for various mobile platforms, as well as scaling methods operational on different physical screen sizes.

7 Usability Engineering

7.1 Usability Engineering Methods

Slide 12-15: Remember Medical Workflows Are Highly Complex...

One of the basic lessons from HCI is, that usability must be considered before prototyping takes place. There are techniques (such as usability context analysis) intended to facilitate such early focus and commitment. When usability inspection, or testing, is first carried out at the end of the design cycle, changes to the interface can be costly and difficult to implement, which in turn leads to usability recommendations. These are often ignored by developers who feel, "We don't have usability problems". The earlier critical design flaws are detected, the more likely they can be corrected. Thus, user interface design should more properly be called user interface development, analogous to software development, since design usually focuses on the synthesis stages, and user interface components include metaphors, mental models, navigation, interaction, appearance, and usability (Holzinger et al. 2005b).

	Inspection Methods			Test Methods		
	Heuristic Evaluation	Cognitive Walkthrough	Action Analysis	Thinking Aloud	Field Observation	Questionnaires
Applicably in Phase	all	all	design	design	final testing	all
Required Time	low	medium	high	high	medium	low
Needed Users	none	none	none	3+	20+	30+
Required Evaluators	3+	3+	1-2	1	1+	1
Required Equipment	low	low	low	high	medium	low
Required Expertise	medium	high	high	medium	high	low
Intrusive	no	no	no	yes	yes	no

Fig. 4 See Slide 12-16

Slide 12-16: Comparison of Usability Engineering Methods

Generally, we can determine between two types of usability engineering methods: Inspection versus Test (Holzinger 2005).

Inspection methods are a set of methods for identifying usability problems and improving the usability of an interface design by checking it against established standards. These methods include heuristic evaluation, cognitive walkthroughs, and action analysis. No end users are needed, and these methods are performed by experts.

Testing with (real) end users is the most fundamental usability method and is in some sense indispensable. It provides direct information about how people use our systems and their exact problems with a specific interface. There are several methods for testing usability, the most common being thinking aloud, field observation, and questionnaires.

7.2 How to Measure Usability?

7.2.1 The System Usability Scale (SUS)

Slide 12-17: The System Usability Scale (SUS)

A rapid evaluation tool is the System Usability Scale (SUS). This ten-item scale was developed by (Brooke 1996) as a "quick and dirty" survey scale that would allow the usability practitioner to quickly and easily assess the

(continued)

(continued)

usability of a given product or service. Although there are a number of other excellent alternatives the SUS has several attributes that make it a good choice for general usability practitioners. The main advantage is, that the survey provides a single score on a scale that is easily understood by the wide range of people (from project managers to computer programmers) who are typically involved in the development of products and services and who may have little or no experience in human factors and usability (Holzinger 2010).

7.2.2 The Software Usability Measurement Inventory (SUMI)

Slide 12-18: The Software Usability Measurement Inventory (SUMI)

According to Kirakowski and Corbett (1993) the assessment of the usability of a computer system should involve measuring not only aspects of users' performance, but also how users subjectively feel about the system. For this purpose the Software Usability Measurement Inventory (SUMI) has been designed in particular to investigate users' *perceptions* of the quality of software systems. SUMI provides a global usability measure, along with five subscale measures and a high level problem diagnosis. There are large samples available, which can be used as benchmarks tested either against generic usability profiles, or against the usability profile of another system. An sample application of a SUMI evaluation can be found in Kosec et al. (2009), a good discussion in Cavallin et al. (2007) and a good source is available here: http://sumi.ucc.ie/sumipapp.html

7.2.3 Usability Measurement Metrics

Slide 12-19: Quantifying Usability Metrics in Software Quality

In this slide we see QUIM: A framework for quantifying usability metrics in software quality models, which is a hierarchical model similar to typical software engineering models (e.g., Boehm model, McCall model, IEEE 1061, ISO 9126, etc.). The difference is that, it distinguishes four levels called factors, criteria, metrics, and data—as can be seen in the slide. The relationship between these layers is an n–m relationship. Factors include effectiveness, efficiency, satisfaction, productivity, safety, internationability (globality); the criteria include attractiveness, consistency, minimal action, minimal memory load, completeness; the metrics include task concordance and visual coherence (Seffah et al. 2001, 2006; Holzinger et al. 2008).

7.3 User-Centered Design and Development

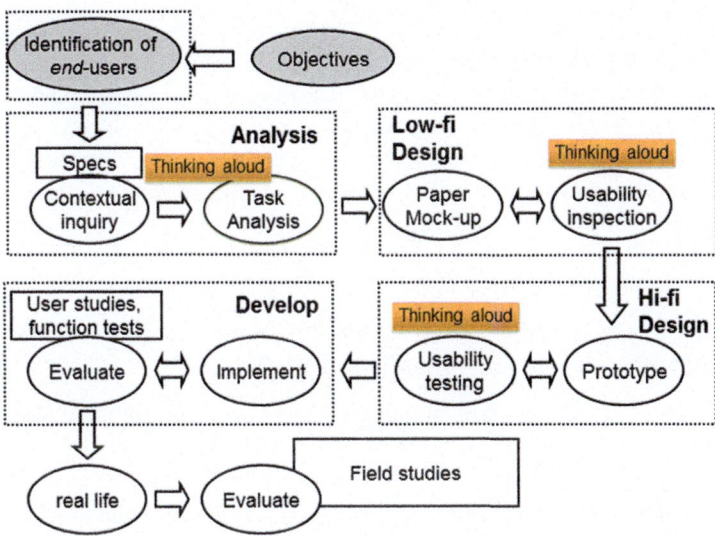

Fig. 5 See Slide 12-20

Slide 12-20: User-Centered Design and Development (UCD)

In contrast to traditional egocentric design, user-centered design and development focuses on the needs, demands, and requirements of the end user.

Note: In software engineering the design is more the "thinking" and problem solving, e.g., the solution of a problem in terms of algorithms and concepts; whereas development includes the implementation of the design. Usually, an engineer performs both: design and development. The emphasis in this model is on the end user. Meanwhile the ISO 13407 standard defines a Human-centred design process, which defines a general process model (similar to the "big picture" in the next slide) but does not define specific methods. In this slide we see a process model, which has been proved in many projects and where for example thinking aloud as a main low-cost method can be applied from the very early stages of the development cycle. The most important step is to **identify the end users** at the very beginning, then to specify the **context of use**, create low-fi design solutions, because they can be redesigned rapidly and with low-cost (Holzinger 2002, 2003; Holzinger et al. 2003).

Slide 12-21: Remember the Big Picture: UCD Process

In this slide we see the "big picture": The UCD process as strategy of the whole development process (Wiklund and Wilcox 2005) includes the concept phase with contextual inquiries and market research; the design requirement phase with task analysis and user profiling, the design specifications phase with the first prototypes a the verification phase with usability testing and finally the validation phase with field studies and evaluation (we come to this in the next slides).

Slide 12-22: The Power of Iteration: A UCD Spiral

The success of Extreme Programming (XP) is based, among other things, on an optimal communication in teams of 6–12 persons, simplicity, frequent releases, and a reaction to changing demands (Beck 1999). Most of all, the customer (and hence end user) is integrated into the development process, with constant feedback. This is very similar to Usability Engineering (UE) which follows a spiral four phase procedure model (analysis, draft, development, test) and a three step (paper mock-up, prototype, final product) production model. In comparison, these phases are extremely shortened in XP; also the ideal team size in UE User-Centered Development is four to six people, including the end user. The two development approaches have different goals but, at the same time, employ similar methods to achieve them. It seems obvious that there must be synergy in combining them. The authors present ideas in how to combine them in an even more powerful development method called Extreme Usability (XU) (Holzinger et al. 2005a; Holzinger and Slany 2006; Hussain et al. 2009a, b).

Slide 12-23: Agility: Make the UCD Spirals as Small as Possible

By making the UCD spirals (Slide 12-22) as small as possible we achieve a series of advantages:

In XP, this danger of dissipating one's energies in details (engineers are particularly susceptible to "featuritis" (Buschmann 2010)) and the client's (end users who are also often overstressed) becomes caught up in the detail is consciously controlled by applying short iterations, frequent replanning and **focusing on simple design**.

Note: Simple things first—they may be the most important ones!

(continued)

(continued)

This enables the client to get a realistic feeling of what can be achieved by the team, if the team implements only what he requested, and what needs to be pushed back to later versions in order to achieve the core functionality needed for the economic success of the project.

In particular, the well-known danger of "featuritis" is harnessed by the conscious decision to avoid thinking about what could happen later and could become meaningful, while being prepared to make extensive adjustments and changes at a later date. Extreme Usability (XU) could become such that all the best practices of UE are kept in the XP process during the planning games, with a restriction of the usability aspects in the next iteration and the equal treatment of usability and functionality. The advantage would be that, with the XP process, the adjustment and gradual improvement until the end of the project is explicitly built into the process, which is very helpful for UE. However, UE can improve the XP development method by focusing on the important aspects of the usability and employing the entire development team to make the customer continually aware of these aspects (by daily inquiry, discussion and testing); also the developers minds will be focused on the most important usability aspects, when at least one developer in the team possesses previous knowledge about UE and by implementing pair-programming, including the complete and frequent mixing of the pairs as well as passing on the on-site customer XP principle. Obviously, UE experience for all developers is an advantage in every project (Holzinger et al. 2005a).

Slide 12-24: Rapid Prototyping: Paper Mock-Ups

A very helpful low-fidelity prototyping method is to use paper mock-ups for rapid prototyping. Using common office supplies (markers, index cards, scissors and transparency film) the engineer can quickly sketch screen contents and each interactive element of the interface (menus, messages) on a separate piece of paper. The paper mock-up is not necessary to be very neat: it may contain hand written text, crooked lines and last minute corrections. It is, however, good enough to show what the screens would look like and provides a good basis for "playing out" some workflows. One developer plays the role of the "computer," simulating the behavior of the software by manipulating the pieces of paper. It is important to ask the end users to perform realistic tasks with the prototype, e.g., "... you are a teacher, set up a theme and create some hours in the catalogue ...". Furthermore, it is important NOT to ask the users for their opinions of the interface—telling them that this is experimental is enough. After every UCD session the team can discuss what they had seen and immediately execute changes to the paper prototype (Holzinger 2004).

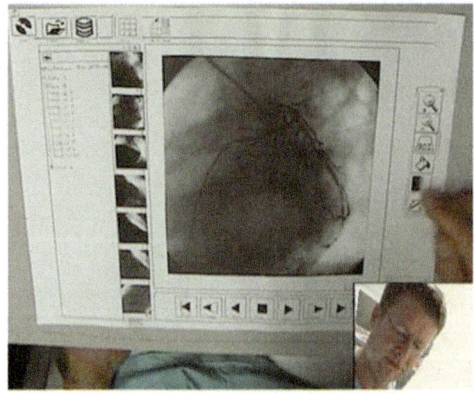

Fig. 6 See Slide 12-25

Slide 12-25: Insight into the End User: Thinking Aloud

From the spectrum of methods in Usability Engineering (review Slide 12-16), one method particularly stands out due to its practical realizability: Thinking aloud (THA). This method originates from early psychological problem-solving research (Duncker 1945) and permits insight into the mental processes: The test person (end user) receives a completely defined set of tasks and is asked to express, out loud, all—also fleeting—ideas and thoughts during the execution of this task. It is advantageous to record this procedure with a video camera because it is then possible to draw conclusions on the work habits from both the verbal and the facial expression and the gestures of the test person, in particular, it is possible to judge their subjective impressions and feelings. The behavior patterns recorded on the video tape and/or the log file analyses usually make it possible to identify where the test person has problems and how, and why, they take certain actions. Additionally, with a behavior observation software (for example INTERACT from the company Mangold Munich, Germany), the video material can also be compared to that of other users in order to find particular behavior patterns. According to Nisbett and Wilson (1977) three to five end users are sufficient to obtain valid statements; however, for scientific studies, it is sometimes necessary to increase the number of test subjects. The principle of breaking off the tests when no further increase in knowledge is effected, has proved satisfactory (Brown and Holzinger 2008; Holzinger and Brown 2008).

7 Usability Engineering

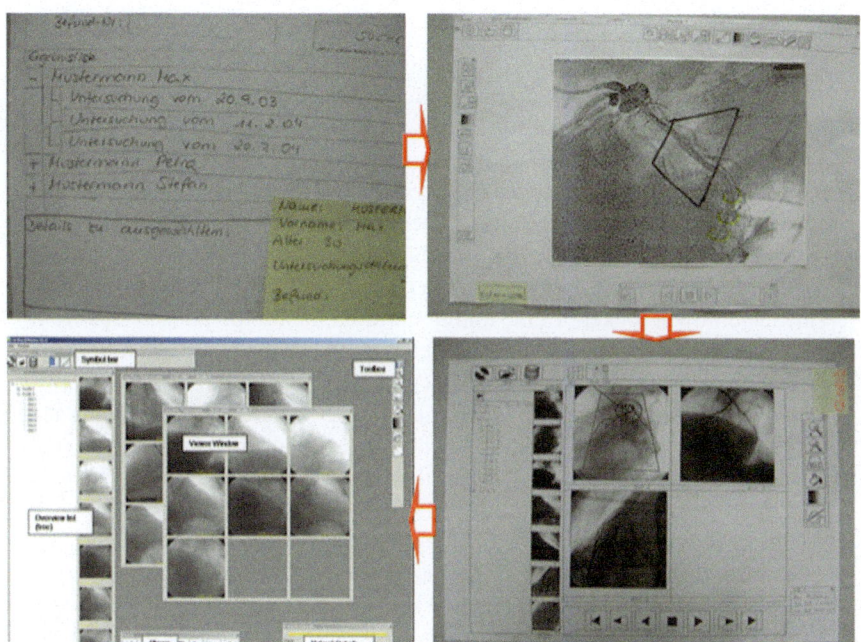

Fig. 7 See Slide 12-26

Slide 12-26: Example UCD Process of Developing a Cardiac Viewer

This practical example from a development project (Holzinger et al. 2005b) shows the four levels of the UCD process:

Level 1 Requirements Analysis: The first goal is to provide specifications of the tasks that the end users must perform in order to support problem solving. The envisioned system shall be discussed **with all the people involved** (not the CEO, who never will do any work with it;-)) in the medical example here it includes the cardiologists, who expressed their demands of what the system must be able to do—which functionalities it should provide and how these should work. During the requirement analysis, which was made with the help of video recordings of clinicians in real work situations, a verbal description of the system emerged.

Level 2 Low-Fi Prototyping (Paper Mock-up): In this project, at first, screen designs and dialogues were sketched on paper. Then a paper mock-up, which can be adjusted whilst working with the cardiologists, was created. The use of paper mock-ups provides a first usability feedback with minimum effort and maximum results.

(continued)

(continued)

Level 3 Hi-Fi Prototyping: Further, a working prototype, for studying the interaction of the end users, was created. During this phase the programmers were able to concentrate on the hi-fi prototype and adapt the choice of software tools to the technological requirements. The advantage of this approach—as opposed to the usual methods—can be seen in the fact that the graphical user interface was available before the full implementation, subsequently the end users know, in advance, exactly what was being provided and how it looked. In the traditional way, the prototype develops from an idea, although the design is predetermined by the data for which the programmer has to provide the interaction. The application consists of two windows; Search Mask and Main Window. The Search Mask, by which the user can find the medical data, is displayed immediately on starting the application. Subsequently, the application switches to the main window to display the data.

The search pop-up is a self-opening search form, in which the user can enter parameters for a search in the data base.

The Viewer is a window containing the actual image data within the main window. An individual viewer window is opened for every patient or patient study.

The Player is a window, within which the selected images can be played, started/stopped, navigated etc.

The Toolbox contains all the tools (functions), which the user requires to manipulate the images.

The Symbol Toolbar is the window beneath the menu. It contains all other tools, which are not in the Toolbox, for example: grid alignment, navigation, etc.

Level 4 Implementation: From the viewpoint of Software Engineering the most essential specifications were: Platform independence (Mac, Windows, Unix, Linux); Support for the most important display formats, including DICOM, BMP, JPEG, GIF, PNG, TIFF, AVI, MPEG 4—including this additional requirements the viewer was developed (and is still in use (November, 2013)), (Holzinger et al. 2005b).

Slide 12-27: Hi-Fi Prototype Allows Low-Level Interaction

It is important to test the functionality with a hi-fi prototype, either by a part implementation or with a simple assistive technology (even for example PowerPoint slides).

8 Evaluation

Slide 2-28: Validation and Verification to Check Quality

Your applications are used by end users—your customers—consequently a solid validation, verification and evaluation, and/or experimental examination is invaluable (for details please refer to Holzinger (2010)).
Let us just clarify some definitions at first:

1. **Validation** is the process of checking if and to what extend your system meets the specifications and therefore fulfils (American English: fulfills) its intended purpose.
2. **Verification** is a quality control process that is used to evaluate whether and to what extend your system complies with official regulations, legal specifications, standards or norms.
3. **Evaluation** is the systematic assessment of your application by use of certain criteria against a defined set of standards.
4. **Experimental Examination** is testing the system against stated Hypotheses (e.g., "By use of the system A the task X is performed in shorter time than by use of system B") either in a laboratory or, better, in the field (real life experiment, field experiment).

Slide 12-29: ISO 13407 Human-Centred Design (1/2)

This brings us back to the ISO 13407 Human-centered design: along with ISO TR 18529 these standards represent a maturing of the discipline of user-centered design. The systems development community sees that Human Factors has processes which can be managed and integrated with existing project processes. This internationally accepted set of processes provides a definition of the capability that an organization must possess in order to implement such user-centered design effectively. It can also be used to assess the extent to which a particular development project employs user-centered design (Earthy et al. 2001).

Slide 12-30: ISO 13407 Human-Centered Design (2/2)

In this slide we see the "big picture"—a good overview on the contents of the ISO 13407 standard.

9 Technology Acceptance

Slide 12-31: Technology Acceptance Model (TAM) by Davis (1989)

Originally, the framework of Shackel (1991) has been one of the most influential paradigms for conceptualizing the acceptability of any given system to its intended end users: He suggested that systems acceptability can be defined as a function of three orthogonal dimensions, which he balanced against cost (Holzinger et al. 2011b):

1. Utility (whether the system does what is needed functionally).
2. Usability (whether and to what extent the users can actually work with the system successfully).
3. Likeability (whether the users feel the system is suitable).

In the slide we see a previous model, proposed to explain and predict user acceptance: the technology acceptance model (TAM) by Davis (1989), confer also to Davis (1993) and Morris and Turner (2001). There have been several theoretical models developed in order to study user acceptance; many of them incorporate perceived ease of use as a determinant of acceptance, the technological acceptance model (TAM) as can be seen in this slide is most widely accepted.

Background: Originally, TAM was adapted from the Theory of Reasoned Action (TRA) by Fishbein and Ajzen (1975) and it proposes that two specific beliefs:

(a) The perceived ease of use.
(b) The perceived usefulness are determining a person's behavioral intention to use technology.

However, the attitude towards using a technology was originally omitted in the final model, due to a partial mediation of the impact of beliefs on intention by attitude, a weak direct link between perceived usefulness and attitude, and a strong direct link between perceived usefulness and intention; this was explained as originating from people intending to use technology, due to it was useful for them even though they did *not* have a positive affect (attitude) towards using. This derives from the Hedonomics-Ergonomics pyramid (see next slide):

Slide 12-32: Ergonomics Versus Hedonomics

Similar to the famous "Maslow Pyramide" (Maslow et al. 1970), Helander and Khalid (2006) proposed a Ergnomics/Hedonomics Pyramide which can

(continued)

(continued)

be seen in this slide: The more to the top, the more individuation, in a sense of personal perfection, takes place. Safety is the basis, followed by functionality, and usability. On these ergonomic "must-haves", there are hedonomics factors including pleasurable experiences.

Slide 12-33: Technology Acceptance in the Clinical Context

In this slide we see a version of the TAM model adapted to the clinical context: Recent empirical research has utilized the TAM to advance the understanding of medical doctors' and nurses' technology acceptance in the clinical workplace. However, the majority of the reported studies are either qualitative in nature or use small samples of medical staff. Additionally, in very few studies moderators are either used or assessed despite their importance in TAM-based research. The study by Melas et al. (2011) focused on the application of TAM in order to explain the intention to use clinical information systems, in a random sample of 604 medical staff (534 physicians) working in 14 hospitals in Greece. The authors introduce physicians' specialty as a moderator in TAM and test medical staff information and communication technology (ICT) knowledge and ICE feature demands, as external variables. The results showed that TAM predicts a substantial proportion of the intention to use clinical information systems (Melas et al. 2011)

A main contribution here is that it is accepted that there are diverse clinical specialists—every of them having different needs, goals and requirements. Note that this work is from 2011 sometimes easy things take very long.

10 Future Outlook

Slide 12-34: Example: Information Retrieval Experience

Along with technological advances, quality of use will become more important in the future including (van der Sluis et al. 2010):

1. Aesthetic and hedonic factors (e.g., beauty, enjoyment, and extending one's personal knowledge and satisfaction).
2. Emotional factors, addressing the antecedents and consequences of, ideally, positive emotions. Although overlapping with the first category,

(continued)

(continued)

   these factors are not seen as a goal on their own; however, they can aid in solving an information need.
3. Experiential factors, combining all contextual and related factors (including e.g., mood, expectations, and active goals) interact with the situation and time in creating the experience.

Slide 12-35: Example: Emotion 2D Measurement Scale

Here we see the factors arousal and pleasure. Difficult to measure (see next slide), but will become more and more important in the future.

Slide 12-36: How to Measure Emotions?

Measuring emotions is not easy and there are three basic approaches (Lopatovska and Arapakis 2011):

1. Neuro-physiological, e.g., brain activity, pulse rate, blood pressure, skin conductance. Can detect short-term changes not measurable by other means; Reliance on non-transparent, invasive sensors; can reduce people's mobility, causing distraction of emotional reactions; prone to noise due to unanticipated changes in physiological characteristics; inability to map data to specific emotions; require expertise and the use of special, often expensive, equipment.
2. Observation, e.g., facial expressions; speech; gestures. Use of unobtrusive techniques for measuring emotion; crosscultural universals; cannot perform context-dependent interpretation of sensory data; highly dependent on environmental conditions (illumination, noise, etc.); some responses can be faked; recognizes the presence of emotional expressions, not necessarily emotions.
3. Self-reporting, e.g., questionnaire, diary; interview; high correlation to neurophysiological evidence; unobtrusive; straightforward and simple—do not require the use of special equipment; Rely on the assumption that people are aware of and willing to report their emotions; subject to the respondent's bias; results of different studies might not be directly comparable.

Slide 12-37: Example Methods for Measuring Emotion

Here just a selection of possible methods:

1. Subjective measures → Kansei Engineering, Semantic scales (e.g., Nagamachi (2001), Helander and Khalid (2006)); Experience sampling method (e.g., Larson and Csikszentmihalyi (1983)); Affect Grid (e.g., Warr (1999), Holzinger et al. (2013)); MACL Checklist (e.g., Nowlis and Greenberg (1979)); PANAS Scale (e.g., Watson et al. (1988)); Philips questionnaire (e.g., Jordan (2000));
2. Objective Measures → Facial action coding system (e.g., Ekman (1982)); Maximally discriminative affect coding system (e.g., Izard (1979)); Facial electromyography (e.g., Davis et al. (1995));
3. Psychogalvanic measures → Galvanic skin response (e.g., McCleary (1950), Westerink et al. (2008)), Wearable sensors (e.g., Picard (2000));
4. Performance measures → Judgment task involving probability estimates (e.g., Larsen and Ketelaar (1991)); Lexical decision task (e.g., Challis and Krane (1988), Niedenthal and Setterlund (1994)).

11 Exam Questions

11.1 Yes/No Decision Questions

Please check the following sentences and decide whether the sentence is true = YES; or false = NO; for each correct answer you will be awarded 2 credit points.

01	Customizing and adaptation of third-party software into the clinical workplace is usually not necessary, they can be better used by the end-users when provided separately.	❏ Yes ❏ No	2 total
02	Time to perform a task is still the major and easy measurable human factor in usability engineering.	❏ Yes ❏ No	2 total
03	Ergonomic aspects of the workplace include: proper distance to the screen, correct table height, proper chair to ensure a comfortable angle from eyes to the screen.	❏ Yes ❏ No	2 total
04	ISO 13407 does not include human aspects into engineering development cycles.	❏ Yes ❏ No	2 total
05	Usability can be defined via a combination of efficiency, effectiveness and satisfaction.	❏ Yes ❏ No	2 total
06	The EU directive 93/42/EEC1 states criteria to define medical devices. For systems and devices that fall under these definitions, the directive states requirements that have to be met.	❏ Yes ❏ No	2 total
07	Neurophysiological measures, e.g. brain activity can be easily achieved by looking at facial expressions.	❏ Yes ❏ No	2 total
08	According to Kirakowski the assessment of the usability of a system should involve measuring not only aspects of performance, but also how users subjectively feel about the system.	❏ Yes ❏ No	2 total
09	Evaluation is the assessment of expected requirements and not are always a nice to have, not necessary for inclusion in the development process.	❏ Yes ❏ No	2 total
10	Validation is the process of checking if and to what extend your system meets the specifications and therefore fulfils (American English: fulfills) its intended purpose.	❏ Yes ❏ No	2 total

Sum of Question Block A (max. 20 points)	

11.2 Multiple Choice Questions (MCQ)

The following questions are composed of two parts: the stem, which identifies the question or problem and a set of alternatives which can contain 0, 1, 2, 3, or 4 correct answers, along with a number of distractors that might be plausible—but are incorrect. Please **select the correct answers** by ticking ☒—and do not forget that it can be none. Each question will be awarded 4 points *only if everything is correct*.

01	A SUS value of 65 means ... ☐ a) ... excellent, no need to look on further usability issues. ☐ b) ... just acceptable – but not good. ☐ c) ... poor – a complete redesign is necessary. ☐ d) ... just on the border between marginal low/high acceptable.	4 total
02	Neuro-physiological measurements include ... ☐ a) ... brain activity. ☐ b) ... pulse rate. ☐ c) ... blood pressure. ☐ d) ... skin conductance.	4 total
03	At the beginning of medical informatics in the 1970ies .... ☐ a) ... the focus was mostly on data acquisition, storage and accounting. ☐ b) ... there was an emerging trend to use Web based health applications. ☐ c) ... personalized medicine was already part of some health strategies. ☐ d) ... the first end-user programmable Mash-ups were introduced.	4 total
04	Hedonomics focuses on ☐ a) ... prevention of pain. ☐ b) ... priority of preference. ☐ c) ... promulgation of process. ☐ d) ... personal perfection.	4 total
05	Thinking aloud can be applied ... ☐ a) ... in the creativity and design phase for finding new solutions. ☐ b) ... in the low-fi design phase for finding usability problems. ☐ c) ... in the development phase for user studies and function tests. ☐ d) ... in the hi-fi design phase for consolidating usability problems.	4 total
06	The Quality Improvement Cycle by Deming follows ... ☐ ... Plan, Do, Check, Act. ☐ ... Plan, Check, Do, Act. ☐ ... Plan, Act, Check, Do. ☐ ... Plan, Do, Study, Act.	4 total
07	ISO 13407 ... ☐ a) ... considers generally Human-Centred Development. ☐ b) ... is the medical device directive. ☐ c) ... contains ISO 9241. ☐ d) ... includes risk management.	4 total
08	Check if the following purposes fall into the medical device directive ... ☐ a) ... control of conception. ☐ b) ... diagnosis, prevention, monitoring, treatment or alleviation of disease. ☐ c) ... diagnosis, monitoring, treatment for an injury or handicap. ☐ d) ... investigation, replacement of the anatomy or a physiological process.	4 total

09	To the system characteristic "classified patient data" the corresponding quality factor is... ☐ a) ... reliability. ☐ b) ... expandability. ☐ c) ... security ☐ d) ... usability.	4 total
10	Heuristic Evaluation ... ☐ a) ... can be applied in all design and development phases. ☐ b) ... needs 3+ users. ☐ c) ... is very time consuming. ☐ d) ... is not intrusive.	4 total

Sum of Question Block B (max. 40 points)

11 Exam Questions

11.3 Free Recall Block

Please follow the instructions below. At each question you will be assigned the credit points indicated if your option is correct (partial points may be given).

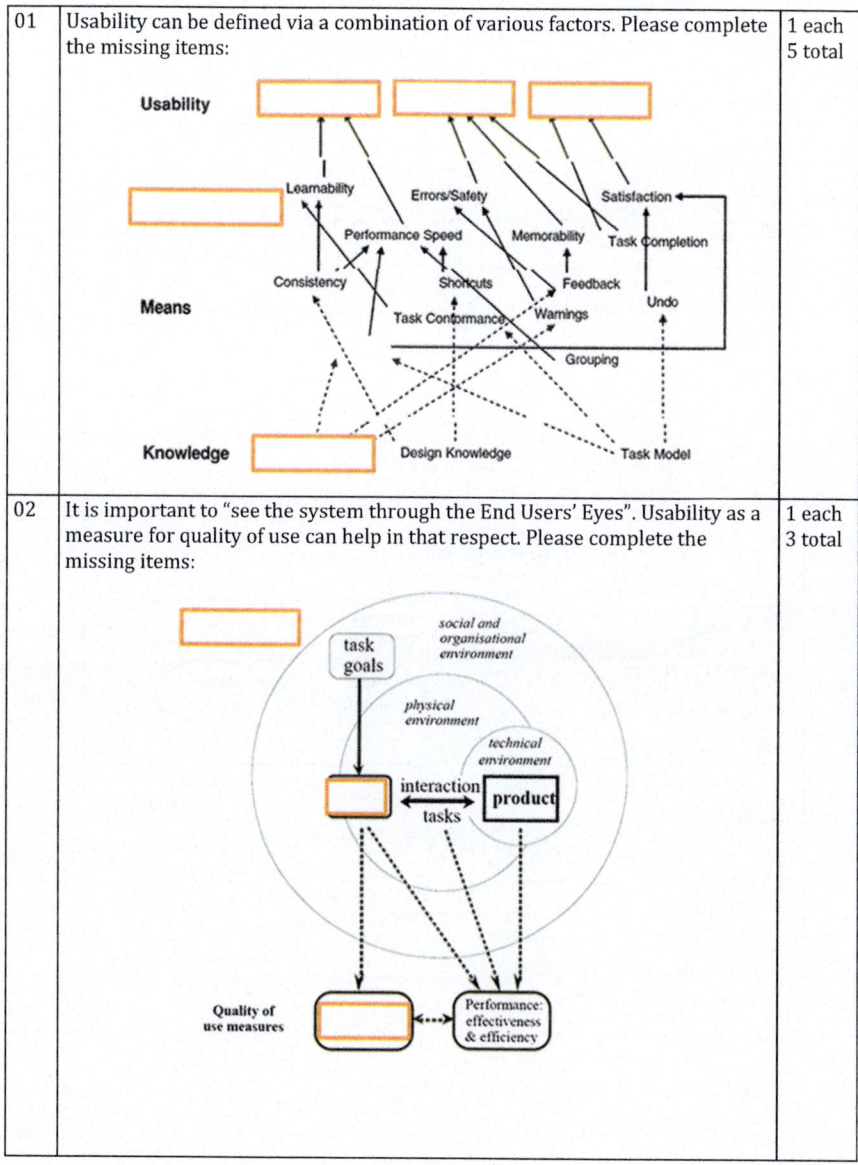

01	Usability can be defined via a combination of various factors. Please complete the missing items:	1 each 5 total
02	It is important to "see the system through the End Users' Eyes". Usability as a measure for quality of use can help in that respect. Please complete the missing items:	1 each 3 total

03 | Usability Engineering Methods can be divided into inspection methods and test methods, each having specific advantages and disadvantages. Please complete the missing items. | 1 each 6 total

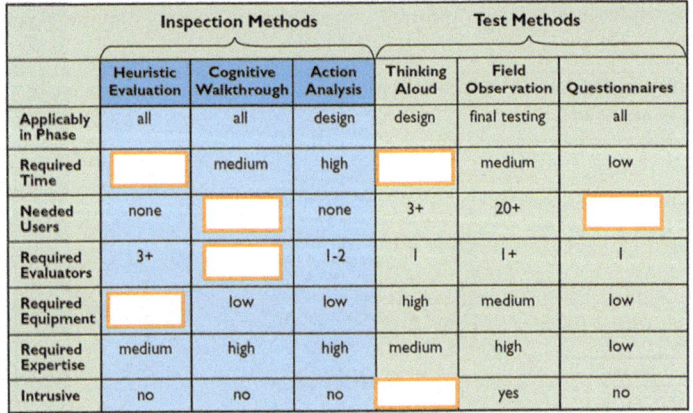

	Inspection Methods			Test Methods		
	Heuristic Evaluation	Cognitive Walkthrough	Action Analysis	Thinking Aloud	Field Observation	Questionnaires
Applicably in Phase	all	all	design	design	final testing	all
Required Time		medium	high		medium	low
Needed Users	none		none	3+	20+	
Required Evaluators	3+		1-2	1	1+	1
Required Equipment		low	low	high	medium	low
Required Expertise	medium	high	high	medium	high	low
Intrusive	no	no	no		yes	no

04 | The UCD- focuses on the needs, demands, requirements of the end user. Please fill in the missing terms in the figure below: | 1 each 4 total

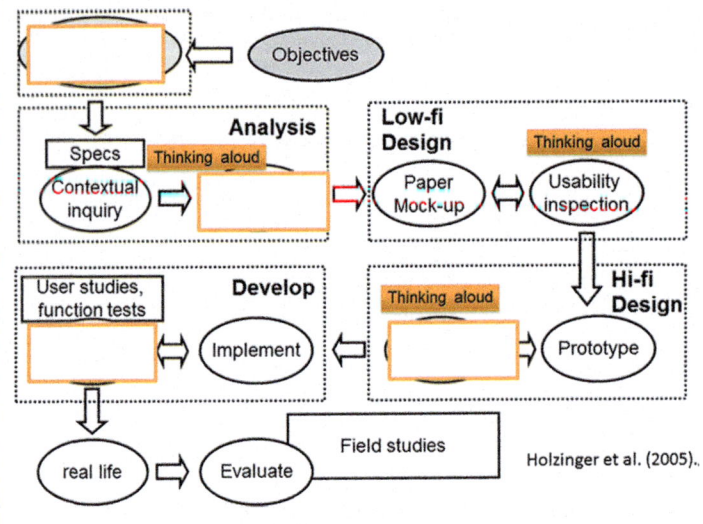

Holzinger et al. (2005).

05	Usability is important, but only a small part. Please identify the factors and assign the correct categories:		1 each 3 total
	accuracy suitability interoperability security	Efficiency	
	time behaviour resource man. utilisation	Functionality	
	maturity fault tolerance recoverability availability	Reliability	

SUM OF QUESTION BLOCK C (MAX. 40 POINTS)

12 Answers

12.1 Answers to the Yes/No Questions

Please check the following sentences and decide whether the sentence is true = YES; or false = NO; for each correct answer you will be awarded 2 credit points.

01	Customizing and adaptation of third-party software into the clinical workplace is usually not necessary, they can be better used by the end-users when provided separately.	☐ Yes ☒ No	2 total
02	Time to perform task is still the major and easy measurable human factor in usability engineering.	☒ Yes ☐ No	2 total
03	Ergonomic aspects of the workplace include: proper distance to the screen, correct table height, proper chair to ensure a comfortable angle from eyes to the screen.	☒ Yes ☐ No	2 total
04	ISO 13407 does not include human aspects into engineering development cycles.	☐ Yes ☒ No	2 total
05	Usability can be defined via a combination of efficiency, effectiveness and satisfaction.	☒ Yes ☐ No	2 total
06	The EU directive 93/42/EEC1 states criteria to define medical devices. For systems and devices that fall under these definitions, the directive states requirements that have to be met.	☒ Yes ☐ No	2 total
07	Neurophysiological measure, e.g. brain activity can be easily achieved by looking at facial expressions.	☐ Yes ☒ No	2 total
08	According to Kirakowski the assessment of the usability of a system should involve measuring not only aspects of performance, but also how users subjectively feel about the system.	☒ Yes ☐ No	2 total
09	Evaluation is the assessment of expected requirements and not are always a nice to have, not necessary for inclusion in the development process.	☐ Yes ☒ No	2 total
10	Validation is the process of checking if and to what extend your system meets the specifications and therefore fulfils (American English: fulfills) its intended purpose.	☒ Yes ☐ No	2 total

Sum of Question Block A (max. 20 points)	

12.2 Answers to the Multiple Choice Questions (MCQ)

01	A SUS value of 65 means ... ☐ a) ... excellent, no need to look on further usability issues. ☒ b) ... just acceptable – but not good. ☐ c) ... poor – a complete redesign is necessary. ☒ d) ... just on the border between marginal low/high acceptable.	4 total
02	Neuro-physiological measurements include ... ☒ a) ... brain activity. ☒ b) ... pulse rate. ☒ c) ... blood pressure. ☒ d) ... skin conductance.	4 total
03	At the beginning of medical informatics in the 1970ies .... ☒ a) ... the focus was mostly on data acquisition, storage and accounting. ☐ b) ... there was an emerging trend to use Web based health applications. ☐ c) ... personalized medicine was already part of some health strategies. ☐ d) ... the first end-user programmable Mash-ups were introduced.	4 total
04	Hedonomics focuses on ☐ a) ... prevention of pain. ☐ b) ... priority of preference. ☐ c) ... promulgation of process. ☒ d) ... personal perfection.	4 total
05	Thinking aloud can be applied ... ☒ a) ... in the creativity and design phase for finding new solutions. ☒ b) ... in the low-fi design phase for finding usability problems. ☐ c) ... in the development phase for user studies and function tests. ☒ d) ... in the hi-fi design phase for consolidating usability problems.	4 total
06	The Quality Improvement Cycle by Deming follows ... ☒ ... Plan, Do, Check, Act. ☐ ... Plan, Check, Do, Act. ☐ ... Plan, Act, Check, Do. ☒ ... Plan, Do, Study, Act.	4 total
07	ISO 13407 ... ☒ a) ... considers generally Human-Centred Development. ☐ b) ... is the medical device directive. ☐ c) ... contains ISO 9241. ☐ d) ... includes risk management.	4 total
08	Check if the following purposes fall into the medical device directive ... ☒ a) ... control of conception. ☒ b) ... diagnosis, prevention, monitoring, treatment or alleviation of disease. ☒ c) ... diagnosis, monitoring, treatment for an injury or handicap. ☒ d) ... investigation, replacement of the anatomy or a physiological process.	4 total
09	To the system characteristic "classified patient data" the corresponding quality factor is... ☐ a) ... reliability. ☐ b) ... expandability. ☒ c) ... security ☐ d) ... usability.	4 total

10	Heuristic Evaluation ... ☒ a) ... can be applied in all design and development phases. ☐ b) ... needs 3+ users. ☐ c) ... is very time consuming. ☒ d) ... is not intrusive.	4 total

Sum of Question Block B (max. 40 points)	

12 Answers

12.3 Answers to the Free Recall Questions

01	Usability can be defined via a combination of various factors. Please complete the missing items:	1 each 5 total
	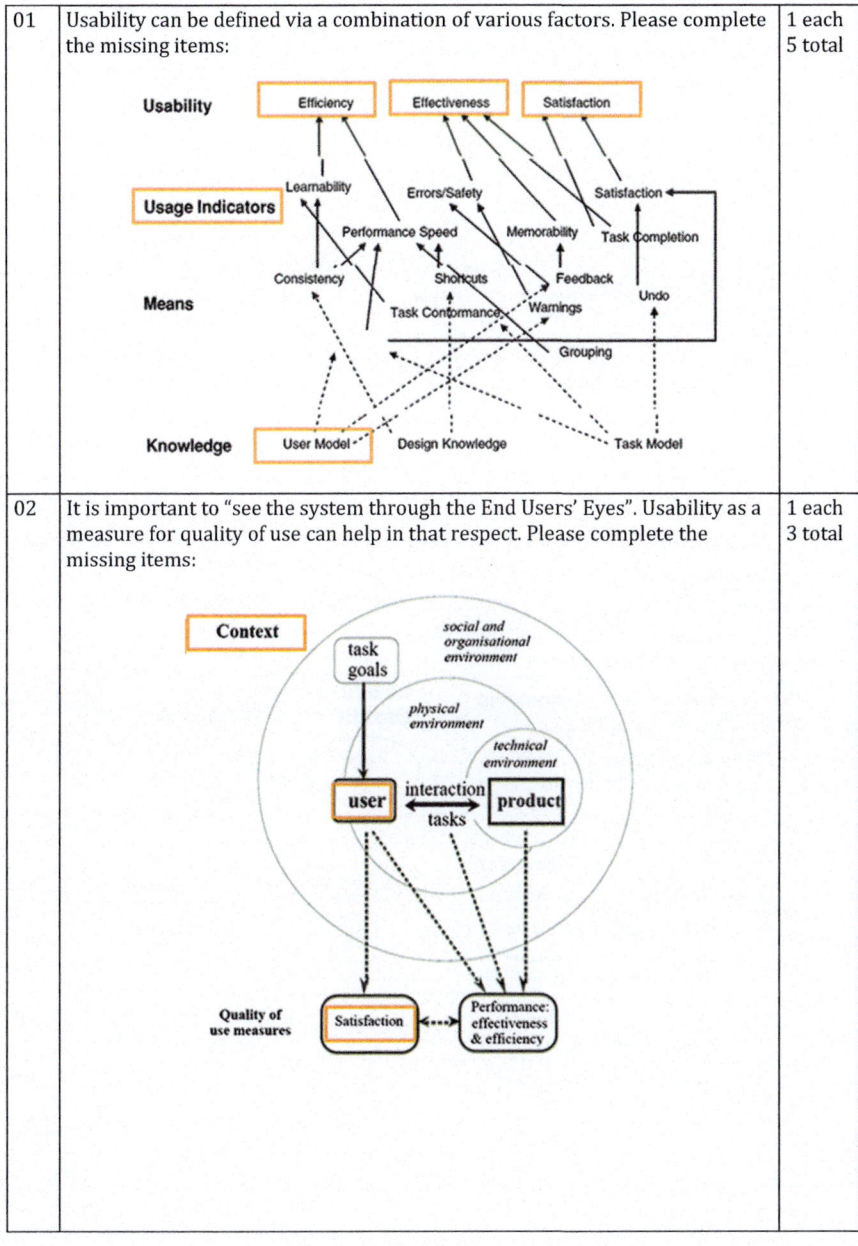	
02	It is important to "see the system through the End Users' Eyes". Usability as a measure for quality of use can help in that respect. Please complete the missing items:	1 each 3 total

03	Usability Engineering Methods can be divided into inspection methods and test methods, each having specific advantages and disadvantages. Please complete the missing items.	1 each 6 total

	Inspection Methods			Test Methods		
	Heuristic Evaluation	Cognitive Walkthrough	Action Analysis	Thinking Aloud	Field Observation	Questionnaires
Applicably in Phase	all	all	design	design	final testing	all
Required Time	low	medium	high	high	medium	low
Needed Users	none	none	none	3+	20+	30+
Required Evaluators	3+	3+	1-2	1	1+	1
Required Equipment	low	low	low	high	medium	low
Required Expertise	medium	high	high	medium	high	low
Intrusive	no	no	no	yes	yes	no

04	The UCD- focuses on the needs, demands, requirements of the end user. Please fill in the missing terms in the figure below:	1 each 4 total

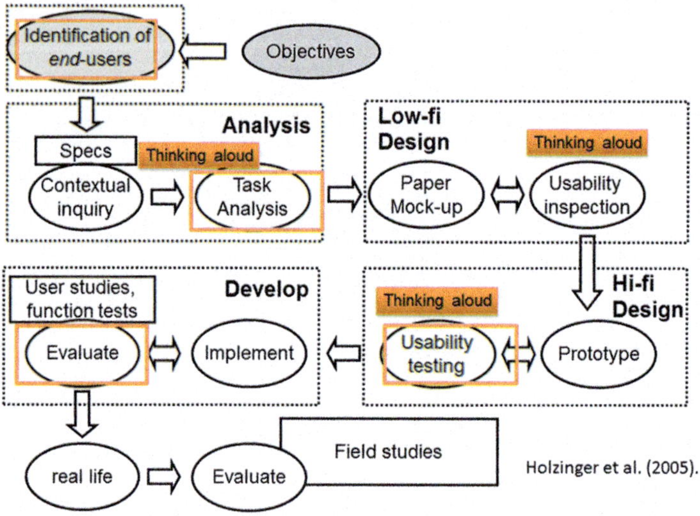

Identification of *end*-users ⇐ Objectives

Analysis
Specs — Thinking aloud
Contextual inquiry ⇒ Task Analysis ⇒

Low-fi Design
Paper Mock-up ⇔ Usability inspection — Thinking aloud

User studies, function tests
Develop
Evaluate ⇐ Implement ⇐ Usability testing ⇔ Prototype — Thinking aloud

Hi-fi Design

real life ⇒ Evaluate — Field studies

Holzinger et al. (2005).

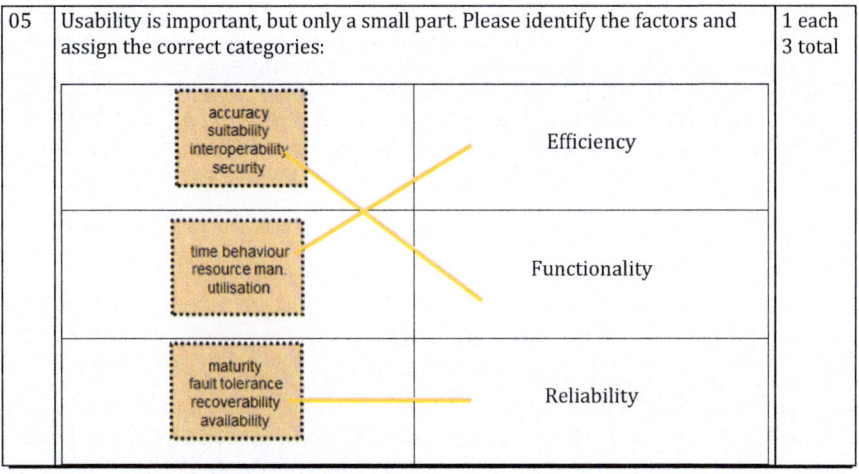

References

Alagoez F, Valdez AC, Wilkowska W, Ziefle M, Dorner S, Holzinger A (2010) From cloud computing to mobile Internet, from user focus to culture and hedonism: the crucible of mobile health care and Wellness applications. 5th International conference on pervasive computing and applications (ICPCA). IEEE, pp 38–45

Beck K (1999) Extreme programming explained: embracing change. Addison Wesley, Boston, MA

Bevan N (1995) Usability is quality of use. In: Anzai Y, Ogawa K, Mori H (eds) 6th International conference on human computer interaction. Elsevier, Tokyo

Bevan N (1997) Quality in use: incorporating human factors into the software engineering lifecycle. 3rd International software engineering standards symposium (ISESS '97), Walnut Creek, CA, 01–06 June 1997, pp 169–179

Bevan N (2009) Extending quality in use to provide a framework for usability measurement. In: Kurosu M (ed) Human centered design HCII 2009, vol 5619, Lecture notes in computer science (LNCS). Springer, Berlin, pp 13–22

Breitwieser C, Terbu O, Holzinger A, Brunner C, Lindstaedt S, Müller-Putz G (2013) iScope—viewing biosignals on mobile devices. In: Zu Q, Hu B, Elçi A (eds) Pervasive computing and the networked world, vol 7719, Lecture notes in computer science (LNCS). Springer, Berlin, pp 50–56

Brooke J (1996) SUS: a "quick and dirty" usability scale. In: Jordan PW, Thomas B, Weerdmeester BA, Mcclelland AL (eds) Usability evaluation in industry. Taylor and Francis, London

Brown S, Holzinger A (2008) Low cost prototyping: Part 1, or how to produce better ideas faster by getting user reactions early and often. In: Abuelmaatti O, England D (eds) Proceedings of HCI 2008. John Moores University, British Computer Society, Liverpool, UK, pp 213–214

Buschmann F (2010) Learning from failure, Part 2: featuritis, performitis, and other diseases. IEEE Softw 27(1):10–11

Catelani M, Ciani L, Diciotti S, Dori F, Giuntini M (2011) ISO 14971 as a methodological tool in the validation process of a RIS-PACS system. IEEE International workshop on medical measurements and applications proceedings (MeMeA) 2011. IEEE, pp 408–412

Cavallin H, Martin WM, Heylighen A (2007) How relative absolute can be: SUMI and the impact of the nature of the task in measuring perceived software usability. AI & Soc 22(2):227–235

Challis BH, Krane RV (1988) Mood induction and the priming of semantic memory in a lexical decision task: asymmetric effects of elation and depression. Bull Psychonomic Soc 26(4):309–312

Cleary BA (1995) Supporting empowerment with Deming's PDSA cycle. Empower Organ 3(2):34–39

Cosgriff P (1994) Quality assurance of medical software. J Med Eng Technol 18(1):1–10

Davis FD (1989) Perceived usefulness, perceived ease of use, and user acceptance of information technology. MIS Q 13(3):319–339

Davis FD (1993) User acceptance of information technology: system characteristics, user perceptions and behavioral impacts. Int J Man Mach Stud 38(3):475–487

Davis W, Rahman MA, Smith LJ, Burns A, Senecal L, Mcarthur D, Halpern JA, Perlmutter A, Sickels W, Wagner W (1995) Properties of human affect induced by static color slides (IAPS): dimensional, categorical and electromyographic analysis. Biol Psychol 41(3):229–253

Deming WE (1994) The new economics. MIT Press, Cambridge, MA

Duncker K (1945) On problem-solving. In: Dashiell JF (ed) Psychological monographs of the American Psychological Association, vol 58. APA, Washington, DC, pp 1–114

Earthy J, Jones BS, Bevan N (2001) The improvement of human-centred processes—facing the challenge and reaping the benefit of ISO 13407. Int J Hum Comput Stud 55(4):553–585

Ekman P, Friesan WV (1982) Felt, false, and miserable smiles. J Nonverbal Behav 6(4):238–252

Fishbein M, Ajzen I (1975) Belief, attitude, intention and behavior: an introduction to theory and research. Addison-Wesley, Reading, MA

Helander MG, Khalid HM (2006) Affective and pleasurable design. In: Salvendy G (ed) Handbook of human factors and ergonomics, 3rd edn. Wiley, Hoboken, NJ

Holzinger A (2002) User-centered interface design for disabled and elderly people: first experiences with designing a patient communication system (PACOSY). In: Miesenberger K, Klaus J, Zagler W (eds) 8th International conference on computer helping people with special needs, ICCHP 2002, vol 2398, Lecture notes in computer science (LNCS). Springer, Berlin, pp 34–41

Holzinger A (2003) Experiences with user centered development (UCD) for the front end of the virtual medical campus Graz. In: Jacko JA, Stephanidis C (eds) Human-computer interaction, theory and practice. Lawrence Erlbaum, Mahwah, NJ, pp 123–127

Holzinger A (2004) Application of rapid prototyping to the user interface development for a virtual medical campus. IEEE Softw 21(1):92–99

Holzinger A (2005) Usability engineering methods for software developers. Commun ACM 48(1):71–74

Holzinger A (2010) Process guide for students for interdisciplinary work in computer science/informatics, 2nd edn. BoD, Norderstedt

Holzinger A (2011) Successful management of research and development. BoD, Norderstedt

Holzinger A, Brown S (2008) Low cost prototyping: Part 2, or how to apply the thinking-aloud method efficiently. In: Abuelmaatti O, England D (eds) Proceedings of HCI 2008. John Moores University, British Computer Society, Liverpool, UK, pp 217–218

Holzinger A, Bruschi M, Eder W (2013) On interactive data visualization of psychological low-cost-sensor data with focus on mental stress. In: Cuzzocrea A, Kittl C, Simos DE, Weippl E, Xu L (eds) Multidisciplinary research and practice for information systems. Springer lecture notes in computer science LNCS 8127. Springer, Heidelberg, pp 469–480

Holzinger A, Dorner S, Födinger M, Valdez A, Ziefle M (2010) Chances of increasing youth health awareness through mobile wellness applications. In: Leitner G, Hitz M, Holzinger A (eds) HCI in work and learning, life and leisure, vol 6389, Lecture notes in computer science (LNCS). Springer, Berlin, pp 71–81

Holzinger A, Errath M, Searle G, Thurnher B, Slany W (2005) From extreme programming and usability engineering to extreme usability in software engineering education. 29th International annual IEEE computer software and applications conference (IEEE COMPSAC 2005), Edinburgh (UK). IEEE, pp. 169–172

Holzinger A, Geierhofer R, Ackerl S, Searle G (2005b) CARDIAC@VIEW: the user centered development of a new medical image viewer. In: Zara J, Sloup J (eds) Central European multimedia and virtual reality conference (available in eurographics library). Czech Technical University (CTU), Prague, pp 63–68

Holzinger A, Kosec P, Schwantzer G, Debevc M, Hofmann-Wellenhof R, Frühauf J (2011a) Design and development of a mobile computer application to reengineer workflows in the hospital and the methodology to evaluate its effectiveness. J Biomed Inform 44(6):968–977

Holzinger A, Leitner H (2005) Lessons from real-life usability engineering in hospital: from software usability to total workplace usability. In: Holzinger A, Weidmann K-H (eds) Empowering software quality: how can usability engineering reach these goals? Austrian Computer Society, Vienna, pp 153–160

Holzinger A, Searle G, Kleinberger T, Seffah A, Javahery H (2008) Investigating usability metrics for the design and development of applications for the elderly. In: Miesenberger K, Klaus J, Zagler W, Karshmer A (eds) 11th International conference on computers helping people with special needs, vol 5105, Lecture notes in computer science (LNCS). Springer, Heidelberg, pp 98–105

Holzinger A, Searle G, Wernbacher M (2011b) The effect of previous exposure to technology (PET) on acceptance and its importance in usability engineering. Univers Access Inform Soc Int J 10(3):245–260

Holzinger A, Slany W (2006) XP + UE - > XU Praktische Erfahrungen mit eXtreme Usability. Informatik Spektrum 29(2):91–97

Holzinger A, Stickel C, Fassold M, Ebner M (2009) Seeing the system through the end users' eyes: shadow expert technique for evaluating the consistency of a learning management system. In: Holzinger A, Miesenberger K (eds) HCI and usability for e-inclusion, vol 5889, Lecture notes in computer science (LNCS). Springer, Berlin, pp 178–192

Holzinger A, Treitler P, Slany W (2012) Making apps useable on multiple different mobile platforms: on interoperability for business application development on smartphones. In: Quirchmayr G, Basl J, You I, Xu L, Weippl E (eds) Multidisciplinary research and practice for information systems, vol 7465, Lecture notes in computer science (LNCS). Springer, Berlin, pp 176–189

Holzinger A, Wascher I, Steinmann C (2003) Design and development of a LO-editor for the virtual medical campus Graz. In: Bode A, Desel J, Rathmayer S, Wessner M (eds) Lecture notes in informatics (LNI), volume P-37, DeLFI 2003. Köllen, Bonn, pp 440–449

Hussain Z, Slany W, Holzinger A (2009a) Current state of agile user-centered design: a survey. HCI and usability for e-inclusion, USAB 2009, vol 5889, Lecture notes in computer science (LNCS). Springer, Berlin, pp 416–427

Hussain Z, Slany W, Holzinger A (2009b) Investigating agile user-centered design in practice: a grounded theory perspective. HCI and usability for e-inclusion, USAB 2009, vol 5889, Lecture notes in computer science, LNCS. Springer, Berlin, pp 279–289

Izard CE (1979) Emotions in personality and psychopathology. Plenum, Oxford

Jordan PW (2000) Inclusive design: an holistic approach. Proceedings of the human factors and ergonomics society annual meeting, SAGE Publications, pp 917–920

Kirakowski J, Corbett M (1993) SUMI: the software usability measurement inventory. Br J Educ Technol 24(3):210–212

Kosec P, Debevc M, Holzinger A (2009) Towards equal opportunities in computer engineering education: design, development and evaluation of video-based e-lectures. Int J Eng Educ 25(4):763–771

Larson R, Csikszentmihalyi M (1983) The experience sampling method. New Dir Meth Soc Behav Sci 15:41–56

Larsen RJ, Ketelaar T (1991) Personality and susceptibility to positive and negative emotional states. J Pers Soc Psychol 61(1):132

Lopatovska I, Arapakis I (2011) Theories, methods and current research on emotions in library and information science, information retrieval and human–computer interaction. Inf Process Manag 47(4):575–592

Maslow AH, Frager R, Fadiman J (1970) Motivation and personality. Harper & Row, New York

McCleary RA (1950) The nature of the galvanic skin response. Psychol Bull 47(2):97–117

Melas CD, Zampetakis LA, Dimopoulou A, Moustakis V (2011) Modeling the acceptance of clinical information systems among hospital medical staff: an extended TAM model. J Biomed Inform 44(4):553–564

Morris MG, Turner JM (2001) Assessing users' subjective quality of experience with the world wide web: an exploratory examination of temporal changes in technology acceptance. Int J Hum Comput Stud 54(6):877–901

Nagamachi M (2001) Kansei engineering: a powerful ergonomic technology for product development. In: Helander MG, Khalid HM, Tham MP (eds) Proceedings of the international conference of affective human factors design, ASEAN Academic Press, London, pp 9–14

Neuhaus C, Polze A, Chowdhuryy MMR (2011) Survey on healthcare IT systems: standards, regulations and security (Technical report). Hasso-Plattner-Institute for Software Engineering, Potsdam

Niedenthal PM, Setterlund MB (1994) Emotion congruence in perception. Pers Soc Psychol Bull 20(4):401–411

Nisbett RE, Wilson TD (1977) Telling more than we can know: verbal reports on mental processes. Psychol Rev 84(3):231–259

Novak J, Ziegler J, Hoppe U, Holzinger A, Heintze C, Böckle M (2012) Mobile Anwendungen für Medizin und Gesundheit. In: Reiterer H, Deussen O (eds) Mensch & computer workshopband: interaktiv informiert—allgegenwärtig und allumfassend!? Oldenbourg, München, pp 227–230

Nowlis DP, Greenberg N (1979) Empirical description of effects of exercise on mood. Percept Mot Skills 49(3):1001–1002

Peischl B, Ferk M, Holzinger A (2013) On the success factors for mobile data acquisition in healthcare. Human aspects in mobile apps engineering at the British HCI '13. Brunel University, British Computer Society, London

Picard RW (2000) Perceptual user interfaces: affective perception. Comm ACM 43(3):50–51

Seffah A, Donyaee M, Kline RB, Padda HK (2006) Usability measurement and metrics: a consolidated model. Softw Qual J 14(2):159–178

Seffah A, Kececi N, Donyaee M (2001) QUIM: a framework for quantifying usability metrics in software quality models. Second Asia-Pacific conference on quality software (APAQS'01), Hong Kong, 10–11 Dec 2001, pp 311–318

Shackel B (1991) Usability—context, framework, definition, design and evaluation. In: Shackel B, Richardson SJ (eds) Human factors for informatics usability. Cambridge University Press, New York, pp 21–37

Thimbleby H (2007) User-centered methods are insufficient for safety critical systems. In: Holzinger A (ed) HCI and usability for medicine and health care: third symposium of the workgroup human-computer interaction and usability engineering of the Austrian computer society, USAB 2007, vol 4799, Lecture notes in computer science (LNCS). Springer, Berlin, pp 1–20

Van Der Sluis F, Van Den Broek EL, Van Dijk B (2010) Information Retrieval eXperience (IRX): towards a human-centered personalized model of relevance. 2010 IEEE/WIC/ACM international conference on web intelligence and intelligent agent technology (WI-IAT), 31 Aug 2010–3 Sept 2010, pp 322–325

Veer GCVD, Welie MV (2004) DUTCH: designing for users and tasks from concepts to handles. In: Diaper D, Stanton N (eds) The handbook of task analysis for human-computer interaction. Lawrence Erlbaum, Mahwah, NJ, pp 155–173

Warr P (1999) Logical and judgmental moderators of the criterion-related validity of personality scales. J Occup Organ Psychol 72(2):187–204
Watson D, Clark LA, Tellegen A (1988) Development and validation of brief measures of positive and negative affect: the PANAS scales. J Pers Soc Psychol 54(6):1063
Westerink JH, Van Den Broek EL, Schut MH, Van Herk J, Tuinenbreijer K (2008) Computing emotion awareness through galvanic skin response and facial electromyography. Probing experience. Springer, Dordrecht, pp 149–162
Wiklund ME, Wilcox SB (2005) Designing usability into medical products. Taylor & Francis, Boca Raton, FL

Index

A
Abduction, 57, 76, 77
Abstraction, 57, 58, 76, 109, 110, 120, 121, 124, 203, 301
Acceptable risk, 459, 469
Accessibility, 58, 63, 66, 73, 129, 137, 162, 169, 253, 254, 283, 393, 464, 475, 483, 501, 502
Accreditation, 422, 437, 501
AceDB, 122, 153
Act, 300, 301, 314, 462, 484, 501, 513, 514, 531, 537
Adverse event, 328, 331, 337, 459, 461, 467, 468, 473, 494
Anonymization, 398, 459, 475, 486, 488, 494
Artifact/surrogate, 58
Artificial neural network (ANN), 177, 251, 264–267, 269, 275, 276, 318
Association rule learning, 251
Audit, 482–484, 501, 502
Authentication, 158, 166, 429, 459, 462, 471, 476, 478, 479

B
Bioinformatics, 1, 11, 19, 20, 63, 136, 165, 168, 210, 211, 214, 233, 379, 399, 421–433
Bioinformatics workflow management system, 421, 432
Biological data visualization, 379
Biological system, 109, 203
Biomarker, 1, 39
Biomedical, 1, 23–25, 27, 29–35, 37, 41, 57, 60, 63, 65, 67, 71, 73, 109, 120–122, 125–127, 135, 136, 151, 153–165, 167, 189, 195, 203, 213, 228, 251, 254, 256, 260, 267–268, 302, 345–376, 386, 396, 401, 405, 421–498
Biomedical data, 1, 31, 33, 34, 60, 459–498
Brute force, 263, 299
Business intelligence (BI), 153, 154, 379, 380
Business process reengineering (BPR), 421

C
Case-based reasoning (CBR), 299, 345, 360–364, 367, 373
Cassandra, 153
Certainty factor model (CF), 345, 352, 365, 366, 371, 372
Certification, 131, 440–442, 502, 512
Cladogram, 153
CLARION, 345
Classical Medicine, 2
Classification, 39, 68, 90, 109–151, 167, 170, 183, 203, 251, 252, 254, 258, 261–263, 266, 276, 282, 284, 295, 300, 318, 323, 357, 358, 360, 379, 405, 418, 423, 462, 466–467
Classification system, 154
Clinical decision support (CDS), 346
Clinical decision support system (CDSS), 35, 276, 346
Clinical pathway, 421
Cloud computing, 154, 157, 165–166, 442
Cluster analysis, 223, 252
Clustering, 3, 4, 10, 14, 64, 154, 156, 165, 170, 171, 187, 216–220, 222, 223, 232, 239, 245, 252, 255, 258, 260–263, 276, 281, 284, 357, 358, 360, 366, 372, 380, 394, 396, 398, 401, 405, 418
Coding, 18, 19, 35, 86, 109–151, 203, 227, 232, 284, 382, 529

Cognition, 11, 31, 261, 299, 305, 380, 381, 393
Cognitive, 2, 32, 74, 80, 91, 114, 126, 160, 172, 256, 257, 259, 281, 299–341, 345, 351, 357, 383, 384, 389, 391, 460, 461, 468, 502, 517
Cognitive load, 91, 299
Cognitive modeling, 327, 334, 340, 502
Cognitive science, 126, 299–341, 351
Cognitive walkthrough, 502
Collective intelligence, 346
Computerized physician order entry (CPOE), 3, 30, 41, 47, 154, 156, 189, 408, 414
Computing, 2, 27–31, 51, 154, 156, 157, 165–166, 188, 189, 224, 255, 256, 280, 300, 354, 408, 412, 414, 423, 425, 428–429, 433, 442–444, 446, 451, 452, 459, 470, 471, 473, 486, 493, 498, 504, 541
Confidentiality, 166, 460, 462, 464, 470, 471, 475, 486, 512
Confounding variable, 300
Consistency, 58, 125, 502, 508, 518
Consistency inspection, 502
Content analytics, 281, 379, 380, 388, 392, 406, 442, 444
Correlation coefficient, 300
Crowd sourcing, 346

D
Data, 1, 57, 109, 153, 203, 251, 300, 346, 379, 421, 459, 504
Data Mart (DM), 154, 163, 164, 170, 179, 253, 258–259, 295, 300, 301, 311–312, 317
Data mining, 2, 74, 92, 170, 172, 206, 223, 259–295, 317, 333, 339, 351, 354, 355, 367, 379, 380, 398, 399, 412
Data model, 110, 171, 204, 210, 403, 407, 413, 477
Data protection, 157, 159, 398, 442, 446, 452, 459, 460, 486
Data quality, 58, 68, 157, 208
Data security, 195, 429, 460
Dataset, 58, 68, 70, 74, 81, 87–89, 115, 170, 258, 263, 264, 273, 276, 321, 404, 405, 438, 486
Data structure, 58, 59, 63, 94, 95, 100, 101, 123, 223
Data visualization, 69, 380, 388–390, 393, 398, 399, 408, 414
Decision making, 24, 31, 131, 259, 260, 270, 283, 299, 300, 302–304, 310–312, 315, 317, 318, 320, 326, 327, 331, 333, 334, 336, 339, 340, 345–376, 381, 407, 413, 424
Decision support system (DSS), 35, 276, 346–360, 366, 367, 370, 372, 375
Deduction, 58, 76, 77, 301, 349, 351, 353
Diagnosis, 2, 34, 35, 37, 110, 118, 130, 158, 204, 232, 299, 300, 302, 303, 312, 317, 320, 321, 327, 346–349, 353, 359, 365, 382, 434, 460, 510, 518, 531, 537
Diagnostic and Statistical Manual (DSM), 110, 136, 204
Differential diagnosis (DDX), 299, 301, 303, 319–320, 330, 336
Digital imaging and communications in medicine (DICOM), 119, 141, 147, 166, 421, 422, 434
DIK Model, 58
DIKW Model, 58
Dirty data, 58
Disparity, 58
Distance matrix method, 154
DSM. *See* Diagnostic and Statistical Manual (DSM)
DXplain, 346

E
Effectiveness, 167, 254, 502, 508, 513, 530, 536
Efficiency, 68, 115, 216, 218, 357, 360, 423, 424, 502, 504, 508, 509, 515
Emotion, 300, 502, 527–529
End user, 28, 31, 34, 41, 47, 58, 73, 116, 135, 159, 165, 172, 173, 189, 195, 259, 279, 304, 354, 367, 372, 387, 390, 403, 406, 408–410, 414, 416, 431, 480, 487, 493, 501–505, 507, 508, 514, 516, 517, 519–526, 530, 531, 533, 534, 536, 537, 539, 540
End-user programming (EUP), 31, 41, 47, 189, 502, 531, 537
EnsEMBL, 15, 22, 41, 47, 95, 101, 154, 165, 171, 188, 195, 276
ENTREZ, 154–156, 167
Entropy, 11, 15, 57–59, 78–90, 95–97, 101, 102, 222, 294
Errors, 2, 29, 58, 68, 86, 92, 115, 123, 208, 264, 265, 275, 299, 302, 303, 317, 326–328, 330, 331, 336, 337, 359, 429, 430, 459, 460, 464–474, 487–489, 494, 503, 504, 508
Evaluation, 57, 67, 77, 91, 93, 160, 261, 264, 266, 312, 355, 382, 431, 501–541

Evidence-based medicine (EBM), 3, 24, 37, 112, 115, 140, 146, 161, 178, 300, 347, 422
Experiment, 58, 81, 82, 90, 168, 169, 525
Expert system, 346, 347, 350, 351, 353, 357, 365, 367, 371, 373, 443, 463
Extensible Markup Language (XML), 68, 75, 110, 119, 120, 128, 141, 147, 156, 169, 175, 204, 207, 209–211, 235, 240, 241, 246, 252, 429, 437, 439
External validity, 300
Extract, transform, and load (ETL), 154

F
Federated database system, 154, 166
Formative Evaluation, 503

G
GALEN, 112, 125, 204
GAMUTS in radiology, 346
Gedanken, 58, 81, 82
Genetic algorithm, 154
Genomics, 1, 2, 12, 20, 32, 60–62, 66, 154, 167, 168, 195, 206, 227, 236, 242, 356
GOLD, 154, 253, 255
Graphical user interface (GUI), 505, 524

H
Hadoop, 154, 155
Hazard, 459, 460, 469
Hbase, 154
Health Level Seven (HL 7), 422, 437, 503
Heart rate variability (HRV), 58, 59
Heuristic evaluation, 503, 517, 532, 538
Hospital information system (HIS), 65, 118, 158–162, 168, 195, 208, 209, 347, 366, 372, 406, 421, 422, 428, 429, 432–434, 436, 437, 440, 442, 444–446, 448, 451, 452, 454, 464, 476
HRV. *See* Heart rate variability (HRV)
Human fallibility, 328, 460, 468, 487, 493
Hypothetico-deductive model (HDM), 299, 300, 312–315

I
ICD. *See* International Classification of Diseases (ICD)
ILIAD, 346
Induction, 58, 76, 77, 94, 100, 263, 320, 325, 345, 353

Information entropy, 58, 81–90
Information Extraction (IE), 155
Information overload, 58
Information quality, 31–33, 58, 433, 501
Information retrieval (IR), 93, 110, 127, 153–199, 204, 251, 527–528
Information visualization, 91, 93, 379–418
Integrating the Healthcare Enterprise (IHE), 422, 442, 443
Internal validity, 300
International Classification of Diseases (ICD), 68, 109, 110, 115, 130–132, 136, 140, 141, 146, 147, 203, 204, 232, 423
IR. *See* Information retrieval (IR)
ISO 9001, 501, 503, 512
ISO 9241, 503, 531, 537
ISO 9241-10, 503
ISO 9241-11, 503
ISO 13407, 503, 519, 523, 525, 530, 531, 536, 537
ISO 13485 (2003), 503, 512, 514
ISO 14971 (2007), 461, 503, 512
ISO 62304 (2006), 503
ISO/HL7, 503
ISO/IEEE, 503
ISO/OECD, 503

K
k-Anonymity, 398, 460
Knowledge, 2, 57, 109, 153, 203, 209, 299, 345, 380, 431, 460, 502
Knowledge Discovery (KD), 7, 35, 36, 73, 74, 92, 159, 170, 172, 187, 251–295, 347, 354, 388, 391, 398, 425, 444
Knowledge Extraction, 252, 357, 358

L
Large data Multidimensional, 59, 91, 95, 101, 165, 170, 251, 252, 264, 405, 442
Learnability, 277, 503, 508, 515
Learning curve, 503

M
MapReduce, 154, 155
Mashup, 31, 155, 480–482, 487, 493
Mash-up, 41, 47, 187, 502–504, 531, 537
Medical Classification, 119, 129–137, 204
Medical error, 299, 327, 328, 460, 464–474, 488, 494
Medical safety design, 504

Medicine, 2, 3, 19, 23–24, 30, 32, 36–39, 109–112, 130, 132–133, 140, 146, 203–205, 234, 304, 327, 346, 347, 398, 422, 436, 465, 473, 474
Medizin Produkte Gesetzn (MPG), 501, 504, 513
MEDLINE, 136, 155, 465
Memorability, 504, 508
Mental model, 79, 80, 95, 101, 166, 380, 385, 391, 407, 408, 413, 414, 502, 504
MeSH, 109, 110, 130, 133–136, 140, 146, 203, 204
Metabolomics, 2, 66
Metadata, 110, 155, 204
Metadata Model, 110, 204
Methodology, 85, 91, 110, 166, 173, 187, 204, 260–280, 284, 327, 330, 334, 336, 340, 485, 501–541
MMMDB, 155
Molecular, 2, 9–13, 15, 16, 19, 23, 31, 34, 36–38, 40, 41, 46, 47, 60, 155, 168, 169, 188, 189, 194, 195, 206, 225, 227, 229, 230, 233, 234, 236, 238, 242, 244, 261, 350, 379, 385, 408, 414, 437
Multidimensional, 59, 63, 67, 69, 74, 93, 94, 100, 110, 204, 279, 380, 382, 395, 406
Multidimensional scaling, 380
Multimedia, 62, 115, 251–295, 383, 421, 434–439, 441
Multi-modality, 59
Multi-variate, 33, 59, 380, 383
MYCIN, 345, 346, 350–353, 365, 366, 371, 372

N
National Electrical Manufacturers Association (NEMA) Paradigm, 111, 119, 422, 438
Natural language processing (NLP), 115, 155, 172, 176, 253, 281
Neural networks, 155, 177, 195, 251, 262, 264–265, 267–268, 270, 284, 366, 372
Nomen Nescio (N.N) patient safety, 460
Non-relational database, 155
Nosography, 110, 204
Nosology, 110, 204

O
Omics data, 2, 36, 38, 65–67, 76, 163, 164
Online Mendelian Inheritance in Man (OMIM), 136, 155, 167

Ontology, 68, 110, 117, 122–130, 136, 138, 140, 141, 146, 147, 185, 204, 207, 232, 388
Ontology engineering, 110, 204

P
Parallel coordinates, 93, 215, 380, 394–398, 411, 415, 417
Participatory design, 504
PCA. See Principal component analysis (PCA)
PDB. See Protein database (PDB)
Perception, 301
Performance, 2, 3, 26–27, 31, 32, 112, 113, 153, 158, 171, 179, 186, 259, 264, 284, 299, 304, 309, 312, 314, 318, 356, 442, 443, 473, 474, 502, 504, 508, 513, 518, 529, 536
Performance measure, 504
Personally identifying information prevention, 460
Pervasive, 2, 27–29, 31, 37–39, 41, 47, 189, 195, 408, 414, 541
Pervasive Health, 2, 28, 37
P-Health Model, 2, 38–39
Phylogenetics, 155, 167
Picture Archiving and Communication System (PACS), 422
Plan-Do-Check-Act (PDCA), 300–301, 312, 314, 514
Principal Component Analysis (PCA), 252, 263, 277, 279, 280
Privacy, 460
Privacy policy, 460, 482–484
PROSITE, 155
Protected health information (PHI), 461, 462
Protein database (PDB), 5, 6, 35, 69, 155, 167, 168, 195
Proteome, 2, 12, 16
Proteomics, 1, 2, 15, 16, 32, 60–62, 66, 95, 101, 167, 261
Pseudonymization, 459, 461, 475–480, 488, 494

Q
Qualitative research, 301
Quantitative research, 301
Quasi-Identifiers, 398, 460, 461

R
RadViz, 379, 380, 394, 399–401, 407, 408, 411, 413, 414, 417
Reasoning, 59, 76, 77, 104, 120, 125–127, 129, 138, 161, 172, 173, 175, 176, 195, 258,

299–304, 311–327, 330, 334, 336, 340, 345–376, 380, 388, 403, 408, 414, 471, 472
Receiver-operating characteristic (ROC), 276, 301, 318, 331, 337
Relational database, 122, 155

S
Safety, 157, 159, 423, 442, 444, 446, 452, 459–498, 504, 506, 509–511, 513, 518, 527
Safety-critical systems research, 461
Safety engineering, 459, 461
Safety risk management, 461
Satisfaction, 430, 504, 507, 508, 518, 527, 530, 536
Security, 37, 157, 159, 195, 423, 429, 442–444, 446, 452, 459–498, 509–512, 515, 537
Semiotic engineering, 380, 386, 387
Semiotics, 380, 386–388, 408, 414, 504
Semi-structured data, 122, 207, 209–211, 237, 242
Sensitive data, 398, 461
Similarity table, 156
SNOMED, 109–111, 130, 132–133, 136, 140, 141, 146, 151, 185, 203–205
SNOP, 20, 110, 132
Software usability measurement inventory (SUMI), 505, 506, 518
Software usability scale (SUS), 505, 506, 517, 518, 531, 537
Space, 2, 4, 14, 40, 46, 63, 70–72, 74, 86, 94, 100, 135, 160, 172, 176–182, 188, 207, 212, 223, 231, 233, 238, 242, 244, 266, 275–278, 280, 380, 381, 393–396, 399, 401, 404, 406, 408, 409, 414, 508
Spatiality, 59
SQL, 156, 163
SRS, 156
Star plot, 379, 380, 395, 401, 403, 408, 410, 414, 416
Structural complexity, 59, 222
Supervised learning, 252, 262–276
Supervised learning algorithm, 252
Support vector machine (SVM), 251, 252, 262, 270, 275–276, 318

Swiss-Cheese model, 328, 461
SWISS-PROT, 156, 169
Symbolic reasoning, 301
System features, 110, 204

T
Task analysis, 505, 520
Technological Performance, 26, 31
Terminology, 68, 110, 115, 123, 124, 127, 129, 132, 136, 184, 204, 437
Thinking aloud, 505, 517, 519, 522, 531, 537
Time dependency Voxel, 59
Translational Medicine, 3
Triage, 301, 311, 330, 336

U
Unified Language System (UMLS), 109, 110, 127, 130, 133, 135–137, 140, 141, 146, 147, 151, 203
UniGene, 156
Unsupervised learning, 252, 262, 263
Usability engineering, 93, 488, 494, 501, 504, 505, 516–524, 530, 534, 540
User interface (UI), 35, 36, 92, 172, 351, 353, 366, 372, 381, 403, 405, 472, 502, 503, 505, 516, 524

V
Validation, 269, 354, 403, 440, 501, 505, 520, 525, 530, 536
Verification, 157, 261, 501, 505, 520, 525
Visual analytics, 379–381, 388, 392, 403–406, 410, 416
Visualization, 8, 9, 74, 91–93, 135, 171, 174, 215, 259, 261, 268, 284, 379–417, 432, 442, 443, 504
Visualization mantra, 381, 404, 408, 414
Von Neumann Computer, 3

W
Workflow, 118, 158–161, 164, 347, 353, 362, 421, 422, 424–434, 436, 438, 441, 507, 516, 521

The manufacturer's authorised representative in the EU is Springer
Nature Customer Service Centre GmbH, Europaplatz 3, 69115 Heidelberg,
Germany. If you have any concerns regarding our products, please
contact ProductSafety@springernature.com

Printed and bound by CPI Group (UK) Ltd, Croydon, CR0 4YY
23/03/2026
02076658-0006